MATHEMATICAL MORPHOLOGY AND ITS APPLICATIONS TO IMAGE AND SIGNAL PROCESSING

Computational Imaging and Vision

Volume 18

MATHEMATICAL MORPHOLOGY AND ITS APPLICATIONS TO IMAGE AND SIGNAL PROCESSING

edited by

John Goutsias
Johns Hopkins University, Baltimore

Luc Vincent
Xerox, Palo Alto

Dan S. Bloomberg
Xerox, Palo Alto

KLUWER ACADEMIC PUBLISHERS
Boston / Dordrecht / London

Distributors for North, Central and South America:
Kluwer Academic Publishers
101 Philip Drive
Assinippi Park
Norwell, Massachusetts 02061 USA
Telephone (781) 871-6600
Fax (781) 681-9045
E-Mail <kluwer@wkap.com>

Distributors for all other countries:
Kluwer Academic Publishers Group
Distribution Centre
Post Office Box 322
3300 AH Dordrecht, THE NETHERLANDS
Telephone 31 78 6392 392
Fax 31 78 6546 474
E-Mail <services@wkap.nl>

Electronic Services <http://www.wkap.nl>

Library of Congress Cataloging-in-Publication Data

Mathematical morphology and its applications to image and signal processing / edited by John Goutsias, Luc Vincent, Dan S. Bloomberg.
p. cm. — (Kluwer series in computational imaging and vision ; CIVI 18)
Proceedings of the 5th International Symposium on Mathematical Morphology and its Applications to Image and Signal Processing, held June 26-28, 2000, Palo Alto, Calif.
Includes bibliographical references and index.
ISBN 0-7923-7862-8 (alk. paper)
1. Image processing—Mathematics—Congresses. 2. Signal processing—Mathematics—Congresses. 3. Digital Filters (Mathematics)—Congresses. I. Goutsias, John. II. Vincent, Luc III. Bloomberg, Dan S. IV. International Symposium on Mathematical Morphology and its Applications to Image and Signal Processing (5th : 2000 : Palo Alto, Calif.) V. Series.

TA1637 .M36 2000
621.36'7'0151—dc21 00-030174

Printed on acid-free paper.
Printed in the United States of America

Contents

Connectivity and Connected Operators

Segmentation

Texture Analysis

Preface

It has been seven years since the first International Symposium on Mathematical Morphology (ISMM) took place in Barcelona, Spain, in May 1993. Since then, meetings were also held in Fontainebleau, France (September 1994), Atlanta, Georgia (May 1996), and Amsterdam, The Netherlands (June 1998). This volume contains an up-to-date sampling of international scientific work on Mathematical Morphology that was presented during the fifth ISMM, held at the Xerox Palo Alto Research Center (PARC), in Palo Alto, California, June 26–29, 2000.

We would like to thank all members of the program committee for their invaluable work in reviewing the submitted papers. A blind review process was used, whereby each paper was reviewed by two to four committee members. This resulted in the acceptance of the 45 papers included in this volume. The tight schedule the ISMM chairs were working with required that committee members review up to ten papers in four weeks, a Herculean effort on their part. We are deeply appreciative of their commitment to making this conference a success.

We are also thankful to Jeanette Figueroa for her amazingly efficient help with all aspects of the organization. The conference would not have been the same without her. We would also like to thank Ulisses Braga Neto and Sinan Batman who handled the rather difficult task of identifying important keywords used to build the subject index at the end of this volume. Finally, we gratefully acknowledge the financial and logistical support provided by Xerox Corporation.

JOHN GOUTSIAS
LUC VINCENT
DAN S. BLOOMBERG

Introduction

In 1975, a seminal book by Georges Matheron, entitled *Random Sets and Integral Geometry*, laid down the foundations of a novel technique for shape processing and analysis known as mathematical morphology. Twenty five years later, mathematical morphology is considered to be a powerful tool for signal and image analysis, in particular for those applications where extraction and analysis of geometric information is of interest. The main idea behind mathematical morphology is to analyze geometric information by "probing" an image with a small geometric template known as the structuring element. This simple idea has lead to a large collection of theoretical results and practical tools and has provided a new approach for processing visual information.

This book contains the proceedings of the fifth *International Symposium on Mathematical Morphology and its Applications to Image and Signal Processing*, held June 26-28, 2000, at Xerox PARC in Palo Alto, California. It provides a broad sampling of the most recent theoretical and practical developments of mathematical morphology and its applications to image and signal processing.

The contributions are classified under several themes, which are briefly introduced below. It should be noted that many papers have aspects belonging to more than one theme.

THEORY. A number of theoretical aspects of mathematical morphology are investigated. These include: an understanding of the relationship between traditional signal processing techniques and mathematical morphology, sequential decompositions of a particular class of morphological operators, decomposition of large concave grayscale structuring elements, minimization of mixed volume functionals of convex polyhedral shapes based on Minkowski addition, investigation of certain topological properties of the discretization of closed sets based on the Hausdorff metric, the introduction of vector levelings and flattenings, and a lattice control model for dynamical systems.

SHAPE ANALYSIS AND INTERPOLATION. This part includes three papers, providing a morphological approach to shape interpolation, a shape deformation technique based on combining morphological interpolation with affine transformations, and the use of affine invariant mathematical morphology for shape recognition.

FILTERING. Morphological filtering is an important subject of mathematical morphology that provides operators capable of removing unwanted structures from images. Here, a method is presented for constructing self-dual morphological filters, a new class of morphological filters is proposed for extracting linear features, aperture filters are applied in several image analysis applications, and a technique, using genetic algorithms, is proposed for the optimization of grayscale soft morphological filters.

CONNECTIVITY AND CONNECTED OPERATORS. Connected operators are non-linear operators that remove connected components from the level zones of a given image. A number of issues regarding connectivity and connected operators are investigated: new insights on the concept of digital connectivity are presented, the concept of a connectivity class is extended to include approximate and multiresolution connectivities, and region-tree pruning is proposed as a tool for constructing connected operators.

SEGMENTATION. Mathematical morphology is instrumental in providing effective solutions to image segmentation problems. Here, a number of papers present morphological approaches for segmenting grayscale, color, 3-D and video images, as well as for extracting semantic video objects. Moreover, flooding, a key step in morphological segmentation, is discussed in the framework of hierarchical morphological segmentation.

TEXTURE ANALYSIS. In this part, morphological tools, like granulometries and antigranulometries, Choquet capacities, erosion and dilation curves, etc. are used for texture synthesis, classification, and mipmapping.

MULTIRESOLUTION TECHNIQUES AND SCALE-SPACES. Multiresolution approaches are extremely useful in many image processing and analysis applications. In this section, morphological pyramids and wavelets are studied on the quincunx lattice, an algebraic framework for morphological scale-space operators is introduced, and an idempotent scale-space approach to segmentation is proposed.

ALGORITHMS. In this part, a number of papers deal with the development of algorithms for the fast implementation of morphological operators. These include algorithms for efficient implementation of standard erosions, dilations, openings, and closings; for attribute openings-closings and change detectors; and for calculating distance transforms, the watershed transform, and the medial axis of 3D objects.

APPLICATIONS. Finally, a number of papers deal with the use of mathematical morphology in a variety of applications. These include: compression of document images, OCR classification, bank check logo segmentation, correction of handwriting baseline skew, segmentation of renaissance books, extraction of linear features in airborne images, face localization, and quantitative description of telecommunication networks.

The papers in this book contribute ideas that are useful to people working in mathematical morphology and other areas of signal and image processing, pattern recognition and computer vision.

A MORPHOLOGICAL VIEW ON TRADITIONAL SIGNAL PROCESSING

RENATO KESHET (KRESCH)
Hewlett-Packard Laboratories - Israel
Technion City, Haifa 32000, Israel.
E-mail: renato@hpli.hpl.hp.com

Abstract. We argue that the fundamentals of mathematical morphology (partial ordered sets, openings, erosions, etc.) could provide a theoretical foundation for signal processing in general. The main observation is that signal processing addresses *simpler versions* of signals (of a given set $\mathcal{S}$), and this actually determines a *partial ordering* on $\mathcal{S}$. Another observation, made in the past by Serra, is that ideal filters are in fact algebraic openings. In this paper, these and other ideas are addressed and developed.

In the first part of this paper, we show that several key signal processing tasks (*linear* filtering, quantization, and decimation) can be seen as particular cases of morphological operators. Specifically, for each of these operators, we show a complete inf-semilattice in which the operator is an erosion. This serves as a background and motivation for investigating the relationship between mathematical morphology and general signal processing.

In the second part, we revisit the foundations of signal processing from the point of view of mathematical morphology. We show that, to every function, one can associate a partial ordering and an ideal filter (algebraic opening in the resulting partial ordered set), which provide a characterization of the "simplification" (information loss) performed by the function. Then, links between classes of signal processing tasks and basic morphological operators are established.

Key words: Mathematical Morphology, Complete Semilattices, Signal Processing, Image Processing.

1. Introduction

1.1. Motivation

Structural and functional similarities between the well-known linear *Laplacian pyramid* and the *morphological skeleton representation*, observed in the past [1] (see also [2, chapter 9]), are the roots of the work presented here. These two image decomposition schemes can be obtained by the same algorithmic structure: *i)* Perform a series of information removal steps, *ii)* at each step perform a restoration operation, and *iii)* take the differences between the restored and the input images at each step. The only difference between the algorithms is that for the Laplacian pyramid the information removal step and restoration operation are *decimation* and *interpolation*, respectively, whereas for the morphological skeleton, these are *erosion* and *dilation*, respectively. As noted in [3] (see also [2, chapter 5]), there are other decomposition schemes that have the same algorithmic structure as above. *Bit-plane decomposition* is obtained if the information removal and restoration operators are set to *point-wise integer division by 2* and *point-wise multiplication by 2*, respectively. *Quadtree decomposition* is obtained for another specific choice of information removal and restoration operators.

Heijmans and Goutsias have recently proposed a common axiomatic framework for these pyramidal schemes [4], deepening the understanding of the similarities between them. On the other hand, one could go forward, and ask if there is a deeper connection between such theoretically different operators as linear decimation, morphological erosion, and integer division.

An interesting result, related to the above question, is given in [3, 2]. It turns out that quadtree and bit-plane decompositions are particular cases of morphological skeletons on complete lattices! I.e., there exists a complete lattice (the usual complete lattice of grayscale images) where point-wise integer division by two is an erosion, and another one (the usual Boolean lattice on $\mathbb{Z}^2$) where the quadtree information removal step is an erosion. It is also shown that the corresponding restoration operators are their adjoint dilations. In this case, could linear decimation be an erosion as well, and the linear interpolation be its adjoint dilation? Unfortunately, there does not seem to exist a complete lattice where this holds.

However, with the extension of MM's theory to complete semilattices [5], we can answer the above question affirmatively. We show here that there exists a *complete inf-semilattice* (CISL) where decimation is indeed an erosion, and interpolation is its adjoint dilation!

The above result triggered a few other interesting results: We also show here that any *linear translation-invariant filter* is an erosion in the same CISL as above. Also, *quantization* is shown to be an erosion in another CISL.

1.2. Analysis

In our opinion, all the above observations are more than just curious remarks. We believe that the very nature of MM constitutes a suitable framework for signal processing (SP) in general, and therefore it is not a coincidence that the information removal steps of all decomposition schemes considered above are erosions. Similarities between MM and other signal processing tasks, like template matching and restoration, have also been reported in the past [8, 2, chapter 9]. Other links are given in [10].

In the second part of this paper, we briefly review signal processing fundamentals from the MM theoretical point of view, in order to justify the above statement. In particular, we argue that the concept of *partial ordering* (the basis of mathematical morphology theoretical framework) could be seen as a formal characterization of the notion of "simpler version" of a signal, which is the heart of signal processing. We show that every operator is an erosion in some partial ordered set (poset), which means that every operator has an underlying simplicity criterion (the partial order) behind it. We suggest that an operator is suitable to some signal processing application only if its underlying simplicity criterion matches the application needs and ones intuition of what "simpler version" means. In this framework, erosions assume the rôle of information removing operators, while their adjoint dilations are the restoration step associated to them. The associated opening represents the ideal filter behind the operation. Interestingly, the closing operation is of no meaning in this framework.

2. Some Traditional SP Tasks Viewed as Morphological Operations

The methodology in this section is as follows: To each example of SP task, we associate a partial ordering that provides the corresponding set of signals with a CISL structure. Then we show that the operation in consideration is an erosion in that CISL. The proof for the first proposition presented here is shown, but that of the others is omitted, for lack of space.

2.1. Scaling and Pyramids

Let us define the partial ordering $\stackrel{\bullet}{\leq}$ on $\ell_2\left(\mathbb{Z}^d\right)$. For all $f, g \in \ell_2\left(\mathbb{Z}^d\right)$:

$$f \stackrel{\bullet}{\leq} g \Leftrightarrow \forall \omega \in [0, 2\pi)^d, \quad \begin{cases} |F(\omega)| \leq |G(\omega)|, \text{ and} \\ \Theta\{F(\omega)\} = \Theta\{G(\omega)\}, \end{cases} \tag{1}$$

where $F(\omega)$ and $G(\omega)$ are the d-dimensional Fourier transforms of $f(x)$ and $g(x)$, respectively, and $\Theta\{\cdot\}$ is the operator that returns the *phase* of a Fourier coefficient.

The partially ordered set $\left(\ell_2\left(\mathbb{Z}^d\right), \stackrel{\bullet}{\leq}\right)$ is actually a CISL, where the least element is the null function. The infimum of a set $\{f_i\} \subseteq \ell_2\left(\mathbb{Z}^d\right)$ is given by the following relation:

$$\stackrel{\bullet}{\wedge}\{f_i\} = \mathcal{F}^{-1}\{W\}, \tag{2}$$

where $\mathcal{F}^{-1}$ denotes inverse d-dimensional Fourier transform, and $W(\omega)$ is defined by:

$$W(\omega) \triangleq \begin{cases} \inf\{|F_i(\omega)|\} \cdot \exp\left(j\Theta\{F_0(\omega)\}\right), & \text{if } \Theta\{F_i(\omega)\} = \Theta\{F_0(\omega)\}, \forall i \\ 0, & \text{otherwise.} \end{cases} \tag{3}$$

Consider the $R:1$ decimation operation $\mathcal{D}_R$, characterized by:

$$\mathcal{D}_R\{f\}(n) = \sum_{i \in \mathbb{Z}^d} h_i \cdot f(nR - i), \tag{4}$$

where $\{h_i\}$ is the set of coefficients of the ideal π/R-cutoff low-pass filter. And consider also the $1:R$ interpolation operation $\mathcal{I}_R$, given by up-sampling, followed by filtering with the same ideal π/R-cutoff low-pass filter $\{h_i\}$ as above.

Proposition 1 *$\mathcal{D}_R$ is an erosion in* $\left(\ell_2\left(\mathbb{Z}^d\right), \stackrel{\bullet}{\leq}\right)$.

Proof: We have to prove that $\mathcal{D}_R\left(\dot{\bigwedge}\{f_i\}\right) = \dot{\bigwedge}\{\mathcal{D}_R(f_i)\}$, for any subset $\{f_i\}$ of $\ell_2\left(\mathbb{Z}^d\right)$.

From signal processing theory, we have:

$$\mathcal{F}\{\mathcal{D}_R(f)\}(\omega) = F(\omega/R). \tag{5}$$

Therefore:

$$\begin{aligned}\mathcal{F}\left\{\mathcal{D}_R\left(\dot{\bigwedge} f_i\right)\right\} &= W(\omega/R) \\ &= \begin{cases} \inf\{|F_i(\omega/R)|\}\exp\left(j\Theta\{F_0(\omega/R)\}\right), \\ \quad \text{if } \Theta\{F_i(\omega/R)\} = \Theta\{F_0(\omega/R)\}, \forall i \\ 0, \text{ otherwise.} \end{cases} \\ &= \mathcal{F}\left\{\dot{\bigwedge}\mathcal{D}_R(f_i)\right\}. \end{aligned} \tag{6}$$

□

Proposition 2 *$\mathcal{I}_R$ is the adjoint dilation of $\mathcal{D}_R$.*

Corollary 1 *The corresponding morphological opening is filtering by $\{h_i\}$.*

Proposition 3 *The Laplacian pyramid, calculated with the decimation and interpolation operations $\mathcal{D}_R$ and $\mathcal{I}_R$, is a morphological skeleton in $\ell_2\left(\mathbb{Z}^d\right)$.*

2.2. Linear Filtering

For simplification, let us consider here only discrete signals, i.e., signals in $\ell_2\left(\mathbb{Z}^d\right)$. The results can be extended to continuous signals as well.

Consider the linear filtering operation $\mathcal{L}_g$, characterized by:

$$\mathcal{L}_g\{f\}(n) = (f * g)(n) = \sum_{i\in\mathbb{Z}^d} g_i \cdot f(n-i), \tag{7}$$

where $\{g_i\}$ is the set of coefficients of an arbitrary filter. The pseudo-inverse $\mathcal{L}_h^{\#}$ of a linear filtering operator $\mathcal{L}_h$ is given by: $\mathcal{L}_h^{\#}(f) \triangleq \mathcal{F}^{-1}\{F(\omega)H^{\#}(\omega)\}$, where

$$H^{\#}(\omega) \triangleq \begin{cases} 1/H(\omega), & |H(\omega)| > 0, \\ 0, & \text{otherwise.} \end{cases} \tag{8}$$

Proposition 4 *$\mathcal{L}_g$ is an erosion in $\left(\ell_2\left(\mathbb{Z}^d\right), \dot{\leq}\right)$.*

Proposition 5 *$\mathcal{L}_g^{\#}$ is the adjoint dilation of $\mathcal{L}_g$.*

Corollary 2 *The morphological opening associated to a given linear filtering operation is an ideal (idempotent) linear filtering operation.*

2.3. Quantization

Consider the complete inf-semilattice $\left(\mathbb{R}, \overset{\mathrm{T}}{\leq}\right)$ of real numbers with the partial ordering:

$$x \overset{\mathrm{T}}{\leq} y \Leftrightarrow 0 \leq x \leq y \text{ or } y \leq x \leq 0, \tag{9}$$

where $\leq$ in the right-hand side denotes the standard scalar ordering. The infimum operation $\overset{\mathrm{T}}{\wedge}$ associated to $\overset{\mathrm{T}}{\leq}$ is given by: $x \overset{\mathrm{T}}{\wedge} y = \text{median}\{x, y, 0\}$.

Now, consider the quantization by truncation with step $q > 0$, $Q_q^{(T)}$, defined by:

$$Q_q^{(T)}(x) = \begin{cases} \lfloor x/q \rfloor, & \text{if } x \geq 0, \\ \lceil x/q \rceil, & \text{if } x < 0. \end{cases} \tag{10}$$

Proposition 6 *For any $q > 0$, $Q_q^{(T)}$ is an erosion in $\left(\mathbb{R}, \overset{\mathrm{T}}{\leq}\right)$. The adjoint dilation of $Q_q^{(T)}$ is the dequantization $D_q^{(T)}$ given by:*

$$D_q^{(T)}(x) \triangleq \begin{cases} q\lceil x \rceil, & \text{if } x \geq 0, \\ q\lfloor x \rfloor, & \text{if } x < 0. \end{cases} \tag{11}$$

Similar results can be obtained for quantization by rounding, $Q_q^{(R)}$, with a more complex partial ordering.

3. Signal Processing on Posets

The various examples in Section 2 motivates us to look at MM theoretical fundamentals as a possible, appropriate framework for signal processing in general. In this section, we investigate signal processing from the point of view of MM on posets. The investigation is limited to deterministic operators.

3.1. Morphology on Posets

The recent extension of MM from complete lattices to complete semilattices [5] is not the most general possible. It turns out, from the work of Heijmans and Ronse [6], that, with the concept of *adjunctions*, MM can be based on general partially ordered sets (posets).

Let $\overset{\mathcal{A}}{\leq}$ and $\overset{\mathcal{B}}{\leq}$ be two partial orderings defined on arbitrary sets $\mathcal{A}$ and $\mathcal{B}$ respectively. Let $\varepsilon : \mathcal{A} \mapsto \mathcal{B}$ and $\delta : \mathcal{B} \mapsto \mathcal{A}$ be two operators. (ε, δ) is called an adjunction between $\mathcal{A}$ and $\mathcal{B}$, if: $\forall a \in \mathcal{A}, b \in \mathcal{B}$, $b \overset{\mathcal{B}}{\leq} \varepsilon(a) \Leftrightarrow \delta(b) \overset{\mathcal{A}}{\leq} a$.

As in MM on complete lattices, operators ε and δ that form an adjunction between posets satisfy distributivity over infimum and supremum, respectively, and therefore will be called *erosion* and *dilation*, respectively.

3.2. Simplification and Partial Orders

Typically, the aim of a signal processing operation is to address a *simplified version* of a given signal. By "addressing a simplified version" we mean either *producing* a simplified version of a given signal, *recovering* a signal from a simplified version of it, or *transforming* the signal in order to ultimately assess a simplified version of it.

The term "simplified version" implies the existence of a partial ordering $\preceq$ on the set of signals $\mathcal{S}$, where a signal $f \in \mathcal{S}$ is a simplified version of another $g \in \mathcal{S}$ iff $f \preceq g$. This partial order constitutes a hierarchy in $\mathcal{S}$.

Often (but not necessarily), one can also associate a simplicity *measure* μ to this hierarchy, which is a non-negative real function that quantifies the simplicity of each signal. Examples of useful simplicity measures of signals include energy, entropy, the inverse of smoothness (roughness), variance, error, area, length, etc. The measure and the partial ordering should be related as follows: $f \prec g \Rightarrow \mu(f) \leq \mu(g)$[1]. Notice that $\mu(f) < \mu(g) \not\Rightarrow f \prec g$, which means that being simpler than another signal does not necessarily means being a simpler *version* of it.

3.3. Ideal Filters and Openings

Simplification of signals are obtained by *filters*. J. Serra compares the functionality of a filter with hand washing [7]: *i)* It does not add dirt to the hands, *ii)* once the hands are clean, further cleaning produces nothing, and *iii)* washing only one hand cleans less than washing both. The above features are mathematically described as *anti-extensivity*, *idempotence*, and *increasingness*, respectively. Serra also points out that these correspond exactly to the definition of *algebraic opening*. The conclusion is that filtering is performing an algebraic opening, and vice-versa.

An interesting and strong aspect of filtering is given by the following proposition.

Proposition 7 *Let $\mathcal{A}$ be an arbitrary set. For every idempotent operator $\gamma : \mathcal{A} \mapsto \mathcal{A}$, there exists a partial ordering $\preceq$, such that γ is an opening in the poset $(\mathcal{A}, \preceq)$.*

Proof: Define $\preceq$ as follows: $a_1 \preceq a_2 \Leftrightarrow a_1 = a_2$ or $a_1 = \gamma(a_2)$. It is easy to show that the above is a partial ordering. In this case, γ is increasing, since $a_1 \preceq a_2 \Rightarrow \gamma(a_1) = \gamma(a_2) \Rightarrow \gamma(a_1) \preceq \gamma(a_2)$. Moreover, γ is idempotent by definition, and anti-extensive, by definition of $\preceq$. Hence, γ is an opening. □

I.e., any idempotent operator can be seen as a filter according to some simplification criterion (partial ordering). However, it is obvious that not every idempotent operator is an appropriate filter for signal processing. Actually, most are not. To be appropriate, the underlying partial ordering of the filter should match one's intuition of what "simpler version" means.

3.4. Information Removal and Openings

Next proposition permits us to conclude that the only way to remove information is by using openings (filters).

---

[1] If $\mathcal{S}$ is discrete, or if the signals f and g are "regular," then $f \prec g \Rightarrow \mu(f) < \mu(g)$

Proposition 8 *Let $\mathcal{A}$ be an arbitrary set. Any operator $\varphi : \mathcal{A} \mapsto \mathcal{A}$ can be decomposed into $\varphi = \varepsilon'\gamma$, where $\gamma : \mathcal{A} \overset{\text{onto}}{\mapsto} \mathcal{A}' \subseteq \mathcal{A}$ is an* idempotent *operator, and $\varepsilon' : \mathcal{A}' \mapsto \mathcal{A}$ is an* invertible *operator.*

Proof: Let $\mathcal{B} \subseteq \mathcal{A}$ be the range of φ. For every $b \in \mathcal{B}$, consider the set $\mathcal{A}_b^\varphi \triangleq \{a \in \mathcal{A} |\ b = \varphi(a)\}$. Construct an operator $\delta : \mathcal{B} \mapsto \mathcal{A}$, by assigning, for each $b \in \mathcal{B}$, an arbitrary element of $\mathcal{A}_b^\varphi$ to $\delta(b)$. Let the range of δ be denoted by $\mathcal{A}'$. Notice that any such construction satisfies $\varphi\delta = id$, and $\delta\varphi$ is an idempotent operator. We then assign $\gamma \triangleq \delta\varphi$. We define ε' as identical to φ, but with domain restricted to $\mathcal{A}'$, i.e., we define $\varepsilon'(a') = \varphi(a')$, for all $a' \in \mathcal{A}'$. Even though φ is not necessarily invertible, ε' is, and its inverse is δ. This is because *i)* $\delta\varepsilon'(a') = \delta\varphi(a') = \gamma(a') = a'$, and *ii)* $\varepsilon'\delta(b) = \varphi\delta(b) = b$. Finally notice that $\varepsilon'\gamma = \varphi\delta\varphi = \varphi$. □

Since ε' is invertible, it does not remove information from a signal. It merely modifies the format of the information. Therefore, if φ is lossy, then the information is lost by means of the idempotent operator γ, which according to Proposition 7 is an opening in some poset. Notice that the decomposition referred to in Proposition 8 is not unique, which means that there may be more than one opening "behind" a given operator, and more than one partial ordering related to an operator. In other words, there may be more than one way to interpret the information loss. But in each case, the latter is caused by an opening.

3.5. INFORMATION REMOVAL AND EROSIONS

Proposition 9 *Let $\gamma : \mathcal{A} \overset{\text{onto}}{\mapsto} \mathcal{A}' \subseteq \mathcal{A}$ be an* idempotent *operator, $\varepsilon' : \mathcal{A}' \overset{\text{onto}}{\mapsto} \mathcal{B} \subseteq \mathcal{A}$ an* invertible *operator, and δ its inverse. Define the operator $\varepsilon \triangleq \varepsilon'\gamma$, as well as the relation $\overset{\bullet}{\preceq}$ in $\mathcal{A}$: $a_1 \overset{\bullet}{\preceq} a_2 \Leftrightarrow a_1 = \delta^n\varepsilon^n(a_2)$ for some non-negative integer[2] n. Then:*

i) The relation $\overset{\bullet}{\preceq}$ is a partial ordering in $\mathcal{A}$.

ii) (ε, δ) is an adjunction between the posets $\left(\mathcal{A}, \overset{\bullet}{\preceq}\right)$ and $\left(\mathcal{B}, \overset{\bullet}{\preceq}\right)$.

iii) γ is the morphological opening associated with the above adjunction.

The proof is omitted because of lack of space. Propositions 8 and 9 together lead to the following even stronger proposition.

Proposition 10 *For every operator $\varphi : \mathcal{A} \overset{\text{onto}}{\mapsto} \mathcal{B} \subseteq \mathcal{A}$ there is a partial ordering $\overset{\bullet}{\preceq}$, for which φ is an erosion between the posets $\left(\mathcal{A}, \overset{\bullet}{\preceq}\right)$ and $\left(\mathcal{B}, \overset{\bullet}{\preceq}\right)$.*

As in the case of openings, the above proposition is relevant only when $\overset{\bullet}{\preceq}$ matches one's intuition of what a "simpler version" is. For instance, one could say that one reason for linear filtering (convolution) to be usually regarded

[2] δ^n and ε^n are the n-fold application of δ and ε, respectively, and $\delta^0 = \varepsilon^0 = id$.

as an appropriate signal processing operation is because it is an erosion with underlying partial ordering (1) associated to *signal energy*[3], which is often recognized as a good measure of signal complexity. On the other hand, in many applications (e.g., segmentation), signal energy is not an appropriate measure, and other (non-linear) methods are usually invoked.

We should also point out that the range $\mathcal{B}$ of the operators φ and ε' in Propositions 7, 8, and 9, are required here to be a subset of $\mathcal{A}$ only for simplicity. One can show that these propositions can be extended also for an arbitrary $\mathcal{B}$, which means that in fact *every* mapping is an erosion between posets.

3.6. Rôles in Signal Processing

Summarizing the above ideas, one identify the following links between signal processing and morphological concepts.

- Simplification criteria for signal processing are partial orderings. In most practical cases, this partial ordering characterizes a CISL, since it is intuitive to expect the existence of a "simplest signal," but not of a "most complicated" one.
- Simplification (information removal) is performed by openings, which are ideal filters.
- Instead of openings, it is nevertheless more usual to use erosions for simplification, which consist in fact of filtering plus an invertible distortion/ transformation. Examples of erosions: linear filtering, decimation, quantization, coding, compression, denoising, pattern detection, etc.
- To each of the above simplification operators, correspond a "pseudo-inverse," the adjoint dilation, which does not add or remove information – just cancels the invertible distortion/transformation performed by the erosion. Examples: Inverse operators in general, interpolation, dequantization, decoding, restoration, pattern reconstruction, enhancement (deblurring), etc.

The above links can be depicted schematically through the diagram in Fig. 1. Two particular examples of SP tasks seen as morphological operations are shown in Fig. 2.

Notice that, in the above scheme, the morphological closing operation has no intrinsic meaning or function. In many cases it is simply equal to the identity operator, and in other cases it is not well defined. In this context, the classical closing on a complete lattice, which usually does have a meaning and function, is seen as an opening in the dual complete lattice. This is because such closing is used for filtering, and therefore function as an opening.

4. Discussion

The ideas described above can develop into two different research approaches: *Analysis*, which aims in providing new insights and understanding of old schemes,

[3] The partial order definition in terms of absolute value of the frequency components in (1) is equivalent to a definition in terms of the energy (square of absolute value) of the frequency components.

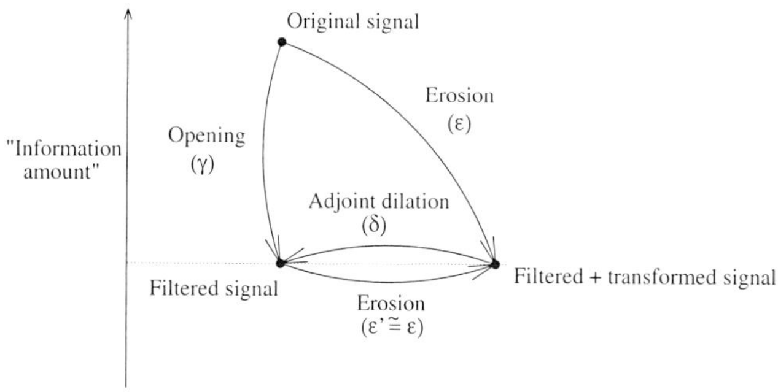

Fig. 1. Schematic description of the links between signal processing tasks and morphological operators.

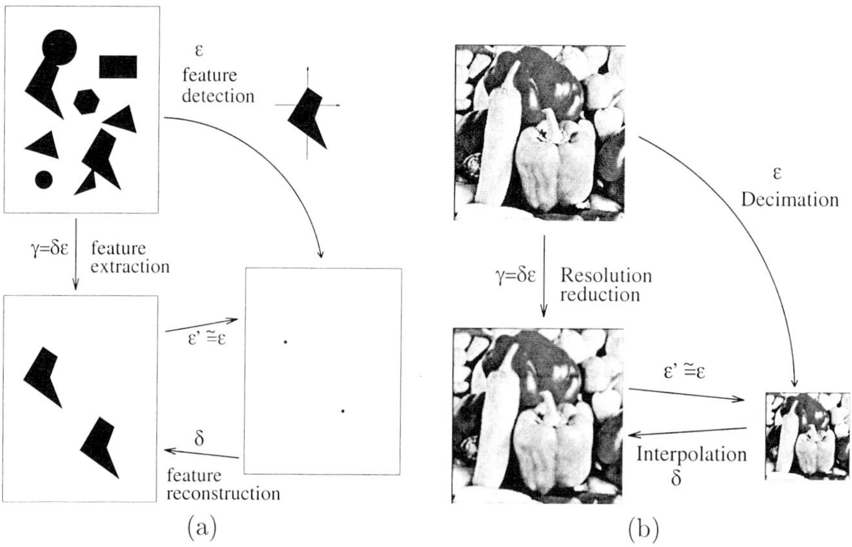

Fig. 2. Examples of signal processing tasks seen as morphological operations. (a) Feature extraction, and (b) resolution reduction.

and *synthesis*, which should result in new signal processing algorithms, tools, etc., or in ways of improving existing ones.

The work described here is purely analytic. It provides observations that lead to a different way of regarding traditional signal processing tasks. Further analysis research should try to answer questions like: "What is the underlying partial ordering (simplification criterion) behind a given signal processing task?" "Does this partial ordering make sense intuitively?" "What is the ideal filter associated to a given signal processing task?" Etc.

Regarding the synthesis approach, no contribution was provided yet, at this point. This approach could work as follows: First, find a partial ordering that is suitable to a specific set of signals and a given application. Then, design algebraic openings on the resulting poset. Finally, design erosions/adjoint-dilations that are associated to the above openings (optional). One possible, interesting candidate for partial ordering is the concept of "leveling," proposed by Meyer (see [9]).

5. Conclusion

The observations presented in this paper suggest that morphological fundamentals (partial ordering, openings, erosions, dilations, adjunctions) on posets could serve as a framework for signal processing in general.

We first showed that some key signal processing tasks can be seen as erosions on CISL's, and we presented underlying partial orderings (simplification criteria) related to them.

We then argued that every operator can be seen as a morphological erosion between posets, and is related to an ideal filtering (the associated opening) and a simplification criterion (the corresponding partial ordering).

We also stressed that the work presented here is purely analytic, and one still has to provide answers to questions like: "Can a MM framework for signal processing provide new algorithms and application, or improve existing ones?" Moreover, extension to stochastic signal processing still has to be investigated.

References

1. P. Salembier and J. Serra, "Pyramids and Multiresolution Decompositions," in *Course on Mathematical Morphology and its Applications to Signal Processing: Lecture notes*, organized by J. Serra and F. Marqués, May 1993, Universitat Politecnica de Catalunya, Barcelona, Spain.
2. R. Kresch, *Morphological Image Representation for Coding Applications*, D.Sc. Thesis (in English), Dept. of Electrical Engineering, Technion, Israel, June 1995. Available at `http://www-sipl.technion.ac.il/publications/thesis/renato/renato.html`.
3. R. Kresch and D. Malah, "Quadtree and Bit-Plane Decomposition as Particular Cases of the Generalized Morphological Skeleton," IEEE Workshop on Non-Linear Signal and Image Processing 1995, Halkidiki, Greece, June 1995.
4. J. Goutsias and H.J.A.M Heijmans, "An Axiomatic Approach to Multiresolution Signal Decomposition," ICIP-98.
5. R. Kresch, "Extension of Morphological Operations to Complete Semilattices and Its Applications to Image and Video Processing," *Mathematical Morphology and its Applications to Image and Signal Processing* (Proc. of ISMM'98), H.J.A.M. Heijmans and J.B.T.M. Roerdink (eds.), pp. 35-42, June 1998, to appear in *Fundamenta Informaticæ*.
6. H.J.A.M. Heijmans and C. Ronse, "The Algebraic Basis of Mathematical Morphology. I. Dilations and Erosions," *Comput. Vis. Graph. Im. Proc.*, Vol. 50, pp. 245-295, 1990.
7. J. Serra, "Morphological Filtering: An Overview", *Proc. First Int. Workshop on Mathematical Morphology and its Applications to Signal Processing*, J. Serra and Ph. Salembier (eds.), Barcelona, Spain, pp.1-10, May 1993.
8. C. Ronse, "A Lattice-Theoretical Morphological View on Template Extraction in Images", *J. Visual Comm. and Image Rep.*, Vol. 7, No. 3, pp.273-295, Sep. 1996.
9. F. Meyer, "The Levelings", *Mathematical Morphology and its Applications to Image and Signal Processing* (Proc. of ISMM'98), H.J.A.M. Heijmans and J.B.T.M. Roerdink (eds.), pp. 35-42, June 1998.
10. P. Maragos, "A Representation Theory for Morphological Image and Signal Processing," *IEEE transactions on Pattern Analysis and Machine Intelligence*, Vol. 11, No. 6, Jun. 1989.

FROM THE SUP-DECOMPOSITION TO A SEQUENTIAL DECOMPOSITION

RONALDO F. HASHIMOTO and JUNIOR BARRERA
Departamento de Ciência da Computação
Instituto de Matemática e Estatística. Universidade de São Paulo.
E-mail: ronaldo@ime.usp.br and jb@ime.usp.br

and

EDWARD R. DOUGHERTY
Texas Center Applied Technology.
Department of Electrical Engineering. Texas A & M University.
E-mail: edward@ee.tamu.edu

Abstract. $W-$operators are discrete set operators that are both translation invariant and locally defined within a finite window W. A particularly interesting property of $W-$operators is that they have a sup-decomposition in terms of a family sup-generating operators, that are parameterized by the operator basis. The sup-decomposition has a parallel structure that usually is not efficient for computation in conventional sequential machines. In this paper, we formalize the problem of transforming sup-decompositions into purely sequential decompositions (when they exist). The techniques were developed for general $W-$operators, specialized for increasing $W-$operators and applied on operators built by alternating compositions of dilations and erosions.

Key words: $W-$Operators, Sup-Decomposition, Sequential Decomposition, Basis.

1. Introduction

$W-$operators are discrete set operators that are both translation invariant and locally defined within a finite window W. Due to their great utility in binary image analysis, they have been intensively studied. One of the most successful approaches to represent $W-$operators is Mathematical Morphology [7, 5, 2].

A particularly interesting property of $W-$operators is that they have sup-decompositions, that is, they can be decomposed in terms of a family of sup-generating operators (i.e., the uniquely intersection of erosions and complemented dilations), that are parameterized by the operator basis [1] (i.e., a collection of maximal intervals). The sup-decomposition has a parallel structure that usually is not efficient for computation in conventional sequential machines.

In this paper, we formalize the problem of transforming the sup-decompositions into purely sequential decompositions (when they exist). The theory proposed consists in the formulation and solution of discrete equations in a lattice space. Due to their discrete nature, the solution of these equations depends on the combinatory techniques that were developed for general $W-$operators, specialized for increasing $W-$operators and applied on operators built by alternating compositions of dilations and erosions.

The results presented here extend to general $W-$operators some results given by Jones [6] for increasing translation invariant operator and introduce some new bounds for the particular case of increasing $W-$operators.

Following this introduction, Sections 2 and 3 reviews some properties of, respectively, the lattice of sets and of collections of maximal intervals. Section 4 introduces and gives bounds for the Minkowski factorization equation. Section 5 specializes the results of Section 4 and introduces a necessary condition for the structuring element. Section 6 gives some properties for collections of $W-$operators. Section 7 shows how to apply the results of Section 5 to compute the structuring elements of morphological operators built by alternating composition of dilations and erosions from their basis. Section 8 gives an application of the proposed method to a simple example for transforming a sup-decomposition into a purely sequential decomposition. Finally, in Section 9, we discuss the results presented and give some future steps of this research.

The proof of the results given in this paper are presented in [4].

2. Lattice of Sets

Let E be a non empty set and let W be a finite subset of E. Let $\mathcal{P}(W)$ denote the power set of W. Elements of $\mathcal{P}(W)$ will be denoted by capital letters $A, B, C, ...$ Let $\subseteq$ be the usual inclusion relation on sets. The pair $(\mathcal{P}(W), \subseteq)$ is a Complete Boolean Lattice [3]. The intersection and union of X and Y in $\mathcal{P}(E)$ are, respectively, $X \cap Y$ and $X \cup Y$. The complementary set of $X \in \mathcal{P}(W)$ with respect to W, denoted X^c_W or, simply, X^c, when no confusion is possible, is $X^c_W = \{x \in W : x \notin X\}$.

Let E be a non empty set, that is an Abelian group with respect to a binary operation denoted by $+$. The zero element of $(E, +)$ is denoted by o.

The *transpose* of a subset $X \in \mathcal{P}(E)$ is the subset X^t, given by $X^t = \{x \in E : -x \in X\}$.

For any $X \in \mathcal{P}(E)$ and $y \in E$, X_y denotes the translation of X by y, that is, $X_y = \{x \in E : x - y \in X\}$.

Let $X, Y \in \mathcal{P}(E)$. The *Minkowski addition* and *subtraction* of X and Y are, respectively, the subsets $X \oplus Y$ and $X \ominus Y$ given by $X \oplus Y = \cup\{X_y : y \in Y\}$ and $X \ominus Y = \cap\{X_{-y} : y \in Y\}$.

Let $X, Y \in \mathcal{P}(E)$. We say Y is an *invariant* of X if and only if (iff) $X = (X \ominus Y) \oplus Y$.

3. Lattice of Collection of Maximal Intervals

Let $\mathcal{P}(\mathcal{P}(W))$ be the collection of all subcollections of $\mathcal{P}(W)$. Elements of $\mathcal{P}(\mathcal{P}(W))$ will be denoted by capital script letters $\mathcal{A}, \mathcal{B}, \mathcal{C}, ...$ If $\subseteq$ is the usual inclusion relation on sets, then the pair $(\mathcal{P}(\mathcal{P}(W)), \subseteq)$ is a Complete Boolean Lattice.

Let $A, B \in \mathcal{P}(W)$ such that $A \subseteq B$. An *interval* of extremities A and B is the subset $[A, B]$ of $\mathcal{P}(W)$ given by $[A, B] = \{X \in \mathcal{P}(W) : A \subseteq X \subseteq B\}$. The sets A and B are called, respectively, the *left* and *right extremities* of the interval $[A, B]$. For all pairs (A, B) such that $A \not\subseteq B$, $[A, B]$ represents the empty collection, also denoted $\emptyset$.

Collections of intervals contained in $\mathcal{P}(W)$ will be denoted by capital bold face letters as $\mathbf{A}_W, \mathbf{B}_W, \mathbf{C}_W, ...,$ or, simply, $\mathbf{A}, \mathbf{B}, \mathbf{C}, ...,$ when no confusion is possible.

An *interval* $[A, B]$ in a collection of intervals $\mathbf{X}$ is called *maximal* iff there does not exist an interval $[A', B']$ in $\mathbf{X}$, distinct of $[A, B]$, such that $[A, B] \subseteq [A', B']$.

The collection of all maximal intervals of $\mathbf{X}$ is denoted $Max(\mathbf{X})$. Of course, if all the intervals in $\mathbf{X}$ are maximal, then $\mathbf{X} = Max(\mathbf{X})$.

Let $\mathcal{X}$ be a subcollection of $\mathcal{P}(W)$. The collection of all maximal intervals contained in $\mathcal{X}$ is denoted $\mathbf{M}(\mathcal{X})$, that is,

$$\mathbf{M}(\mathcal{X}) = Max(\{[A, B] \subseteq \mathcal{P}(W) : [A, B] \subseteq \mathcal{X}\}).$$

We denote by $\cup\mathbf{X}$ the collection of all elements of $\mathcal{P}(W)$ that are elements of intervals in $\mathbf{X}$, that is, $\cup\mathbf{X} = \{X \in \mathcal{P}(W) : X \in [A, B],\ [A, B] \in \mathbf{X}\}$.

Let Π_W denote the set $\{\mathbf{M}(\mathcal{X}) : \mathcal{X} \subseteq \mathcal{P}(W)\}$. We will define the partial order $\leq$ on the elements of Π_W by setting, for all $\mathbf{X}, \mathbf{Y} \in \Pi_W$,

$$\mathbf{X} \leq \mathbf{Y} \quad \Leftrightarrow \quad \forall [A, B] \in \mathbf{X},\ \exists [A', B'] \in \mathbf{Y} : [A, B] \subseteq [A', B'].$$

The poset $(\Pi_W, \leq)$ constitutes a Complete Boolean Lattice [2]. The supremum and infimum operations in the lattice $(\Pi_W, \leq)$ are given, respectively, by for any $\mathbf{X}, \mathbf{Y} \in \Pi_W$, $\mathbf{X} \sqcup \mathbf{Y} = \mathbf{M}((\cup\mathbf{X}) \cup (\cup\mathbf{Y}))$ and $\mathbf{X} \sqcap \mathbf{Y} = \mathbf{M}((\cup\mathbf{X}) \cap (\cup\mathbf{Y}))$.

Given two collections of maximal intervals $\mathbf{X}_{W_1} \in \Pi_{W_1}$ and $\mathbf{Y}_{W_2} \in \Pi_{W_2}$, we say that $\mathbf{X}_{W_1} = \mathbf{Y}_{W_2}$ iff $W_1 = W_2$, (i.e., $\Pi_{W_1} = \Pi_{W_2}$) and $\mathbf{X}_{W_1} \leq \mathbf{Y}_{W_1}$ and $\mathbf{Y}_{W_1} \leq \mathbf{X}_{W_1}$.

Let $W', W \in \mathcal{P}(E)$ such that $W' \supseteq W$. We define the set $\Pi_{W'/W} \subseteq \Pi_{W'}$ as $\Pi_{W'/W} = \{\mathbf{X} \in \Pi_{W'} : A' \subseteq W \text{ and } B' - W = W^c,\ \forall\, [A', B'] \in \mathbf{X}\}$.

Proposition 1 *Let $W', W \in \mathcal{P}(E)$ such that $W' \supseteq W$. The mapping $\mathcal{W}(\cdot)$, from $(\Pi_W, \leq)$ to $(\Pi_{W'/W}, \leq)$, defined by*

$$\mathcal{W}(\mathbf{X}_W) = \{[A', B'] \subseteq \mathcal{P}(W') : A' = A \text{ and } B' = B \cup W^c,\ [A, B] \in \mathbf{X}_W\}$$

constitutes a lattice isomorphism between the lattices $(\Pi_W, \leq)$ and $(\Pi_{W'/W}, \leq)$. The inverse of the mapping $\mathcal{W}(\cdot)$ is the mapping $\mathcal{W}^{-1}(\cdot)$, from $\Pi_{W'/W}$ to Π_W, defined by

$$\mathcal{W}^{-1}(\mathbf{X}_{W'}) = \{[A, B] \subseteq \mathcal{P}(W) : A = A' \text{ and } B = B' \cap W,\ [A', B'] \in \mathbf{X}_{W'}\}.$$

As a consequence of Proposition 1, if $W \subseteq W'$ we can change of representation of a collection of maximal intervals $\mathbf{X}_W \in \Pi_W$ to $\mathbf{X}_{W'} \in \Pi_{W'/W} \subseteq \Pi_{W'}$, and vice-versa.

For any $\mathbf{X} \in \Pi_W$ and $y \in E$, $\mathbf{X}_y \in \Pi_{W_y}$ denotes the translation of all intervals of $\mathbf{X}$ by y, that is, $\mathbf{X}_y = \{[A_y, B_y] \in \Pi_{W_y} : [A, B] \in \mathbf{X}\}$.

Let $\mathbf{X}_W \in \Pi_W$ and $C \in \mathcal{P}(E)$. The *Minkowski addition* and *subtraction* of $\mathbf{X}_W$ by C are, respectively, the collection of maximal intervals $\mathbf{X}_W \oplus C \in \Pi_{W \oplus C}$ and $\mathbf{X}_W \ominus C \in \Pi_{W \oplus C^t}$ given by $\mathbf{X}_W \oplus C = \sqcup\{(\mathbf{X}_h)_{W \oplus C} : h \in C\}$ and $\mathbf{X}_W \ominus C = \sqcap\{(\mathbf{X}_{-h})_{W \oplus C^t} : h \in C\}$.

4. Minkowski Factorization Equation

Problem *Given a collection of maximal intervals $\mathbf{Y}_{W'} \in \Pi_{W'}$ and a set $C \in \mathcal{P}(E)$, find all collections of maximal intervals $\mathbf{X}_W \in \Pi_W$ such that*

$$\mathbf{X}_W \oplus C^t = \mathbf{Y}_{W'}. \tag{1}$$

Since $\mathbf{X}_W \oplus C^t \in \Pi_{W \oplus C^t}$ and $\mathbf{Y}_{W'} \in \Pi_{W'}$, then, by Equation (1) the windows W and W' satisfy $W \oplus C^t = W'$.

The next result is a property of collection of maximal intervals.

Proposition 2 *Let $\mathbf{X}_W \in \Pi_W$, $C \in \mathcal{P}(E)$ and $\mathbf{Y}_{W'} \in \Pi_{W'}$. If $\mathbf{X}_W \oplus C^t = \mathbf{Y}_{W'}$, then, for any $h \in E$, $(\mathbf{X}_h)_{W_h} \oplus (C_h)^t = \mathbf{Y}_{W'}$.*

4.1. Upper Bound for W

The next result states an upper bound for the window W in Equation (1) and is a direct consequence of the adjunction relation given in [5, p. 84, Eq. 4.41].

Proposition 3 *Let $W, C, W' \in \mathcal{P}(E)$. If $W \oplus C^t = W'$, then $W \subseteq W' \ominus C^t$.*

Given a set $C \in \mathcal{P}(E)$ and a collection of maximal intervals $\mathbf{Y}_{W'}$, by Propositions 3 and 1, the collections $\mathbf{X}_W \in \Pi_W$ that satisfy Equation (1) can have their representations changed to $\mathbf{X}_{W' \ominus C^t} \in \Pi_{W' \ominus C^t / W}$. Therefore, it is sufficient to consider that $W = W' \ominus C^t$.

4.2. Upper Bound for $\mathbf{X}_W$

In this section, we state an upper bound for $\mathbf{X}_W$. Let $W, C \in \mathcal{P}(E)$. An immediate consequence of the Minkowski addition is that $o \in C^t \oplus C$. So, we can easily see that $W \subseteq W \oplus (C^t \oplus C)$. Thus, by the associative property [5, p. 82, Eq. 4.29] of the Minkowski addition, $W \subseteq (W \oplus C^t) \oplus C$. Hence, by Proposition 1, we can change the representation of any collection of maximal intervals $\mathbf{X}_W \in \Pi_W$ to $\mathbf{X}_{W''} \in \Pi_{W''}$, where $W'' = (W \oplus C^t) \oplus C$.

The following theorem [4] gives an upper bound for all $\mathbf{X}_{W''} \in \Pi_{W''}$.

Theorem 1 *Let $\mathbf{Y}_{W'} \in \Pi_{W'}$ and $C \in \mathcal{P}(E)$. For any $\mathbf{X}_W \in \Pi_W$ such that $\mathbf{X}_W \oplus C^t = \mathbf{Y}_{W'}$, we have that $\mathbf{X}_{W''} \leq \mathbf{Y}_{W'} \ominus C^t$, where $W'' = (W \oplus C^t) \oplus C$.*

Let $\mathbf{X}_W$ be a solution of Equation (1). Thus, $\mathbf{X}_{W''} \in \Pi_{W''/W}$. By Theorem 1, $\mathbf{X}_{W''} \leq \mathbf{Y}_{W'} \ominus C^t$. So, we get an upper bound for $\mathbf{X}_{W''}$. Now, we will show an upper bound for $\mathbf{X}_W$. Since $\mathbf{X}_{W''} \leq \mathbf{Y}_{W'} \ominus C^t$, then, for any $[A, B] \in \mathbf{X}_{W''}$ there exists $[L, R] \in \mathbf{Y}_{W'} \ominus C^t$ such that $[A, B] \subseteq [L, R]$. As $\mathbf{Y}_{W'} \ominus C^t \in \Pi_{W''}$, then $L \subseteq A$ and $B \subseteq R \subseteq W''$. Since $[A, B] \in \mathbf{X}_{W''}$ and $\mathbf{X}_{W''} \in \Pi_{W''/W}$, then $A \subseteq W$ and $B - W = W^c$. So, $L \subseteq A \subseteq W$ and $W^c = B - W \subseteq R - W \subseteq W'' - W = W^c$. Thus, $R - W = W^c$, and, consequently, $[L, R] \in \Pi_{W''/W}$. Hence, for any interval $[A, B] \in \mathbf{X}_{W''}$, there exists an interval $[L, R] \in \mathbf{Y}_{W'} \ominus C^t$, such that $[A, B] \subseteq [L, R]$ and $[L, R] \in \Pi_{W''/W}$. Therefore, if $\mathbf{U}_{W''} = \{[L, R] \in \mathbf{Y}_{W'} \ominus C^t : [L, R] \in \Pi_{W''/W}\}$ and $\mathbf{U}_W = \mathcal{W}^{-1}(\mathbf{U}_{W''})$, then, for all $\mathbf{X}_W$ that satisfy Equation (1), we have that $\mathbf{X}_W \leq \mathbf{U}_W$. Consequently, $\mathbf{U}_W$ is an upper bound for $\mathbf{X}_W$.

4.3. Lower Bounds for $\mathbf{X}_W$

Given an interval $[A', B'] \subseteq \mathcal{P}(W')$ and a set $C \in \mathcal{P}(E)$, we define the collections of intervals $\mathcal{L}_W^{[A',B'],C}$ and $\mathcal{R}_W^{[A',B'],C}$, contained in $\mathcal{P}(\mathcal{P}(W))$, where $W = W' \ominus C^t$, as

$$\mathcal{L}_W^{[A',B'],C} = \{[A'_{-x}, B \cap W] : A'_{-x} \subseteq W,\ B \subseteq B'_{-x},\ B_{-x} - W = W^c,\ x \in C^t\}$$

and

$\mathcal{R}_W^{[A',B'],C} = \{[A, B'_{-x} \cap W] : A'_{-x} \subseteq A,\ A'_{-x} \subseteq W,\ B_{-x} - W = W^c,\ x \in C^t\}$.

We define the set $\mathcal{H}_W^{[A',B'],C}$ as $\mathcal{H}_W^{[A',B'],C} = \{\{[P,Q]\} \sqcup \{[R,S]\} : [P,Q] \in \mathcal{L}_W^{[A',B'],C}, [R,S] \in \mathcal{R}_W^{[A',B'],C}\}$.

Let $I = \{1, 2, 3, \cdots, n\}$ be a set of indices. Let $\mathbf{Y}_{W'} = \{[A'_i, B'_i] : i \in I]\} \in \Pi_{W'}$ and $C \in \mathcal{P}(E)$. We define the set $\mathcal{S}_W^{\mathbf{Y}_{W'},C} = \mathcal{H}_W^{[A'_1,B'_1],C} \times \cdots \times \mathcal{H}_W^{[A'_n,B'_n],C}$.

Given a collection of maximal intervals $\mathbf{Y}_{W'} \in \Pi_{W'}$ and a subset $C \in \mathcal{P}(E)$, let us define the set of collection of intervals $\Theta_W^{\mathbf{Y}_{W'},C}$, where $W = W' \ominus C^t$, by $\Theta_W^{\mathbf{Y}_{W'},C} = \{\mathbf{Z}_W \in \Pi_W : \mathbf{Z}_W = \sqcup\{\mathbf{S}_W^i,\ i \in I\},\ (\mathbf{S}_W^1, \mathbf{S}_W^2, \cdots, \mathbf{S}_W^n) \in \mathcal{S}_W^{\mathbf{Y}_{W'},C}\}$.

The next result states lower bounds for $\mathbf{X}_W$ in Equation (1).

Theorem 2 *Let $C \in \mathcal{P}(E)$ and $\mathbf{Y}_{W'} \in \Pi_{W'}$. For all $\mathbf{X}_W \in \Pi_W$ such that $\mathbf{X}_W \oplus C^t = \mathbf{Y}_{W'}$, there exists $\mathbf{Z}_W \in \Theta_W^{\mathbf{Y}_{W'},C}$ such that $\mathbf{Z}_W \leq \mathbf{X}_W$.*

5. Fixed Right Extremity Simplification

Let us define the set $\mathcal{I}_W \subseteq \Pi_W$ as the set of all collections of maximal intervals such that the right extremity of any interval is the window W. We call any collection in $\mathcal{I}_W$ by *right window collection*.

Now, consider the problem, presented in Section 4, restricted to $\mathcal{I}_W$ and $\mathcal{I}_{W'}$, that is, given a collection $\mathbf{Y}_{W'} \in \mathcal{I}_{W'}$ and a set $C \in \mathcal{P}(E)$, find all collections of maximal intervals $\mathbf{X}_W \in \mathcal{I}_W$ such that $\mathbf{X}_W \oplus C^t = \mathbf{Y}_{W'}$.

For simplicity, where there is no risk of confusion, we denote the intervals $[A, W]$ of $\mathbf{X} \in \mathcal{I}_W$ by $[A]$.

5.1. Lower Bound Simplification

Given an interval $[A'] \subseteq \mathcal{P}(W')$ and a subset $C \in \mathcal{P}(E)$, we define the sets $\mathcal{L}_W^{[A'],C} = \{[A'_{-x}, W] :\ x \in C^t\}$ and $\mathcal{H}_W^{[A'],C} = \{\{[P,W]\} : [P,W] \in \mathcal{L}_W^{[A'],C}\}$.

Let $I = \{1, 2, 3, \cdots, n\}$ be a set of indices. Let $\mathbf{Y}_{W'} = \{[A'_i] : i \in I]\} \in \mathcal{I}_{W'}$ and $C \in \mathcal{P}(E)$. We define the set $\mathcal{F}_W^{\mathbf{Y}_{W'},C} = \mathcal{H}_W^{[A'_1],C} \times \cdots \times \mathcal{H}_W^{[A'_n],C}$.

Given a collection of maximal intervals $\mathbf{Y}_{W'} \in \mathcal{I}_{W'}$ and a subset $C \in \mathcal{P}(E)$, let us define the set of collection of intervals $\Phi_W^{\mathbf{Y}_{W'},C}$, where $W = W' \ominus C^t$, by $\Phi_W^{\mathbf{Y}_{W'},C} = \{\mathbf{Z}_W \in \mathcal{I}_W : \mathbf{Z}_W = \sqcup\{\mathbf{F}_W^i,\ i \in I\},\ (\mathbf{F}_W^1, \mathbf{F}_W^2, \cdots, \mathbf{F}_W^n) \in \mathcal{F}_W^{\mathbf{Y}_{W'},C}\}$.

Note that, by definition of the set $\Theta_W^{\mathbf{Y}_{W'},C}$ in Section 4.3, if the right extremity of the intervals in $\mathbf{Y}_{W'}$ and $\mathbf{Z}_W$ are, respectively, the windows W' and W, then $\Theta_W^{\mathbf{Y}_{W'},C}$ is reduced to the set $\Phi_W^{\mathbf{Y}_{W'},C}$. Thus, we can easily see that, $\Phi_W^{\mathbf{Y}_{W'},C} \subseteq \Theta_W^{\mathbf{Y}_{W'},C}$.

The following result, that is a particular case of Theorem 2, states lower bounds for $\mathbf{X}_W$ in Equation (1) for right window collections.

Theorem 3 *Let $C \in \mathcal{P}(E)$ and $\mathbf{Y}_{W'} \in \mathcal{I}_{W'}$. For all $\mathbf{X}_W \in \mathcal{I}_W$ such that $\mathbf{X}_W \oplus C^t = \mathbf{Y}_{W'}$, there exists $\mathbf{Z}_W \in \Phi_W^{\mathbf{Y}_{W'},C}$ such that $\mathbf{Z}_W \leq \mathbf{X}_W$ and $\mathbf{Z}_W \oplus C^t = \mathbf{Y}_{W'}$.*

As a consequence of Theorem 3, all lower bounds for $\mathbf{X}_W$ in Equation (1) are in $\Phi_W^{\mathbf{Y}_{W'},C}$. In fact, each $\mathbf{Z}_W \in \Phi_W^{\mathbf{Y}_{W'},C}$ such that $\mathbf{Z}_W \oplus C^t = \mathbf{Y}_{W'}$ is a lower bound for $\mathbf{X}_W$.

Let $W', W \in \mathcal{P}(E)$ such that $W' \supseteq W$. We define the set $\mathcal{I}_{W'/W} \subseteq \mathcal{I}_{W'}$ as $\mathcal{I}_{W'/W} = \{\mathbf{X} \in \mathcal{I}_{W'} : A' \subseteq W, \forall [A'] \in \mathbf{X}\}$. Note that $\mathcal{I}_{W'/W} \subseteq \Pi_{W'/W}$.

Now, we present an algorithm for solving the Equation (1) in the case $\mathbf{Y}_{W'} \in \mathcal{I}_{W'}$.

Algorithm SEARCH $(C, \mathbf{Y}_{W'})$:

Input: A set $C \in \mathcal{P}(E)$ and a collection of intervals $\mathbf{Y}_{W'} \in \mathcal{I}_{W'}$.

Output: The collections $\mathbf{X}_W \in \mathcal{I}_W$, where $W = W' \ominus C^t$, such that $\mathbf{X}_W \oplus C^t = \mathbf{Y}_{W'}$.

```
begin
  W'' ← (W ⊕ C^t) ⊕ C;  U_W'' ← {[L] ∈ Y_W' ⊖ C^t : [L] ∈ I_W''/W}.
  for each Z_W ∈ Φ_W^{Y_W',C} do
    if Z_W ⊕ C^t = Y_W' then
      for each X_W such that Z_W ≤ X_W ≤ U_W
        if X_W ⊕ C^t = Y_W' then
          output X_W;
end.
```

5.2. FEASIBLE SETS C

In Equation (1), a subset $C \in \mathcal{P}(E)$ is a fixed parameter. Obviously, given a collection $\mathbf{Y}_{W'} \in \mathcal{I}_{W'}$, there exist some subsets in $\mathcal{P}(E)$ such that Equation (1) has no solution. The subsets in $\mathcal{P}(E)$ such that Equation (1) has at least one solution are called *feasible sets.*

In this section, we study some properties of Equation (1) in order to give a necessary condition for the existence of feasible sets. Observe that, by Proposition 2, if a subset $C \in \mathcal{P}(E)$ is feasible, then, so is C_h, for any $h \in E$.

Let $[A] \in \mathbf{X} \in \mathcal{I}_W$. We say that the left extremity A is *minimal* in $\mathbf{X}$ iff $|A| \leq |B|$, for any interval $[B] \in \mathbf{X}$. Clearly, if $|A| = |B|$, then the extremities of $[A]$ and $[B]$ are minimal.

Let $\mathbf{Z} \in \mathcal{I}_W$. Let us denote by $\mathbf{Min}(\mathbf{Z})$ the set of all intervals in $\mathbf{Z}$ such that its left extremity is minimal in $\mathbf{Z}$, that is, $\mathbf{Min}(\mathbf{Z}) = \{[A] \in \mathbf{Z} : A \text{ is minimal in } \mathbf{Z}\}$.

Given a collection of maximal intervals $\mathbf{Z} \in \mathcal{I}_W$, for each set $A \in \mathcal{P}(E)$, let us define the set $S_A^{\mathbf{Z}} \in \mathcal{P}(E)$ as $S_A^{\mathbf{Z}} = \{h \in E : [A_{-h}] \in \mathbf{Z}\}$.

The next result gives a necessary condition for feasible sets.

Theorem 4 *Let $\mathbf{Y}_{W'} \in \mathcal{I}_{W'}$ and $C \in \mathcal{P}(E)$. If C is a feasible set, then, for any $[A'] \in \mathbf{Min}(\mathbf{Y}_{W'})$, there exists $a \in E$ such that $C_a \subseteq S_{A'}^{\mathbf{Y}_{W'}}$ and C is an invariant of $S_{A'}^{\mathbf{Y}_{W'}}$.*

Given a collection of maximal intervals $\mathbf{Y}_{W'} \in \mathcal{I}_{W'}$, as a consequence of Proposition 2, if C is feasible, then so is C_h, for any $h \in E$. By Theorem 4, if C is feasible, then, for any $[A'] \in \mathbf{Min}(\mathbf{Y}_{W'})$, a translation of C, say C_a, is a subset of $S_{A'}^{\mathbf{Y}_{W'}}$. Since C_a is also feasible, then the feasible sets can be found by searching $C \subseteq S_{A'}^{\mathbf{Y}_{W'}}$ such that C is an invariant of $S_{A'}^{\mathbf{Y}_{W'}}$.

Now, given a collection of maximal intervals $\mathbf{Y}_{W'} \in \mathcal{I}_{W'}$, we present an algorithm that outputs pairs $(C, \mathbf{X}_W) \in \mathcal{P}(E) \times \mathcal{I}_W$ such that $\mathbf{X}_W \oplus C^t = \mathbf{Y}_{W'}$.

Algorithm SEARCH_ALL $(\mathbf{Y}_{W'})$:

Input: A collection of intervals $\mathbf{Y}_{W'} \in \Pi_{W'}$.

Output: The pairs $(C, \mathbf{X}_W) \in \mathcal{P}(E) \times \mathcal{I}_W$, with $W = W' \ominus C^t$, such that $\mathbf{X}_W \oplus C^t = \mathbf{Y}_{W'}$.

```
begin
  let [A'] ∈ Min(Y_{W'}) such that |S_{A'}^{Y_{W'}}| is minimum.
  for each C ⊆ S_{A'}^{Y_{W'}} such that C is an invariant of S_{A'}^{Y_{W'}} do
    begin
      let {X_1, X_2, ···, X_n} be the output of SEARCH (C, Y_{W'});
      for i = 1, 2, ···, n output the pair (C, X_i);
    end
end.
```

6. Lattice of W–Operators

A mapping from $\mathcal{P}(E)$ to $\mathcal{P}(E)$ is called an *operator*. The operators will be denoted by lower case Greek letters $\alpha, \beta, \gamma, \ldots$ The set of all operators will be denoted by Ψ. The set Ψ inherits the Complete Boolean Lattice structure of $(\mathcal{P}(E), \subseteq)$ by setting, for any $\psi_1, \psi_2 \in \Psi$,

$$\psi_1 \leq \psi_2 \quad \Leftrightarrow \quad \psi_1(X) \subseteq \psi_2(X) \quad (X \in \mathcal{P}(E)).$$

The supremum and infimum of two operators ψ_1 and ψ_2 of Ψ verify, respectively, $(\psi_1 \vee \psi_2)(X) = \psi_1(X) \cup \psi_2(X)$ and $(\psi_1 \wedge \psi_2)(X) = \psi_1(X) \cap \psi_2(X)$, for any $X \in \mathcal{P}(E)$.

Let $C \in \mathcal{P}(E)$. The *dilation* and *erosion* by C are the operators δ_C and ε_C given by, for any $X \in \mathcal{P}(E)$, $\delta_C(X) = X \oplus C$ and $\varepsilon_C(X) = X \ominus C$. Usually, the subset C is called *structuring element.*

The operator $\nu \in \Psi$ defined by $\nu(X) = X^c$, for any $X \in \mathcal{P}(E)$, is called *negation operator.*

The *dual operator* of the operator ψ, denoted ψ^*, is given by $\psi^* = \nu\psi\nu$.

An operator ψ is called *translation invariant* (t.i.) iff, for any $x \in E$ and $X \in \mathcal{P}(E)$, $\psi(X_x) = \psi(X)_x$.

Let W be a finite subset of E. An operator ψ is called *locally defined* within W iff, for any $x \in E$ and $X \in \mathcal{P}(E)$, $x \in \psi(X) \Leftrightarrow x \in \psi(X \cap W_x)$.

An operator ψ is called a *W–operator* iff it is both t.i. and locally defined within W. The set of all W–operators will be denoted by Ψ_W. The pair $(\Psi_W, \leq)$ constitutes a sublattice of the lattice $(\Psi, \leq)$ [2].

The *kernel* of an operator $\psi \in \Psi_W$ is the set $\mathcal{K}_W(\psi)$ given by $\mathcal{K}_W(\psi) = \{X \in \mathcal{P}(W) : o \in \psi(X)\}$.

Barrera and Salas [2] stated the following lattice isomorphism between the complete lattices $(\Pi_W, \leq)$ and $(\Psi_W, \leq)$.

Proposition 4 *The mapping $\mathbf{M}(\mathcal{K}_W(\cdot))$ from $(\Psi_W, \leq)$ to $(\Pi_W, \leq)$ constitutes a lattice isomorphism between the lattices $(\Psi_W, \leq)$ and $(\Pi_W, \leq)$. The inverse of the mapping $\mathbf{M}(\mathcal{K}_W(\cdot))$ is the mapping $\mathcal{K}_W^{-1}(\cup(\cdot))$, where $\mathcal{K}_W^{-1}(\cdot)$ is defined by $\mathcal{K}_W^{-1}(\mathcal{X})(X) = \{x \in E : (X - x) \cap W \in \mathcal{X}\}$.*

For any operator $\psi \in \Psi_W$, the *basis* of ψ, denoted $\mathbf{B}_W(\psi)$, or simply $\mathbf{B}(\psi)$ when no confusion is possible, is the collection of all maximal intervals contained in $\mathcal{K}_W(\psi)$, that is, $\mathbf{B}_W(\psi) = \mathbf{M}(\mathcal{K}_W(\psi))$.

Given $\psi \in \Psi_W$ and $C \in \mathcal{P}(E)$, the operators $\delta_C\psi$ and $\varepsilon_C\psi$ are locally defined, respectively, within $W \oplus C^t$ and $W \oplus C$. The following proposition [2] shows how to build the basis of $\delta_C\psi$ and $\varepsilon_C\psi$ from the basis of ψ.

Proposition 5 *If $\psi \in \Psi_W$ and $C \in \mathcal{P}(E)$, then*

$$\mathbf{B}_{W \oplus C^t}(\delta_C\psi) = \mathbf{B}_W(\psi) \oplus C^t \text{ and } \mathbf{B}_{W \oplus C}(\varepsilon_C\psi) = \mathbf{B}_W(\psi) \ominus C^t.$$

The following result is a consequence of Proposition 5 and the fact that $(\varepsilon_{C^t}\psi)^* = \delta_C\psi^*$ [4].

Corollary 1 *Let* $\psi' \in \Pi_{W'}$, $C \in \mathcal{P}(E)$ *and* $\psi \in \Pi_W$. *Then,* $\psi' = \varepsilon_{C^t}\psi$ *iff* $\mathbf{B}_{W'}((\psi')^*) = \mathbf{B}_W(\psi^*) \oplus C^t$.

An operator ψ is called *increasing* iff $\forall X, Y \in \mathcal{P}(E)$, if $X \subseteq Y$, then $\psi(X) \subseteq \psi(Y)$. We denote by Ω_W the set of all increasing $W-$operators.

A very interesting property of basis of increasing $W-$operators is that $\psi \in \Psi_W$ is an increasing operator iff for any interval $[A, B] \in \mathbf{B}_W(\psi)$, $B = W$. Thus, we can easily prove that, ψ is increasing iff $\mathbf{B}_W(\psi) \in \mathcal{I}_W$.

Given $\psi \in \Psi_W$, ψ^* is locally defined within W [2]. In addition, if $\psi \in \Pi_W$ is increasing, then so is ψ^* [5, p. 46].

Let $I = \{1, 2, 3, \cdots, n\}$ be a set of indices. Given the basis of an increasing $W-$operator ψ, the next result shows how to build the basis of ψ^* from the basis of ψ. This proposition is a particular case of the result stated in [2].

Proposition 6 *If* ψ *is an increasing* $W-$*operator with basis* $\mathbf{B}_W(\psi) = \{[A_i] : i \in I\}$, *then the basis of its dual operator* ψ^* *is*

$$\mathbf{B}_W(\psi^*) = \sqcap\{[\{a\}] : a \in A_i^c, i \in I\}.$$

7. Compositions of Erosions and Dilations

We denote by Υ_W the set of all $W-$operators that may be built by an alternating compositions of erosions and dilations. Note that the set of all alternating sequential filters [8], locally defined within a window W, is a subset of Υ_W. Given the basis of an operator $\psi \in \Upsilon_W$, we describe how to find a representation for ψ.

If ψ is an operator in Υ_W, then an representation of ψ may start by a dilation or an erosion, that is, ψ may be rewritten by $\delta_C\psi_1$ or $\varepsilon_C\psi_1$.

Given the basis of an operator $\psi \in \Upsilon_W$ that starts by a dilation (respectively, erosion), then, by Proposition 5 (respectively, by Corollary 1), we can find a representation of ψ applying the procedure SEARCH_ALL for $\mathbf{B}_W(\psi)$ (respectively, for $\mathbf{B}_W(\psi^*)$). If $(C, \mathbf{X})$ is an output of the procedure SEARCH_ALL $(\mathbf{B}_W(\psi))$ (respectively, $\mathbf{B}_W(\psi^*)$), then ψ can be rewritten by $\psi = \delta_C\psi_1$ (respectively, by $\psi = \varepsilon_{C^t}\psi_1$), where ψ_1 is an increasing $W-$operator and $\mathbf{B}(\psi_1) = \mathbf{X}$ (respectively, $\mathbf{B}(\psi_1^*) = \mathbf{X}^*$).

Thus, given the basis of an operator $\psi \in \Upsilon_W$, we will construct the tree that represents the space of all possible representations of ψ in the following way. The root is the basis of ψ. A node is a collection of maximal intervals $\mathbf{Y}_W \in \mathcal{I}_W$. If $\mathbf{Y}_W = \{[\{o\}]\}$, then $\mathbf{Y}_W$ has no descendants. If $\mathbf{Y}_W = \{[\{a\}]\}$, $a \neq o$, then the descendant of $\mathbf{Y}_W$ is $\{[\{o\}]\}$ and the edge that joins $\mathbf{Y}_W$ and its descendant is labeled $\delta_{\{-a\}}$. In any other case, compute $\mathbf{Y}_W^*$ and apply the procedure SEARCH_ALL for $\mathbf{Y}_W$ and $\mathbf{Y}_W^*$. If $(C, \mathbf{X})$ is an output of SEARCH_ALL $(\mathbf{Y}_W)$, then $\mathbf{X}$ is a descendant of $\mathbf{Y}_W$ and the edge that joins $\mathbf{Y}_W$ to $\mathbf{X}$ is labeled δ_C. If $(C, \mathbf{X})$ is an output of the procedure SEARCH_ALL $(\mathbf{Y}_W^*)$, then $\mathbf{X}^*$ is a descendant of $\mathbf{Y}_W$ and the edge that joins $\mathbf{Y}_W$ to $\mathbf{X}^*$ is labeled ε_{C^t}. This tree is called *Representation Tree.*

Note that, given the basis of an operator $\psi \in \Upsilon_W$, the labels of the edges on the path from the root to a node $\mathbf{Y}_W = \{[\{o\}]\}$ forms a representation for ψ.

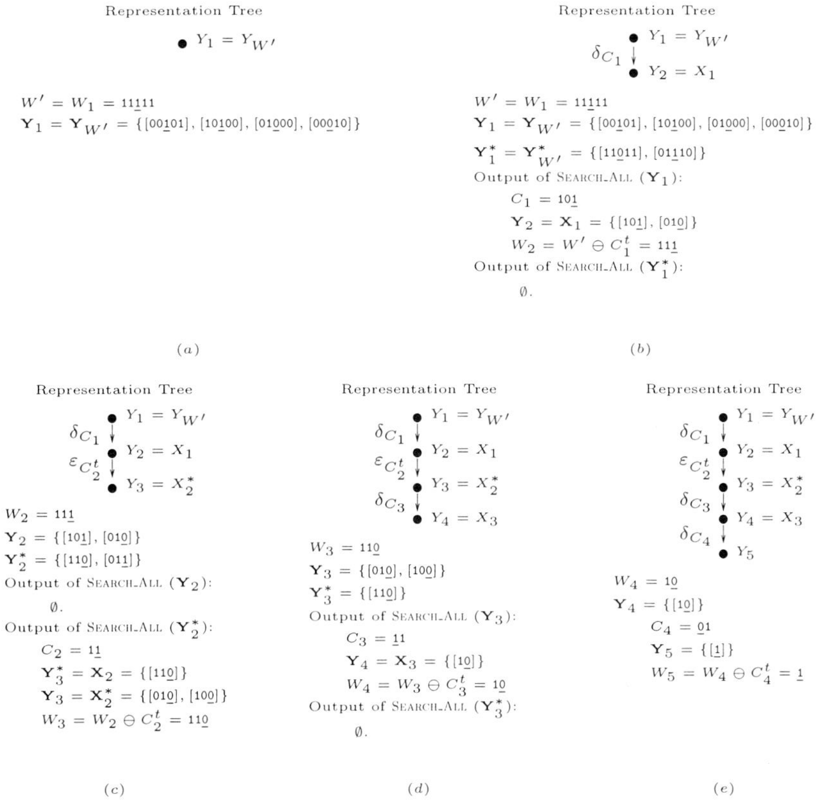

Fig. 1. Application of the proposed method to a simple example.

8. Application Example

We have implemented the method proposed in Section 7 for transforming a sup-decomposition into a sequential decomposition. In this section, we show an application of this algorithm to a simple example. For simplicity of notation, we represent the subsets of a window W by strings of 0's and 1's, where 0 means that the point does not belong to the subset and 1 means that it does. Moreover, the origin is represented by an underline character in the string description of a subset of W. For example, if W is the set $\{(-1,0),(0,0),(1,0)\}$, the subset $\{(0,0),(1,0)\}$ is represented by $0\underline{1}1$.

In Figure 1, we show a simple example for finding a sequential representation of an operator ψ in $\Upsilon_{W'}$. In this example, the basis of ψ is the collection of maximal intervals $\mathbf{Y}_1 = \mathbf{Y}_{W'}$ presented in Figure 1a. The root of the representation tree is the input basis, that is, $\mathbf{Y}_1$.

Figure 1b shows that the outputs of procedure SEARCH_ALL when applied for $\mathbf{Y}_1$ and $\mathbf{Y}_1^*$ are, respectively, $(C_1, \mathbf{X}_1)$ and $\emptyset$. Thus, the descendant of the root $\mathbf{Y}_1$ is $\mathbf{Y}_2 = \mathbf{X}_1$ and the edge that joins $\mathbf{Y}_1$ and $\mathbf{Y}_2$ is labeled δ_{C_1} (see Figure 1b).

In Figure 1c, we see that the outputs of SEARCH_ALL when applied for $\mathbf{Y}_2$ and $\mathbf{Y}_2^*$ are, respectively, $\emptyset$ and $(C_2, \mathbf{X}_2)$. Thus, the descendant of the node $\mathbf{Y}_2$

is $\mathbf{Y}_3 = \mathbf{X}_2^*$ and the edge that joins $\mathbf{Y}_2$ and $\mathbf{Y}_3$ is labeled $\varepsilon_{C_2^t}$, as it is shown in Figure 1*c*.

In Figure 1*d*, we show that the outputs of SEARCH_ALL when applied for $\mathbf{Y}_3$ and $\mathbf{Y}_3^*$ are, respectively, $(C_3, \mathbf{X}_3)$ and $\emptyset$. Thus, the descendant of the node $\mathbf{Y}_3$ is $\mathbf{Y}_4 = \mathbf{X}_3$ and the edge that joins $\mathbf{Y}_3$ and $\mathbf{Y}_4$ is labeled δ_{C_3}, as it is shown in Figure 1*d*.

Finally, in Figure 1*e*, $\mathbf{Y}_4 = \{[\{a\}]\}$, where $a = (-1, 0)$. Since $a \neq o$, then the descendant of $\mathbf{Y}_4$ is $\mathbf{Y}_5 = \{[\{o\}]\}$ and the edge that joins $\mathbf{Y}_4$ and $\mathbf{Y}_5$ is labeled δ_{C_4} with $C_4 = \{-a\}$.

The labels of the edges on the path from the root to the node $\mathbf{Y}_5$ forms a sequential representation of ψ, i.e., $\delta_{C_1}\varepsilon_{C_2^t}\delta_{C_3}\delta_{C_4}$ is a sequential representation of ψ.

As we can see in this example, we transform the sup-decomposition of ψ (that has a parallel structure) into a purely sequential representation of ψ. The advantage of the sequential representation over sup-decomposition is that the sequential representation is usually more efficient for computation in conventional sequential machines. For this example, the number of shifts, unions and intersections for computation of ψ by using its sup-decomposition is 9; if its sequential decomposition is used, this number is decreased to 7.

9. Conclusion

In this paper, we have studied the problem of transforming the basis representation of morphological operators into more efficient representations (when they exist). The solution of this problem depends on the solution of Minkowski factorization equation, that is a hard combinatorial problem.

We have given new bounds for the space of solutions of Minkowski factorization equation and showed how to apply them to build sequential representations from the basis of sequences of dilations and erosions.

The next steps of this research are improvements on the implemented algorithm for the proposed technique and study of more restrict bounds for the family of alternating sequential filters.

References

1. G. J. F. Banon and J. Barrera. Minimal Representations for Translation-Invariant Set Mappings by Mathematical Morphology. *SIAM J. Appl. Math.* 51(6):1782–1798, December 1991.
2. J. Barrera and G. P. Salas. Set Operations on Collections of Closed Intervals and their Applications to the Automatic Programming of Morphological Machines. *Journal of Electronic Imaging*, 5(3):335-352, July 1996.
3. G. Birkhoff. *Lattice Theory*. American Mathematical Society, Providence, Rhode Island, 1967.
4. R. F. Hashimoto and J. Barrera. Finding Solutions for the Dilation Factorization Equation. Technical Report RT-MAC-9914, Departamento de Ciência da Computação, IME-USP, November, 1999.
5. H. J. A. M. Heijmans. *Morphological Image Operators*. Academic Press, 1994.
6. R. Jones. The Transformation of the Basis Representation into Cascaded Representation. In *Mathematical Morphology and its Applications to Signal Processing*, pages 239-244. Barcelona, Spain, 1993.
7. J. Serra. *Image Analysis and Mathematical Morphology*. Academic Press, 1982.
8. J. Serra and L. Vincent. An Overview of Morphological Filtering. *Circuits Systems Signal Process*, 11(1):47-108, 1992.

DECOMPOSITION OF SEPARABLE CONCAVE STRUCTURING FUNCTIONS

R. VAN DEN BOOMGAARD, E.A. ENGBERS and A.W.M. SMEULDERS
Intelligent Sensory Information Systems, University of Amsterdam

Abstract. This paper presents a decomposition scheme for a large class of greyscale structuring elements from mathematical morphology. In contrast with many existing decomposition schemes, our method is valid in the continuous domain. Conditions are given under which this continuous method can be properly discretized. The class of functions that can be decomposed with our method contains the class of quadratic functions, that are of major importance in, for instance, distance transforms and morphological scale space. In the continuous domain, the size of the structuring elements resulting from the decomposition, can be chosen arbitrarily small. For functions from the mentioned class, that can be separated along the standard image axes, a discrete decomposition in 3×3 elements can be guaranteed.

Key words: Mathematical Morphology, Structuring Element Decomposition, Concave Structuring Elements.

1. Introduction

In this paper we study the decomposition of a large class of structuring functions often used in morphological image processing. We study decomposition of continuous functions and prove that a subset of the class of concave functions can be decomposed into a dilation sequence of functions with finite effective domain. Only then we show what are the requirements for proper discretization of the proposed decomposition scheme.

In mathematical morphology [18, 7], concave structuring functions play an important role. Matheron [14] already showed that convexity of structuring sets is needed to axiomatize the concept of size. This analysis lead to the notion of granulometries. Concave functions are needed to extend the notion of granulometries to grey level functions [11]. In [10], Xu presents a decomposition scheme for *convex* polygon-shaped structuring elements in binary morphology.

Concave structuring functions play an important role in the calculation of (Euclidean) distance transforms [2]. Sternberg [19] showed that the Euclidean distance transform can be calculated by dilating the indicator function of a set with a cone shaped structuring function. The cone function $c(x,y) = -\sqrt{x^2+y^2}$ encodes the distance to its center. The infinite support function c can be decomposed into the sequence $c = t \oplus t \oplus \cdots$ where t is the "top" of the cone $t(x,y) = c(x,y)$ for $x^2+y^2 \leq 1$ and $t(x,y) = -\infty$ elsewhere. Unfortunately this decomposition cannot be properly discretized, i.e. discretizing the tops and then dilating leads to a different result then first dilating and then discretizing. Therefore a *chamfer distance transform* [1, 26] can only be an approximation of the Euclidean distance transform.

Dilating the indicator function of a set with the square of the cone function leads to the square of the distance transform [22] of that set. Huang[8] showed that the discretized *squared* cone *can* be decomposed into a sequence of finite support discrete structuring functions. In this paper we will give a geometrical *continuous* construction that is a generalization of this result in the sense that our result can be used for a larger class of (continuous) concave functions (not only the parabola).

The parabolic function is not only important because it can be used to calculate the distance transform. It has been shown by van den Boomgaard [24] and Jackway [9] that the parabola in a very specific sense is the morphological analogue of the Gaussian function as used in linear convolutions [22, 24].

Decomposition of structuring elements ([28]) has a lot of literature devoted to it. Some of the reported results deal with continuous structuring elements. For example the decomposition of convex symmetric polygons into the dilation of the edge line segments [18] is simple to prove for polygons in $\mathbb{R}^2$. Also the logarithmic decomposition of a convex set into the dilation of the set with its extreme elements (the vertices of a polygon) [15, 25] is proved for continuous sets (and the conditions for proper discretization are given in [21]). Most of the results on decomposition, however, are focused on the decomposition of discrete sets [5, 20, 12, 16, 6]. Whereas the decomposition of continuous sets tend to be of a geometrical nature, the proofs for the decomposition of discrete sets tend to be of an algebraic nature [17].

Decomposition into small structuring elements is important from a practical point of view. Even for the Euclidean distance transform, that can be implemented quite efficiently using the dimensional decomposition (see [23]), the decomposition into small 3×3 elements is profitable as it allows for inhomogeneous distance transforms [26]. These are for instance needed in the watershed algorithm while keeping track of the distance traveled from the starting marker points [27].

2. Decomposition

Consider the dilation of a function f with respect to the structuring function g:

$$(f \oplus g)(x) = \bigvee_{y \in \mathbb{R}^n} f(x-y) + g(y).$$

The effective domain of the structuring function g is the set of points y where $g(y) \neq -\infty$. Note that only points in the effective domain of g need to be considered in calculating the dilation result.

The structuring functions considered in this paper are all concave. Concave structuring functions generalize the notion of convex sets to the domain of grey value images. A function $g : \mathbb{R}^n \to \mathbb{R}$ is concave if $g(tx + (1-t)y) \geq tg(x) + (1-t)g(y)$ for $x, y \in \mathbb{R}^n$ and $0 \leq t \leq 1$. A concave function is *proper concave* if $g(x) > -\infty$ for at least one x and $g(x) < \infty$ for all x.

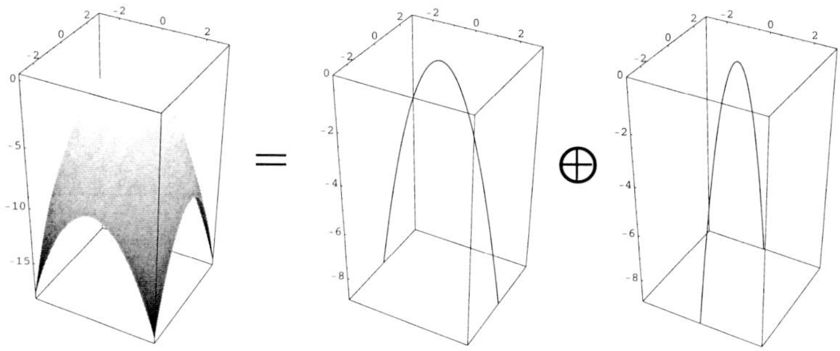

Fig. 1. Example of a separation of a parabola g into $\chi_{\vec{v}} g_{\vec{v}}$ and $\chi_{\vec{w}} g_{\vec{w}}$

We restrict ourselves to structuring functions g that can be separated in one-dimensional proper concave functions $g_{\vec{v}}$ and $g_{\vec{w}}$. The separation process makes use of the embedding operator $\chi_{\vec{v}}$, to embed one-dimensional functions into two dimensional space (see [23]):

Definition 1 *Let l be a one dimensional real function. The operator $\chi_{\vec{v}}$ (with $\vec{v}$ a direction vector, $\|\vec{v}\| = 1$) embeds the function l into 2-dimensional space, resulting in the function $\chi_{\vec{v}} l : \mathbb{R}^2 \to \mathbb{R}$:*

$$\chi_{\vec{v}} l(\vec{x}) = \begin{cases} l(\|\vec{x}\|) & , \textit{if } \vec{x} \textit{ parallel to } \vec{v} \\ -\infty & , \textit{otherwise} \end{cases} \tag{1}$$

The structuring function g is called separable if it can be separated in functions $g_{\vec{v}}$ and $g_{\vec{w}}$ such that:

$$g = \chi_{\vec{v}} g_{\vec{v}} \oplus \chi_{\vec{w}} g_{\vec{w}}, \tag{2}$$

(see figure 1 for an example)

Notice that when we take two arbitrary proper concave functions $g_{\vec{v}}$ and $g_{\vec{w}}$ the resulting function $g(\vec{x}) = \chi_{\vec{v}} g_{\vec{v}} \oplus \chi_{\vec{w}} g_{\vec{w}}$ is always a proper concave function.

The vectors $\vec{v}$ and $\vec{w}$ in the separation must be linearly independent. Functions $f : \mathbb{R}^2 \to \mathbb{R}$ which can be written as $f(x, y) = f_1(x) + f_2(y)$, with $f_1 : \mathbb{R} \to \mathbb{R}$ and $f_2 : \mathbb{R} \to \mathbb{R}$ are a subclass of the functions that can be separated cf. equation 2.

2.1. The decomposition scheme

In theorem 1 we will prove that a one dimensional proper concave function g can be decomposed in two concave functions u and r, such that $f = u \oplus r$ (under certain conditions). This function u always has a finite effective domain. Following this theorem allows for the decomposition of $g_{\vec{v}}$ and $g_{\vec{w}}$:

$$g_{\vec{v}} = (u_{\vec{v}} \oplus r_{\vec{v}}), \qquad g_{\vec{w}} = (u_{\vec{w}} \oplus r_{\vec{w}}).$$

Now dilation of image I with function g is $I \oplus g = I \oplus (\chi_{\vec{v}} g_{\vec{v}} \oplus \chi_{\vec{w}} g_{\vec{w}})$, with $\chi_{\vec{v}} g_{\vec{v}} \oplus \chi_{\vec{w}} g_{\vec{w}} = \chi_{\vec{v}}(u_{\vec{v}} \oplus r_{\vec{v}}) \oplus \chi_{\vec{w}}(u_{\vec{w}} \oplus r_{\vec{w}})$. The embedding operator can be distributed over the dilation, so $g = (\chi_{\vec{v}} u_{\vec{v}} \oplus \chi_{\vec{v}} r_{\vec{v}}) \oplus (\chi_{\vec{w}} u_{\vec{w}} \oplus \chi_{\vec{w}} r_{\vec{w}})$, and since

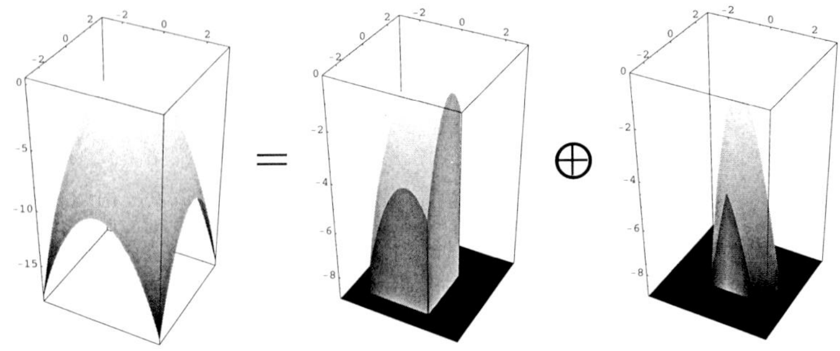

Fig. 2. Example of a decomposition of a parabola g into u_1 and r_1

dilation is commutative and associative $g = \chi_{\vec{v}} u_{\vec{v}} \oplus \chi_{\vec{w}} u_{\vec{w}} \oplus \chi_{\vec{v}} r_{\vec{v}} \oplus \chi_{\vec{w}} r_{\vec{w}}$. If we define $u_1 = \chi_{\vec{v}} u_{\vec{v}} \oplus \chi_{\vec{w}} u_{\vec{w}}$ and $r_1 = \chi_{\vec{v}} r_{\vec{v}} \oplus \chi_{\vec{w}} r_{\vec{w}}$, then $I \oplus g = (I \oplus u_1) \oplus r_1$. (See for an example figure 2.) Since $r_{\vec{v}}$ and $r_{\vec{w}}$ are proper concave functions, the same process can be repeated and the dilation with g can be carried out as follows $I \oplus g = I \oplus u_1 \oplus u_2 \oplus r_2$. For n decomposition steps this results in

$$I \oplus g = I \oplus u_1 \oplus \ldots \oplus u_n \oplus r_n. \tag{3}$$

Only in case the function g has a finite effective domain, it can be decomposed in a finite number of functions $u_1 \ldots u_n$, such that r_n is the pulse function. For functions with an infinite effective domain, the rest function r_n always has an infinite effective domain.

2.2. Decomposition of one dimensional concave structuring functions

The main theorem of this section gives a decomposition of one dimensional proper concave functions f into two proper concave functions u and r, such that $f = u \oplus r$. In this decomposition u always has a finite effective domain, while r only has a finite effective domain if f has a finite effective domain. To simplify the decomposition, we assume that $f(x) < 0$, except for $x = 0$, where $f(0) = 0$. For proper concave functions this comes down to translating the function such that the maximum is obtained in the origin. Since the decomposed function is used for dilation, this only results in a simple translation of the result.

The function u is constructed from f as follows

$$u(x) = \begin{cases} f(x) & \text{, if } x_1^* \leq x \leq x_2^* \\ -\infty & \text{, otherwise} \end{cases}, \tag{4}$$

where x_1^* and x_2^* are chosen such that $f(x_1^*) = t_1$ and $f(x_2^*) = t_2$ for some real numbers $t_1 \leq 0$, $t_2 \leq 0$ and $x_1^* \leq 0$, $x_2^* \geq 0$, see figure 3. The function r is constructed from f as follows

$$r(x) = \begin{cases} f(x + x_1^*) - t_1 & \text{, if } x < 0 \\ f(x + x_2^*) - t_2 & \text{, otherwise.} \end{cases} \tag{5}$$

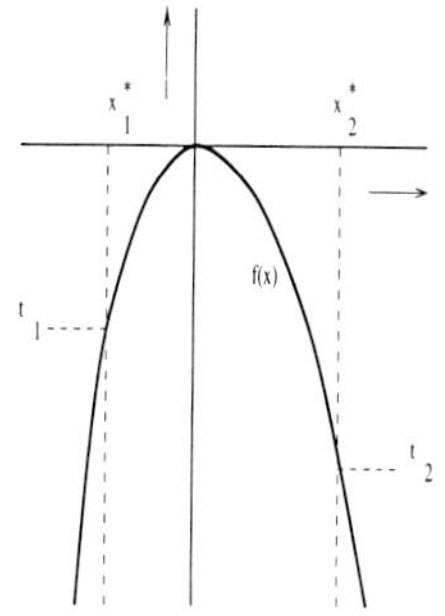

Fig. 3. Visualization of the symbols used in the decomposition.

The terms $-t_1$ and $-t_2$ translate the maximum of r to the origin so that the dilation $u \oplus r$ will not be a translated version of f. r is again a proper concave function, so the same decomposition scheme can be used to decompose r (with possible other values of x_1^* and x_2^*).

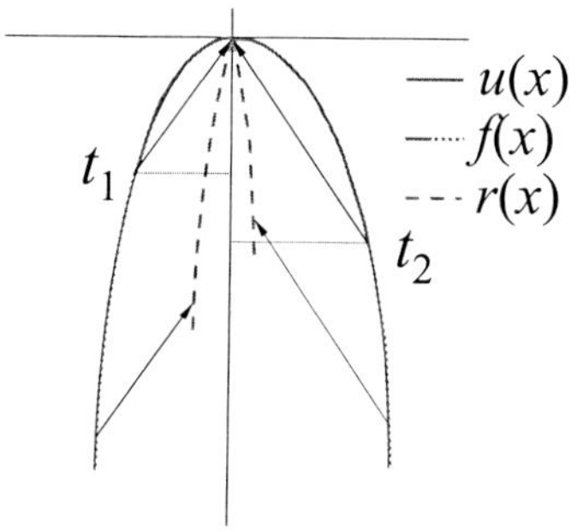

Fig. 4. Decomposition of f into u and r.

A non-trivial decomposition exists in case the effective domain of the concave function is of the form (a, b) where $a < 0$ and $b > 0$. Then t_1 and t_2 can always be found such that there exist $x_1^* \leq 0$ and $x_2^* \geq 0$, with $f(x_1^*) = t_1$ and $f(x_2^*) = t_2$.

Theorem 1 ensures that the decomposition given by equations 4 and 5 indeed results in a decomposition of f such that $f = u \oplus r$. The proof of this theorem, and all other proofs omitted in this paper, can be found in [4].

Theorem 1 *Let $f(x) : I\!R \to I\!R$ be a one dimensional proper concave function with $f(x) < 0$ except for $x = 0$, where $f(0) = 0$. Assume that there exist $x_1^* \leq 0$ and $x_2^* \geq 0$ such that $f(x_1^*) = t_1$ and $f(x_2^*) = t_2$ for some real numbers $t_1 \leq 0$, $t_2 \leq 0$. Now define*

$$u(x) = \begin{cases} f(x) & , \text{ if } x_1^* \leq x \leq x_2^* \\ -\infty & , \text{ otherwise} \end{cases}$$

and

$$r(x) = \begin{cases} f(x + x_1^*) - t \ , & \textit{if } x < 0 \\ f(x + x_2^*) - t \ , & \textit{otherwise.} \end{cases}$$

Then $u \oplus r(x) = f(x)$.

3. Discretization of the Decomposition Scheme

In this section we discuss the conditions under which the decomposition process can be properly discretized, i.e. the conditions under which dilating with the original discretized function is the same as dilation with the discretized results of the decomposition. First the definition of the discretization operator $\Delta_{\vec{c},\vec{d}}$. The vectors $\vec{c}$ and $\vec{d}$ generate the sampling points in the discretization grid.

Definition 2 *The discretization operator* $\Delta_{\vec{c},\vec{d}}$ *constructs a function* $\Delta_{\vec{c},\vec{d}}f : \mathbb{Z}^2 \to \mathbb{R}$ *from a function* $f : \mathbb{R}^2 \to \mathbb{R}$ *as follows:*

$$(\Delta_{\vec{c},\vec{d}}f)(p,q) = f(p\vec{c} + q\vec{d}).$$

Likewise, the discretization $\Delta_k f$ of a function $f : \mathbb{R} \to \mathbb{R}$ is constructed by $(\Delta_k f)(p) = f(pk)$.

Now the term "properly discretizable" of the decomposition $f = u \oplus r$ can be formalized as:

$$\Delta_{\vec{c},\vec{d}}f = \Delta_{\vec{c},\vec{d}}u \oplus \Delta_{\vec{c},\vec{d}}r.$$

In general the decomposition process can not be guaranteed to be properly discretizable.

The following lemma gives the conditions under which one dimensional decomposable functions can be properly discretized.

Lemma 1 *Let the decomposition for a function* f *be given by the parameters* x_1^* *and* x_2^* *and let the discretization be given by parameter* d. *Then*

$$\Delta_d f = \Delta_d u \oplus \Delta_d r$$

iff $\forall k \in \mathbb{Z}_0^-,\ kd \le d\lceil \frac{x_1^*}{d} \rceil$:

$$f(kd) = f(kd - (d\lceil \frac{x_1^*}{d} \rceil - x_1^*)) + f(d\lceil \frac{x_1^*}{d} \rceil) - f(x_1^*)$$

and $\forall k \in \mathbb{Z}_0^+,\ kd \ge d\lfloor \frac{x_2^*}{d} \rfloor$:

$$f(kd) = f(kd + (x_2^* - d\lfloor \frac{x_2^*}{d} \rfloor)) + f(d\lfloor \frac{x_2^*}{d} \rfloor) - f(x_2^*).$$

For two-dimensional decomposable functions f that can be separated into $f_{\vec{v}}$ and $f_{\vec{w}}$, the question whether it can be discretized properly with parameters $\vec{v}$ and $\vec{w}$ boils down to the question whether $f_{\vec{v}}$ can be discretized with parameter $|\vec{v}|$ and $f_{\vec{w}}$ can be discretized with parameter $|\vec{w}|$.

If the discretization vectors $\vec{c}$ and $\vec{d}$ do not correspond with the separation parameters $\vec{v}$ and $\vec{w}$, it is still possible for the discretization of the decomposition scheme to be proper.

In the following theorem we use $[\vec{v}\vec{w}]$ to denote the 2×2 matrix whose column are the vectors $\vec{v}$ and $\vec{w}$.

Theorem 2 *Let f be a two-dimensional decomposable function that can be separated into $f_{\vec{v}}$ and $f_{\vec{w}}$, for which $\Delta_{|\vec{v}|} f_{\vec{v}} = \Delta_{|\vec{v}|} u_{\vec{v}} \oplus \Delta_{|\vec{v}|} r_{\vec{v}}$ and $\Delta_{|\vec{w}|} f_{\vec{w}} = \Delta_{|\vec{w}|} u_{\vec{w}} \oplus \Delta_{|\vec{w}|} r_{\vec{w}}$. Then if $\vec{v}$ and $\vec{w}$ are two points from the lattice formed by $\vec{c}$ and $\vec{d}$ and there exists an integral matrix U such that $det(U) = \pm 1$ and $[\vec{v}\vec{w}] = [\vec{c}\vec{d}]U$, then*

$$\Delta_{\vec{c},\vec{d}} f = \Delta_{\vec{c},\vec{d}} u \oplus \Delta_{\vec{c},\vec{d}} r.$$

The well-known example decomposing a (discrete) parabola $g(p,q) = -(o^2 + q^2)$ that also follows directly from the theory presented in this paper is:

$$g = \left\{ \begin{array}{ccc} -2 & -1 & -2 \\ -1 & \underline{0} & -1 \\ -2 & -1 & -2 \end{array} \right\} \oplus \left\{ \begin{array}{ccc} -6 & -3 & -6 \\ -3 & \underline{0} & -3 \\ -6 & -3 & -6 \end{array} \right\} \oplus \cdots$$

Here we use the notation that within the curly brackets the values of a structuring function in the discrete sampling points are given. The origin is marked with an underscore. All the values that are not shown are implicitly assumed to be equal to $-\infty$.

The second example illustrates that non axes aligned quadratic structuring functions (QSF's) are also decomposable using our approach. Consider the QSF $g(p,q) = -p^2 + 2pq - 2q^2$. This QSF is decomposed as:

$$g = \left\{ \begin{array}{ccccc} & & -2 & -1 & -2 \\ & -1 & \underline{0} & -1 & \\ -2 & -1 & -2 & & \end{array} \right\} \oplus \left\{ \begin{array}{ccccc} & & -6 & -3 & -6 \\ & -3 & \underline{0} & -3 & \\ -6 & -3 & -6 & & \end{array} \right\} \oplus \cdots$$

4. Computational Complexity

In general, the dilation with a structuring function with effective domain nQ where Q is convex is of complexity $\mathcal{O}(M^2 n^2 |Q|)$, where $|Q|$ is the number of points in Q, and the dimension of the image is $M \times M$. If the structuring function nQ can be decomposed such that for the effective domain of nQ holds that

$$\underbrace{Q \oplus Q \oplus \ldots Q}_{n \text{ times}},$$

the complexity is reduced to $\mathcal{O}(M^2 n |Q|)$.

Due to the nature of our separation process, the decomposition scheme always returns structuring functions with a parallelogram shaped effective domain. Since the size of the effective domains of the resulting functions of our decomposition scheme can be chosen at will, the domains can be chosen equal to Q, where the effective domain of the original structuring function is nQ. The decomposition scheme then reduces the complexity by a factor n.

5. Conclusions

In this paper we have presented a decomposition scheme for a large class of concave structuring functions. The results are valid in the continuous domain, but we have proved the requirements for proper discretization as well. The important class of quadratic structuring functions prove to be decomposable into a sequence of dilations with structuring functions restricted to a finite (small) effective domain. For the axis aligned parabola (an element of the class of quadratic functions) our proof is a generalization of the decomposition presented by Huang [8]. The main difference with existing approaches is that we have chosen a continuous geometrical view on decomposition instead of a discrete algebraic approach.

We have restricted our proofs to the functions that are separable by dimension. This is somewhat of a limitation that may well be eliminated using more elaborate proofs using the slope transform description of morphological operators [3] or equivalently (for concave functions) using the Fenchel conjugate functions (or the upper and lower slope transform) from convex analysis [13, 3]. Such an extension of the theory is left to future work.

We believe that the presented approach for decomposition of concave structuring functions provides an intuitive feeling for the decomposition (being a "cut-and-paste" procedure as illustrated in figure 4) that is fruitful for a deeper understanding of morphological operators modifying and probing the geometry of visual observations.

References

1. G. Borgefors. Chamfering: A fast method for obtaining approximations of the Euclidian distance in n dimensions. In *Third Scandinavian Conference on Image Analysis*, Copenhagen, 1983.
2. P. Danielsson. Euclidian distance mapping. *Computer Graphics and Image Processing*, 14:227–248, 1980.
3. L. Dorst and R. van den Boomgaard. Morphological signal processing and the slope transform. *Signal Processing*, 38:79–98, 1994.
4. E. Engbers, R. van den Boomgaard, and A. Smeulders. Decomposition of separable concave structuring functions. Technical report, ISIS, University of Amsterdam, 1999.
5. P. Gader. Separable decompositions and approximations of greyscale morphological templates. *CVGIP: Image Understanding*, 53(3):288–296, 1991.
6. P. Gader and S. Takriti. Decomposition techniques for grey-scale morphological templates. In P. Gader, editor, *Image Algebra and Morphological Image Processing*, Proceedings of SPIE, pages 431–442, 1990.
7. H. J. A. M. Heijmans. *Morphological Image Operators*. Academic Press, Boston, 1994.
8. C. Huang and O. Mitchell. A Euclidean distance transform using greyscale morphology decomposition. *IEEE PAMI*, 16(4):443–448, 1994.
9. P. Jackway and M. Deriche. Scale-space properties of the multiscale morphological dilation-erosion. *IEEE PAMI*, 18:38–51, 1996.
10. J.Xu. Decomposition of convex polygonal morphological structuring elements into neighborhood subsets. *IEEE PAMI*, 13(2):153–162, 1991.
11. E. J. Kraus, H. J. A. M. Heijmans, and E. R. Dougherty. Gray-scale granulometries compatible with spatial scalings. *Signal Processing*, 34:1–17, 1993.
12. D. Li and G. Ritter. Decomposition of separable and symmetric convex templates. In P. Gader, editor, *Image Algebra and Morphological Image Processing*, Proceedings of SPIE, pages 408–418, 1990.

13. P. Maragos. Max-min difference equations and recursive morphological systems. In *Mathematical Morphology and its Applications to Signal Processing*, Barcelona, 1993.
14. G. Matheron. *Random Sets and Integral Geometry.* Academic Press, London, 1975.
15. J. Pecht. Speeding up successive Minkowski operations. *Pattern Recognition Letters*, 3(2):113–117, 1985.
16. P.Sussner and G. Ritter. Decomposition of gray-scale morphological templates using the rank method. *IEEE PAMI*, 19(6):1–10, 1997.
17. G. Ritter, J. Wilson, and J. Davidson. Image algebra: An overview. *CVGIP: Image Understanding*, 49:297–331, 1990.
18. J. Serra. *Image Analysis and Mathematical Morphology.* London: Academic, 1982.
19. S. Sternberg. Language and architecture for parallel image processing. In E. Gelsema and L. Kanal, editors, *Pattern Recognition in Practice.* North Holland Publishing Company, 1980.
20. S. Takriti and P. Gader. Local decomposition of greyscale morphological templates. *Journal of Mathematical Imaging and Vision*, 2:39–50, 1992.
21. R. van den Boomgaard. *Mathematical Morphology: Extensions towards Computer Vision.* PhD thesis, University of Amsterdam, 1992.
22. R. van den Boomgaard. The morphological equivalent of the Gauss convolution. *Nieuw Archief voor Wiskunde (in English)*, 38(3):219–236, 1992.
23. R. van den Boomgaard, L. Dorst, S. Makram Ebeid, and J. Schavemaker. Quadratic structuring functions in mathematical morphology. In R. S. P. Maragos and M. Butt, editors, *Mathematical Morphology and its Applications in Image and Signal Processing*, Proceedings of the 3rd International Symposium on Mathematical Morphology (ISMM '96), pages 147–154, 1996.
24. R. van den Boomgaard and A. W. M. Smeulders. The morphological structure of images, the differential equations of morphological scale-space. *IEEE PAMI*, 16(11):1101–1113, 1994.
25. R. van den Boomgaard and D. Wester. Logarithmic shape decomposition. In *Aspects of Visual Form Processing*, Capri, 1994.
26. B. Verwer. *Distance Transform: Metrics, Algorithms and Applications.* PhD thesis, Delft University of Technology, 1991.
27. L. Vincent. Morphological algorithms. In E. R. Dougherty, editor, *Mathematical Morphology in Image Processing*, chapter 8, pages 255–288. Marcel Dekker, 1993.
28. X. Zhuang and R. Haralick. Morphological structuring element decomposition. *CVGIP: Image Understanding*, 35:370–382, 1986.

MINKOWSKI SUM VOLUME MINIMIZATION FOR CONVEX POLYHEDRA*

ALEXANDER V. TUZIKOV and STANISLAV A. SHEYNIN
Institute of Engineering Cybernetics
Academy of Sciences of Republic Belarus, Minsk, Belarus
{tuzikov,sheynin}@mpen.bas-net.by

Abstract. The paper considers a problem of minimization the volume of Minkowski sum or mixed volume functionals for convex polyhedral shapes. It is assumed that one of the shapes can be rotated about arbitrary axis through arbitrary angle. This problem is of interest for some approaches developed for shape pose determination, invariant shape comparison, shape symmetry analysis. It is shown that the problem can be solved efficiently, i.e. there exists a finite number of rotation axes and rotation angles which are candidates for the best solution. Some implementations problems of the developed algorithm and results of experiments are discussed.

Key words: Minkowski Addition, Convex Polyhedron, Volume, Mixed Volume.

1. Introduction

Minkowski addition of two sets $A, B \subseteq \mathbf{R}^3$ is defined by

$$A \oplus B = \{a + b \mid a \in A,\ b \in B\}.$$

Denote by $V(A)$ the volume of the set $A \subset \mathbf{R}^3$. Given convex sets $A, B \subset \mathbf{R}^3$ and $\alpha, \beta \geq 0$ it is known that the following relation is true [3, p.353]:

$$V(\alpha A \oplus \beta B) = \alpha^3 V(A) + 3\alpha^2 \beta V(A, A, B) + 3\alpha\beta^2 V(A, B, B) + \beta^3 V(B). \quad (1)$$

Here $V(A, A, B)$ and $V(A, B, B)$ are called *mixed volumes*.

It is well-known [10] that every convex body A is uniquely determined by its *support function* given by:

$$h(A, u) = \sup\{\langle a, u\rangle \mid a \in A\}, \qquad u \in S^2.$$

Here $\langle a, u\rangle$ is the inner product of vectors a and u, and S^2 denotes the unit sphere in $\mathbf{R}^3$. It is also known [10] that

$$h(A \oplus B, u) = h(A, u) + h(B, u), \qquad u \in S^2.$$

The *support set* $F(A, u)$ of A at $u \in S^2$ consists of all points $a \in A$ for which $\langle a, u\rangle = h(A, u)$. Support sets can be of dimension $0, 1, 2$. If A is a convex polyhedron, then its support sets of dimension $0, 1, 2$ are vertices, edges and facets of A, respectively.

* The authors were supported by the INTAS grant N 96-785.

It is known from Minkowski's existence theorem [10, p. 390] that a convex polyhedron is uniquely determined by areas and normal vector directions of its facets.

If P is a convex polyhedron with facets F_i and corresponding outward unit normal vectors u_i, $i = 1, \dots, k$, then

$$V(A, P, P) = \frac{1}{3} \sum_{i=1}^{k} h(A, u_i) S(F_i),$$

where $S(F_i)$ is the area of the facet F_i of P.

Let ℓ be an axis passing through the coordinate origin and $r_{\ell,\alpha}$ be the rotation in $\mathbb{R}^3$ about ℓ by an angle α in a counter-clockwise direction. Suppose that P and Q are convex polyhedra in $\mathbb{R}^3$.

The problem addressed in this report is how to compute efficiently the minimum of the functionals

$$V(P, P, r_{\ell,\alpha}(Q)), \; V(P, r_{\ell,\alpha}(Q), r_{\ell,\alpha}(Q)) \; \text{ and } \; V(P \oplus r_{\ell,\alpha}(Q))$$

for all rotations $r_{\ell,\alpha}$ in $\mathbb{R}^3$. Below we refer to these functionals as *objective functionals.*

What is the motivation to study this problem? First point out the following inequalities known from Brunn-Minkowski theory (Brunn-Minkowski and Minkowski inequalities [5, 10]).

For two convex compact sets $A, B \subset \mathbb{R}^3$ it holds:

$$V(A \oplus B)^{\frac{1}{3}} \geq V(A)^{\frac{1}{3}} + V(B)^{\frac{1}{3}},$$

$$V(A, A, B)^3 \geq V(A)^2 V(B)$$

with equality if and only if A and B are homothetic modulo translation.

Suppose that the set B is a rotated, scaled and shifted version of the set A. Then the minimization of the objective functionals under all possible rotation $r_{\ell,\alpha}(B)$ of the set B allows to define the orientation of the set B respective to the set A. Here it was used also the translation invariance of Minkowski addition as well as volume and mixed volume functionals. This property was utilized in [2, 7, 8, 9] to develop procedures for attitude determination of 3D shapes. It also allowed to develop recently an approach for comparing convex shapes and computation their symmetry measures [6, 11, 12]. However the paper [11] presents only a theoretical framework for comparing convex shapes and contains a complexity analysis of the solution but does not consider implementations. The aim of this report is to discuss some implementation problems and to provide the results of experiments.

To describe the developed procedure we use the slope diagram representation (SDR) of convex polyhedra [4]. According to this representation, facets, edges and vertices of a polyhedron are given by points, spherical arcs and convex spherical polygons of the unit sphere S^2.

- *Facet representation.* A facet F_i of a polyhedron which is orthogonal to the unit vector u_i is represented on the sphere S^2 by the end point of this vector;

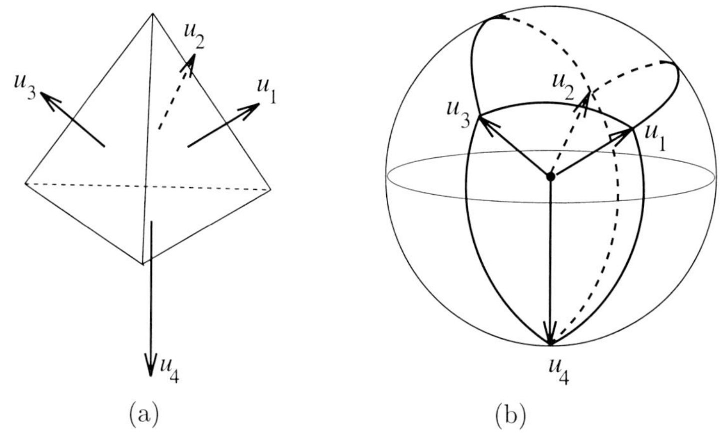

Fig. 1. Polyhedron (a) and its slope diagram representation (b).

- *Edge representation.* Each edge is represented by the arc of the great circle (spherical arc) joining two points corresponding to the two adjacent facets of the edge;
- *Vertex representation.* The region (called the spherical polygon) of the sphere bounded by spherical arcs corresponding to the edges which are adjacent to a polyhedral vertex, represents this vertex on the sphere S^2. The spherical arcs are included in the region.

Also, weights of spherical points and spherical arcs are used. The weight of a spherical point or arc equals the area of the corresponding polyhedral facet, or the length of the corresponding polyhedral edge, respectively.

2. Minimization for a Fixed Rotation Axis

Let ℓ be a *fixed* rotation axis.

While rotating the slope diagram of polyhedron Q about a fixed axis a finite number of situations arise when spherical points of the rotated $SDR(Q)$ intersect spherical arcs or points of $SDR(P)$ or vice versa. These rotations of Q are called ℓ-critical in that sense that they are candidates for local minima of the considered functionals. To be more precise the volume and the mixed volume functionals are concave at any interval between two neighbor critical rotations.

This property is illustrated in Fig. 3. The formal definition of critical rotations and the description of properties of considered functionals are discussed in [11].

Thus the following proposition is true [11].

Proposition 1 *Given an axis of rotation ℓ, the mixed volume $V(P, r_{\ell,\alpha}(Q), r_{\ell,\alpha}(Q))$ of convex polyhedra P and Q is a function of α which is piecewise concave on $[0, 2\pi)$, i.e. concave on every interval $(\alpha'_k, \alpha'_{k+1})$, for $k = 1, 2, \ldots, N$ and $\alpha'_{N+1} = \alpha'_1$.*

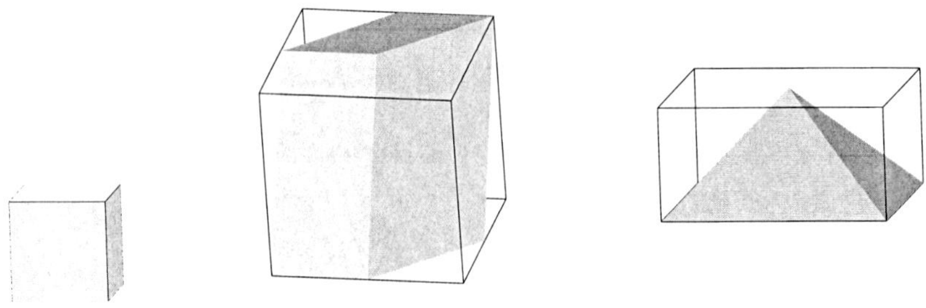

(a) P_1 (b) P_2 (c) P_3

Fig. 2. (a),(b),(c) Convex polyhedra P_1, P_2 and P_3.

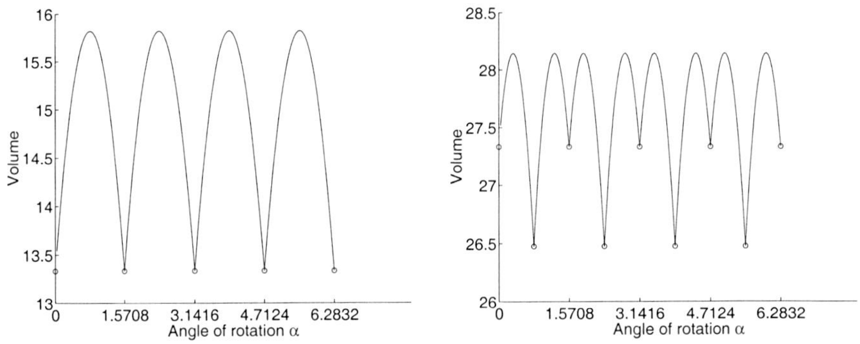

(a) (b)

Fig. 3. (a),(b) Functionals $V(P_i \oplus r_{u,\alpha}(P_3))$, $u = (0,0,1)$, $i = 1,2$ are piecewise concave. The $\circ$s indicate the values of the functionals at critical rotations.

Here $0 \leq \alpha_1' < \alpha_2' < \ldots < \alpha_N' < 2\pi$ are the ℓ-critical angles of Q with respect to P, i.e angles for which spherical points of rotated $SDR(Q)$ intersect spherical arcs or points of $SDR(P)$ (further called *ℓ-critical angles for mixed volume*).

The similar proposition is also true for volume $V(P \oplus r_{\ell,\alpha}(Q))$ but with a larger set of critical rotations including both ℓ-critical angles of Q with respect to P and ℓ-critical angles of P with respect to Q (further called *ℓ-critical angles for volume*).

These results allow us to compute efficiently the minimum of functionals $V(P, r_{\ell,\alpha}(Q), r_{\ell,\alpha}(Q))$, $V(P, P, r_{\ell,\alpha}(Q))$ and $V(P \oplus r_{\ell,\alpha}(Q))$ for any *fixed* rotation axis ℓ and all rotation angles $\alpha \in [0, 2\pi)$. It is sufficient to compute them only at a finite number of corresponding ℓ-critical angles.

3. Minimization for all Rotation Axes

Since our interest is to find the minimum of objective functionals for all possible rotation axes, we have to know which axes have to be checked. Fortunately it is possible to reduce essentially the set of such rotation axes. This reduction is based on the following property of functionals under consideration. If for a fixed position of polyhedron $Q' = r_{\ell_1,\alpha_1}(Q)$ there exists an axis ℓ' such that Q' is not ℓ'-critical w.r.t. P, then the value $V(P, Q', Q')$ is not a local minimum of the mixed volume functional $V(P, r_{\ell,\alpha}(Q), r_{\ell,\alpha}(Q))$. A smaller value of the functional can be found by rotating polyhedron Q' about the axis ℓ'. A similar property is true for other objective functionals as well.

In [11] it was proved that for minimizing the mixed volume functional $V(P, r_{\ell,\alpha}(Q), r_{\ell,\alpha}(Q))$ the following critical rotations $r_{\ell,\alpha^\star}$ of Q with respect to P are only of interest (denote $Q' = r_{\ell,\alpha^\star}(Q)$):

1. it is fulfilled simultaneously that one spherical point of $SDR(Q')$ coincides with a spherical point of $SDR(P)$ and another spherical point of $SDR(Q')$ intersects a spherical arc of $SDR(P)$;
2. three spherical points of $SDR(Q')$ intersect three spherical arcs of $SDR(P)$ simultaneously.

A similar result (but with a larger set of critical rotations defined by spherical points and spherical arcs of both $SDR(P)$ and $SDR(Q')$) is true for the volume functional.

Item 1 means that one facet of Q' is parallel to some facet of P and another facet of Q' is parallel to some edge of P. It is clear that there exists only a finite number of critical rotations corresponding to this item. Given any two spherical points from $SDR(P)$ and one spherical point and one spherical arc from $SDR(Q)$, the corresponding rotation vector and critical angle making them coinciding can be easily computed (see Section 4 for details).

Finding critical rotations corresponding to item 2 is a more complicated problem. Suppose that for given spherical points a, b, c and spherical arcs with normal vectors l, m, n there exists a rotation r such that rotated spherical points belong to the above spherical arcs. Then it is true

$$\begin{cases} \langle l, r(a) \rangle = 0, \\ \langle m, r(b) \rangle = 0, \\ \langle n, r(c) \rangle = 0. \end{cases}$$

The set of all 3D rotations can be parameterized by homogeneous coordinates. A rotation corresponding to homogeneous coordinates (t, u, v, w) has the following matrix [7, ch.18]

$$\frac{1}{t^2+u^2+v^2+w^2}\begin{pmatrix} t^2+u^2-v^2-w^2 & 2(uv-tw) & 2(tv+uw) \\ 2(uv+tw) & t^2-u^2+v^2-w^2 & 2(vw-tu) \\ 2(uw-tv) & 2(tu+vw) & t^2-u^2-v^2+w^2 \end{pmatrix}$$

Therefore we obtain a system of 3 homogeneous algebraic equation of order 2. By other words, this problem is equivalent to finding intersection points of

3 quadratic surfaces in projective space. Given triples of spherical points and arcs this system of equations is equivalent to some equation of degree 8 and can be solved numerically. We do not discuss this case and refer to [1] for a detailed discussion. However it seems that this case is of theoretical value rather than of practical one.

4. Experiments

In this section, we apply the above results for the minimization of Minkowski sum volume of convex polyhedra given in Fig. 2. For every pair of these polyhedra the volume functional was minimized under all critical rotations for volume of the first type (it corresponds to item 1 in section 3). No computations were performed for critical rotations of the second type which corresponds to item 2. Therefore the obtained final results give a good local minimum of the functional but there is no guarantee in fact that this local minimum equals to the global one.

Note that the set of critical rotations for volume is larger than for mixed volume. It includes all rotations of Q for which a spherical point u of $SDR(Q')$ coincides with a spherical point of $SDR(P)$ (i.e. two facets are parallel), and simultaneously another spherical point from SDR of one polyhedron (P or Q') intersects a spherical arc from SDR of the second polyhedron while rotating Q' about the axis defined by point u.

To find all such rotations we should perform at first for every pair of spherical points $u \in SDR(P)$ and $v \in SDR(Q)$ a rotation of Q making these points coinciding. But for simplification we use rotations of both polyhedra transferring points u and v to the point $(0,0,1)$. Then we find all critical rotations for the fixed axis passing through the point $(0,0,1)$ (see section 2). The latter procedure is reduced to solving the following equation system

$$\begin{cases} x^2 + y^2 = 1 - z_0^2 \\ n_1 x + n_2 y + n_3 z_0 = 0 \end{cases}$$

for every spherical point (x_0, y_0, z_0) and spherical arc with normal vector (n_1, n_2, n_3). The solution of this system gives us the position (x, y, z_0) of the spherical point after rotation. This rotation is critical if the point belongs to the spherical arc (not to its exterior).

This algorithm was implemented using MatLab system. The full complexity of it is $O(n^4)$, where n is the number of polyhedra facets.

Table I shows the volumes of original polyhedra given in Fig. 2. For every pair of original polyhedra the minimum (in fact the local one) of the volume functional and the corresponding critical rotation of the second polyhedron were computed. The obtained results are given in Table II. Note that for polyhedra P_1 and P_3 the minimum is achieved for their initial positions, therefore the rotation angle is 0 and the rotation axis is undefined. Fig. 4 presents the corresponding Minkowski sums having minimum volume found.

For comparison we show also in Fig. 5 polyhedra $P_1 \oplus P_2$ and $P_2 \oplus P_3$. Volumes of these polyhedra (equal 18 and 27.33) are greater than volumes of

TABLE I
Volumes of polyhedra given in Fig. 2.

Polyhedron	P_1	P_2	P_3
Volume of polyhedron	1	2	1.333

TABLE II
The found minimal values of the volume functional for pairs of polyhedra given in Fig. 2.

Pairs of polyhedra	P_1 and P_2	P_1 and P_3	P_2 and P_3
Minimum of the volume functional	17.899	13.333	18.513
Axis of rotation	$(0.863, 0.357, 0.357)$	-	$(0.465, 0.465, -0.753)$
Angle of rotation	4.5654	0	1.8508

polyhedra shown in Fig. 4 (a),(c).

References

1. H. Bekker and J. B. T. M. Roerdink. Calculating critical orientations of polyhedra for similarity measure evaluation. In *Proc. 2nd Annual IASTED International Conference on Computer Graphics and Imaging*, Palm Springs, California, USA, Oct. 25–27 1999. to be published.
2. P. Brou. Using the Gaussian image to find the orientation of objects. *International Journal of Robotics Research*, 3(4):89–125, 1984.
3. R. Gardner. *Geometric Tomography.* Cambridge University Press, 1995.
4. P. K. Ghosh and R. M. Haralick. Mathematical morphological operations of boundary-represented geometric objects. *Journal of Mathematical Imaging and Vision*, 6(2/3):199–222, 1996.
5. H. Hadwiger. *Vorlesungen über Inhalt, Oberfläche und Isoperimetrie.* Springer, Berlin, 1957.
6. H. J. A. M. Heijmans and A. Tuzikov. Similarity and symmetry measures for convex shapes using Minkowski addition. *IEEE Transactions on Pattern Analysis and Machine Intelligence*, 20(9):980–993, 1998.
7. B. K. P. Horn. *Robot Vision.* MIT Press, Cambridge, 1986.
8. Y. Li and R. J. Woodham. Orientation-based representation of 3D shape. In *Proc. IEEE Computer Society Conf. on Computer Vision and Pattern Recognition*, pages 182–187, Seattle, Washington, June 21–23 1994.
9. J. J. Little. Determining object attitude from extended Gaussian images. In *Proc. of the Ninth Int. Joint Conf. on Artificial Intel.*, volume 2, pages 960–963, California, August 18–23 1985.
10. R. Schneider. *Convex Bodies. The Brunn-Minkowski Theory.* Cambridge University Press, Cambridge, 1993.
11. A. Tuzikov, J. B. T. M. Roerdink, and H. J. A. M. Heijmans. Similarity measures for convex polyhedra based on Minkowski addition. *Pattern Recognition*, 2000. to appear.
12. A. V. Tuzikov, J. B. T. M. Roerdink, H. J. A. M. Heijmans, and S. A. Sheynin. Measures and indices of reflection symmetry for convex polyhedra. In *Mathematical Morphology and its Application to Image and Signal Processing*, pages 59–66, 1998.

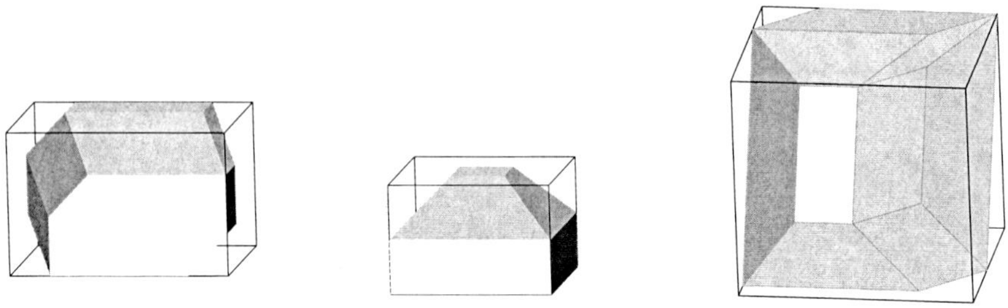

(a) P_1 and P_2 (b) P_1 and P_3 (c) P_2 and P_3

Fig. 4. Minkowski sums for which minimum in Table II is achieved.

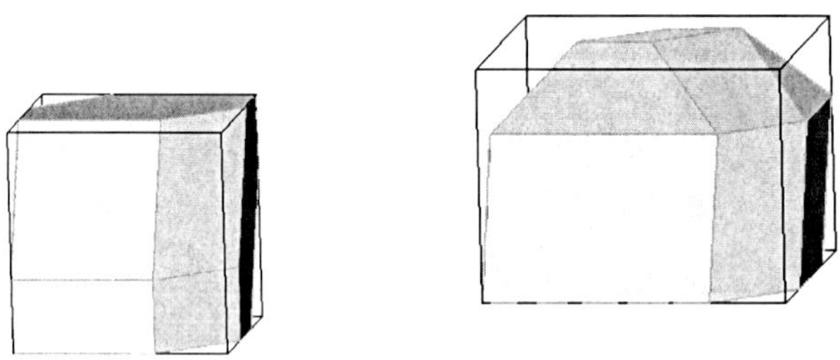

(a) $P_1 \oplus P_2$ (b) $P_2 \oplus P_3$

Fig. 5. Polyhedra $P_1 \oplus P_2$, $P_2 \oplus P_3$.

TOPOLOGICAL PROPERTIES OF HAUSDORFF DISCRETIZATIONS

MOHAMED TAJINE and CHRISTIAN RONSE
Laboratoire des Sciences de l'Image, de l'Informatique et de la Télédétection (LSIIT, UPRES-A CNRS 7005)
Boulevard Sébastien Brant, F-67400 Illkirch
France
email: {tajine, ronse}@dpt-info.u-strasbg.fr

Abstract. We study a new framework for discretization of closed sets based on Hausdorff metric as described in [15, 16, 23, 24]. Let F be a non-empty closed subset of $\mathbb{R}^n$, $S \subseteq \mathbb{Z}^n$ is a Hausdorff discretization of F if it minimizes the Hausdorff distance to F. We study the properties of Hausdorff discretization for homogeneous metrics. For such metrics the popular covering discretizations are Hausdorff discretizations. We also study some topological properties of Hausdorff discretizations. Actually, a Hausdorff discretization of a connected closed set is 8-connected, its maximal Hausdorff discretization is 4-connected, and a Hausdorff discretization "preserves" the homotopy for a class of closed sets and a class of homogeneous metrics. Under some general condition, a Hausdorff discretization is "homeomorphic" to the original set.

Key words: Connected, n-Connected, Covering Discretization, Hausdorff Metric, Homeomorphic, Homogeneous Metric, Homotopically Equivalent.

1. Introduction

The problem of discretizing shapes, images, and image processing operations has been extensively studied in mathematical morphology [10, 18]. Although the importance of the Hausdorff metrics in image analysis is recognized [18], current discretization theories are not based on a metric approach. This drawback has been the starting point of our study of Hausdorff discretization [15, 16, 19, 22, 23, 24]. The basic idea is to select as possible discretizations of a Euclidean set F all discrete set S such that the Hausdorff distance between F and S is minimal. This leads to several possible choices for such a discretization.

Our framework is as follows: given a metric d on $\mathbb{R}^n$, let F be a non-empty closed subset of $\mathbb{R}^n$; then $S \subseteq \mathbb{Z}^n$ is a Hausdorff discretization of F if it minimizes the Hausdorff distance $H_d(S, F)$ to F. We characterize the set $\mathcal{M}_{H_d}(F)$ of subsets of $\mathbb{Z}^n$ which are a Hausdorff discretization of a non-empty closed set F. We have proved that a Hausdorff discretization of a non-empty closed set F converges to F when the resolution of the discrete space tends to 0, as in [3]. We refine the study of Hausdorff discretization for the class of homogeneous metrics. Actually, we investigate the relationship between Hausdorff discretizations and covering discretizations. We also study some topological properties of Hausdorff discretizations for homogeneous metrics. We have proved that a Hausdorff discretization of a connected closed set is 8-connected, its maximal Hausdorff discretization is 4-connected, and that Hausdorff discretization "preserves" the homotopy for r-convex closed sets for a subclass of homogeneous metrics [19]. Under some general condition, a Hausdorff discretization is "homeomorphic" to the original set. This new result generalizes the one in [12].

This paper is divided into six sections. In the second section we briefly recall classical notions of metric space and Hausdorff space. In the third section, we introduce the Hausdorff discretization. Section 4 deals with homogeneous metrics, while Section 5 considers topological properties of Hausdorff discretization. The last section is a conclusion.

The proofs are not given here, most of them can be found in [19, 21, 23].

2. Some Metric Notions and Hausdorff Metric

We assume that the reader is familiar with classical notions of topological space, metric space and normed space, see for example [2, 9, 11]. We introduce here our notations, most of them are recalled in [19, 22, 23, 24].

Definition 1 *Let $(\mathcal{E}, d)$ be a metric space, and let $p \in \mathcal{E}$ and $r \in I\!R^+$,*

$$\mathcal{B}_r^d(p) = \{x \in \mathcal{E} \mid d(x,p) \leq r\}.$$

$\mathcal{B}_r^d(p)$ is called the ball of center p and of radius r. *Let $E \subseteq \mathcal{E}$.*
• *$int(E) = \{x \in E \mid \exists r > 0,\ \mathcal{B}_r^d(x) \subseteq E\}$, $int(E)$ is called the* interior *of E.*
• *$cl(E)$ is the intersection of all closed set containing E, $cl(E)$ is called the* closure *of E.*

In all the following all topological notions in a metric space $(\mathcal{E}, d)$ are considered relatively to the topology induced by d. All metrics used in this paper are induced by norm. So if N is a norm over $\mathcal{E}$, then the function d_N such that: $\forall x, y \in \mathcal{E},\ d_N(x,y) = N(x-y)$ is called the metric over $\mathcal{E}$ induced by N.

Example:
Let $x = (x_1, x_2, ..., x_n) \in I\!R^n$. Then $\forall p \geq 1$, $|x|_p = \sqrt[p]{|x_1|^p + ... + |x_n|^p}$ and $|x|_\infty = max_{1 \leq i \leq n}|x_i|$ are a norms over $\mathcal{E} = I\!R^n$. The metrics induced by these norms are denoted d_p and d_∞ respectively.

2.1. Hausdorff metric

The definitions and results presented in this subsection can be found in [2, 9].

Definition 2 *Let $(\mathcal{E}, d)$ be a metric space; $\mathcal{H}(\mathcal{E})$ is the set of non-empty compact subsets of $\mathcal{E}$, $\mathcal{F}(\mathcal{E})$ the set of closed subsets of $\mathcal{E}$, and $\mathcal{F}'(\mathcal{E})$ the set of non-empty closed subsets of $\mathcal{E}$.*

On $\mathcal{H}(\mathcal{E})$, we will define a metric H_d, such that if $(\mathcal{E}, d)$ is a complete metric space then $(\mathcal{H}(\mathcal{E}), H_d)$ is a complete metric space.

Definition 3 *Let $(\mathcal{E}, d)$ be a metric space and let $A, B \in \mathcal{H}(\mathcal{E})$. We define* the oriented Hausdorff metric *from A to B by $h_d(A,B) = max_{a \in A}(d(a,B))$ where $d(a,B) = inf_{b \in B}(d(a,b))$.*

Definition 4 *Let $(\mathcal{E}, d)$ be a metric space. The* Hausdorff distance *between two compact sets $A,\ B \in \mathcal{H}(\mathcal{E})$ is defined by $H_d(A, B) = max(h_d(A, B), h_d(B, A))$.*

Remark:
Let $\mathcal{F}'(\mathcal{E})$ be the set of non empty closed set of $\mathcal{E}$. Then, the functions h_d and H_d can be extended in natural way as function from $\mathcal{F}'(\mathcal{E}) \times \mathcal{F}'(\mathcal{E})$ to $I\!R^+ \cup \{+\infty\}$. H_d is a "generalized metric" in the sense that it satisfies the axioms of metric, but can take infinite values.

3. Hausdorff Discretization

In this section, we present our framework of discretization based on Hausdorff metric. Our result are proved in [19]. In the rest of this paper we assume that we have as metric space $(I\!R^n, d)$, where d is a metric induced by a norm on $I\!R^n$, and as a discrete space $\mathcal{D}_\rho = \rho \mathbb{Z}^n$, for $\rho > 0$. So

$$\forall x \in I\!R^n, \forall r \geq 0,\ \mathcal{B}^d_r(x) \cap \mathcal{D}_\rho\ is\ a\ finite\ set,$$

and for such distance, $M \subseteq \mathcal{D}_\rho$ implies that $M \in \mathcal{F}(I\!R^n)$.

Definition 5 *Let d be a metric on $I\!R^n$ and $\rho > 0$. The* covering radius *of the metric d in $\mathcal{D}_\rho$ is*

$$r_c(d, \rho) = sup_{x \in I\!R^n}(d(x, \mathcal{D}_\rho)).$$

Let F be a non-empty closed subset of $I\!R^n$; $M \subseteq \mathcal{D}_\rho$ is a Hausdorff discretization of F in $\mathcal{D}_\rho$ if it minimizes the Hausdorff distance to F. In this section, we study the properties of Hausdorff discretizations. In [15, 16, 23, 24], we have studied the Hausdorff discretizations when F is a compact set.

Definition 6 *Let $F \in \mathcal{F}'(I\!R^n)$.*

- *A set $M \subseteq \mathcal{D}_\rho$ is a* Hausdorff discretization *of F in $\mathcal{D}_\rho$ if $H_d(F, M) = inf(\{H_d(F, S) \mid S \subseteq \mathcal{D}_\rho\})$.*
- *$\mathcal{M}_{H_d}(F, \rho) = \{M \subseteq \mathcal{D}_\rho \mid H_d(F, M) = inf_{S \subseteq \mathcal{D}_\rho}(H_d(F, S))\}$ is the set of Hausdorff discretizations of F.*
- *$\Delta_{H_d}(F, \rho) = (\bigcup_{M \in \mathcal{M}_{H_d}(F,\rho)} M)$ is called* the maximal Hausdorff discretization *of F in $\mathcal{D}_\rho$.*
- *The value $r_H(F, d, \rho) = inf_{x \in F}(d(x, \mathcal{D}_\rho))$ is called* the Hausdorff radius *of F in $\mathcal{D}_\rho$ for the metric d.*

We characterize now the Hausdorff discretization.

Theorem 1 *Let $F \in \mathcal{F}'(I\!R^n)$; then:*

- *$\mathcal{M}_{H_d}(F, \rho)$ is nonvoid and for $M \in \mathcal{M}_{H_d}(F, \rho)$, $H_d(F, M) = r_H(F, d, \rho)$;*
- *for a family $(M_i)_{i \in I}$ of members of $\mathcal{M}_{H_d}(F, \rho)$, $\bigcup_{i \in I} M_i \in \mathcal{M}_{H_d}(F, \rho)$ and so $\Delta_{H_d}(F, \rho) \in \mathcal{M}_{H_d}(F, \rho)$;*
- *if $(M_n)_{n \in I\!N}$ is a decreasing sequence in $\mathcal{M}_{H_d}(F, \rho)$ (relatively to the set inclusion) then $\bigcap_{n \in I\!N} M_n \neq \emptyset$ and $\bigcap_{n \in I\!N} M_n \in \mathcal{M}_{H_d}(F, \rho)$;*
- *$\Delta_{H_d}(F, \rho) = \{p \in \mathcal{D}_\rho \mid d(p, F) \leq r_H(F, d, \rho)\}$;*

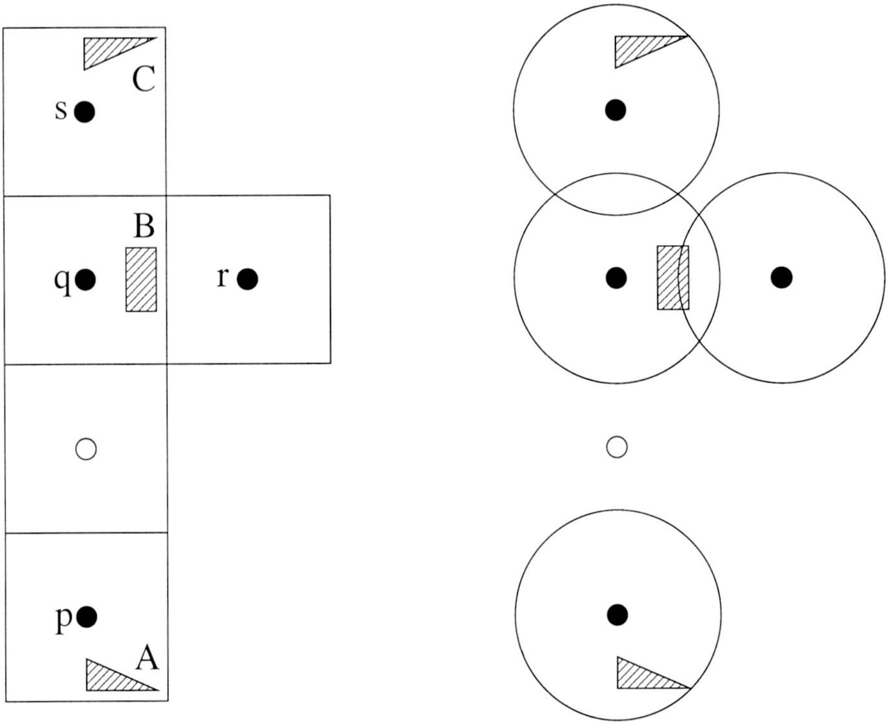

Fig. 1. Left: A compact set $K = A \cup B \cup C$ overlayed with discrete points p, q, r, s and their square cells $C(p,1), C(q,1), C(r,1), C(s,1)$. Right: For $d = d_2$ (the Euclidean distance), the maximal Hausdorff discretization of K is $\{p, q, r, s\}$; indeed, we show the circles of radius $r_H(K, d, \rho)$ centered about these points. The unique other Hausdorff discretizing set of K is $\{p, q, s\}$.

- *$\mathcal{M}_{H_d}(F, \rho)$ is the set of all $M \in \mathcal{D}_\rho$ such that $M \subseteq \Delta_{H_d}(F, \rho)$ and F is included in the union of balls of radius $r_H(F, d, \rho)$ centered about points of M;*
- *$r_H(F, d, \rho) \leq r_c(d, \rho)$.*

Remark:
In [15, 23] we have proved that, if $K \in \mathcal{H}(I\!R^n)$, then $\mathcal{M}_{H_d}(K, \rho)$ is finite and $\forall M \in \mathcal{M}_{H_d}(K, \rho)$, M is finite. Actually, if $r = Sup_{x \in K}(d(O, x)) + r_H(K, \rho)$ *(K being a compact set implies that r is finite)*, then $\Delta_{H_d}(K, \rho) \subseteq \mathcal{B}^d_r(O) \cap \mathcal{D}_\rho$, which is a finite set.

In Figure 1 we illustrate the construction of Hausdorff discretizations for a compact set K: computing the Hausdorff radius (maximal distance from points of K to the discrete space), one takes for $\Delta_{H_d}(K, \rho)$ all discrete points p such that the ball of center p and Hausdorff radius intersect K; any subset M of $\Delta_{H_d}(K, \rho)$ such that the corresponding balls for $p \in M$ cover K, will be a Hausdorff discretization.

In the following proposition, we see that, the "digital geometry" converges relatively to Hausdorff metric to the "Euclidean geometry", as in [3], by using lattices with increasing resolution. In the following proposition we assume that the distance d verify the condition :

$$\lim_{\rho \to 0} r_c(d, \rho) = 0$$

This condition is verified by all distances induced by norms [16].

Proposition 1 *Let $F \in \mathcal{F}'(I\!R^n)$, then for any choice of $M^\rho \in \mathcal{M}_{H_d}(F, \rho)$,*

$$\lim_{\rho \to 0} H_d(F, M^\rho) = 0.$$

Remark:
Let $BH_d : \mathcal{F}'(I\!R^n) \times \mathcal{F}'(I\!R^n) \to I\!R^+$ defined by:

$$BH_d(F, F') = Sup_{x \in I\!R^n}(|d(x, F) - d(x, F')|e^{-d(O,x)}) \ for\ all\ F, F' \in \mathcal{F}'(I\!R^n)$$

where O is any fixed point in $I\!R^n$ for example the 0 vector. Then, BH_d is a metric on $\mathcal{F}'(I\!R^n)$; BH_d is called the Busemann-Hausdorff metric [2, 9, 10, 13] . The topology induced by BH_d is the classical hit-or-miss topology if for example all balls *(relatively to the metric d)* are compact sets.

We have, $\forall F, F' \in \mathcal{F}'(I\!R^n), BH_d(F, F') \leq H_d(F, F')$. So for any choice $M^\rho \in \mathcal{M}_{H_d}(F, \rho)$, $\lim_{\rho \to 0} BH_d(F, M^\rho) = 0$, which implies that a Hausdorff discretization M^ρ of the closed set F converges to F for the hit-or-miss topology when the resolution ρ of the discrete space converge to 0.

Let us briefly compare our approach to the morphological one [10, 18]. Here one chooses a structuring element $A(\rho)$ at each resolution ρ, which satisfies the *covering assumption* $\mathcal{D}_\rho \oplus A(\rho) = I\!R^n$. One takes then the *discretization of F by dilation by $A(\rho)$* [15], consisting of all points $p \in \mathcal{D}_\rho$ such that $A_p \cap F \neq \emptyset$. Making some technical assumptions on A, these authors prove the convergence to F, when the resolution ρ of the discrete space tends to 0, not of the discretization of F, but of a "reconstruction" of that discretization. In fact, we have shown [15, 20] that under the covering assumption, the distance between a closed set and its discretization by dilation is bounded by the *radius* of the structuring element $A(\rho)$, namely the supremum of distances of points of A to the origin. We can choose $A(\rho) = \rho A(1)$, and then for a distance induced by a norm, the radius of $A(\rho)$ is proportional to ρ, so it tends to 0 when ρ tends to 0. Hence we get here for the discretization by dilation a similar Hausdorff metric convergence of the discretization itself (not a "reconstruction") to the original set when the resolution tends to 0. See [15, 20] for more details.

4. Homogeneous Metric and Hausdorff Discretization

We present some properties of a homogeneous metric and we refine the characterization of Hausdorff discretizations for such metrics. Actually, we study the relationship between Hausdorff discretizations and covering discretizations. Again, results of this section are proved in [19].

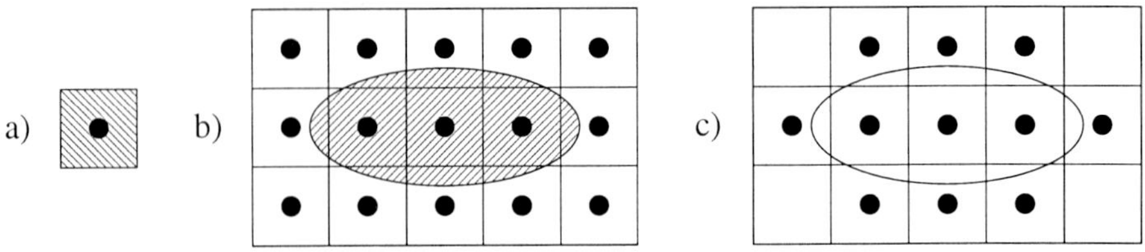

Fig. 2. **(a)** The structuring element A is $C(o,1)$. **(b)** A Euclidean set X overlayed with the discrete points p and their cells $C(p,1)$. **(c)** $\Delta_{SC}(X,1)$.

Definition 7 *Let $p \in \mathcal{D}_\rho$, we define the* the cell *of center p as*

$$\mathcal{C}(p,\rho) = \{x \in I\!R^n \mid d_\infty(x,p) \leq \frac{\rho}{2}\}$$

Note that cells are closed and overlap only at their boundaries.

Definition 8 *A metric d over $I\!R^n$ is called* cellular *if $\forall x \in I\!R^n$, $\forall p,\ q \in Z\!\!Z^n$, $x \in \mathcal{C}(p,1) \Longrightarrow d(p,x) \leq d(q,x)$. In particular, if $x \in \mathcal{C}(p,1) \cap \mathcal{C}(q,1)$, then $d(p,x) = d(q,x)$.*

Definition 9 *A norm N on $I\!R^n$ is* homogeneous *if for every $(x_1,...,x_n) \in I\!R^n$, $(\varepsilon_1,...,\varepsilon_n) \in \{-1,1\}^n$, and for every permutation σ of $\{1,...,n\}$, we have $N(\varepsilon_1 x_{\sigma(1)},...,\varepsilon_n x_{\sigma(n)}) = N(x_1,...,x_n)$. A metric induced by homogeneous norm is called a* homogeneous *metric.*

Theorem 2 *Let d be a homogeneous metric induced by a norm N, then:*

- *d is cellular,*
- *$r_c(d,1) = \frac{1}{2}N(1,...,1)$ and*
- *$\mathcal{B}^{d_\infty}_{\frac{1}{2}}(O) \subseteq \mathcal{B}^{d}_{r_c(d)}(O) \subseteq \mathcal{B}^{d_1}_{\frac{n}{2}}(O)$.*

Example:

In $I\!R^n$, $\forall p \geq 1$, d_p is a homogeneous metric and thus, $r_c(d_p,1) = (\frac{n^{\frac{1}{p}}}{2})$, and $r_c(d_\infty,1) = \frac{1}{2}$.

Definition 10 *Let $E \subseteq I\!R^n$, a subset $S \subseteq \mathcal{D}_\rho$ is called a* covering discretization *of E in $\mathcal{D}_\rho$, if $\forall p \in S$, $E \cap \mathcal{C}(p,\rho) \neq \emptyset$ and $E \subseteq \bigcup_{p \in S} \mathcal{C}(p,\rho)$*

An example of covering discretization is the *supercover discretization* Δ_{SC} which associates to every $X \in \mathcal{P}(I\!R^n)$ the set of all $p \in \mathcal{D}$ such that $C(p,1)$ intersects X (see Figure 2):

$$\forall X \subseteq I\!R^n, \qquad \Delta_{SC}(X,\rho) = \{p \in \mathcal{D} \mid C(p,\rho) \cap X \neq \emptyset\}.$$

Theorem 3 *Let d be a cellular metric and let $F \in \mathcal{F}'(I\!R^n)$. If S is a covering discretization of F in $\mathcal{D}_\rho$, then $S \in \mathcal{M}_{H_d}(F,\rho)$.*

5. Topological Properties of Hausdorff Discretizations in the Plane

In this section, we study some topological properties of Hausdorff discretizations for homogeneous metrics. First we recall some results of [19]: 1) every Hausdorff discretization of a connected closed set is 8-connected, and its maximal Hausdorff discretization is 4-connected; 2) every Hausdorff discretization "preserves" the homotopy for a "r-convex" closed set and for a subclass of homogeneous metrics. The proofs of theses results is given in [19]. Finally, we give a new result: using a stronger condition than "r-convexity", we prove that for a subclass of homogeneous metrics, every Hausdorff discretization is "homeomorphic" with the original closed set. The proof of this new result will be given in [21] (working document in preparation).

Notations:
Let d be a metric on $I\!R^2$, and let $p \in \mathcal{D}_\rho$.
- $\mathcal{V}_8(p) = \{q \in \mathcal{D}_\rho \mid d_\infty(p,q) = \rho\}$ and
- $\mathcal{V}_4(p) = \{q \in \mathcal{D}_\rho \mid d_1(p,q) = \rho\}$.

Property 1 *Let d be a homogeneous metric, and let $F \in \mathcal{F}'(I\!R^n)$. If F is connected then $\forall M \in \mathcal{M}_{H_d}(F)$, M is 8-connected, and $\Delta_{H_d}(F)$ is 4-connected.*

The converse of the last property is not true for every closed sets. Consider for example the set $E = \{(x, \frac{1}{x}) \mid x \in]0,1]\} \cup \{(-x, \frac{1}{x}) \mid x \in]0,1]\}$ which is a closed set of $I\!R^2$ relatively to any distance induced by a norm. It is easy to see that, $\forall \rho > 0, \forall M^\rho \in \mathcal{M}_{H_d}(E,\rho)$, M^ρ is 8-connected, but E is not connected.

In the following we show that the last property has a converse if F is a compact set.

Property 2 *Let d be a homogeneous metric, and let $K \in \mathcal{H}(I\!R^2)$, and assume that there exists $\rho_0 > 0$ such that for every $\rho < \rho_0$ there exists $M^\rho \in \mathcal{M}_{H_d}(K,\rho)$ such that M^ρ is 8-connected; then K is connected.*

Definition 11 *Let d be a homogeneous metric and let $F \in \mathcal{F}(I\!R^n)$; F is called* r-convex relatively to d *if $\forall x \in I\!R^n, \forall r' \leq r$, $\mathcal{B}^d_{r'}(x) \cap F$ is a connected subset of F, and if $d(x,F) \leq r$, then there exists an unique point in F denoted $\pi(x)$ such that $d(x,F) = d(x,\pi(x))$.*

Remark:
If F is a finite set of points and if $\alpha = min_{x \neq y \in F}(d(x,y))$ then F is r-convex relatively to d for every $r < \frac{\alpha}{2}$.

Definition 12 *Let X, Y be metric spaces. Two continuous maps $f, g : X \to Y$ are* homotopic *if there exists a continuous map $H : X \times [0,1] \to Y$ such that*

$$\forall x \in X,\ f(x) = H(x,0) \quad and \quad g(x) = H(x,1).$$

The homotopic relation is an equivalence relation.

Definition 13 *Two metric spaces X, Y are* homotopically equivalent *if there exist maps $f : X \to Y$ and $g : Y \to X$ such that $g \circ f$ and Id_X are homotopic, and $f \circ g$ and Id_Y are homotopic, where Id_X and Id_Y are the identity maps of X and Y respectively.*

The homotopic equivalence is an equivalence relation. The following theorem, proved in [19], states that for a subclass of homogeneous metrics, if M is a Hausdorff discretization in $\mathcal{D}_\rho$ of a r-convex closed set F, then $\bigcup_{m \in M} \mathcal{C}(m, \rho)$ and F are homotopically equivalents for $\rho < \rho_0$. This result generalizes the results of homotopy proved in [12].

Definition 14 *A metric d on $\mathbb{R}^2$ is called* strictly homogeneous *if d is homogeneous and $\mathcal{B}^d_{r_c(d,1)}(0,0) \cap \mathcal{B}^d_{r_c(d,1)}(1,1) = \{(\frac{1}{2}, \frac{1}{2})\}$.*

In other words, the balls of covering radius centered about diagonally adjacent discrete points intersect only at their corners. For example d_p is strictly homogeneous for all $p > 1$ and for $p = \infty$.

Theorem 4 *Let d be a strictly homogeneous metric. Let F be r-convex relatively to d, $\rho \leq \frac{r}{3r_c(d,1)}$ and $M \in \mathcal{M}_{H_d}(F, \rho)$. Then the set $\bigcup_{p \in M} \mathcal{C}(p, \rho)$ is homotopically equivalent to F.*

Definition 15 *Two topological spaces X, Y are called* topologically equivalent *or* homeomorphic *if there exists a bijection $f : X \to Y$ such that both f and its inverse function f^{-1} are continuous.*

Definition 16 *A topological space X is called a* bordered 2D manifold *if every point of X has a neighbourhood homeomorphic to a relatively open subset of a closed half-plane. A connected component of a 2D bordered manifold is called a* bordered surface.

Definition 17 *Let $r > 0$. A closed set $F \in \mathcal{F}'(\mathbb{R}^2)$ is called* strictly r-convex *relatively to a metric d, if F is r-convex relatively to d, and $\forall P \in F$ there exists $Q \in F$ such that $P \in \mathcal{B}^d_r(Q)$ and $\mathcal{B}^d_r(Q) \subseteq F$.*

Definition 18 *Let $r > 0$. A closed set $F \in \mathcal{F}'(\mathbb{R}^2)$ is called r-regular relatively to a metric d, if F and $cl(\mathbb{R}^2 \setminus F)$ are both strictly r-convex relatively to d.*

Theorem 5 *[1, 4, 5, 8, 14]*
Two bordered 2D surfaces are homeomorphic iff they agree in character of orientability, number of contours, and Euler characteristic.

Definition 19 *A subset $S \subset \mathcal{D}_\rho$ is called singular if there exists $P \in S$ such that*

- *$P + \rho(1,1) \in S$, but $P + \rho(0,1) \notin S$ and $P + \rho(1,0) \notin S$; or*
- *$P + \rho(1,-1) \in S$, but $P + \rho(0,-1) \notin S$ and $P + \rho(1,0) \notin S$.*

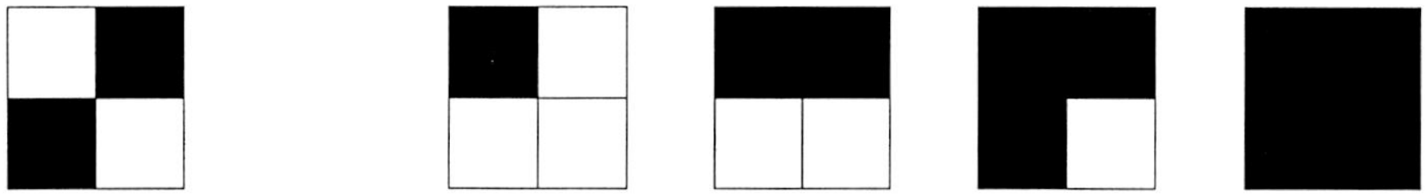

(a) (b)

Fig. 3. (a) Singular configuration. (b) Non singular configurations.

The two different configurations characterizing the singular sets are illustrated in Figure 3(a) by considering alternatively the white and the black pixels as the foreground.

We present now our new main result, which will be proved in [21] (working document in preparation).

Lemma 1 *Let F be a closed r-regular set in $\mathbb{R}^2$ and let $\rho \leq \frac{r}{2\sqrt{2}r_c(d,1)}$. Then $\forall M^\rho \in \mathcal{M}_{H_d}(F,\rho)$, M^ρ is not singular.*

Remarks:
Let F be a closed r-regular set in $\mathbb{R}^2$ and let $\rho \leq \frac{r}{2\sqrt{2}r_c(d,1)}$. If $M^\rho \in \mathcal{M}_{H_d}(F,\rho)$ and $P \in M^\rho$, then all possible configurations at P are represented in Figure 3(b) (modulo a reflection and/or a 90° rotation).

Corollary 1 *Let F be a closed r-regular set in $\mathbb{R}^2$ and let $\rho \leq \frac{r}{2\sqrt{2}r_c(d,1)}$. Then $\forall M^\rho \in \mathcal{M}_{H_d}(F,\rho)$, the set $\bigcup_{P\in M^\rho} \mathcal{C}(P,\rho)$ is a bordered 2D manifold.*

Theorem 6 *Let d be a strictly homogeneous metric and K be a r-regular compact subset of $\mathbb{R}^2$ such that K is a bordered 2D manifold. Let $\rho \leq \frac{r}{2\sqrt{2}r_c(d,1)}$. Then $\forall M^\rho \in \mathcal{M}_{H_d}(F,\rho)$, the set $\bigcup_{P\in M^\rho} \mathcal{C}(P,\rho)$ and K are homeomorphic.*

Let us mention related results from the literature: [17] showed that under certain conditions on a Euclidean set X, in the supercover discretization $\Delta_{SC}(X,\rho)$ there are points that can be removed in such a way that for the remaining subset S of points, $\bigcup_{p\in S} C(p,\rho)$ is homotopic to X. On the other hand [6, 7, 12] gave sufficient conditions under which $\bigcup_{p\in\Delta_{SC}(X,\rho)} C(p,\rho)$ is homeomorphic to X, but our result is more general.

6. Conclusion

Throughout several papers we have introduced a new framework for the discretization of a non-empty closed set, based on the Hausdorff distance. We have the convergence *(in Hausdorff metric sense)* of the discretization to the original object when the resolution of the discrete space converges to zero. We refine the study of the Hausdorff discretizations for homogeneous metrics and we study the topological properties of Hausdorff discretizations.

We intend to do further investigations on:

- the differential properties of Hausdorff discretization;
- its extension to grey-level images;
- the discretizations of basic geometrical and morphological operators: rotation, symmetry, dilation, erosion, opening and closing by a structuring element.

In [19], we have already shown that the Minkowski addition is discretizable in our framework.

References

1. L. V. Ahlfors and L. Sario. *Riemann surfaces.* Princeton University Press, 1960.
2. H. Busemann *The geometry of geodesics.* Academic Press, New York, 1955.
3. S. Duval and M. Tajine. Digital geometry and Fractal Geometry. *DGCI'95*, pp. 93–106, 1995.
4. P. A. Friby and C. F. Gardiner. *Surface Topology.* John Wiley and Sons, 1982.
5. H. B. Griffiths. *Surfaces.* Cambridge University Press, 1981.
6. A. Gross, and L.J. Latecki. Digitizations preserving topological and differential geometric properties. *Computer Vision & Image Processing*, 62(3):370–381, Nov. 1995.
7. A. Gross, and L.J. Latecki. Digital geometric methods in document image analysis. *Pattern Recognition*, 32(3):407–424, Mar. 1999.
8. J. L. Gross and T. W. Tucker *Topological Graph Theory* John Wiley and Son, 1987.
9. F. Hausdorff. *Set Theory.* Chelsea, New York, 1962.
10. H. J. A. M. Heijmans. *Morphological Image Operators.* Academic Press, 1994.
11. J.G. Hocking and G.S. Young *Topology.* Dover Publications Inc., New York, 1988.
12. L. J. Latecki, C. Conrad, and A. Gross. Preserving topology by a digitalization process. *Journal of Mathematical Imaging and Vision*, 8:131–159, 1998.
13. G. Matheron. *Random sets and integral geometry.* John Wiley and Sons, New York, 1975.
14. E. E.Moise *Geometric topology in dimensions 2 and 3.* Springer-Verlag, New York, 1977.
15. C. Ronse and M. Tajine. Discretization in Hausdorff Space. *Journal of Mathematical Imaging & Vision*, to appear, 2000.
16. C. Ronse and M. Tajine. Hausdorff discretization for cellular distances, and its relation to cover and supercover discretization. *Submitted*, 1999.
17. M. Schmitt. Digitization and connectivity. In H. Heijmans & J. Roerdink, editors, *International Symposium on Mathematical Morphology 1998. Mathematical morphology and its applications to image and signal processing IV*, pp. 91–98, Kluwer Academic Publishers, June 1998.
18. J. Serra. Image analysis and mathematical morphology. *Academic Press*, London, 1982.
19. M. Tajine and C. Ronse. Preservation of topology by Hausdorff discretization and comparison to other discretization schemes. *Submitted*, 1999.
20. M. Tajine and C. Ronse. Hausdorff sampling of closed sets in a boundedly compact space. *In preparation*, 2000.
21. M. Tajine and C. Ronse. Topological properties of Hausdorff discretizations. *Working document in preparation*, 2000.
22. M. Tajine, D. Wagner and C. Ronse. Hausdorff discretization and its comparison with other discretization schemes. *DGCI'99*, Paris, LNCS Springer-Verlag, Vol. 1568, pp. 399–410, 1999.
23. D. Wagner. Distance de Hausdorff et problème discret-continu. Mémoire de D.E.A. (M.Sc. Dissertation), Université Louis Pasteur, Strasbourg (France), June 1997. URL: ftp://dpt-info.u-strasbg.fr/pub/recherche/Vision/DEA_97_Wagner.ps.gz
24. D. Wagner, M. Tajine and C. Ronse. An approach to discretization based on the Hausdorff metric. In H. Heijmans & J. Roerdink, editors, *International Symposium on Mathematical Morphology 1998. Mathematical morphology and its applications to image and signal processing IV*, pp. 67–74, Kluwer Academic Publishers, June 1998.

VECTORIAL LEVELINGS AND FLATTENINGS

FERNAND MEYER
Centre de Morphologie Mathématique - Ecole des Mines de Paris
35, rue Saint-Honoré - 77305 Fontainebleau (FRANCE)
Email: meyer@cmm.ensmp.fr

Abstract. It is possible to level a colour image by applying to each of its coordinates a separate grey-tone leveling. Interpreting this operation in a vectorial space gives a better insight into levelings and flattenings, interpreted as Minkowski subtractions.

Key words: Levelings, Flattenings, Vectorial Images, Minkowski Subtraction.

1. Introduction

Levelings and flattenings have been defined and used for grey-tone images. Colour images may be leveled, by leveling independently each colour component. Applying a scalar leveling or flattening to each component of a vector image will be called vectorial leveling. In the present paper, we give a geometrical interpretation of vectorial levelings and flattenings in the vectorial space itself. From this analysis, a deeper insight may be gained in levelings and flattenings, from which new vector operators may be derived; in particular, a way to produce vectorial levelings independent of the coordinate axis is presented. As it appears that levelings are Minkowski subtractions and flattenings are erosions, we will examine whether it is possible to construct dilations and openings. The present study also leads to more synthetic algorithms for vector levelings, in which all coordinates are treated at the same time. Levelings and flattenings have been introduced by F. Meyer [3], [4]. An extensive algebraic study has been made by G. Matheron [2]. Binary levelings have been studied by J. Serra [6].

2. Separating Functions

2.1. Notations

In this paper, we study vectorial functions defined on a discrete grid G. The neighboring relations are defined by a planar graph. The vector function maps the grid G into a vector space R^n, on which an orthonormal basis is defined g: $p \rightarrow g_p = (g_p.x,\ g_p.y,\ g_p.z, \dots.)$.

Most often n is equal to 2 or 3. It may be three colour coordinates or the coordinates of a vector field for instance Figure 1 presents an example of leveling of a motion field; the initial field presents a number of spurious vectors; a marker is generated and the original field is leveled with this marker. The illustrations in the present paper all use $n = 2$. In what follows, we identify the points p of R^n with the vectors of origin O and extremity p.

Given two distinct vectors a and b and one of the vectors i of the basis we will define:

1. HalfSpace(a, b, i) as the half space passing by a, containing b and orthogonal to i. The limiting hyperplane is parallel to the axis of coordinates other than i. If the vector i is orthogonal to the vector $\overrightarrow{ab}$, or if $a = b$, then HalfSpace(a, b, i) is the plane containing a and b and orthogonal to i.
2. The quadrangular zone with corner a and with faces parallel to the axis of the coordinates: Quad(a, b) = HalfSpace$(a, b, x) \cap$ HalfSpace$(a, b, y) \cap$ HalfSpace$(a, b, z) \cap \ldots$.
3. The parallelepiped, with faces parallel to the coordinate axis and with opposite corners a and b Box(a, b) = Quad$(a, b) \cap$ Quad(b, a).
4. More generally, Box$(a_1,\ a_2,\ a_3,\ \ldots., a_n)$ is the smallest parallelepiped parallel to the coordinate axis and containing the family (a_i). This box always exists and is unique. However, not all a_i are summits of this box.

Given a parallelepiped H and a corner a of H, the opposite corner b will be written Opposite$(a,\ H)$. The segment ab will be a maximal diagonal of the parallelepiped. By convention, if the box H is empty, then Opposite$(a, \emptyset) = a$.

2.2. Class of Functions Separating two Functions

Matheron in [2] introduces the following notions for scalar functions, from a set E into a complete lattice T.

Definition 1 *For $g, h, f \in T^E$, we say that h separates g and f, and we write (ghf) or equivalently $(f\ h\ g)$, if and only if, for any $p \in E$, the series $(g_p\ h_p\ f_p)$ is monotonous, i.e. $\forall p \in E : g_p \leq h_p \leq f_p$ or $g_p \geq h_p \geq f_p$.*

We extend now the previous definitions to the vectorial case.

Definition 2 *For $g, h, f \in (R^n)^G$, we say that h separates g and f, and we write $(g\ h\ f)$ or equivalently $(f\ h\ g)$, if and only if, for any $p \in G$, the point h_p belongs to* Box(g_p, f_p).

Obviously, the relation $(g\ h\ f)$ holds if and only if for each coordinate direction i the relation $(g.i\ \ h.i\ \ f.i)$ holds.

Definition 3 *We call* Inter(g, f) *the set of functions separating g and f.*

2.3. The order $g >_f h$

In the scalar case, Serra in [5] defined the activity order between two functions g and h with respect to a reference function f as: g is more active than $h \Leftrightarrow g \wedge f \leq h \wedge f$ and $g \vee f \geq h \vee f$. This activity order extends to the vectorial case as follows:

Definition 4 *We say that g is more far away from f than h, or that g is bigger than h, in the order f, and we write $g >_f h$, if and only if h separates g and $f : g >_f h \Leftrightarrow (g\ h\ f)$.*

Proposition 1 *$>_f$ is an order relation on $(R^n)^G$, and moreover $g >_f h \Leftrightarrow f >_g h$.*

2.4. Existence of a supremum of a family of functions

The family of vectorial functions is not a complete lattice for the order relation $>_f$ since there is in general no supremum for a family of functions (h^i). A function k is larger than each function in a family (h^i) for $>_f$ iff for each pixel p and each index i we have $\mathrm{Box}(h^i_p, f_p) \subset \mathrm{Box}(k_p, f_p)$. This will be the case if and only if $\mathrm{Box}(f_p, h^1_p,\ h^2_p, ..., h^n_p) \subset \mathrm{Box}(k_p, f_p)$. But such an inclusion can only become true if f_p is a summit of $\mathrm{Box}(f_p, h^1_p,\ h^2_p, ..., h^n_p)$, which generally is not the case. There are particular situations however where it is the case. We will evocate two of them.

a) There exists a function g such that for each function of the family (h^i) and each pixel p we have $\mathrm{Box}(f_p, h^i_p) \subset \mathrm{Quad}(f_p, g_p)$.

b) As a particular case: $\mathrm{Box}(f_p, h^i_p) \subset \mathrm{Box}(f_p, g_p)$ for all h^i. In other words, all functions h^i belong to $\mathrm{Inter}\,(g, f)$.

In both cases, there exists a supremum, i.e. the family of elements which are superior to all functions of the family (h^i) has a smallest element, which is defined as $\left[\bigvee_f (h^i)\right]_p = \mathrm{Opposite}(f_p,\ \mathrm{Box}(f_p, h^1_p,\ h^2_p, ..., h^n_p))$. The supremum of two functions h^1 and h^2 is illustrated in fig. 2, with a first example (E) where the supremum exists and a second example (F) where it does not exist.

Remark 1 *The RGB color representation uses only positive or null values for each component. Hence the condition a) for the existence of a supremum is satisfied.*

2.5. The inf-semi lattice of functions

If (h^i) is a family of functions, then a function k separates the function f and each function h^i if and only if for each pixel p, we have $k_p \in \bigcap_i \mathrm{Box}(h^i_p, f_p)$. But $\bigcap_i \mathrm{Box}(h^i_p, f_p)$ is a rectangular box itself; hence $\mathrm{Opposite}(f_p,\ \bigcap_i \mathrm{Box}(h^i_p, f_p))$ is the largest function for $<_f$which separates the function f and each function h^i. This function is the infimum of the family (h^i) for the order relation $>_f$. Its expression at pixel p is given by $[\bigwedge_i h_i]_p = \mathrm{Opposite}(f_p,\ \bigcap_i \mathrm{Box}(h^i_p, f_p))$.

The infimum of two functions h^1 and h^2 is illustrated with three different configurations in fig. 2(E,F,G). Since any family of functions of $(R^n)^G$ has an infimum for $<_f$ but not necessarily a supremum, it is called an inf-semi-lattice.

3. Levelings = Minkowski Subtraction

3.1. Definition of levelings

Scalar levelings have been defined in [4] and [2]: a scalar function a is a leveling of a scalar function b if and only if $a = (b \wedge \alpha a) \vee \beta a = (b \vee \beta a) \wedge \alpha a$ where α is an extensive dilation and β the adjunct anti-extensive erosion. In this section, we will consider the extensive morphological dilation by a flat structuring element B which contains the origin (and the adjunct erosion). We call V the structuring element made of a central pixel and all its neighbors on the grid. The same structuring elements, without the central pixel, will be called $\overset{o}{B}$ and $\overset{o}{V}$. The corresponding morphological dilations and erosions will be written as Minkowski addition: $\alpha a = a \oplus B = \bigcup_{t \in B} s_{.+t}$ and Minkowski subtraction $\beta a = a \ominus B = \bigcap_{t \in B} s_{.+t}$ (for the structuring element V one classically uses: $\delta a = a \oplus V = \bigcup_{t \in V} a_{.+t}$; $\varepsilon a = a \ominus B = \bigcap_{t \in V} a_{.+t}$) where $s_{.+t}$ is the translation of s by the vector t.

Examining the expression $\beta a = a \ominus B = \bigcap_{t \in B} a_{.+t}$, we see that the Minkowski subtraction is nothing but the infimum of translated copies of the function a, by all translations defined the structuring element. Since there exists an infimum for the order relation $>_f$, we may similarly define in this new lattice a Minkowski subtraction by a structuring element B of a vectorial function g:

$$g \ominus_f B = \bigwedge_f^{t \in B} g_{.+t}.$$

The explicit formulation is the following:

$$(g \ominus_f B)_p = \text{Opposite}(f_p, \bigcap_{t \in B} \text{Box}(f_p, g_{p+t})).$$

The construction is illustrated, with three different configurations, by figures 2H, I, and J for a structuring element made of two pixels $(1, 2)$ with the centre at 1.

Definition 5 *A function g is a B-leveling of a function f iff g is invariant by the Minkowski subtraction for the structuring element B: $g = g \ominus_f B$.*

The function g is a B-leveling of a function f iff each coordinate $g.x$, $g.y$, $g.z, ...$ is a scalar B-leveling of the corresponding coordinates $f.x$, $f.y$, $f.z, ...$ of the function f.

If we now take a function h that is not a B-leveling of the function f, it is possible by repeating upon convergence the operation $g = g \ominus_f B$ on a copy g of h; we then obtain the largest B-leveling of f verifying: $f <_f g <_f h$.

3.1.1. *Basic properties*

Curiously, this Minkowski subtraction is not increasing: $g <_f h \nRightarrow (g \ominus_f B) <_f (h \ominus_f B)$, and does not commute with the infimum: $(g \wedge_f h) \ominus_f B \neq (g \ominus_f B) \wedge_f$

$(h \ominus_f B)$; hence, it is not an erosion. It is interesting to analyse why it is not increasing. Let g and h be two functions such that $g <_f h$. This means that, for any pixel p, $\text{Box}(f_p, g_p) \subset \text{Box}(f_p, h_p)$. But for $t \neq 0$, we do not necessarily have $\text{Box}(f_p, g_{p+t}) \subset \text{Box}(f_p, h_{p+t})$ and hence $(g \ominus_f B) \not<_f (h \ominus_f B)$.

G. Matheron in [2] has shown that the flattenings are the smallest increasing (for $<_f$) operators greater than the levelings. We will give a direct proof in the section of flattenings below and show how to construct them.

The following relation $(g \ominus_f B) \ominus_f B = g \ominus_f 2B$ also is not true.

3.1.2. *V-Levelings (levelings based on the structuring element V) are connected operators*

The formula for levelings $g_p = (g \ominus_f V)_p = \text{Opposite}(f_p, \bigcap_{t \in V} \text{Box}(f_p, g_{p+t}))$ implies the following property:

$$\forall (p, q) \text{ neighboring pixels, } \text{Box}(f_p, g_p) \subset \text{Box}(f_p, g_q).$$

Exchanging the roles of p and q yields $\text{Box}(f_q, g_q) \subset \text{Box}(f_q, g_p)$. In the case where $f_p = f_q$ these two relations have as immediate consequence that $g_p = g_q$. Hence, levelings are connected operators.

Let us now analyse in the vector space why V-levelings produce flat zones.

Proposition 2 *Let g be a V-leveling of f and (p, q) be two neighboring pixels. Then,* $\begin{matrix} f_p \in \text{Quad}(g_q, f_q) \\ f_q \in \text{Quad}(g_p, f_p) \end{matrix} \Rightarrow g_p = g_q$

Proof. From $\text{Box}(f_p, g_p) \subset \text{Box}(f_p, g_q)$ we derive that

$$\text{Quad}(g_p, f_p) \subset \text{Quad}(g_q, f_p).$$

On the other hand $f_p \in \text{Quad}(g_q, f_q)$ implies $\text{Quad}(g_q, f_p) \subset \text{Quad}(g_q, f_q)$; putting everything together yields $\text{Quad}(g_p, f_p) \subset \text{Quad}(g_q, f_q)$. Exchanging the roles of p and q we also obtain $\text{Quad}(g_q, f_q) \subset \text{Quad}(g_p, f_p)$. Hence

$$\text{Quad}(g_p, f_p) = \text{Quad}(g_q, f_p)$$

implying that $g_p = g_q$. □

3.2. Supremum of Minkowski subtractions

The result of a Minkowski subtraction always belongs to $\text{Inter}(g, f)$. If we consider the results of several Minkowski subtractions with different structuring elements B_i, then the family $(g \ominus_f B_i)$ has a supremum $\bigvee_f (g \ominus_f B_i)$ which also belongs to $\text{Inter}(g, f)$. As an example, we may construct a leveling for which the elementary step is the supremum of three Minkowski subtractions with segments as structuring elements.

4. The Flattenings

Let g and h be two functions such that $g <_f h$. This means that, for any pixel p, $\text{Box}(f_p, g_p) \subset \text{Box}(f_p, h_p)$. But for $t \neq 0$, we do not necessarily have $\text{Box}(f_p, g_{p+t}) \subset \text{Box}(f_p, h_{p+t})$ and hence $(g \ominus_f B) \not<_f (h \ominus_f B)$. For this reason, the levelings are not increasing. The inclusion $\text{Box}(f_p, g_{p+t}) \subset \text{Box}(f_p, h_{p+t})$ is false in general but may become true if we increase the size of each box. Since $g <_f h$, we have $\text{Box}(f_{p+t}, g_{p+t}) \subset \text{Box}(f_{p+t}, h_{p+t})$ hence $\text{Box}(f_p, f_{p+t}, g_{p+t}) \subset \text{Box}(f_p, f_{p+t}, h_{p+t})$. The preceding relation is true for all vectors t belonging to a given neighborhood V. Furthermore $\text{Box}(f_p, g_p) \cap_{t \in \overset{o}{B}} \text{Box}(f_p, f_{p+t}, g_{p+t})$ defines a box for which f_p is a summit. Hence the following operator, called flattening step, defined for each pixel p, is increasing for $>_f$: $g_p \to \text{Opposite}(f_p,\ \text{Box}(f_p, g_p) \cap_{t \in \overset{o}{B}} \text{Box}(f_p, f_{p+t}, g_{p+t}))$.

This operator is the basic step of a flattening, which we write: $g \oslash_f B = \text{Opposite}(f,\ \text{Box}(f, g) \cap_{t \in \overset{o}{B}} \text{Box}(f, f_{.+t}, g_{.+t}))$ and is illustrated with three different configurations in fig. 2 (K,L,M), for a structuring element made of two pixels $(1, 2)$ with the centre at 1.

If we remark that $\text{Box}(f, g) = \text{Box}(f, g) \cap \text{Quad}(f, g)$, we may write

$$g \oslash_f B = \text{Opposite}\left[f, \text{Box}(f, g) \cap_{t \in \overset{o}{B}} (\text{Quad}(f, g) \cap \text{Box}(f, f_{.+t}, g_{.+t}))\right] = \bigwedge_f^{t \in B} g^t,$$

where the operator T: $g \to g^t = \text{Opposite}\left[f, \text{Quad}(f, g) \cap \text{Box}(f, f_{.+t}, g_{.+t})\right]$ may be interpreted as a translation. Thus the flattening step appears as a Minkowski subtraction in the $\wedge_f - \textit{lattice}$ but the translation is rather baroque. In contrast, the elementary leveling step is a Minkowski subtraction also in the $\wedge_f - \textit{lattice}$ based on the usual translation $g_{.+t}$.

The scalar version of the "translation" T has been introduced by Kresh in [1]. Contrarily to the ordinary translation, this translation commutes with $\wedge_f$ and under some restrictive conditions with $\vee_f$.

Lemma 1 *The operator T is increasing: $g <_f h \Rightarrow g^t <_f h^t$ and commutes with $\wedge_f$: $(g \wedge_f h)^t = g^t \wedge_f h^t$. If functions g and h possess a supremum, i.e. $\text{Quad}(f, g) = \text{Quad}(f, h)$, then $(g \vee_f h)^t = g^t \vee_f h^t$.*

As an immediate consequence of the commutativity of the operator $\wedge_f$ and the fact that it commutes with T we obtain:

Proposition 3 *The elementary flattening $g \oslash_f B$ commutes with the $\wedge_f$ infimum, hence it is an erosion. It is the smallest increasing operator bigger than the elementary leveling $g \ominus_f B$.*

Definition 6 *A function g is a B-flattening of a function f iff g is invariant by the flattening step with the structuring element B: $g = g \oslash_f B$.*

Obviously, we have: $g \ominus_f B <_f g \oslash_f B <_f g$. Hence, if g is a leveling of f, i.e. $g \ominus_f B = g$, then $g \oslash_f B = g$, meaning that g also is a flattening of f. Hence, any leveling of f also is a flattening of f.

Contrarily to classical erosions, it is not true that: $(g \oslash_f B) \oslash_f B = g \oslash_f 2B$.

4.0.1. *Characterisation of V-flattenings (flattenings based on the structuring element V)*
If g is a V-flattening of a function f, then for the same neighboring pixels (p,q) : $\text{Box}(f_p, g_p) \subset \text{Box}(f_p, f_q, g_q)$. It is easy to check that these two relations imply, for each coordinate, the classical characterization of flattenings: $f_p.x \wedge f_q.x \wedge g_q.x \leq g_p.x \leq f_p.x \vee f_q.x \vee g_q.x$.

4.1. An increasing Minkowski addition

On a complete lattice, an adjunct dilation may be associated to each erosion ([5]). As we only have a semi-lattice, we could follow Kresh and define the following adjunct dilation ([1]): $\delta_f(g) = \bigwedge\limits_f \{h \mid g <_f h \oslash_f B\}$. This dilation is increasing, but its expression does not lead to a practical construction. For this reason we prefer to construct an explicit increasing Minkowski addition, based on the increasing "translation" T: which for functions that possess a supremum behaves as a dilation

$$g \otimes_f B = \bigvee_f^{t \in B} g^t = \text{Opposite}(f,\ \text{Quad}(f,g) \cap \text{Box}(f, g, ..., f_{-t_1},\ g_{-t_1}, ...)).$$

Using the restricted commutation of T with $\vee_f$ we obtain:

Proposition 4 *If a family (h^i) of functions is comparable, i.e. for $i \neq j$, $\text{Quad}(f_p, h^i) = \text{Quad}(f_p, h^j)$, then the increasing Minkowski addition commutes with the supremum of the family, hence it is a dilation.*

Due to the fact that the operator T has not all properties of the usual translation, in particular $(g^t)^{-t} \neq g$, we do not have complete adjunction properties between the operators $\oslash_f$ and $\otimes_f$. We only managed to show the following, for structuring elements which are couple of pixels (o,t):

$$h \otimes_f (0,t) <_f g \Rightarrow h^{-t} <_f g \oslash_f (0,t)$$

$$h <_f g \oslash_f (0,t) \Rightarrow h \otimes_f (0,t) <_f g^{-t}.$$

Unfortunately we cannot have total adjunction, as shows the following counter-example. Let us consider in 1-dimension three periodic scalar functions functions f, g and h where $f_{2k+1} = f_1$ and $f_{2k} = f_2$. Their values are distributed as follows: $f_2 > (h_1 = g_2) > (h_2 = g_1) > f_1$. Since $\text{Box}(f_1, g_1) \subset \text{Box}(f_1, h_1) \cap \text{Box}(f_1, f_2, h_2)$ and $\text{Box}(f_2, g_2) \subset \text{Box}(f_2, h_2) \cap \text{Box}(f_2, f_1, h_1)$ we indeed have $g <_f (h \oslash_f B)$. However, we do not have an adjunction as it is not true that $(g \otimes_f B) <_f h$ as $\text{Quad}(f_2, g_2) \cap \text{Box}(f_1, f_2, g_1, g_2) = \text{Box}(f_1, f_2) \not\subseteq \text{Box}(f_2, h_2)$.

4.2. Pseudo-openings and closings

Pseudo-openings will be obtained by applying, in sequence, a flattening (increasing Minkowski subtraction) by a structuring element B followed by an increasing Minkowski addition by the symmetrical structuring element B' :

$((g \oslash_f B) \otimes_f B')$. Pseudo-closings will be obtained by applying, in sequence, an increasing Minkowski addition by a structuring element B followed by a flattening (increasing Minkowski subtraction) by the symmetrical structuring element $B' : ((g \otimes_f B) \oslash_f B')$.

5. Conclusion

The levelings and flattenings presented in this paper may be called separable, as they may be obtained by processing each coordinate axis separately. This feature is reflected by the fact that the basic elements $\mathrm{Quad}(a, b)$ and $\mathrm{Box}(a, b)$ are rectangular shapes, with the faces parallel to the axis of coordinates. This separability may be considered as an advantage, in terms of computation, as a processor may be affected to each component, and as there is no cross-effects between coordinates. On the other hand, it is also a disadvantage, as the result relies heavily on the choice of coordinate axis. For this reason, it may be useful to construct vector levelings and flattenings, less dependant, or independent of this choice of axis. In order to get an independence of the coordinate axis, we use disks in the place of boxes and adopt the following definition for the vectorial activity order relation, on which the construction of levelings would apply without problems.

Definition 7 *For $g, h, f \in (R^n)^G$, we say that h separates g and f, and we write $(g\ h\ f)$, or equivalently $(f\ h\ g)$, if and only if for any $p \in G$, the point h_p belongs to the disk $\mathrm{Disk}(g_p, f_p)$ for which the points g_p and f_p are extremities of diameter.*

For constructing the infimum of a family of functions $(g_1, g_2, ..., g_n)$ at a pixel p we have to construct the intersection of all disks $\mathrm{Disk}(g_p^k, f_p)$ which is rather tedious in the initial space, but becomes simple after transforming the space by an inversion centered in f_p. Each of the disks is transformed into a half-plane not containing the point f_p. The intersection of all these halfplanes forms a convex body, on which we have to project the point f: this projection always exists is unique but may be at infinity; after inversion, the image of this projected point will be the farthest point in the intersection of disks $\mathrm{Disk}(g_p^k, f_p)$. The inversion of the point at infinity gives f_p. The construction is illustrated by fig. 3.

References

1. R. Kresh, "Extension of morphological operations to complete semi-lattices and its applications to image and video processing." In *Mathematical Morphology and its Applications to Image and Signal Processing*, H. Heijmans and J. Roerdink, editors, pp. 35-42, Kluwer, 1998.
2. G. Matheron, "Les nivellements," *Technical Report*, CMM, 1997.
3. F. Meyer, "From connected operators to levelings." In *Mathematical Morphology and its Applications to Image and Signal Processing*, H. Heijmans and J. Roerdink, editors, pp. 191-199, Kluwer, 1998.
4. F. Meyer, "The levelings." In *Mathematical Morphology and its Applications to Image and Signal Processing*, H. Heijmans and J. Roerdink, editors, pp. 199-207, Kluwer, 1998.

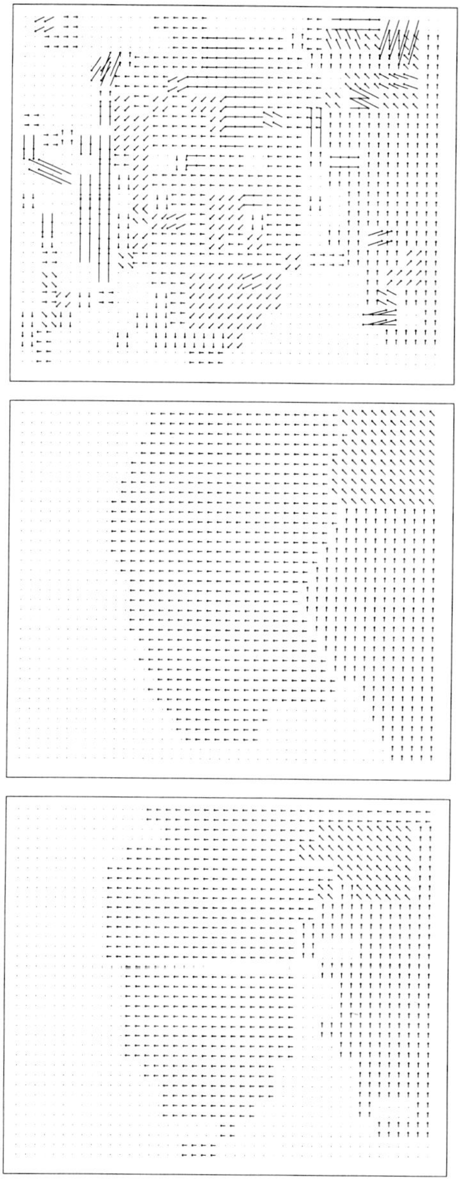

Fig. 1. Original motion vector field, marker, and vectorial leveling.

5. J. Serra, ed., *Image Analysis and Mathematical Morphology. Volume 2: Theoretical Advances.* London, England: Academic Press, 1988.
6. J. Serra, "Quelques propriétés des nivellements," *Technical Report 30/98/MM*, CMM, 1998.

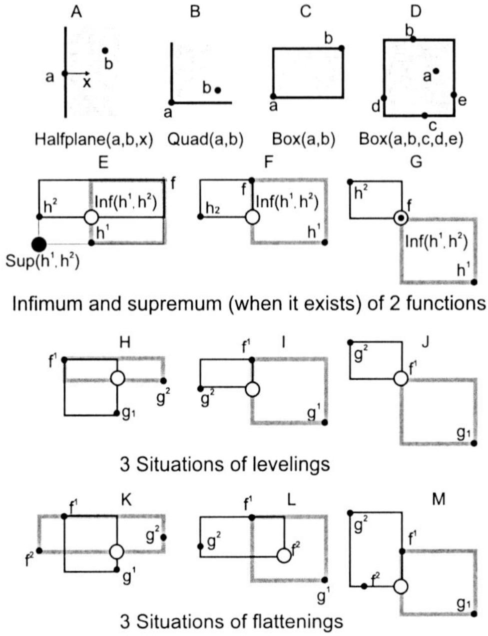

Fig. 2.

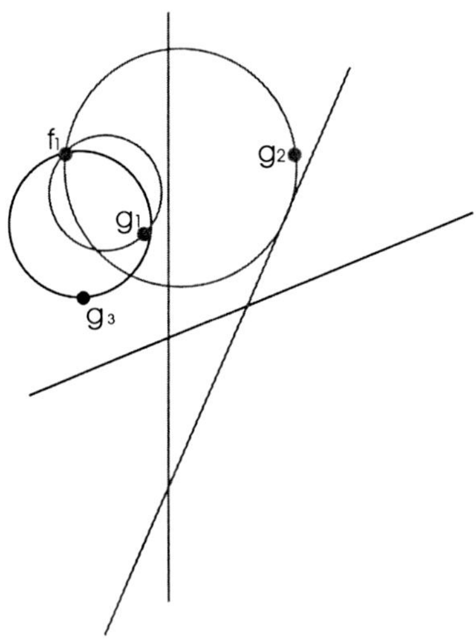

Fig. 3. An inversion of centre f_1 transforms each of the disks (f_1, g_k) into a halfplane. The intersection of the disks becomes by inversion the intersection of the halfplanes. The farthest points in the intersection of the disks corresponds to the closest points in the intersection of the halfplanes.

A LATTICE CONTROL MODEL OF FUZZY DYNAMICAL SYSTEMS IN STATE-SPACE*

PETROS MARAGOS, GIORGOS STAMOU and
SPYROS TZAFESTAS
National Technical University of Athens,
Dept. of Electrical & Computer Engineering,
Zografou, 15773 Athens, Greece.

Abstract. Lattice control can unify nonlinear control systems where the basic vector and signal superpositions or transformations are based on the lattice supremum and infimum. In this paper we introduce a special case of lattice control that can model fuzzy dynamical systems in state space. Vector and signal transformations are represented as lattice dilations or erosions. The state and output responses are computed via supremal convolutions based on fuzzy norms. Causality and stability issues are studied. Finally, solutions to the controllability and observability problem are found using lattice adjunctions.

Key words: Nonlinear control, lattice morphology, fuzzy systems, minimax algebra.

1. Lattice Model for Max and Min Control

In [7, 8, 6] a unified model was proposed based on lattice theory for large classes of nonlinear control systems, such as discrete event dynamical systems, recursive morphological filters, and fuzzy dynamical systems. Lattice morphology [9, 3] is ideally suited to studying such systems because all vector and signal operations and mappings involved can be expressed as morphological operators, and solutions to important control issues such as responses, stability and controllability are obtained using simple lattice-theoretic morphological concepts. In this paper we examine a special case of lattice control systems applicable to fuzzy dynamical systems.

In classical linear control the state vectors, the input/output signals, and the system matrices take values from the field of reals equipped with standard addition and multiplication. In lattice control we take the set $\mathbb{V}$ of *scalars* to be a complete sublattice of $\overline{\mathbb{R}}$ and equip it with the standard real number ordering $\leq$ and four binary operations:

(A). A generalized **'addition'**, which will be the supremum $\vee$ on reals.

(A′). A **'dual addition'**, which will be the infimum $\wedge$ on reals.

(M). A commutative generalized **'multiplication'** $\star$ under which: (i) $\mathbb{V}$ is a monoid (i.e., semigroup possessing an identity) with identity V_{id} and null element $V_{\text{inf}} = \bigwedge \mathbb{V}$, and (ii) $\star$ is a scalar dilation, i.e., distributes over any supremum.

(M′). A commutative **'dual multiplication'** $\star'$ under which: (i) $\mathbb{V}$ is a monoid

* This research was supported by the Greek Secretariat for Research and Technology under Grants $E\Pi ET - 98\Gamma T26$ and $\Pi ENE\Delta - 99E\Delta 164$.

with identity $V_{\rm id}'$ and null element $V_{\sup} = \bigvee \mathbb{V}$, and (ii) $\star'$ is a scalar erosion, i.e., distributes over any infimum.

We group the above requirements into three assumptions:
(LCA1). $(\mathbb{V}, \vee, \wedge)$ is a complete infinitely-distributive lattice.
(LCA2). $(\mathbb{V}, \star)$ is a commutative monoid, and $\star$ is a dilation.
(LCA3). $(\mathbb{V}, \star')$ is a commutative monoid, and $\star'$ is an erosion.

Under the above assumptions $(\mathbb{V}, \vee, \wedge, \star, \star')$ becomes a commutative *complete lattice-ordered double monoid (CLODUM)*. This will be the most general and minimally required algebraic structure we consider for the set of scalars. In our model, all the vectors/matrices/signals take values from $\mathbb{V}$, and their 'addition' is done via pointwise sup or inf. The most important abstraction is 'multiplication' of two matrices.[1] Thus, the generalized max-$\star$ 'product' of a matrix $\mathbf{Q} = [q_{ij}] \in \mathbb{V}^{m\times\ell}$ with a matrix $\mathbf{R} = [r_{ij}] \in \mathbb{V}^{\ell\times n}$ yields a matrix $\mathbf{P} = [p_{ij}] \in \mathbb{V}^{m\times n}$ defined by:

$$\mathbf{P} = \mathbf{Q} \boxast \mathbf{R} \quad , \quad p_{ij} = \bigvee_{k=1}^{\ell} q_{ik} \star r_{kj} \tag{1}$$

The state equations of the **max control** model are:

$$\begin{aligned} \mathbf{x}(k+1) &= \mathbf{A} \boxast \mathbf{x}(k) \;\vee\; \mathbf{B} \boxast \mathbf{u}(k) \\ \mathbf{y}(k) &= \mathbf{C} \boxast \mathbf{x}(k) \;\vee\; \mathbf{D} \boxast \mathbf{u}(k) \end{aligned} \tag{2}$$

where k is a discrete time index. We assume a n-dimensional state vector $\mathbf{x} = [x_1, x_2, ..., x_n]^t \in \mathbb{V}^n$, a p-dimensional input $\mathbf{u} \in \mathbb{V}^p$, and an r-dimensional output $\mathbf{y} \in \mathbb{V}^r$. Therefore, the four matrices have the following sizes: $\mathbf{A} \in \mathbb{V}^{n\times n}, \mathbf{B} \in \mathbb{V}^{n\times p}, \mathbf{C} \in \mathbb{V}^{r\times n}$, and $\mathbf{D} \in \mathbb{V}^{r\times p}$. By replacing $\vee$ with $\wedge$ and $\boxast$ with a dual matrix 'product' $\boxast'$, where a row and a column vector are 'multiplied' via a min-$\star'$ operation, we obtain a *dual* model that describes the state-space dynamics of **min control** systems.

By specifying the scalar 'multiplication' $\star$ and its dual $\star'$, we obtain a large variety of classes of nonlinear dynamical systems that are described by the above unified lattice control model. Two such choices are:

(1) *Max-Sum Control* where $\mathbb{V} = \overline{\mathbb{R}}$ and $\star = +$. Such systems (with special choices of $\mathbf{A}, \mathbf{B}, \mathbf{C}, \mathbf{D}$) have been used in [1, 4, 8] to model the dynamics of certain classes of *discrete event dynamical systems (DEDS)* as applied to material flow in manufacturing systems and related scheduling problems. The underlying nonlinear matrix operations are the basis of the *minimax algebra* [2], which has found numerous applications in DEDS and operations research. In typical applications of DEDS, the states $x_i(k)$ may represent the start-up or completion time of the k-th cycle of machine i, the input $\mathbf{u}$ represents availability times of parts, $\mathbf{y}$ represents exit times, and the elements of the matrices $\mathbf{A}, \mathbf{B}, \mathbf{C}, \mathbf{D}$ represent service/delay times or activity durations. Further, the max-sum control model can also capture the dynamics of *recursive morphological filters* described by max-sum difference equations [8].

[1] *Notation:* If $\mathbf{M} = [m_{ij}]$ is a matrix, its (i,j)th element is also denoted as $\{\mathbf{M}\}_{ij} = m_{ij}$. Similarly, if $\mathbf{x} = [x_i]$ is a vector, its $i^{\rm th}$ element is denoted as $\{\mathbf{x}\}_i$ or simply x_i.

(2) *Max-Tnorm Control:* There are many types of nonlinear control systems where the elements of the state, input and output vectors represent fuzzy set memberships [12], possibilities, or probabilities. Examples include *fuzzy* control systems, *probabilistic* automata, fuzzy classifiers, as well as certain types of neural nets with nonlinear combinations of inputs; surveys of such fuzzy systems can be found in [5]. Fuzzy state-space models can be useful for qualitative modeling of problems with large number of states where quantitative modeling is impossible. The dynamics of large classes of such systems can be described via the lattice control model by restricting the set of scalars to be $\mathbb{V} = [0,1]$ and using a fuzzy intersection norm (a.k.a. 'triangular-norm') $T(a,b) \stackrel{\triangle}{=} a{\star}b$ as the scalar 'multiplication'. This paper deals with this special case of lattice control.

2. Max-Tnorm Control

A **fuzzy intersection norm**, in short a Tnorm, is a binary operation T: $[0,1] \to [0,1]$ that satisfies the following conditions [5]: For all $a, b, c \in [0,1]$

F1. $T(a,1) = a$ and $T(a,0) = 0$ (boundary conditions).
F2. $T(a, T(b,c)) = T(T(a,b), c)$ (associativity).
F3. $T(a,b) = T(b,a)$ (commutativity).
F4. $b \leq c \Longrightarrow T(a,b) \leq T(a,c)$ (increasing).

For the Tnorm to satisfy the general algebraic conditions we require from the set of scalars, it must also satisfy the following:

F5. T is a continuous function.

Conditions F1-F3 make $([0,1], T)$ a commutative monoid with identity $V_{\text{id}} = 1$ and null $V_{\text{inf}} = 0$. Conditions F4-F5 suffice to make T a scalar dilation with respect to any argument, as proven next.

Proposition 1 . *Let T be a continuous fuzzy intersection norm. Then, the operator $x \mapsto T(x,a)$, for any arbitrary fixed $a \in [0,1]$, is a dilation.*

Proof: Consider a (finite or infinite) collection $\{x_j : j \in J\}$ of points in $[0,1]$ with $x = \bigvee_{j \in J} x_j$. Since $[0,1]$ is compact, we can find an increasing subsequence $\{x_k = x_{j_k}\}$, such that $x_k \leq x_{k+1}$ and $x = \bigvee_k x_k = \lim_{k\to\infty} x_k$. Since T is increasing, $T(x_k, a)$ is also an increasing sequence that converges to its supremum $\bigvee_k T(x_k, a)$. Further, $\bigvee_k T(x_k, a) \leq \bigvee_j T(x_j, a) \leq T(x, a)$. Finally, since T is continuous, we have $\bigvee_k T(x_k, a) = \lim_{k\to\infty} T(x_k, a) = T(x, a)$. This yields

$$T(\bigvee_j x_j, a) = \bigvee_j T(x_j, a)$$

which proves that T is a scalar dilation. □

As a 'dual multiplication' we may use a **fuzzy union norm** $T'(a,b) \stackrel{\triangle}{=} a{\star}'b$, where T' satisfies F2-F5 and a dual boundary condition:

F1$'$. $T'(a,0) = a$ and $T'(a,1) = 1$ (dual boundary conditions).

Clearly, $([0,1], T')$ is a commutative monoid, and T' is an erosion. Choosing in the lattice control model the above set of scalars and 'multiplications' among them creates the case of *max-T and min-T′ control* systems, obtained by replacing in the general state equations of lattice control the general matrix 'product' $\boxplus$ and its dual $\boxplus'$ with the following max-T and min-T' versions:

$$\mathbf{P} = \mathbf{Q} \,\square_T\, \mathbf{R} \ , \ p_{ij} = \bigvee_k T(q_{ik}, r_{kj}) \tag{3}$$

$$\mathbf{P} = \mathbf{Q} \,\square'_T\, \mathbf{R} \ , \ p_{ij} = \bigwedge_k T'(q_{ik}, r_{kj}) \tag{4}$$

The most obvious choice for the T norm and its dual norm T' are the min and max, respectively. But there are also numerous other choices [5].

3. Vector and Signal Lattice Operators

The space of vectors and the space of signals with values from the lattice $\mathbb{V}$ are special cases of **function lattices**. The underlying set of these lattices is the set $\mathcal{L} = \mathbb{V}^E$ of all functions mapping an arbitrary nonempty set E into $\mathbb{V}$. In particular, if $E = \{1, 2, ..., n\}$, then $\mathcal{L}$ becomes the set of all n-dimensional vectors (n-tuples) $[x_1, ..., x_n]^t$ with elements from $\mathbb{V}$. If $E = \mathbb{Z}$, then $\mathcal{L}$ becomes the set of all discrete-time signals with values from $\mathbb{V}$. The set $\mathcal{L}$ becomes a complete infinitely distributive lattice if we define on it the standard pointwise partial ordering $\leq$, supremum $\vee$, and infimum $\wedge$ induced by $\mathbb{V}$.

Pointwise 'multiplication' of a lattice element $F \in \mathcal{L} = \mathbb{V}^E$ by a scalar $a \in \mathbb{V}$ yields elementary dilations on $\mathcal{L}$ that are called **translations** $\tau_a(F)(x) \triangleq T(F(x), a)$, $x \in E$. An operator ψ on $\mathcal{L}$ is called **translation invariant** iff it commutes with any translation, i.e., $\psi\tau = \tau\psi$ for all τ. All the above concepts apply as well for the *dual translations* $\tau'_a(F)(x) \triangleq T'(F(x), a)$, which are elementary erosions on $\mathcal{L}$.

More general dilations and erosions on the function lattice $\mathcal{L} = \mathbb{V}^E$ can be decomposed into suprema and infima of scalar dilations and erosions on $\mathbb{V}$, respectively.

Proposition 2 ([3]). *Let* $\mathbb{V}$ *be a complete lattice and* E *an arbitrary nonempty set. The pair* (ε, δ) *is an adjunction on the function lattice* $\mathbb{V}^E$ *iff for every* $x, y \in E$ *there exists an adjunction* $(\varepsilon_{x,y}, \delta_{x,y})$ *on* $\mathbb{V}$ *such that*

$$\delta(F)(y) = \bigvee_{x \in E} \delta_{x,y}(F(x)) \quad , \quad \varepsilon(G)(x) = \bigwedge_{y \in E} \varepsilon_{y,x}(G(y)) \tag{5}$$

If we define the *impulse functions* q and their duals q'

$$q_{z,v}(x) = \begin{cases} v, & x = z \\ 0, & x \neq z \end{cases} \quad , \quad q'_{z,v}(x) = \begin{cases} v, & x = z \\ 1, & x \neq z \end{cases} \tag{6}$$

we can enable the decomposition (5) by defining the scalar dilations to be

$$\delta_{x,y}(v) = \delta(q_{x,v})(y), \quad v \in \mathbb{V} \tag{7}$$

and $\varepsilon_{y,x}$ to be the adjoint erosion of $\delta_{x,y}$.

3.1. VECTOR LATTICE

Consider now the vector space $\mathcal{L} = \mathbb{V}^n$, equipped with the partial ordering $\mathbf{x} \leq \mathbf{y}$, which means $x_i \leq y_i$ $\forall i$, the supremum $\mathbf{x} \vee \mathbf{y} = [x_i \vee y_i]$ and the infimum $\mathbf{x} \wedge \mathbf{y} = [x_i \wedge y_i]$ between any vectors $\mathbf{x}, \mathbf{y} \in \mathcal{L}$. Then, $(\mathcal{L}, \vee, \wedge)$ is a complete infinitely distributive lattice. Elementary vector dilations are the *vector translations* $\tau_a(\mathbf{x}) = [T(x_i, a)]$ and their duals $\tau'_a(\mathbf{x}) = [T'(x_i, a)]$. By defining as 'impulse functions' the unit vectors $\mathbf{e}$ and their duals $\mathbf{e}'$

$$\mathbf{e}_i \stackrel{\triangle}{=} [0, ..., 0, 1, 0, ..., 0]^t, \quad \mathbf{e}'_i \stackrel{\triangle}{=} [1, ..., 1, 0, 1, ..., 1]^t$$

each vector $\mathbf{x} = [x_1, ..., x_n]^t$ can be represented as a max of translated impulse vectors or as a min of dual-translated dual impulse vectors

$$\mathbf{x} = \bigvee_{i=1}^{n} \tau_{x_i}(\mathbf{e}_i) = \bigwedge_{i=1}^{n} \tau'_{x_i}(\mathbf{e}'_i) \tag{8}$$

More general forms of vector dilation (δ_M) and erosion (ε_M) are, respectively, the max-T and min-T' 'product' of a matrix $\mathbf{M}$ with an input vector:

$$\delta_M(\mathbf{x}) \stackrel{\triangle}{=} \mathbf{M} \,\Box_T\, \mathbf{x}, \quad \varepsilon_M(\mathbf{x}) \stackrel{\triangle}{=} \mathbf{M} \,\Box'_T\, \mathbf{x} \tag{9}$$

A vector operator ψ on $\mathcal{L}$ is (dual-)translation invariant iff it commutes with any vector (dual-)translation.

Theorem 1 ([6]). *(a) Any translation invariant dilation δ on the vector lattice $\mathcal{L} = [0, 1]^n$ can be represented as a matrix-based dilation δ_M where $\mathbf{M} = [m_{ij}]$ with $m_{ij} = \{\delta(\mathbf{e}_j)\}_i$, and vice-versa.*
(b) Any dual-translation invariant erosion ε on $\mathcal{L}$ can be represented as a matrix-based erosion $\varepsilon_{M'}$ where $\mathbf{M}' = [m'_{ij}]$ with $m'_{ij} = \{\varepsilon(\mathbf{e}'_j)\}_i$, and vice-versa.

Given a vector dilation $\delta(\mathbf{y}) = \mathbf{M} \,\Box_T\, \mathbf{y}$ with $\mathbf{M} = [M_{ij}]$, what is its *adjoint* erosion ε? The **scalar adjoint erosion** stems from a binary operation ξ : $[0, 1]^2 \to [0, 1]$ defined by

$$\xi(w, a) \stackrel{\triangle}{=} \sup\{v : T(v, a) \leq w\} \tag{10}$$

For example, the adjoints of the minimum and product Tnorms are:

$$T(v, a) = \min(v, a) \implies \xi(w, a) = \frac{1 + w}{2} + \left(\frac{1 - w}{2}\right) \text{sign}(w - a) \tag{11}$$

$$T(v, a) = av \implies \xi(w, a) = \min(\frac{w}{a}, 1) \tag{12}$$

where $\text{sign}(r) = 1$ if $r \geq 0$ and -1 else. If we consider the scalar dilations $\delta_{ij}(v) = \{\delta(\mathbf{e}_{i,v})\}_j = T(v, m_{ji})$ then their adjoint scalar erosions are $\varepsilon_{ji}(w) = \xi(w, m_{ji})$. Thus, according to the decomposition (5), the adjoint vector erosion is

$$\varepsilon(\mathbf{x}) = \mathbf{M}^t \,\Box_\xi\, \mathbf{x}, \quad \{\mathbf{M}^t \,\Box_\xi\, \mathbf{x}\}_i \stackrel{\triangle}{=} \bigwedge_j \xi(x_j, m_{ji}) \tag{13}$$

where $(\cdot)^t$ denotes matrix transposition.

3.2. Signal Lattice

Consider the set $\mathbb{V}^{\mathbb{Z}}$ of all discrete-time signals $f : \mathbb{Z} \to \mathbb{V}$ with values form $\mathbb{V}$. Equipped with pointwise sup $\vee$ and inf $\wedge$, this becomes a complete infinitely distributive lattice $\mathcal{L}$ with partial order the pointwise signal relation $\leq$. The signal translations are the operators $\tau_{i,v}(f)(k) = T(f(k-i), v)$, where $(i, v) \in \mathbb{Z} \times \mathbb{R}$ and $f(k)$ is an arbitrary input signal. A signal operator on $\mathcal{L}$ is called *translation invariant* iff it commutes with any such translation. Consider now two elementary signals, called the *impulse* q and the *dual impulse* q':

$$q(k) \triangleq \begin{cases} 1, & k = 0 \\ 0, & k \neq 0 \end{cases}, \quad q'(k) \triangleq \begin{cases} 0, & k = 0 \\ 1, & k \neq 0 \end{cases}$$

Then every signal f can be represented as a sup of translated impulses or as inf of dual-translated dual impulses:

$$f(k) = \bigvee_i T[f(i), q(k-i)] = \bigwedge_i T'[f(i), q'(k-i)]$$

General signal dilation and erosion can result, respectively, from the sup-T convolution $\bigcirc_T$ and the inf-T' convolution $\bigcirc'_T$ of two signals f and g defined by

$$f \bigcirc_T g(k) \triangleq \bigvee_i T[f(i), g(k-i)], \quad f \bigcirc'_T g(k) \triangleq \bigwedge_i T'[f(i), g(k-i)] \qquad (14)$$

The following theorem characterizes all translation invariant signal dilation or erosion systems as nonlinear convolutions of the above type.

Theorem 2 ([6]). *(a) An operator Δ on the signal lattice $[0,1]^{\mathbb{Z}}$ is a translation invariant dilation iff it can be represented as the sup-T convolution of the input signal with the system's impulse response $h = \Delta(q)$.*

(b) An operator E on the signal lattice $[0,1]^{\mathbb{Z}}$ is a dual-translation invariant erosion iff it can be represented as the inf-T' convolution of the input signal with the system's dual impulse response $h' = E(q')$.

Given a signal dilation $\Delta(f) = f \bigcirc_T h$ and its representation via scalar dilations as

$$\Delta(f)(\ell) = \bigvee_{k \in \mathbb{Z}} \Delta_{k,\ell}(f(k)), \quad \Delta_{k,\ell}(v) = T[v, h(\ell - k)]$$

it follows from the decomposition (5) that its adjoint signal erosion is

$$E(g)(k) = \bigwedge_{\ell \in \mathbb{Z}} \xi[g(\ell), h(\ell - k)] \qquad (15)$$

4. State and Output Responses

The basic state-space model of a max-T control system can now be represented via matrix-based dilations:

$$\begin{aligned} \mathbf{x}(k+1) &= \delta_A[\mathbf{x}(k)] \ \vee \ \delta_B[\mathbf{u}(k)] \\ \mathbf{y}(k) &= \delta_C[\mathbf{x}(k)] \ \vee \ \delta_D[\mathbf{u}(k)] \end{aligned} \tag{16}$$

Solving the state equations by using induction on k yields the *state response*:

$$\begin{aligned} \mathbf{x}(k) &= \mathbf{A}^{(k)} \square_T \mathbf{x}(0) \ \vee \ \left(\bigvee_{i=0}^{k-1} \mathbf{A}^{(k-1-i)} \square_T \mathbf{B} \square_T \mathbf{u}(i) \right) \\ &= \delta_A^k[\mathbf{x}(0)] \ \vee \ \left(\bigvee_{i=0}^{k-1} \delta_A^{k-1-i} \delta_B[\mathbf{u}(i)] \right) \end{aligned} \tag{17}$$

where $\mathbf{A}^{(k)}$ denotes the k-fold max-T matrix 'product' of $\mathbf{A}$ with itself for $k \geq 1$ and $\mathbf{A}^{(0)} = \mathbf{I}_n$ where $\mathbf{I}_n$ is the $n \times n$ identity matrix.

The above result yields in turn the *output response:*

$$\mathbf{y}(k) = \underbrace{\delta_C \delta_A^k[\mathbf{x}(0)]}_{\text{'zero'-input resp.}} \ \vee \ \underbrace{\left(\bigvee_{i=0}^{k-1} \delta_C \delta_A^{k-1-i} \delta_B[\mathbf{u}(i)] \right) \vee \delta_D[\mathbf{u}(k)]}_{\mathbf{y}_{zs}(k) \stackrel{\triangle}{=} \text{'zero'-state resp.}} \tag{18}$$

Thus, the output response is found to consist of two parts: (i) the 'zero'-input response which is due only to the initial conditions $\mathbf{x}(0)$ and assumes an input equal to 0, and (ii) the 'zero'-state response which is due only to the input $\mathbf{x}(0)$ and assumes initial conditions equal to 0.

For *single-input single-output* systems the mapping $u(k) \mapsto y_{zs}(k)$ can be viewed as a translation invariant dilation system Δ. Hence, the 'zero'-state response can be found as the sup-T convolution of the input with the system's impulse response $h = \Delta(q)$:

$$y_{zs}(k) = \Delta(u)(k) = \bigvee_{\ell} T[u(\ell), h(k-\ell)] \tag{19}$$

Assuming the system is initially at rest, its impulse response is found to be

$$h(k) = \begin{cases} 0, & k < 0 \\ \mathbf{D}, & k = 0 \\ \mathbf{C} \square_T \mathbf{A}^{(k-1)} \square_T \mathbf{B}, & k > 0 \end{cases} \tag{20}$$

The last two results can be easily extended to multi-input multi-output systems.

5. Causality, Stability

A max-T control system initially at rest can be viewed as a translation invariant dilation system Δ mapping the input u to the output y. (Assume for brevity single-input single-output systems.) Let $h = \Delta(q)$ be the impulse response of

Δ. A useful bound for signals $f(k)$ processed by such systems is their global supremum

$$||f||_\vee \triangleq \bigvee_{k\in\mathbb{Z}} f(k)$$

which can be viewed as a semi-norm. Such systems are called bounded-input bounded-output (BIBO) *stable* iff a bounded input yields a bounded output, i.e., if $||u||_\vee < 1 \Longrightarrow ||y||_\vee < 1$. The following theorem provides us with simple algebraic criteria for checking the causality and stability of max-T control systems based on their impulse response.

Theorem 3 ([6]). *Consider a max-Tnorm control system Δ initially at rest and let $h = \Delta(q)$ be its impulse response. (a) The system is causal iff $h(k) = 0$ for all $k < 0$. (b) The system is BIBO stable iff $||h||_\vee < 1$.*

6. Controllability, Observability

A max-T control system is *controllable* if the following system of nonlinear equations can be solved and provide the vector $\mathbf{u} = [u(0), u(1), ..., u(N-1)]^t$ of input values required to drive the system from the initial state $\mathbf{x}(0)$ to any desired state $\mathbf{x}(N)$ in N steps:

$$\mathbf{x}(N) = \underbrace{\begin{bmatrix} x_1(N) \\ \vdots \\ x_n(N) \end{bmatrix}}_{\mathbf{x}} = \mathbf{A}^{(N)} \Box_T \mathbf{x}(0) \vee \left(\underbrace{[\mathbf{A}^{(N-1)} \Box_T \mathbf{B}, \cdots, \mathbf{B}]}_{\mathcal{C}} \Box_T \underbrace{\begin{bmatrix} u(0) \\ \vdots \\ u(N-1) \end{bmatrix}}_{\mathbf{u}} \right) \tag{21}$$

Assuming that the input is dominating the initial conditions, i.e., the second term $\mathcal{C} \Box_T \mathbf{u}$ is not smaller than the first term $\mathbf{A}^{(N)} \Box_T \mathbf{x}(0)$ of the right hand side, which is true if $\mathbf{x}(0) = \mathbf{0}$, we can rewrite the above as

$$\mathcal{C} \Box_T \mathbf{u} = \mathbf{x} \tag{22}$$

Equations of the form (22) have been studied in [10, 11] in the context of fuzzy relations. The next lemma provides a sufficient condition for their solvability. Let $\mathbb{N}_n = \{1, 2, ..., n\}$.

Lemma 1 ([10, 11]). *Equation (22) has a solution if for any $i \in \mathbb{N}_n$ there exists $j \in \mathbb{N}_n$ such that $\{\mathcal{C}\}_{ij} = 1$ and $\{\mathcal{C}\}_{kj} = 0 \; \forall \; k \neq i$.*

In general, the set of solutions of (22) forms a sup-semilattice. The *greatest solution* is given by

$$\mathbf{u} = \mathcal{C}^t \Box_\xi \mathbf{x} \tag{23}$$

where ξ is the adjoint scalar erosion of the dilation T defined in (10). In certain applications the conditions of Lemma 1 can be restrictive. An important aspect in such cases is finding the **reachable set** $\mathcal{R}$, i.e., the set of state vectors $\mathbf{x}$ for which (22) is solvable. Let us first recall some definitions given in [11]. With the aid of ξ and a related binary operation

$$\zeta(a,b) \triangleq \begin{cases} 0, & a < b \\ \inf\{x : T(a,x) = b\}, & a \geq b \end{cases} \tag{24}$$

we define the following three *solution matrices*:

$$\{\hat{\Gamma}\}_{ij} \triangleq \xi(\{\mathcal{C}^t\}_{ij}, x_j),\ \{\overline{\Gamma}\}_{ij} \triangleq \zeta(\{\mathcal{C}^t\}_{ij}, x_j),\ \{\check{\Gamma}\}_{ij} \triangleq \zeta_{\min}\left(\bigwedge_k \{\hat{\Gamma}\}_{ik}, \{\overline{\Gamma}\}_{ij}\right)$$

where $\zeta_{\min}$ corresponds to the case where T is the min norm. Given the above matrices, the *greatest* solution is the row-wise infimum of $\hat{\Gamma}$, the *mean* solution is the row-wise supremum of $\check{\Gamma}$, and the *minimal* solutions also result from $\check{\Gamma}$. Note that the greatest solution that results from the solution matrix $\hat{\Gamma}$ is identical to the one provided by the adjoint vector erosion in (23).

The reachable set $\mathcal{R}$ can be found via the following Lemma:

Lemma 2 ([10, 11]). *The state vector* $\mathbf{x}$ *belongs to the reachable set* $\mathcal{R}$ *iff for any* $i \in N_n$ *such that* $x_i \neq 0$ *the* i*th column of* $\check{\Gamma}$ *is not equal to* $\mathbf{0}$.

Hence, if a desired state $\mathbf{x}$ belongs to $\mathcal{R}$, the control $\mathbf{u}$ given by (23) drives the system to this state $\mathbf{x}$.

If $\mathbf{x}$ does not belong to $\mathcal{R}$, then it may be sufficient to solve an approximate controllability problem that has some optimality aspects. Specifically, consider the problem of finding an optimal input vector $\mathbf{u}$ as solution to the following optimization problem:

$$\text{Minimize } ||\mathbf{x} - \mathcal{C} \,\Box_T\, \mathbf{u}|| \text{ subject to } \mathcal{C} \,\Box_T\, \mathbf{u} \leq \mathbf{x} \tag{25}$$

where the norm $||\cdot||$ is either the ℓ_∞ or the ℓ_1 norm. The optimal controllability solution is actually a lattice erosion $\mathbf{u} = \varepsilon(\mathbf{x}) = \mathcal{C}^t \,\Box_\xi\, \mathbf{x}$ identical to the greatest solution (23). ε is the adjoint erosion of the dilation $\delta(\mathbf{y}) = \mathcal{C} \,\Box_T\, \mathbf{y}$. Its optimality can be proven simply by noting that (ε, δ) forms a lattice adjunction, and hence $\delta\varepsilon$ is an *opening* operator. Opening is always anti-extensive, and hence $\delta(\varepsilon(\mathbf{x})) \leq \mathbf{x}$. Therefore, $\mathbf{u} = \varepsilon(\mathbf{x})$ is the largest solution with $\delta(\mathbf{u}) \leq \mathbf{x}$.

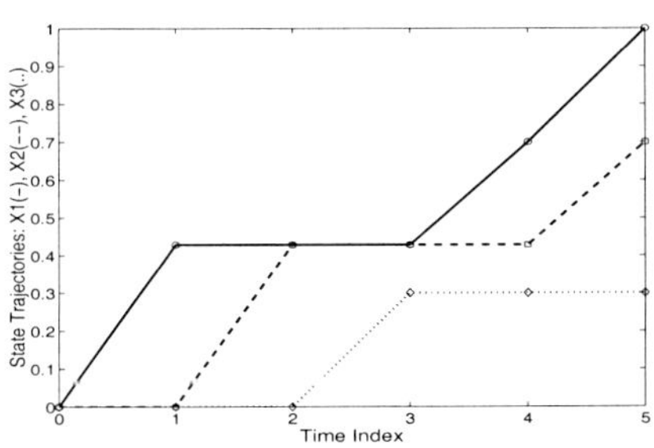

Fig. 1. State trajectories from initial state at $k = 0$ to a desired state at $k = 5$.

Example: Consider a max-T control system with $T(a,b) = ab$, $n = 3$ states, $p = 1$ input, and

$$\mathbf{A} = \begin{bmatrix} 0.2 & 0.8 & 0.1 \\ 1 & 0.3 & 0.2 \\ 0 & 0.7 & 1 \end{bmatrix}, \quad \mathbf{B} = \begin{bmatrix} 1 \\ 0 \\ 0 \end{bmatrix}, \quad \mathbf{x}(0) = \begin{bmatrix} 0 \\ 0 \\ 0 \end{bmatrix}, \quad \mathbf{x}(N) = \begin{bmatrix} 1 \\ 0.7 \\ 0.3 \end{bmatrix}$$

This system can drive the initial state $\mathbf{x}(0)$ to the desired $\mathbf{x}(N)$ in $N = 5$ steps by using the following scalar control signal

$$[u(0), u(1), u(2), u(3), u(4)]^t = [0.4286, 0.4286, 0.4286, 0.7, 1.0]^t$$

Figure 1 shows the state trajectories. □

The above ideas on the controllability problem can also be applied to the observability problem. A max-T control system is *observable* if we can estimate the initial state by observing a sequence of output values. This can be done if the following system of nonlinear equations can be solved:

$$\begin{bmatrix} y(0) \\ \vdots \\ y(n-1) \end{bmatrix} = \underbrace{\begin{bmatrix} \mathbf{C} \\ \vdots \\ \mathbf{C} \,\square_T\, \mathbf{A}^{(n-1)} \end{bmatrix}}_{\mathcal{O}} \square_T\, \mathbf{x}(0) \vee [h(n-1), \cdots, h(0)] \,\square_T \begin{bmatrix} u(0) \\ \vdots \\ u(n-1) \end{bmatrix} \tag{26}$$

This max-T matrix equation can be solved either exactly or approximately by using the same methods as for the controllability equation.

References

1. G. Cohen, P. Moller, J.P. Quadrat and M. Viot, "Algebraic Tools for the Performance Evaluation of Discrete Event Systems", *Proc. IEEE*, vol. 77, pp. 39–58, Jan. 1989.
2. R. Cuninghame-Green, *Minimax Algebra*, Springer-Verlag, New York, 1979.
3. H.J.A.M. Heijmans, *Morphological Image Operators*, Acad. Press, 1994.
4. E. W. Kamen, "An Equation-based Approach to the Control of Discrete Event Systems with Applications to Manufacturing", in *Proc. Int'l Conf. on Control Theory & its Applications*, Israel, Oct. 1993.
5. G. J. Klir and B. Yuan, *Fuzzy Sets and Fuzzy Logic: Theory and Applications*, Prentice-Hall, 1995.
6. P. Maragos, "Lattice Control: General Theory", manuscript in preparation.
7. P. Maragos and S. Tzafestas, "A Lattice Calculus Unification of Min–Max Control Systems of the Morphological and Fuzzy Type", *Proc. Int'l Symp. on Soft Computing in Engineering Applications (SOFTCOM–98)*, Athens, pp. 247-248, June 1998.
8. P. Maragos and S. Tzafestas, "Min–Max Control Systems with Application to Discrete Event Dynamic Systems", in *Advances in Manufacturing: Decision, Control, and Information Technology*, (S.G. Tzafestas, Ed.,) Springer-Verlag, London, 1999, pp. 217–230.
9. J. Serra, Ed., *Image Analysis and Mathematical Morphology, Vol.2: Theoretical Advances*, NY: Acad. Press, 1988.
10. G. Stamou and S. Tzafestas, "Fuzzy Relation Equations and Fuzzy Inference Systems: An Inside Approach", *IEEE Trans. Syst. Man & Cybernetics: Part B*, Dec. 1999.
11. G. Stamou and S. Tzafestas, "Resolution of Composite Fuzzy Relation Equations Based on Archimedean Triangular Norms", *Fuzzy Sets and Systems*, to appear.
12. L. A. Zadeh, "Fuzzy Sets", *Information and Control*, vol. 8(3), pp. 338–353, 1965.

A MORPHOLOGICAL INTERPOLATION APPROACH - GEODESIC SET DEFINITION IN CASE OF EMPTY INTERSECTION

I. GRANADO, N. SIRAKOV and F. MUGE
CVRM / Centro de Geo-Sistemas, Instituto Superior Técnico
Av. Rovisco Pais, 1049-001 Lisboa, PORTUGAL

Abstract. This article presents a morphological approach for the generation of intermediate 2D objects, using others 2D objects, as initial ones. The approach is useful for a smooth realistic 3D object's visualisation, in case of shortage input information. The study is based on mathematical morphology concepts, such as geodesic distance and geodesic set definition, dealing with the case of empty intersection between the objects, in a orthogonal projection over a plane. A classical approach is used to define spatial polynomial curves, interpolating the extreme left and right sets of visible essential border points of the initial objects. Further, the arc of each curve bordered between a couple of initial 2D objects is orthogonal projected over the lower plane together with the upper 2D object. The projections and the object, all over the same plane, are used to create the geodesic set. Then, a set of intermediate sections, between each couple of initial objects, is defined by the application of a morphological linear interpolation. Experiments were performed in order to validate the theory. Real data obtained by well logs performed in Vale de Milhaços (Setúbal - Portugal) was used for this propose, and the results are given in the article.

Key words: Mathematical Morphology, Binary Interpolation, Polynomial Interpolation, Geodesic Distance, 3D Smooth Visualisation.

1. Introduction

This paper presents the development of an approach for the interpolation of binary images, using as initial information 2D objects with empty intersection. The interpolation developed is twofold: firstly is used a polynomial one to define the geodesic set within it the interpolations will be defined, and secondly it is used a linear one to generate such intermediate sections. This method is helpful for smooth realistic 3D object's visualisation in case of shortage input information. It may lead to advantages in 3D subsurface object's visualisation, since the prospective drills or well logs are the only data source for it. Moreover, a shortage of information usually exists, due to the expensive costs associated with the performance of enough number of drills, or difficulties related with the possibility of visiting all the necessary locations. The innovation introduced in here is the creation of the geodesic set between two consecutive objects, whenever exists empty intersection among them, in a top view projection. The linear interpolation applied consists in a generalisation of the one proposed by [3], who have developed an approach for objects with non-empty intersection. Later, improved by [2] the interpolation methodology was enlarged to objects with empty intersection, using a geodesic set containing both initial and final, defined by the object's connection through a straight segment. In this article it is proposed a method to define such geodesic set, in a more precise way, through the employment of a polynomial interpolation. Until now, the morphological interpolation methodology was only applied to binary cases, but it can also be extended to grey level images. The methodology was applied to the generation of an impermeable groundwater unit, using data from its respective well logs.

2. Basic Notions

Hereafter are presented some basic notions used for the method implementation on morphological interpolation, such as geodesic distance, geodesic dilation, geodesic erosion and reconstruction. Let X be a set, x and y being two points of X. The geodesic distance $(d_X(x,y))$ is the length of the shortest path included in X and linking the two points. Geodesic dilation of a marker set Y by an elementary structuring element B in the geodesic set X can be thus defined as:

$$D_X^B(Y) = (Y \oplus B) \cap X \tag{1}$$

where $\oplus$ is Minkowski addition. If this operation is performed until idempotence, we get what is called the reconstruction set by the marker Y which is made of all the connected components of X marked by Y. This transformation is called reconstruction $(R_X(Y))$ and can be written as:

$$R_X(Y) = D_X^{\infty} = \lim(D_X^B ... D_X^B)(Y) \tag{2}$$

This transformation is increasing, anti-extensive and idempotent and thus it is an opening. Because it preserves the connectivity of X, it can be called a connected opening. The dual transformation is the geodesic erosion that can be defined for the complementary sets X^c and Y^c:

$$E_{X^c}^B(Y^c) = (Y^c \ominus B) \cup X^c \tag{3}$$

where $\ominus$ is Minkowski subtraction. The geodesic transformation is repeated until idempotence and the dual transformation $(R_{X^c}^*(Y^c))$ of the reconstruction $R_X(Y)$ is obtained:

$$R_{X^c}^*(Y^c) = (R_X Y)^c = E^{\infty} X^c = \lim(E_{X^c} ... E_{X^c})(Y^c) \tag{4}$$

This transformation is increasing, extensive and idempotent and therefore it is a closing, that can be called connected closing, because the connectivity is preserved. The basic notions just described were used in the interpolation methodology implementation, which is presented in the next item.

3. Interpolation Approach

Reference [3] introduces an interpolation procedure between sets of non-empty intersection, adapted from the one developed by [5], and [1], for reconstructing a topographical surface from its contour lines. Meyer uses the notion of geodesic distance between sets, in which the respective intersection plays a major role. To transform a set X into a set Y, X will shrink and become $X \cap Y$, and at the same time $X \cap Y$ will grow and become Y. Thus, this interpolation algorithm only works when there is a non-empty intersection between the sets. The geodesic set in which Meyer's algorithm works is the union of the two non-disjoint sets X and Y, and each transformation $X \Rightarrow (X \cap Y)$ and $(X \cap Y) \Rightarrow Y$ taken separately consists in finding all the interpolations between a set and another one contained in it. The proposed algorithm follows basically the same procedure of Meyer's algorithm - two different interpolations between increasing or self-contained sets using geodesic distance functions - but enlarging the geodesy to a mask containing the two sets. In order to illustrate the interpolation methodology sets, a geodesic one defined by the union of straight segments between the objects, is used in figure 1 just as an example. Let X and Y being two sets we want to interpolate and Z a geodesic set containing both sets X and Y: $(X \cup Y) \subset Z$ (figure 1). The algorithm calculates two interpolation functions defined as follows:

$$\underset{X}{\overset{Z}{Int}}(x) = \begin{cases} 0 \; on \; X \; and \; 1 \; on \; the \; boundary \; of \; Z \\ d_X/(d_X+d_Z)\,, on \; Z/X \\ +\infty \; on \; Z^c \end{cases} \tag{5}$$

$$\underset{Y}{\overset{Z}{Int}}(x) = \begin{cases} 0 \; on \; Y \; and \; 1 \; on \; the \; boundary \; of \; Z \\ d_Y/(d_Y+d_Z)\,, on \; Z/Y \\ +\infty \; on \; Z^c \end{cases} \tag{6}$$

Where d_X and d_Y are the geodesic distances of a point x to the sets X and Y, respectively, and d_Z is the respective geodesic distance to the boundary of the set Z.

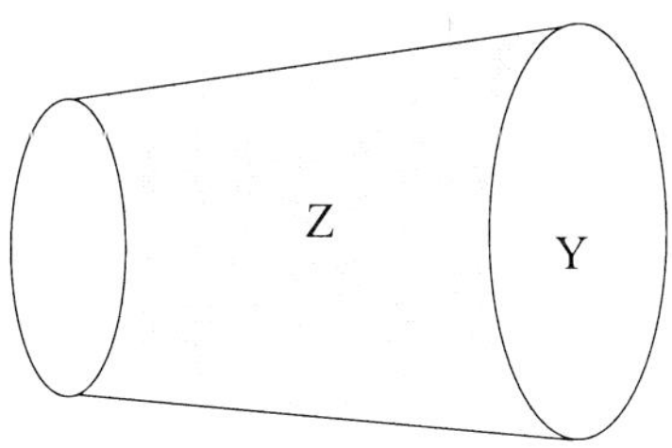

Fig. 1. Sets X, Y and geodesic set Z.

The interpolated sets between X and Z, and between Y and Z, are obtained by a single threshold between 0 and 1:

$$\underset{X}{\overset{Z}{Int}}(\alpha) = \left\{ x \mid \underset{X}{\overset{Z}{Int}}(x) \leq \alpha \right\} \tag{7}$$

$$\underset{Y}{\overset{Z}{Int}}(\alpha) = \left\{ x \mid \underset{Y}{\overset{Z}{Int}}(x) \leq \alpha \right\} \tag{8}$$

Once performed these two interpolations, the interpolation set between X and Y, i.e., the sets that are at a distance (α) from X and $(1-\alpha)$ from Y, can be obtained by the intersection of two partial interpolated sets:

$\underset{X}{\overset{Z}{Int}}(\alpha)$ - interpolated set at a distance α between X and Z;

$\underset{Y}{\overset{Z}{Int}}(\alpha)$ - interpolated set at a distance α between Y and Z.

Thus:

$$\underset{X}{\overset{Y}{Int}}(\alpha) = \underset{X}{\overset{Z}{Int}}(\alpha) \cap \underset{Y}{\overset{Z}{Int}}(1-\alpha) \tag{9}$$

In conclusion, through the application of this expression it is possible to generate intermediate interpolated objects, for certain α values, where $\alpha \in [0, 1]$. Until this phase it has been considered a general geodesic set containing both initial and final objects, in order to describe the morphological interpolation methodology. In the next item it is proposed a technique to define the geodesic set using the orthogonal projection of polynomial spatial curves between objects, which interpolate the set of extreme left and extreme right border points. This way the geodesic set definition is improved, and thus the morphological interpolation.

3.1. GEODESIC SET DEFINITION

Let us consider "similar" 2D-objects [7], situated on a set of plane sections, which will be used to create the surface of a single 3D object. Let the coordinate system Oxz, in each finite plane be connected with the upper left corner (the axis Ox^+ is oriented toward the upper right, and the axis Oz^+ toward the down left corner). And, the spatial axis Oy^+ is connected with the same point of the first plane, being oriented from the first to the last plane section. Assume that the point A belongs to a 2D object's border. We shall call A an extreme left (right) visible with respect to Oz^-, if its x coordinate is minimal (maximal). Therefore, for a certain set of 2D objects situated on the space, two sets of extreme points are defined: set of extreme left, and set of extreme right visible points. Further, a classical approach is employed in order to define the polynomial curve interpolating each set. Let us consider a set of spatial points $N = \{(x_i, y_i, z_i), i = 1, ..n\}$. The polynomial curve L, interpolating the points may be represented by the next parametric equations:

$L : x = f(t), y = g(t), z = h(t)$, where t is a parameter, and

$$\begin{aligned} x = f(t) &= a_{(n-1)}.t^{(n-1)} + a_{(n-2)}.t^{(n-2)} + ... + a_0 \\ y = g(t) &= b_{(n-1)}.t^{(n-1)} + b_{(n-2)}.t^{(n-2)} + ... + b_0 \\ z = h(t) &= c_{(n-1)}.t^{(n-1)} + c_{(n-2)}.t^{(n-2)} + ... + c_0 \end{aligned} \tag{10}$$

Let us denote by: $A_n = \begin{pmatrix} a_{n-1} \\ a_{n-2} \\ \cdots \\ a_0 \end{pmatrix}$, $B_n = \begin{pmatrix} b_{n-1} \\ b_{n-2} \\ \cdots \\ b_0 \end{pmatrix}$, $C_n = \begin{pmatrix} c_{n-1} \\ c_{n-2} \\ \cdots \\ c_0 \end{pmatrix}$,

$$X_n = \begin{pmatrix} x_{n-1} \\ x_{n-2} \\ \dots \\ x_0 \end{pmatrix}, Y_n = \begin{pmatrix} y_{n-1} \\ y_{n-2} \\ \dots \\ y_0 \end{pmatrix}, Z_n = \begin{pmatrix} z_{n-1} \\ z_{n-2} \\ \dots \\ z_0 \end{pmatrix}$$

If one choose for each $i = 1, ..., n$ a fixed value of $t = t_i$, then can be constructed the following matrix:

$$T_{(n\times n)} = \begin{pmatrix} t_1{}^{(n-1)} & t_1{}^{(n-2)} & \dots & 1 \\ t_2{}^{(n-1)} & t_2{}^{(n-2)} & \dots & 1 \\ \dots & \dots & \dots & \dots \\ t_n{}^{(n-1)} & t_n{}^{(n-2)} & \dots & 1 \end{pmatrix}$$

This way the equation system (10) may be presented in the following matrix form:

$$T_{(n\times n)}.A_n = X_n \qquad T_{(n\times n)}.B_n = Yn \qquad T_{(n\times n)}.C_n = Z_n \tag{11}$$

When the coordinates of n spatial points are given, we may calculate the corresponding vectors A_n, B_n and C_n for each fixed $T_n(t_1, ..., t_n)$, through the resolution of the equation systems (11). Thus, will be defined the curve L interpolating the points. Employing the upper approach over the sets of extreme left and right border points, the polynomial curve interpolating each set may be defined. Further, the arc of each curve bordered between a couple of initial 2D objects is orthogonal projected over the lower plane together with the upper 2D object. The projections and the objects, over the same plane, are used to create the geodesic set needed for the morphological approach generating the intermediate 2D objects between the initial couple of 2D objects.

3.2. GEODESIC DISTANCE ALGORITHM

In order to obtain the geodesic distances from each 2D object, was developed a program in C to automatically execute several commands that are described hereafter. Two images are considered, one with the object to which the distances are concerned - object X, and another one with the geodesic set (containing both objects, but with zero value over X). Next it is applied a dilation to object X, followed by its intersection with the geodesic set. With this operation it is obtained the object X surrounded by pixels with value 1, because it was made 1 dilation. To obtain the pixels with geodesic distance 2, we will have to apply another geodesic dilation over the previous one, followed by an XOR operation with the results of geodesic distance 1, assigning value 2 to such difference. Repeating these procedures for object X until idempotence, in the respective geodesic set, it will be obtained the various geodesic distances. For the application of the interpolation methodology presented previously, the geodesic distances from each object, within the respective geodesic set are needed. So, all those distances were calculated as described above, resulting in several images such as d_x - geodesic distance to object X calculated in Z/X, and d_y - geodesic distance to object Y calculated in Z/Y, both generated for each interval between two consecutive sections.

4. Case Study

As it has been mentioned, the morphological interpolation method developed so far was applied for the generation of intermediate 2D objects, in order to provide a smooth realistic 3D reconstruction and visualisation, in case of shortage input information. Using input data obtained by well logs from Vale de Milhaços (Setúbal península in Portugal) six images were generated employing a morphological approach developed by [4]. A set of sections cutting a single impermeable unit (figure 2) was derived by the application of a recognition approach based on regularity's definition [6], and notion of morphological similarity [7]. Those sections are then considered parallel to the plane Oxz.

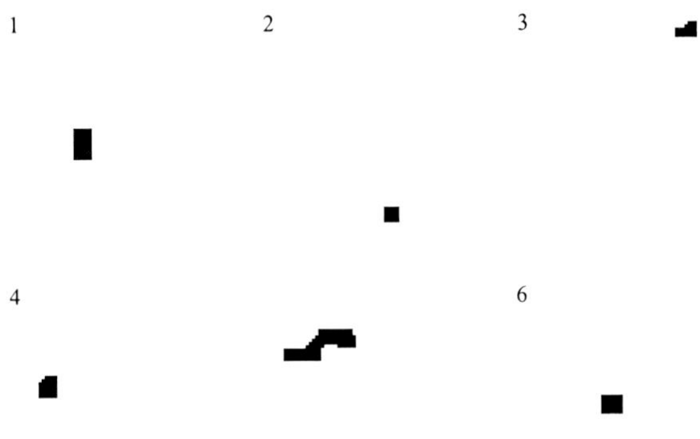

Fig. 2. Initial sections used for the generation of intermediate ones.

Observing them from a top view can be concluded that they do not intersect each other (figure 2). Therefore, the original Meyer's method [3] could not be applied for the generation of intermediate sections, leading to the enlargement of the interpolation context by [2]. The geodesic set definition consists in an important step due to its influence on the subsequent interpolations, noting that in Meyer's method the geodesic set is given by the object's union, and in here such set is defined. Thus, the polynomial interpolation presented in paragraph 3.1 was applied to all sections, obtaining two spatial curves connecting them: one corresponding to the connection of the extreme left points of each section, and the other line connecting its extreme right visible points. The equation systems describing those curves are given below:

- for the right one:

$$\begin{aligned} X &= f(t) = -7.4 * t^5 + 128.5 * t^4 - 827 * t^3 + 2408.5 * t^2 - 3085.6 * t + 1500 \\ Z &= g(t) = -1.9 * t^5 + 35.70 * t^4 - 239 * t^3 + 688.79 * t^2 - 813.58 * t + 534 \\ Y &= h(t) = t \end{aligned} \tag{12}$$

- for the left one:

$$X = u(t) = -5.01 * t^5 + 92 * t^4 - 609.1 * t^3 + 1809 * t^2 - 2337.7 * t + 1152$$

$$Z = v(t) = -2.33 * t^5 + 42.12 * t^4 - 275 * t^3 + 780.3 * t^2 - 917 * t + 576$$
$$Y = w(t) = t \quad (13)$$

Based on the polynomial functions it was calculated the orthogonal projection's location of three points, respectively at distance $t = 0.25$, $t = 0.50$ and $t = 0.75$ from each initial spatial section (considering that the distance between two consecutive sections is set to one). All the triples of points used to bind the different geodesic sets are given below (1 and 2 represents couples of sections used to define left-l and right-r sets of points):

TABLE I
Points coordinates for geodesic sets definitions.

x_{1r}	y_{1r}	x_{1l}	y_{1l}	x_{2r}	y_{2r}	x_{2l}	y_{2l}
82	207	76	207	160	280	140	289
94	223	87	223	171	274	155	283
128	242	115	244	194	227	178	231
197	272	175	274	249	115	225	111
214	278	191	280	272	83	243	76
216	275	193	277	286	76	254	67
177	250	162	249	283	128	250	120
146	229	134	227	265	175	234	170
116	206	103	205	237	224	210	222
78	163	54	165	161	294	142	296
79	148	43	153	121	301	105	303
95	140	44	148	85	284	72	286
158	146	79	161	50	201	38	199
184	161	103	175	65	161	51	157
183	180	119	190	112	154	95	150

In the next figure is represented an example of such point's location corre spondent to the fifth and sixth sections.

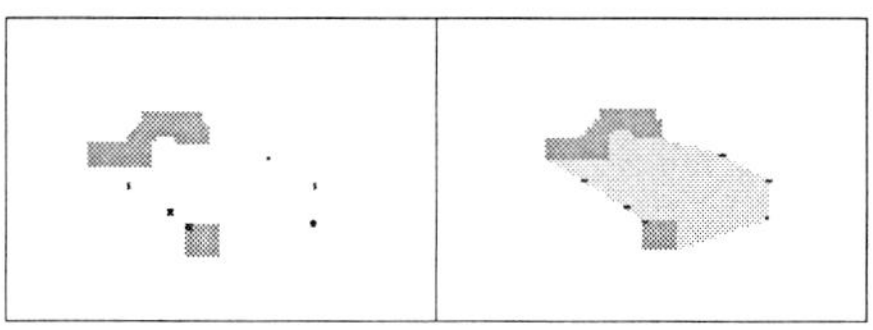

Fig. 3. (A) Fifth and sixth sections, and the points obtained from the polynomial curves projection; (B) Geodesic set definition by linking the extreme left and right points, with the 2D objects.

Then, after the geodesic set's definition for each two consecutive sections, it was applied the morphological interpolation method, in order to obtain the intermediate ones. The result is presented in figure 4, together with the initial sections, displayed at the left and right columns. In the next phase the

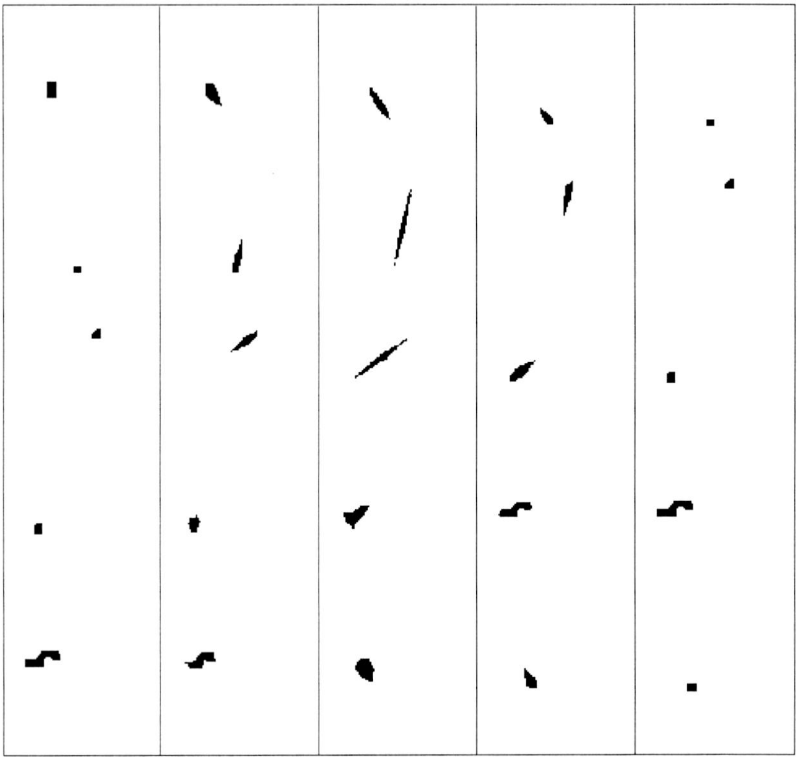

Fig. 4. Interpolated sections and initial sections (left and right columns).

interpolated sections are introduced in a software tool - SURFDRIVE (demo version), in order to provide the impermeable unit surface 3D visualisation. In figure 5A are given the initial sections situated according to their spatial distribution. And the same sections combined with the interpolated ones are shown in figure 5B.

The surface obtained in figure 6B using the geodesic set approach by a polynomial function to generate intermediate sections, is a smoother result, and thus more accurate than the one obtained in figure 6A.

5. Conclusions

With the results obtained may be concluded that the interpolation method enables to provide smooth 3D reconstruction and visualisation, when the initial 2D transversal sections do not overlap each other (orthogonal projection) and their number is not enough. These advantages are due to the following points:

- definition of the geodesic set using polynomial interpolation in case of empty intersections between the initial 2D objects;

- generation of intermediate objects between each couple of initial 2D objects;

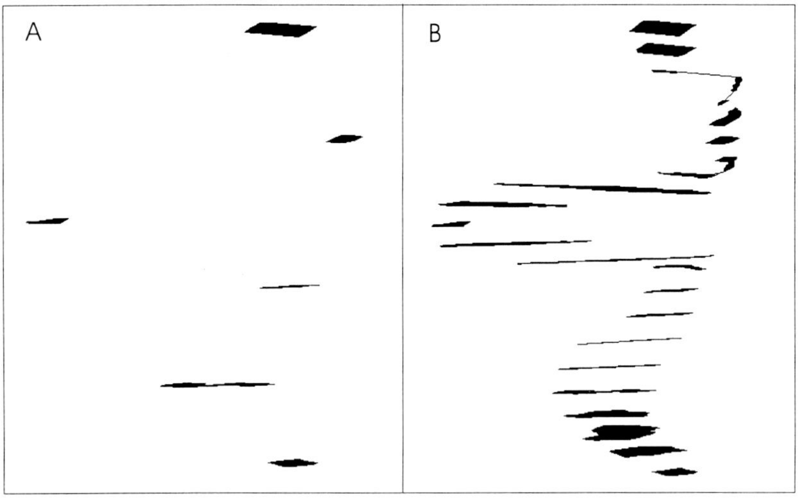

Fig. 5. (A) Initial sections spatially distributed; (B) Interpolated and initial sections, spatially distributed.

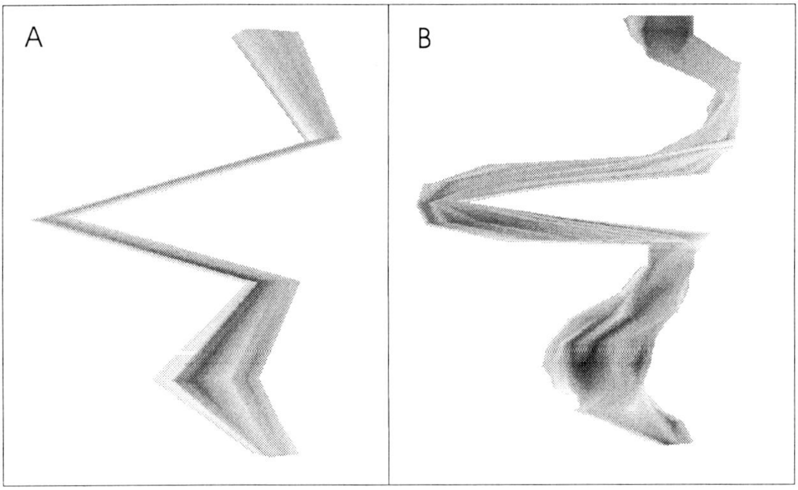

Fig. 6. (A) Aquifer 3D visualisation based in initial sections; (B) Aquifer 3D visualisation based in initial sections and interpolated ones.

- improvement of data usage, needed for a smooth realistic 3D visualisation in case of shortage input information.

The interpolation method for grey level images is still under development, but we believe that this process can also be extended to that context, with positive results, especially to study the topological/spatial relations between various permeable types of subsurface objects.

Acknowledgements

Thanks to CVRM - IST, and NATO grant CP (BU) /08/99/P0.

References

1. I. Granado, F. Muge. Estimation and Comparison of Digital Terrain Models. In 8Th Portuguese Conference on Pattern Recognition RECPAD'96, from the 21st to the 22nd of March of 1996, Guimarães, 395–402.
2. I. Granado, F. Muge. A New Method for the Interpolation of Binary and Grey Scale Images. 10th Portuguese Conference on Pattern Recognition RECPAD'98, from the 26th to 27th of March of 1998, Lisbon, 273–283.
3. F. Meyer. A Morphological Interpolation Method for Mosaic Images. International Symposium on Mathematical Morphology ISMM'96, Atlanta - Georgia, May of 1996, 337–344.
4. F. Muge, L. Ribeiro, P. Pina, V. Oliveira. Morphological characterization of the spatial variability of hydrostratigraphic units. In: Geostatistics Wollongong'96, (ed. by E. Y. Baafi & N. A. Schofield), Kluwer Academic Publishers, 2, 1134–1145, (1997).
5. P. Soille. Morphologie Mathématique: du relief a la Dimensionalité - algoritmes et méthodes. Université Catholique de Louvain, 28th February 1992.
6. N. Sirakov, F. Muge. Comparing 2D Borders Using Regular Structures. Proceedings of RecPad'94. Lisbon, Portugal, 169–176.
7. N. Sirakov. Automatic Reconstruction of 3D Branching Objects, In 13th ICPR, TU Vienna, Austria, August 25-29, 1996. Pattern Recognition and Signal Analysis, IEEE Computer Society, Vol. II, 620–624.

THE MORPHOLOGICAL-AFFINE OBJECT DEFORMATION

MARCIN IWANOWSKI* and JEAN SERRA
Centre de Morphologie Mathematique, Ecole des Mines de Paris
35, rue Saint-Honore, 77305 Fontainebleau, France
E-mail: iwanowski@cmm.ensmp.fr, serra@cmm.ensmp.fr

Abstract. The combination of morphological interpolation and affine transformation is presented. The proposed approach unites the advantages of both methods: the displacement is performed by using affine transformation, and the shape deformation by morphological interpolation. It allows the transformation of one binary set into another in semi-automatic or fully-automatic way.

Key words: Mathematical Morphology, Interpolation, Affine Transformation, Objects Deformation.

1. Introduction

The current paper describes the combination of morphological interpolation and affine transformation and its application to the deformation of binary image objects. The morphological interpolation (section 2) allows one to *deform* the object's shape in a very elegant and robust way. The disadvantage is that the result of interpolation of distant sets is either not possible to obtain or not very realistic, depending on the method applied. On the other hand the affine transformation (section 3) is an ideal solution for *displacements* like translation and rotation. The change of shape however, could be performed to a very restricted extent: only by rescaling and shearing, which does not allow the modification of a shape into another (as morphological interpolation does).

The approach presented in the current paper unites the advantages of both methods. the displacement is performed by using an affine transformation and the shape deformation by a morphological interpolation. The method proposed (section 4) consists of three major steps. At the beginning an affine transformation is applied to each of the sets in order to place them in the central position. In the next step, the morphological interpolation is performed. Finally, in the third step, the interpolated set is moved to its final position by once again using an affine transformation. Contrary to morphological interpolation, the affine transformation requires some input parameters. The methods of calculation of these parameters are described in section 5. Section 6 contains the results and conclusions.

* Also affiliated to: Institute of Control and Industrial Electronics, Warsaw University of Technology, ul.Koszykowa 75, 00-662 Warszawa, POLAND.

2. Morphological Interpolation of Sets

Morphological interpolation is a recent subject. The main papers on it were presented during the last two ISMM symposiums: in 1996 by F. Meyer [4, 5], and in 1998 by J. Serra [6, 7] and S. Beucher [1, 2]. The interpolation method applied here is based on the interpolation of the intermediary sets between two sets with non-empty intersection. Two approaches to this issue have been proposed. The first one is based on the median set [1, 2, 6, 7], the second one - on the geodesic distance functions [4, 5]. The median set of two binary sets with non-empty intersection is an influence zone of the intersection of both in the union of them. It could also be expressed by using the basic morphological operators: dilation and erosion as follows [6, 7]:

$$M(P,Q) = IZ_{(P \cup Q)}(P \cap Q) = \cup\{((P \cap Q) \oplus \lambda B) \cap ((P \cup Q) \ominus \lambda B), \lambda \geq 0\} \quad (1)$$

where P, Q ($P \cap Q \neq \emptyset$) are the initial sets and $M(P,Q)$ is the median set. The interpolated set at a given level could be obtained by the successive generation of medians ([1, 2]). This approach, however, is not very fast. A faster solution based on geodesic distance functions, which produces the same results, has been proposed in [4, 5]. It allows one to obtain the interpolated set at a given level by a simple thresholding of two interpolation functions, namely $int^P_{P\cap Q}$, which interpolates between $P \cap Q$ and P; and $int^Q_{P\cap Q}$, which interpolates between $P \cap Q$ and Q. The interpolated set at level $0 \leq k \leq 1$, is the union of two cross-sections of the interpolation function:

$$Z_k = Thr_k(int^P_{P\cap Q}) \cup Thr_{(1-k)}(int^Q_{P\cap Q}) \quad (2)$$

where Thr_k operator represents the thresholding at level k, and Z_k is the interpolated set at that level. The interpolation function $int^P_{P\cap Q}$ is obtained as a combination of two geodesic distance functions: d_1 and d_2. The first one is the distance from $P \cap Q$ to P and is obtained by the successive dilations of $P \cap Q$ within mask P. The function d_2 represents the geodesic distance function from $\overline{P}$ to $\overline{(P \cap Q)}$, obtained by successive geodesic dilations of $\overline{P}$ with mask $\overline{(P \cap Q)}$. Finally, the interpolation function $int^P_{P\cap Q}$ equals $\frac{d_1}{d_1+d_2}$. Function $int^Q_{P\cap Q}$ is obtained in a similar way.

3. Affine Transformation

An affine transformation allows one to *translate*, *rotate*, *rescale* and *shear* an image. In the case described in the current paper, this operation is applied to the transformation of binary sets. We match two sets as well as possible, leaving the more precise change of shape to the morphological interpolation.

3.1. General form

The transformation can be explained in terms of the general transformation matrix [8, 3] which in the case of an affine transformation can be expressed as follows:

$$[x, y, 1] = [u, v, 1] \cdot A \, ; \; A = \begin{bmatrix} a_{11} & a_{12} & 0 \\ a_{21} & a_{22} & 0 \\ a_{31} & a_{32} & 1 \end{bmatrix} \tag{3}$$

The affine transformation considered here includes three kinds of operations: translation, rotation and scaling[1]; the transformation matrices of which are, respectively, the following:

$$T(t_x, t_y) = \begin{bmatrix} 1 & 0 & 0 \\ 0 & 1 & 0 \\ t_x & t_y & 1 \end{bmatrix} ; \; R(\theta) = \begin{bmatrix} \cos\theta & \sin\theta & 0 \\ -\sin\theta & \cos\theta & 0 \\ 0 & 0 & 1 \end{bmatrix} ; \; S(s_x, s_y) = \begin{bmatrix} s_x & 0 & 0 \\ 0 & s_y & 0 \\ 0 & 0 & 1 \end{bmatrix} \tag{4}$$

The equations above describe the *forward* mapping [8, 3]. It means that we calculate the coordinates of the final image for every pixel from the initial one. In order to obtain an appropriate matrix for the *inverse* mapping (which characterizes the calculation of the coordinates on the initial image for every pixel of the final one), the matrix A from Eq. 3 should be inverted[2]. In general, if we consider the affine transformation consisting of n basic operations, translations, rotations, scalings or shearings:

$$A = A_1 \cdot A_2 \cdot \ldots \cdot A_n \tag{5}$$

the inverse transformation matrix A' will be equal to:

$$A' = A^{-1} = A_n^{-1} \cdot \ldots \cdot A_2^{-1} \cdot A_1^{-1} \tag{6}$$

The inverse matrices of the three basic transformations introduced above (defined as in Eq. 4) are respectively:

$$T^{-1}(t_x, t_y) = T(-t_x, -t_y); \; R^{-1}(\theta) = R(-\theta); \; S^{-1}(s_x, s_y) = S(\frac{1}{s_x}, \frac{1}{s_y}) \tag{7}$$

4. Combination of Morphological Interpolation and Affine Transformation

The proposed approach combines both methods described above. In the first step an affine transformation is performed on each of the two sets under study; this results in locating both (modified) sets in a central position. It is shown in Fig. 1(b) (initial sets: (a) and (c)). This first transformation consists of translation, rotation and scaling. In the second step the morphological interpolation using the interpolation function is performed. Finally the morphologically interpolated set is affine-transformed to its final position(s) (Fig. 1(d,e,f)). In fact, the transformation to the final position could be done directly from both initial sets, and the morphological interpolation could be performed there without transforming them into the central position. But the proposed solution is

[1] The fourth kind of the affine transformation: shearing is not taken into consideration.

[2] The inverse matrix describes also an affine transformation.

faster with regard to the speed of the calculations of the morphologically interpolated set. The interpolation functions are generated only once, and the final interpolated sets at different levels are obtained by the appropriate thresholding and one affine transformation to the final position for each level. The particular steps of treatment are described in the following sections.

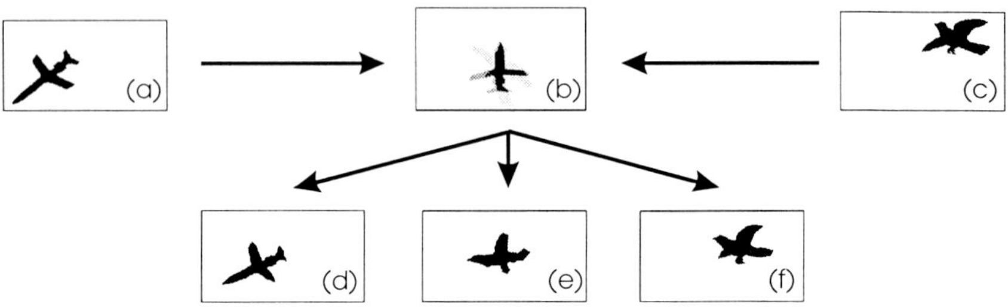

Fig. 1. Generation of the interpolated set: (a),(c)-initial sets, (b)-central position, (d,e,f)-interpolated sets.

4.1. Putting both sets in the central position

The first step of treatment is performed separately for two sets P and Q. In order to perform the affine transformation the following auxiliary data sets must be associated with them: the *middle point* of the set (for the set P: (x_{P0}, y_{P0}), and (x_{Q0}, y_{Q0}) for Q), and the *proper angle* (α_P and α_Q). The first data is used twice: firstly for the translation to the central position, and secondly as a center of the rotation. The second data - the proper angle, describes the orientation of the set and is the angle between the x-axis and the line indicating the characteristic direction of the set. There exists also the third input data, which depend on the relation between both sets: the *scaling coefficients* s_x, s_y. They describe the proportion ratio between the sets. We will consider now all these data as known a priori. How to obtain them is described in the next section. The transformation which moves the initial sets P to the central position is the following:

$$A_P = T(-x_{P0}, -y_{P0}) \cdot R(-\alpha_P) \cdot T(x_C, y_C) \tag{8}$$

It rotates the set around the point (x_{P0}, y_{P0}) and translates it in such a way that the middle point is placed in the central point (x_C, y_C). The appropriate transformation of the set Q contains one additional step, rescaling:

$$A_Q = T(-x_{Q0}, -y_{Q0}) \cdot R(\alpha_Q) \cdot S(s_x, s_y) \cdot T(x_C, y_C) \tag{9}$$

The only difference with the former transformation is that between the rotation and the translation, the set Q is rescaled by using the coefficients s_x, s_y[3]. Both

[3] In order to match both sets, only one of them - in our case Q - has to be rescaled.

equations shown above tell us what has to be done with the initial set in order to obtain the transformed one - they define the forward mapping. In this case for each point from the initial image (containing the initial set), its new coordinates on the final one (which contains the set in the central position) are calculated. This approach, however, has one important disadvantage. Due to the numerical inaccuracies, it can happen that not all the points on the final image are filled by the values of the appropriate points from the initial one and some holes are present. In order to avoid it, the inverse mapping is applied. In this case, according to the equations 6 and 7, the transformation matrices are respectively the following:

$$\begin{aligned} A_P^{-1} &= T(-x_C, -y_C) \cdot R(\alpha_P) \cdot T(x_{P0}, y_{P0}) \\ A_Q^{-1} &= T(-x_C, -y_C) \cdot R(-\alpha_Q) \cdot S(\tfrac{1}{s_x}, \tfrac{1}{s_y}) \cdot T(x_{Q0}, y_{Q0}) \end{aligned} \tag{10}$$

By using both equations indicated above one transform both sets to the central position (see Fig. 2(b)). Now the distance functions: $int^{P'}_{P' \cap Q'}$ and $int^{Q'}_{P' \cap Q'}$ (where P' and Q' represent the transformed sets) are produced. They allow one to later obtain the morphologically interpolated set at given level. All the steps described in this section are performed only once, even if the interpolated sets on different levels have to be produced.

4.2. Interpolated set at given level

Let $0 \leq k \leq 1$ be the level on which the interpolated set is calculated (for $k = 0$ it is equal to P and for $k = 1$, to Q). The interpolation functions calculated in the previous step are thresholded according to Eq. 2. In order to place it in the final position the affine transformation is applied once again. This time, however, new transformation parameters are calculated (all of them depend on k): final middle point $(x_F(k), y_F(k))$, rotation angle $\beta(k)$ and scaling coefficient $s'_x(k), s'_y(k)$. The appropriate matrix of the affine transformation is (in case of forward mapping):

$$A_{int}(k) = T(-x_C, -y_C) \cdot S(s'_x(k), s'_y(k)) \cdot R(-\beta(k)) \cdot T(x_F(k), y_F(k)) \tag{11}$$

The transformation parameters are calculated by using the following equations:

$$s'_x(k) = \frac{1}{(1 + k(s_x - 1))} \; ; \; s'_y(k) = \frac{1}{(1 + k(s_y - 1))} \tag{12}$$

$$\beta(k) = \alpha_P + k \cdot (\alpha_Q - \alpha_P) \tag{13}$$

$$x_F(k) = x_{P0} + k \cdot (x_{Q0} - x_{P0}) \; ; \; y_F(k) = y_{P0} + k \cdot (y_{Q0} - y_{P0}) \tag{14}$$

As in the previous case, instead of the forward mapping, the reverse one is applied, the matrix of which is the following:

$$A_{int}^{-1}(k) = T(-x_F(k), -y_F(k)) \cdot R(-\beta(k)) \cdot S(\frac{1}{s'_x(k)}, \frac{1}{s'_y(k)}) \cdot T(x_C, y_C) \tag{15}$$

5. Calculation of the Parameters of the Affine Transformation

For calculating the parameters of the affine transformation four methods have been developed - one semiautomatic and three fully automatic ones.

5.1. Semiautomatic by the triplet of points

In the current method a triplet of control points is associated with each of the two sets. The affine transformation can transform any triple of non collinear points into any other in the image space by using the combination of all four base transformations. In the current paper we neglect shearing, which reduces the triplets which can be transformed. Let $\{p_1 = (x_{P1}, y_{P1}), p_0 = (x_{P0}, y_{P0}), p_2 = (x_{P2}, y_{P2})\}$ and $\{q_1 = (x_{Q1}, y_{Q1}), q_0 = (x_{Q0}, y_{Q0}),\ q_2 = (x_{Q2}, y_{Q2})\}$ be the triplets of control points of respectively the sets P and Q. The necessary condition is that they must be the vertices of a right-angled triangle, which can be expressed by using the following equation (the right angle is indicated by a pixel with index 0): $x_0(x_0 - x_1 - x_2) + y_0(y_0 - y_1 - y_2) = y_1 y_2 - x_1 x_2$. Example sets and the triplets are shown on Fig. 2. The appropriate points are the ends of the gray straight lines.

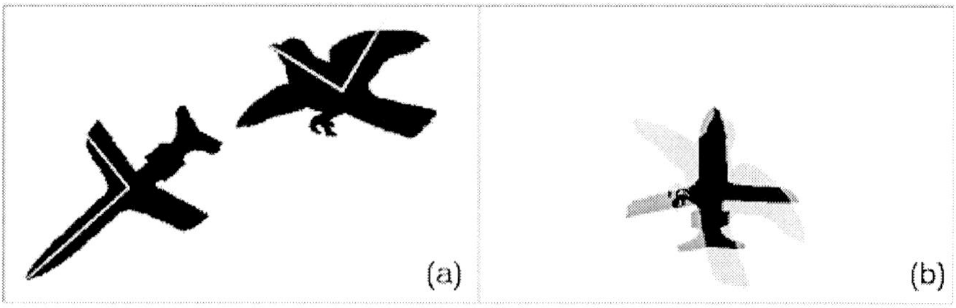

Fig. 2. On the left: two initial sets (with the control points), on the right: transformed set in the central position.

Both sets are transformed using the affine transformation in such a way that their appropriate triplets of points superimpose (see Fig. 2(b)). The middle points are already known and they are equal to p_0 for the set P and q_0 for the set Q. The proper angle α_P[4] is defined as an angle between the line passing through the points p_0 and p_1 and the x-axis of the base coordinate system:

$$\alpha = sgn(y_{P1} - y_{P0}) \cdot \arccos\left(\frac{x_{P1} - x_{P0}}{\sqrt{(x_{P1} - x_{P0})^2 + (y_{P1} - y_{P0})^2}}\right)\ ;\ -\frac{\pi}{2} < \alpha \leq \frac{\pi}{2} \tag{16}$$

Scaling coefficients are calculated by considering the lengths of segments $p_1 p_0$, $p_2 p_0$ and their relationship to the lengths of segments $q_1 q_0$, $q_2 q_0$:

[4] When the equation considers only one set, it means that the appropriate equation for the second one is to be obtained in exactly the same way.

$$s_x = \sqrt{\frac{(x_{P1} - x_{P0})^2 + (y_{P1} - y_{P0})^2}{(x_{Q1} - x_{Q0})^2 + (y_{Q1} - y_{Q0})^2}} \; ; \; s_y = \sqrt{\frac{(x_{P2} - x_{P0})^2 + (y_{P2} - y_{P0})^2}{(x_{Q2} - x_{Q0})^2 + (y_{Q2} - y_{Q0})^2}} \tag{17}$$

5.2. Automatic without rescaling

In the automatic method, at first the middle point is calculated as a *center of gravity* of the set. In order to obtain the proper angle α, one measures the intercepts $h_{\lambda,x}$ of the set P at every point $x \in P$ and for every angle λ. Let l be the maximum lengths of these intercepts:

$$l = sup\{h_{\lambda,x}(P)\, ,\, x \in P\, ,\, \lambda \in [0, \pi]\} \tag{18}$$

then λ is the direction associated with l - the proper angle of P[5].

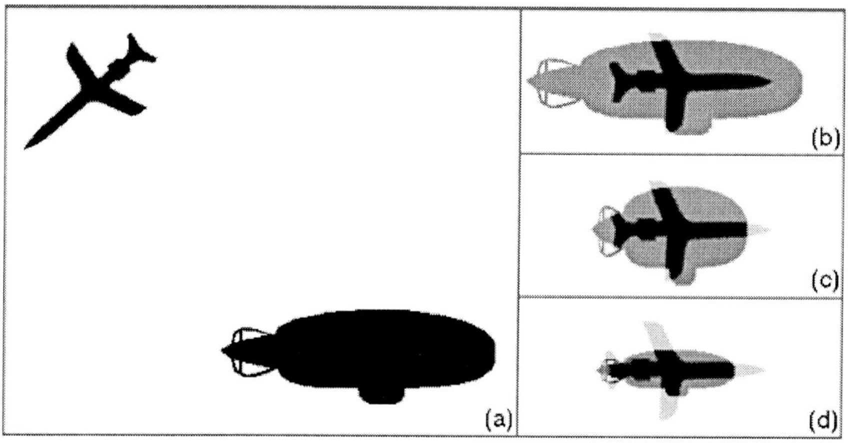

Fig. 3. The initial sets (a) and the automatic matching obtained by using different criteria: (b) - without rescaling, (c) - by the smallest rectangle matching, (d) - with the areal criterion.

The scaling coefficients are not calculated here ($s_x = s_y = 1$). The change of size is performed only by means of the morphological interpolation.

5.3. Automatic by smallest rectangle matching

The middle point and the proper angle is calculated as in the previous method. In order to obtain the scaling coefficients the size of the smallest rectangle containing the set is considered. The coefficients are than calculated by using the extreme values of the coordinates of the points belonging to the set:

$$s_x = \frac{x_{Pmax} - x_{Pmin}}{x_{Qmax} - x_{Qmin}} \; ; \; s_y = \frac{y_{Pmax} - y_{Pmin}}{y_{Qmax} - y_{Qmin}} \tag{19}$$

[5] Of course a set can have more proper angles λ - not only one.

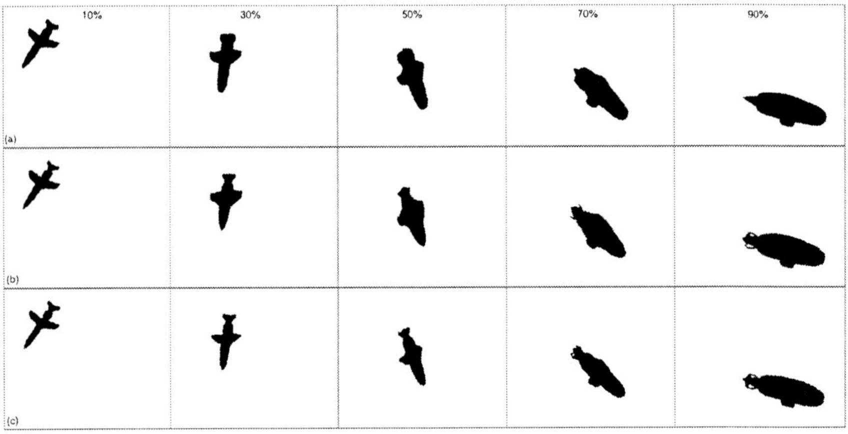

Fig. 4. The sequence generated by using the automatic calculation of parameters: (a) - without rescaling, (b) - by the smallest rectangle matching, (c) - with the areal criterion. Initial images are shown on Fig. 3.

where indices *max* and *min* stand for the extremal coordinate values along the appropriate axes. These values can be computed either before or after performing the rotation - in the result shown in the next section they are calculated after the rotation.

5.4. Automatic by using the areal criterion

In this approach, the middle point and the proper angle are calculated in the same way as in the former ones. The difference lies in the calculation of the scaling coefficients. They are calculated according to the relation between the areas of both sets - the areal criterion:

$$s_x = s_y = \sqrt{\frac{area(P)}{area(Q)}} \tag{20}$$

where $area(P)$ and $area(Q)$ represent the area of respectively P and Q.

6. Results and Conclusions

The results of the automatic methods are presented in Fig. 4. The automatically computed parameters are the following: $x_{P0} = 133$, $y_{P0} = 184$, $x_{Q0} = 353$, $y_{Q0} = 273$, $\alpha_P = -140^o$ and $\alpha_Q = 0^o$. The result of the first method (a) shows that the absence of the rescaling leaves the entire change of size to the morphological interpolation, the results of which is not as precise as in the combination with the rescaling. Next two methods contain the rescaling, the automatically computed parameters of which are following for (b): $s_x = 0.56$, $s_y = 1.05$; for (c):$s_x = s_y = 0.5$. If we compare both methods, the second one (c), with the areal criterion produces finer interpolated sets

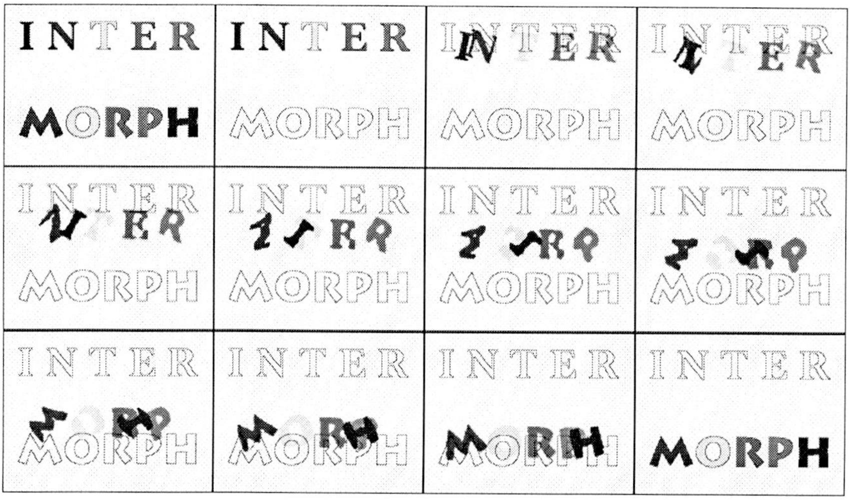

Fig. 5. An animation obtained automatically by using the affine-morphological deformation.

(because the transformed sets in the central position are smaller). The initial sets and the matched sets in the central position are shown in Fig. 3.

The proposed method deals with objects. The objects, however, are always a part of a particular image. They are represented as connected components. If one considers images, one cannot avoid the question of noise-sensitivity. In case of noisy image one have to filter the noise before the interpolation starts. Input images for the interpolation must be noise-free and contain only the objects.

The method of set deformation proposed in the current paper combines the advantages of the affine transformation and the morphological interpolation. It allows one to transform one binary set into another one in a semi-automatic or fully-automatic way. The shape of the interpolated sets looks natural and the transition is performed smoothly. It can be applied in different areas of image processing. One of the possible application areas is the animation. It could be applied to animate the titles, graphics, or other objects on the image. The example of the animation obtained by using the proposed method is presented on Fig. 5. The first word 'INTER' is transformed into another one: 'MORPH'. Each frame of the animation has been obtained as a superposition of the interpolations of single letters. Each letter was transformed by using the automatic method consisting of translation and rotation (rescaling has not been performed). In order to improve the smoothness of the final animated frame, an additional morphological filtering (closing of size 1) has been applied.

References

1. Serge Beucher. Interpolation of sets, of partitions and of functions In H.Heijmans and J.Roerdink *Mathematical morphology and its application to image and signal processing*, Kluwer, 1998.

2. Serge Beucher. *Interpolation d'ensembles, de partitions et de fonctions* Tech.Rep .N-18/94/MM Centre de Morphologie Mathematique, Ecole des Mines de Paris.
3. James D. Foley, Andries van Damm, Steven K.Feiner, John F.Hughes and Richard L.Philips. *Introduction to computer graphics* Addison-Wesley, 1994,1990.
4. Fernand Meyer. Morphological interpolation method for mosaic images In P. Maragos, R.W.Schafer, M.A.Butt *Mathematical morphology and its application to image and signal processing*, Kluwer, 1996.
5. Fernand Meyer. *Interpolations* Tech.Rep. N-16/94/MM Centre de Morphologie Mathematique, Ecole des Mines de Paris.
6. Jean Serra. *Interpolations et distance de Hausdorff* Tech.Rep. N-15/94/MM Centre de Morphologie Mathematique, Ecole des Mines de Paris.
7. Jean Serra. Hausdorff distance and interpolations In H.Heijmans and J.Roerdink *Mathematical morphology and its application to image and signal processing*, Kluwer, 1998.
8. George Wolberg. *Digital Image Warping* IEEE Computer Society Press, Los Alamos CA, 1990.

AFFINE INVARIANT MATHEMATICAL MORPHOLOGY APPLIED TO A GENERIC SHAPE RECOGNITION ALGORITHM

J.L. LISANI*, L. MOISAN*, P. MONASSE* and J.M. MOREL*
CMLA ENS-Cachan, 61 av du Pdt Wilson, 94235 Cachan Cedex, France.

Abstract. We design a generic contrast and affine invariant planar shape recognition algorithm. By generic, we mean an algorithm which delivers a list of all shapes two digital images have in common, up to any affine transform or contrast change. We define as "shape elements" all pieces of level lines of the image. Their number can be drastically reduced by using affine and contrast invariant smoothing Matheron operators, which we describe as alternate affine erosions-dilations. We then discuss an efficient local encoding of the shape elements. We finally show experiments. Applications aimed at include image registration, image indexing, optical flow.

Key words: Shape Recognition, Contrast and Affine Invariance, Partial Occlusion.

1. Introduction

Recently, various strategies to rigorously define distances between shapes have been proposed [25]. This distance method allows large nonparametric deformations. In this communication, we shall restrict ourselves to the case where perturbations boil down to contrast changes, planar affine transforms and occlusions. This restrictive framework is just sufficient to recognize an image which has undergone a Xerox copy or a photograph (if it is a painting) and is thereafter subject to contrast changes and an arbitrary framing (occlusion on the boundary). The affine invariant framework is a well acknowledged topic [3, 4, 12, 13].

The restrictions we are taking are not arbitrary, but result from a hopefully rigorous invariance analysis. We first argue that the local contrast invariant information of an image is completely contained in its level lines ([5, 6]), which turn out to be Jordan curves. In order to overcome the occlusion phenomena, we wish to have an encoding as local as possible. The locality is obtained by segmenting each level line into its smallest meaningful parts which must finally be described by small codes. The curve segmentation-encoding process must therefore be itself invariant.

Moreover, the description of the curves must involve some smoothing since level lines are influenced by the quantization process. Thus, smoothing must be performed in order to get rid of this influence. Another reason to smooth shapes, is given by the "scale space ideology" [24]. Indeed, many of the fine scale oscillations of the shapes may be parts of the shape; the analysis of the

* {lisani,moisan,monasse,morel}@cmla.ens-cachan.fr

shape would be lost in those details.

Following [1], the only contrast invariant, local, smoothing and affine invariant scale space leads to a single PDE,

$$\frac{\partial u}{\partial t} = |Du|\mathrm{curv}(\mathrm{u})^{\frac{1}{3}}, \tag{1}$$

where Du is the gradient of the image, curv(u) the curvature of the level line and t the scale parameter. This equation is equivalent to the "affine curve shortening" ([22])

$$\frac{\partial x}{\partial t} = |\mathrm{Curv}(\mathrm{x})|^{\frac{1}{3}}\vec{\mathrm{n}}, \tag{2}$$

where x denotes a point of a level line, Curv(x) its curvature and $\vec{n}$ the signed normal to the curve, always pointing towards the concavity.

This equation is the only possible smoothing under the invariance requirements mentioned above. This gives a helpless bottleneck to the local shape recognition problem, since it is easily checked ([1]) that no further invariance requirement is possible. Despite some interesting attempts [10], there is no way to define a projective invariant local smoothing. The use of curvature-based smoothing for shape analysis is not new [2, 14, 9].

The contrast invariance requirement leads us to describe the shapes in terms of mathematical morphology [23]. In [7], connected components of level sets are proven to be invariant under contrast changes and [6] proposed to take as basic elements of an image the boundaries of the level sets (the so called **level lines**), a complete representation of the image which they call **topographic map**. A fast algorithm for the decomposition of an image into connected components of level lines is described in [20] and its application to a semi-local scale-space representation in [21]. Each one of these connected components is a closed Jordan curve and in many cases, we shall identify the term "shape" with these Jordan curves.

In Section 2, a fast algorithm to perform equation (2) is derived by going back to the mathematical morphology formalism ([23, 16]) and defining first an affine distance and then affine erosions and dilations. This leads us to an axiomatic justification for a fast algorithm introduced by Moisan ([17, 18]). This presentation follows the general line of a book in preparation [11].

In Section 3, we explain how to segment the smoothed curves into affine invariant parts and how these pieces of level lines can be encoded in an efficient way for matching. Section 4 gives a first account of what can be done with the generic algorithm.

2. Affine Invariant Mathematical Morphology and PDE's

We first define an "affine invariant distance" which will be a substitute to the classical Euclidean one. We consider shapes X, subsets of $I\!R^2$. Let $x \in I\!R^2$ and Δ an arbitrary straight line passing by x. We consider all connected components of $I\!R^2 \setminus (X \cup \Delta)$. If $x \notin \bar{X}$, exactly two of them contain x in their boundary. We denote them by $\mathrm{CA}_1(\mathrm{x}, \Delta, \mathrm{X})$, $\mathrm{CA}_2(\mathrm{x}, \Delta, \mathrm{X})$ and call them the "chord-arc sets" defined by x, Δ and X, and we order them so that $\text{area}(\text{CA}_1(\text{x}, \Delta, \text{X})) \le \text{area}(\text{CA}_2(\text{x}, \Delta, \text{X}))$.

Definition 1 *Let X be a "shape" and $x \in \mathbb{R}^2, x \notin \bar{X}$. We call affine distance of x to X the (maybe infinite) number $\delta(x, X) = \inf_\Delta \text{area}(\text{CA}_1(\text{x}, \Delta, \text{X}))^{1/2}$, $\delta(x, X) = 0$ if $x \in X$.*

Definition 2 *For $X \subset \mathbb{R}^2$. We call affine a-dilate of X the set $\tilde{D}_a X = \{x, \delta(x, X) \le a^{1/2}\}$. We call affine a-eroded of X the set $\tilde{E}_a X = \{x, \delta(x, X^c) > a^{1/2}\} = (\tilde{D}_a X^c)^c$.*

Proposition 1 *$\tilde{E}_a$ and $\tilde{D}_a$ are special affine invariant (i.e., they commute with area preserving affine maps) and monotone operators.*

Proof: It is easily seen that if $X \subset Y$, then for every x, $\delta(x, X) \ge \delta(x, Y)$. From this, we deduce that $X \subset Y \Rightarrow \tilde{D}_a X \subset \tilde{D}_a Y$. The monotonicity of $\tilde{E}_a$ follows by the duality relation $\tilde{E}_a X = (\tilde{D}_a X^c)^c$. The special affine invariance of $\tilde{D}_a$ and $\tilde{E}_a$ follows from the fact that $\det \text{A} = 1$ implies that $\text{area}(\text{X}) = \text{area}(\text{AX})$. □

Remark 1 *One can show that $\tilde{E}_a$ and $\tilde{D}_a$ are affine invariant in the sense of Definition 14.19, in [11] that is, for every linear map A with $\det A > 0$, $A\tilde{E}_{(\det A)^{1/2} a} = \tilde{E}_a A$.*

We shall now use Matheron Theorem (Theorem 6.2 in [11]) in order to give a standard form to $\tilde{E}_a$ and $\tilde{D}_a$.

Definition 3 *We say that B is an affine structuring element if 0 is in the interior of B, and if there is some $b > 1$ such that for every line Δ passing by 0, both connected components of $B \setminus \Delta$ containing 0 in their boundary have an area larger or equal to b. We denote the set of affine structuring elements by $\mathbb{B}_{\text{aff}}$.*

Proposition 2 *For every set X,*

$$\tilde{E}_a X = \bigcup_{B \in \mathbb{B}_{\text{aff}}} \bigcap_{y \in a^{1/2} B} X - y = \{x, \exists B \in \mathbb{B}_{\text{aff}}, x + a^{1/2} B \subset X\}$$

Proof: We simply apply Matheron theorem. The set of structuring elements associated with $\tilde{E}_a$ is $\mathbb{B} = \{X, \tilde{E}_a X \ni 0\}$. Now,

$$\tilde{E}_a X \ni 0 \Leftrightarrow \delta(0, X^c) > a^{1/2} \Leftrightarrow \inf_\Delta \text{area}(\text{CA}_1(0, \Delta, X))^{1/2} > a^{1/2}$$

This means that for every Δ, both connected components of $X \setminus \Delta$ containing 0 have area larger than some number $b > a$. Thus, X belongs to $a^{1/2}\mathbb{B}_{\text{aff}}$ by definition of $\mathbb{B}_{\text{aff}}$. □

By Proposition 2, x belongs to $\tilde{E}_a X$ if and only if for every straight line Δ, chord-arc sets containing x have an area strictly larger than a. Conversely we can state :

Corollary 1 *$\tilde{E}_a X$ is obtained from X by removing, for every straight line Δ, all chord-arc sets contained in X which have an area smaller or equal than a.*

2.1. Application to curve affine erosion/dilation schemes

Let c_0 be a Jordan curve, boundary of a simply connected set X. Iterating affine erosions and dilations on X gives a numerical scheme that computes the affine shortening c_T of c_0 at a given scale T. In general, the affine erosion of X is not simple to compute, because it can be strongly non local. However, if X is convex, then it has been shown in [18] that it can be exactly computed in linear time. In practice, c will be a polygon and the exact affine erosion of X —whose boundary is made of straight segments and pieces of hyperbolae— is not really needed; numerically, a good approximation by a new polygon is enough. Now the point is that we can approximate the combination of an affine erosion plus an affine dilation of X by computing the affine erosion of each *convex component* of c, provided that the erosion/dilation area is small enough. The algorithm consists in the iteration of a four-steps process:

1. **Break the curve into convex components.**
2. **Sample each component.**
3. **Apply discrete affine erosion to each component.**
4. **Concatenate the pieces of curves obtained at step 3.**

• **Discrete affine erosion**. This is the main step of the algorithm: compute quickly an approximation of the affine erosion of scale σ of the whole curve. The first step consists in the calculus of the "area" A_j of each convex component $\mathcal{C}^j = P_0^j P_1^j ... P_{n-1}^j$, given by $A_j = \sum_{i=1}^{n-2} \left[P_0^j P_i^j, P_0^j P_{i+1}^j \right] /2$. Then, the effective area used to compute the affine erosion is $\sigma_e = \max \{\sigma/8, \min_j A_j\}$. We restrict the erosion area to σ_e because the simplified algorithm for affine erosion may give a bad estimate of the continuous affine erosion+dilation when the area of one component is less than the erosion parameter. The term $\sigma/8$ is rather arbitrary and guarantees an upper bound to the number of iterations required to achieve the final scale. The discrete erosion of each component is defined as the succession of each middle point of each segment $[AB]$ such that

1. A and B lie on the polygonal curve
2. A or B is a vertex of the polygonal curve
3. the area enclosed by $[AB]$ and the polygonal curve is equal to σ_e

• **Iteration of the process**. To iterate the process, we use the fact that if E_σ denotes the affine erosion plus dilation operator of area σ, and $h = (h_i)$ is a subdivision of the interval $[0, H]$ with $H = T/\omega$ and $\omega = \frac{1}{2}\left(\frac{3}{2}\right)^{2/3}$, then

$$E_{(h_1-h_0)^{3/2}} \circ E_{(h_2-h_1)^{3/2}} \circ ... \circ E_{(h_n-h_{n-1})^{3/2}} \left(c_0\right) \longrightarrow c_T$$

as $|h| = \max_i h_{i+1} - h_i \to 0$, where c_T is the affine shortening of c_0 described above by (2).

The algorithm has linear complexity in time and memory, and its stability is ensured by the fact that each new curve is obtained as the set of the middle points of some chords of the initial curve, defined themselves by an integration process (an area computation). Hence, no derivation or curvature computation appears in the algorithm.

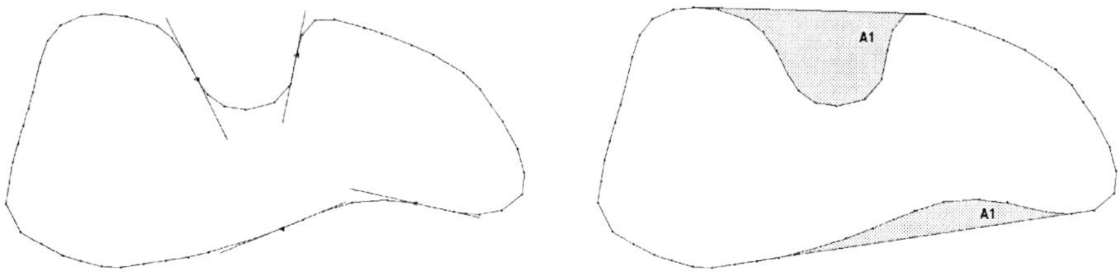

Fig. 1. Inflexion points (marked with small triangles) and bitangents of a closed curve. The area defined by each bitangent and the original curve is marked ($A1$).

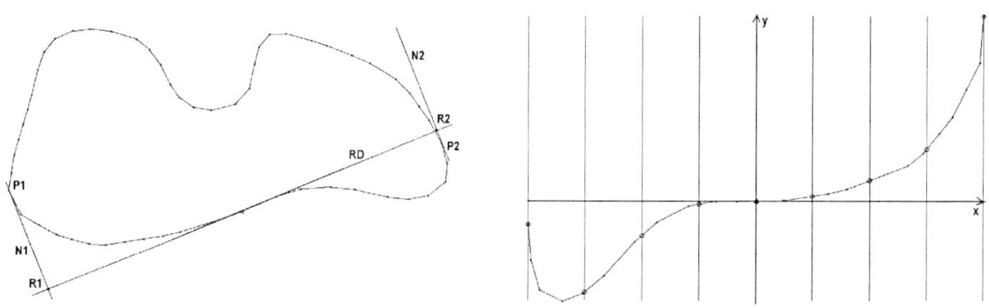

Fig. 2. Left: Local reference system for similarity invariant normalization: reference direction (RD), normal directions (N1, N2) and reference points (R1, R2). The portion of the curve normalized with this reference system starts at P1 and ends at P2, passing through the inflexion point. Right: Similarity invariant normalization. The y-ordinate of the marked points is used to encode the piece of curve.

3. Algorithms for the Description of Shapes in Images.

3.1. Similarity Invariant Description of Curves

In the search for an invariant description of a curve, the starting point for the sampling must be invariant, and so must be the sampling mesh. Typically, inflexion points have been chosen because they are affine invariant. Now, since the curve is almost straight at inflexion points, their position is not robust, but the direction of the tangent to the curve passing through them is. Another affine invariant robust semilocal descriptor is given by the lines which are bitangent to the curve (see Fig. 1).

Our reference system is formed by such a line, and the next and previous tangents to the curve which are orthogonal to it (see Fig. 2). The intersections of each one of these lines with the reference line provide two reliable points independent of the discretization of the curve. The portion of the curve to be normalized is limited by these points. Normalization consists in a similarity transform that maps the reference line to the x-axis and that sets the distance between the two reference points to 1. We discretize each one of the normalized portions of the curve with a fixed number n of points, and we store, for each

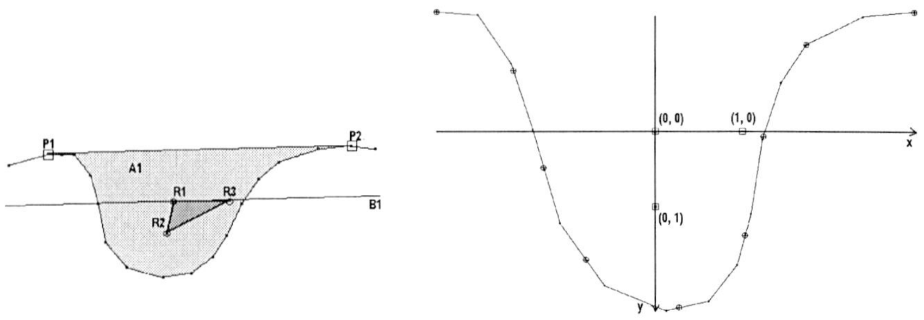

Fig. 3. Left: Local reference system for affine invariant normalization: reference points (R1, R2, R3). The portion of the curve to be encoded has endpoints P1 and P2. Right: Affine invariant normalization. The length of the normalized piece of curve together with the x and y coordinates of the marked points are used to locally encode the curve.

discretized point, its y coordinate (see Fig. 2). This set of n values is used to compare portions of curves.

3.2. Affine invariant description of curves

If we look at Fig. 1, we can observe that the portion of the curve between the points defining the bitangent, together with the bitangent itself, define an area (*A1*), from which further invariant features can be computed. In particular, we can compute the barycenter of this area, an affine invariant reference point. We compute then the line *B1* parallel to the bitangent and passing through the barycenter. *B1* divides the initial area into two parts and we compute the barycenter of the part which does not contain the bitangent (see Fig. 3). This second barycenter is a second reference point. Finally a point in line *B1* such that the area of the triangle formed by this point and the two preceding barycenters is a fixed fraction of the initial area *A1* is a third reference point (see Fig. 3). We therefore obtain three nonaligned points, that is an affine reference system. This strategy is related to [8]. Some discretization points are taken at uniform intervals of length on the normalized curve and they are used to compare portions of curves.

4. Experimental Results

Figure 4 displays a picture of a man and the same picture after an occlusion of the face with his forearm and their level lines after smoothing with the iterative scheme described in section 2. Clearly some level lines have suffered a significant occlusion, and, even if some parts of the level line remain unchanged, registration methods based on global matching would fail in detecting those lines. In Figure 5, we show the result of the matching of several pieces of an occluded level line with other pieces of level lines in the second image.

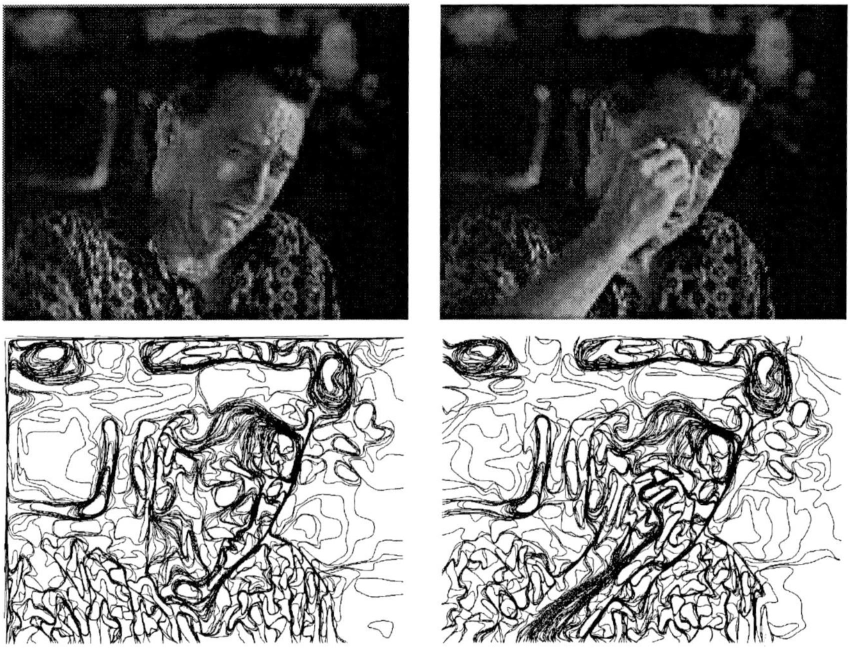

Fig. 4. Up: Original images (from the film 'Analyze This' (Warner Bros)). Down: their smooth level lines (smoothing method of section 2).

References

1. L. Alvarez, F. Guichard, P.L. Lions, J.M. Morel, *Axioms and Fundamental Equations on Image Processing*, Technical Report 9231, CEREMADE, 1992 and Arch. for Rat. Mech. Anal. 16(9), 200-257, 1993.
2. H. Asada and M. Brady. *The curvature primal sketch.* IEEE PAMI, 8(1), 2-14, 1986.
3. K. Astrom. *Affine and projective normalization of planar curves and regions*, ECCV 94, pp. B :439-448, 1994.
4. S.K. Bose, K.K. Biswas and S.K. Gupta. Model-based object recognition : the role of affine invariants. AIEng, 10(3) : 227-234, 1996.
5. V. Caselles, B. Coll, and J.M. Morel. *A Kanisza programme.* Progress in Nonlinear Differential Equations and their Applications, 25, 1996.
6. V. Caselles and B. Coll and J.M. Morel, *Topographic maps*, preprint CEREMADE, 1997. To appear in I.J.C.V.
7. V. Caselles and J.L. Lisani and G. Sapiro and J.M. Morel, *Shape preserving histogram modification*, IEEE Trans. on Image Processing, February 1999.
8. Cohignac, T. and Lopez, C. and Morel, J.M., *Integral and Local Affine Invariant Parameter and Application to Shape Recognition*, ICPR pp. A:164-168, 1994.
9. G. Dudek and J.K. Tsotsos. *Shape representation and recognition from multiscale curvature*, CVIU, 68(2), pp. 170-189, 1997.
10. O. Faugeras and R. Keriven, *Some recent results on the projective evolution of 2D curves.* In Proc. IEEE ICIP, vol. 3, pp. 13-16, Washington, October 1995.
11. F. Guichard and J.M. Morel, *Image iterative filtering and PDE's*, Preprint, 1999. Book in preparation.
12. R.A. Hummel, H.J. Wolfson. *Affine invariant matching.* DARPA88, pp. 351-364, 1988.
13. P. Kempenaers, L. Van Gool, and A. Oosterlinck. *Shape recognition under affine distortions.* VF91, pp. 323-332, 1991.

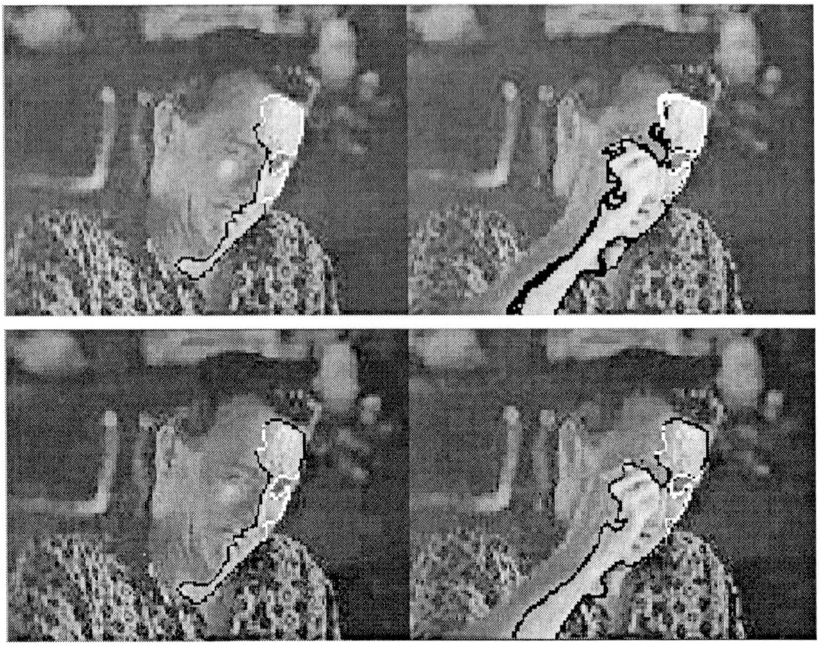

Fig. 5. Some correspondences (marked in white, the remaining parts of the curves in black) between some pieces of one level line in the first image (left) and some pieces of other level lines in the second image (right). Up: By using the similarity invariant registration method based on inflexion points. Down: By using the affine invariant registration method based on bitangents.

14. A. Mackworth and F. Mokhtarian, *A theory of multiscale, curvature-based shape representation for planar curves*, IEEE PAMI, 14 : 789-805, 1992.
15. G. Koepfler, L. Moisan, *Geometric Multiscale Representation of Numerical Images*, Proc. of the Second Int. Conf. on Scale-Space Theories in Computer Vision, in Springer Lecture Notes in Computer Science, vol. 1682, pp. 339-350, 1999.
16. G. Matheron, *Random Sets and Integral Geometry*, John Wiley, N.Y., 1975.
17. L. Moisan, *Traitement numérique d'images et de films: équations aux dérivées partielles préservant forme et relief*, PhD dissertation, Université Paris-Dauphine, France, 1997.
18. L.Moisan, *Affine Plane Curve Evolution : a Fully Consistent Scheme*, IEEE Transactions On Image Processing, vol. 7:3, pp. 411-420, 1998.
19. P. Monasse, *Contrast Invariant Image Registration*, Proc. of I. V. Conf. on Acoustics, Speech and Signal Processing, Phoenix, Arizona, vol 6, 1999, pp. 3221-3224.
20. P. Monasse and F. Guichard, *Fast Computation of a Contrast-Invariant Image Representation*, to appear in IEEE Transactions on Image Processing, 1998, preprint CMLA 9815, available at *http://www.cmla.ens-cachan.fr*
21. P. Monasse and F. Guichard, *Scale-Space from a Level Lines Tree*, Proc. of 2nd Int. Conf. on Scale-Space Theories in Computer Vision, Corfu, Greece, 1999, pp. 175-186.
22. G. Sapiro and A. Tannenbaum, *Affine Invariant Scale Space*, IJCV, 11 (1), 25-44, 1993.
23. J.Serra, *Image Analysis and Mathematical Morphology*, Academic Press, 1982.
24. A.P. Witkin, *Scale space filtering*, Proc. IJCAI, 1019-1023, 1983.
25. L. Younes *Computable elastic distances between shapes*, SIAM J. of Ap. Maths., 1998.

FOLDING INDUCED SELF-DUAL FILTERS

ANDREW J.H. MEHNERT and PAUL T. JACKWAY
Department of Computer Science and Electrical Engineering,
The University of Queensland, Brisbane 4072 Australia.
Fax: +61-7-3365-3684
e-mail: {A.Mehnert,P.Jackway}@cssip.uq.edu.au

Abstract. In this paper we present a method for constructing self-dual grey-scale image operators from arbitrary morphological operators defined on what we call *fold-space*. We call this class of self-dual operators *folding induced self-dual filters* (FISFs). We show examples of their application to noise filtering.

Key words: Chain, Complete Lattice, Folded Ordering, Partial Ordering, Quasi-Ordering, Self-Duality.

1. Introduction

In image analysis self-dual filters are of interest because they treat both the light and dark areas of an image in an equivalent manner. They are well suited to situations where we desire to "separate two components, one of which is sometimes lighter and sometimes darker than the other" [11]. Examples of such situations include the elimination of salt-and-pepper noise, and the filtering of images—e.g. natural scenes, textures, bipolar radar images—for which there is no distinction between foreground and background.

Non-linear self-dual filters offer several advantages over their linear counterparts. They may be designed: (i) such that they do not reduce the dynamic range and high frequencies in the image [11]; (ii) to be independent of monotone changes in intensity called *anamorphoses* [10, 11]; (iii) such that no new intensity values are introduced in the image; and (iv) to be idempotent but not induce the *ringing* degradation [4] that is characteristic of ideal linear filters. Self-duality is a property possessed by convolution and therefore by all linear filters [11]. It is, however, a property possessed by only some non-linear filters (the median filter is perhaps the best known example). It is therefore desirable to know how to construct non-linear filters that are self-dual. This has been the motivation for much of the research on non-linear self-dual filters since the late 1980s [11, 12, 8, 5, 3, 6]. Recently Evans, Svalbe and Jones [3] introduced the idea of imposing an *alternative ordering* (which they call *folded ordering*) on the grey-scale (intensity) values in an image such that the application of a single idempotent morphological closing yields a self-dual filter. Unfortunately, they note that the imposed reordering seriously inhibits the ability of the resulting filter to attenuate impulse noise. They subsequently abandon this ordering in favour of another that induces only *approximate* self-duality. In this paper we present a theoretical framework for the construction of self-dual operators from arbitrary morphological operators defined on what we call *fold-space.* Fold-space generalises the notion of folded ordering (hereinafter referred to as the ESJ folded ordering). We call the class of self-dual operators that can be constructed from fold-space operators *folding induced self-dual filters* (FISFs). The folded closing (hereinafter referred to as the ESJ folded closing) proposed by Evans, Svalbe, and Jones [3] represents but one example from this class. We show that other FISFs can be designed that do not suffer the limitations of the ESJ folded closing when filtering impulse noise.

The remainder of this paper is organised as follows. In the next section we briefly review ordering in sets, and the complete lattice of grey-scale functions (this serves as our model for grey-scale images). In Section 3 we introduce the space of folded grey-scale functions (fold-space) and show that it is a complete lattice. In Section 4 we define and characterise FISFs, give examples of how they may be constructed, and address issues with regard to implementation. Finally in Section 5 we provide a short discussion and conclusion.

2. Sets, Ordering, and Complete Lattices

The concept of ordering in sets is the central idea in this paper. Moreover it is a fundamental concept in mathematical morphology. The algebraic structure used in the formal definition and study of mathematical morphology is the *complete lattice* which is in essence an ordered set equipped with a supremum and infimum. Hence, in this section we briefly review several definitions and results pertaining to sets, ordering, and complete lattices that will be needed for the material presented in subsequent sections.

Let $\mathcal{S}$ be an arbitrary set and let $\mathcal{R}$ be a subset of $\mathcal{S} \times \mathcal{S}$. The set $\mathcal{R}$ is called a *binary relation* on $\mathcal{S}$. If the ordered pair $(X, Y) \in \mathcal{R}$ then we write $X\mathcal{R}Y$ and say that "the relation $\mathcal{R}$ holds between X and Y".

Definition 1 (properties of binary relations) *A binary relation $\mathcal{R}$ defined on a set $\mathcal{S}$ is said to be*

1. reflexive *if $X\mathcal{R}X$ for all $X \in \mathcal{S}$;*
2. transitive *if $X\mathcal{R}Y$ and $Y\mathcal{R}Z \Rightarrow X\mathcal{R}Z$ for all $X, Y, Z \in \mathcal{S}$;*
3. symmetric *if $X\mathcal{R}Y \Rightarrow Y\mathcal{R}X$ for all $X, Y \in \mathcal{S}$; and*
4. anti-symmetric *if $X\mathcal{R}Y$ and $Y\mathcal{R}X \Rightarrow X = Y$ for all $X, Y \in \mathcal{S}$.*

A binary relation that is both reflexive and transitive is called a relation of *quasi-ordering*. A binary relation that is reflexive, transitive, and symmetric is called an *equivalence* relation. A binary relation that is reflexive, transitive, and anti-symmetric is called a *partial order* relation.

Theorem 1 (Schröder's theorem [2]) *Let $\mathcal{R}$ be a relation of quasi-ordering defined on a set $\mathcal{S}$. The relation $\mathcal{E}$ defined*

$$X\mathcal{E}Y \text{ if and only if } X\mathcal{R}Y \text{ and } Y\mathcal{R}X$$

is an equivalence relation. Moreover, if the equivalent *elements are identified, $\mathcal{R}$ becomes a partial order relation.*

If $\mathcal{R}$ is a partial order relation defined on a set $\mathcal{S}$ then the pair $(\mathcal{S}, \mathcal{R})$ is called a partially ordered set or poset. In addition if $X\mathcal{R}Y$ or $Y\mathcal{R}X$ or both for all $(X, Y) \in \mathcal{S}$ then the poset is said to be *totally* or *linearly* ordered and is called a *chain*. A partially ordered set $(\mathcal{S}, \mathcal{R})$ is said to be a *complete lattice* if every non-empty subset of $\mathcal{S}$ (finite or not) has both a *supremum* and an *infimum*.

Let $(\mathcal{L}, \leq)$ be a complete lattice and let $\mathcal{O}(\mathcal{L})$ be the set of all operators $\psi : \mathcal{L} \to \mathcal{L}$. An operator $\psi \in \mathcal{O}(\mathcal{L})$ is said to be a *negation* if it is a bijection (one-to-one and onto), reverses the ordering (i.e. $X \leq Y \Rightarrow \psi(Y) \leq \psi(X)\ \forall X, Y \in \mathcal{L}$), and satisfies $\psi(\psi(X)) = X\ \forall X \in \mathcal{L}$. A complete lattice may possess no negations, a single negation, or multiple negations [7]. An operator $\psi \in \mathcal{O}(\mathcal{L})$ is said to be *self-dual* if $\psi(X) = [\psi(X^*)]^*\ \forall X \in \mathcal{L}$, where $*$ denotes a negation.

2.1. Complete Lattice of Grey-scale Functions

Mathematically grey-scale images can be represented as functions $f : E \to G$. Although the domain space E may be an arbitrary set, it is usually taken to be either $\mathbb{R}^n$ or $\mathbb{Z}^n$ (representing pixel coordinates). The set G defines the set of grey-values. It is essential that G is a complete lattice. For example, in image processing G is typically one of the infinite sets $\overline{\mathbb{R}}$ or $\overline{\mathbb{Z}}$, or the finite set $\{0, 1, \ldots, m\}$, each of which is a complete lattice for the usual partial order relation $\leq$. Let $\mathcal{F}$ be the set of functions $f : E \to G$. The partial order relation $\leq$ on G can be used to define a partial ordering on the set $\mathcal{F}$ as follows:

$$f \leq g \;\Leftrightarrow\; f(x) \leq g(x)\ \forall x \in E.$$

The pair $(\mathcal{F}, \leq)$ is a complete lattice. When G is the finite set $\{0, 1, \ldots, m\}$, the lattice possesses the unique negation $f^*(x) = m - f(x)$. When G is one of the infinite sets $\overline{\mathbb{R}}$ or $\overline{\mathbb{Z}}$, the lattice possesses multiple negations [7] (in grey-scale morphology special attention is paid to the negation $f^*(x) = -f(x)$ wherein it is seen as the counterpart to set complementation on a boolean lattice).

3. Fold-Space and the Folding Operator

The ESJ folded ordering is a *distance ordering* [1] defined on the finite grey-value set $G = \{0, \ldots, m\}$. It is an ordering of the grey values in relation to their absolute deviation or *distance* from some reference value. The ESJ folded ordering in fact defines an equivalence relation on the set $G = \{0, \ldots, m\}$, namely

$$a \equiv b \text{ if and only if } a \text{ and } b \text{ are equidistant from the reference value,}$$

where $a, b \in G$. When the reference value is the median of the set G then the ordering can be used to construct a self-dual filter on the lattice $(\mathcal{F}, \leq)$ from a filter that is not self-dual. The ESJ folded closing [3], for example, is constructed as follows: (i) the image grey-values are folded about the median (the locations of affected pixels are recorded in a template image so that the process can be inverted), (ii) a morphological closing with a flat structuring element

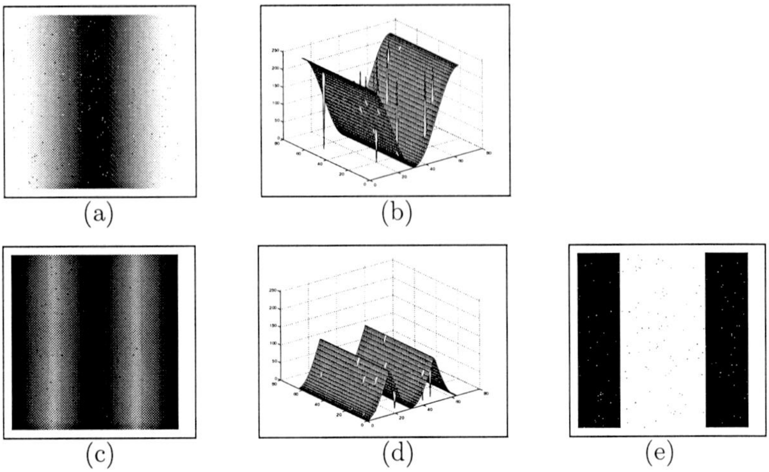

Fig. 1. Illustration of folded ordering. (a),(b) Original 8-bit grey-scale image (sine-wave corrupted with salt-and-pepper noise) and its rendering as a surface. (c),(d) The folding of the original image about the median and its rendering as a surface. (e) Template image needed to invert folding.

is applied, and (iii) the folding is inverted (see Figure 1). We note in passing that this idea bears some similarity to the concept of a *difference semilattice* recently introduced by Kresch [9]. The elements of a difference semilattice are difference functions (i.e. the difference between two real functions $f, g \in \mathcal{F}$). The null element of the lattice is the function $o(x) = 0$ (the lattice has no universal element). The infimum operation between two difference functions h and k is defined point-wise as follows:

$$(h \wedge k)(x) = \begin{cases} \min(h(x), k(x)), & \text{if } h(x) > 0 \text{ and } k(x) > 0 \\ \max(h(x), k(x)), & \text{if } h(x) < 0 \text{ and } k(x) < 0 \\ 0, & \text{otherwise.} \end{cases}$$

If we consider difference functions of the form $f(x) - c$, where $f \in \mathcal{F}$ is an integer valued function and c is the median of G, then we have an operation akin to folding (the sign of the difference values plays the same role as the template image). However this is where the similarity ends because the infimum operation defined for folded grey-values cannot be expressed as an infimum in the difference semilattice framework.

In order to formalise the idea of folded ordering, and to extend its definition to the infinite chains $\overline{\mathbb{R}}$ and $\overline{\mathbb{Z}}$ (and any chains that are isomorphic to them) we now introduce the concept of *fold-space* and the *folding operator*.

In the remainder of this paper it is necessary to assume that the set of grey-values G is, in addition to being a complete lattice, totally ordered (which is the case for the usual sets $\overline{\mathbb{R}}$, $\overline{\mathbb{Z}}$, and $\{0, 1, \ldots, m\}$). Let $\mathcal{H}$ be the set of all functions $\widetilde{f} : E \to \widetilde{G}$ where $\widetilde{G} = G \times \{-1, 0, 1\}$ (the set $\{-1, 0, 1\}$ is arbitrary in the sense that it can be any chain of three elements—they are indicator values). We call $\mathcal{H}$ the space of folded grey-scale functions or simply fold-space.

Definition 2 (folding operator) *Let $\sigma : \mathcal{F} \to \mathcal{H}$ be the* folding operator *defined point-wise as follows:*

$$\sigma(f)(x) = \begin{cases} (f(x), 1), & \text{if } f(x) < f^*(x) \\ (f(x), 0), & \text{if } f(x) = f^*(x) \\ (f^*(x), -1), & \text{if } f(x) > f^*(x) \end{cases}$$

This operator maps a function $f \in \mathcal{F}$ into a pair $\widetilde{f} = (f_1, f_2) \in \mathcal{H}$ comprising the folded function f_1 and an indicator function f_2. The folding operator is a one-to-one mapping of $\mathcal{F}$ on to $\mathcal{H}$; i.e. the image of every distinct element f of $\mathcal{F}$ is a distinct element $\widetilde{f}$ of $\mathcal{H}$, and in addition each element of $\mathcal{H}$ is an image. Consequently the folding operator has an inverse $\sigma^{-1} : \mathcal{H} \to \mathcal{F}$ which is defined point-wise as follows:

$$\sigma^{-1}\left(\widetilde{f}\right)(x) = \begin{cases} f_1(x), & \text{if } f_2(x) = 1 \\ c, & \text{if } f_2(x) = 0 \\ f_1^*(x), & \text{if } f_2(x) = -1 \end{cases},$$

where $\widetilde{f} = (f_1, f_2)$, $f_1 \in \mathcal{F}$, $f_2 : E \to \{-1, 0, 1\}$, and $c \in G$ such that $c = c^*$. The constant c is called the *crease* and its existence and value are solely determined by the negation operator; e.g. when G is the infinite set $\overline{\mathbb{R}}$, then the negation $f^*(x) = -f(x) + k$ on the lattice $(\mathcal{F}, \leq)$ prescribes the value of the crease to be $k/2$ (see also the comments in Section 4.2).

From the definition of the folding operator, it is easy to prove the following properties.

Proposition 1 (properties of the folding operator)

1. *If $\sigma(f) = (f_1, f_2)$ then $\sigma(f^*) = (f_1, -f_2)$ where $f \in \mathcal{F}$.*
2. *$\sigma^{-1}((f_1, f_2)) = \left[\sigma^{-1}((f_1, -f_2))\right]^*$ where $\widetilde{f} = (f_1, f_2) \in \mathcal{H}$.*

3.1. Complete Lattice of Folded Grey-scale Functions

In keeping with the underlying idea of the ESJ folded ordering we can define the following equivalence relation on the set $\widetilde{G}$:

$$a \equiv b \text{ if and only if } a_1 = b_1$$

where $a = (a_1, a_2)$ and $b = (b_1, b_2)$ are elements of the set $\widetilde{G}$. This relation is defined in terms of the equality relation defined on the first component of the elements of $\widetilde{G}$, i.e. the folded grey-values. We can think of these folded grey-values as being elements of the complete chain $(G, \leq)$. The partial order relation $\leq$ on G can be used to define the following order relation on $\widetilde{G}$:

$$a \ll b \text{ if and only if } a_1 \leq b_1.$$

Unfortunately, however, the relation $\ll$ is not itself a partial order relation. It is only a relation of quasi-ordering because although it is reflexive and transitive, it is not anti-symmetric; e.g. if $G = \{0, 1, \ldots, m\}$ then it is clear that $(2, 0) \ll$

$(2,1)$ and $(2,1) \ll (2,0)$ but $(2,0) \neq (2,1)$. Consequently, $\left(\widetilde{G}, \ll\right)$ is not a complete lattice. However, the definition of the equivalence relation above can equally be defined

$$a \equiv b \text{ if and only if } a \ll b \text{ and } b \ll a.$$

The relation induces a partition of $\widetilde{G}$ into the subsets (equivalence classes) $X = \{(x,-1),(x,0),(x,1)\}$, $Y = \{(y,-1),(y,0),(y,1)\}$, ... where $x, y, \ldots \in G$. If we let $\mathcal{S}$ be the set of equivalence classes $X, Y, \ldots$ then the relation $\preceq$ defined

$$X \preceq Y \text{ if and only if } a \ll b,$$

for some $a \in X$ and $b \in Y$, is a partial order relation (by Theorem 1). It follows that the pair $(\mathcal{S}, \preceq)$ is a complete lattice. In fact $\preceq$ defines a total ordering and so $(\mathcal{S}, \preceq)$ is a complete chain.

Rather than explicitly forming the set $\mathcal{S}$ of equivalence classes, it is possible to work with the set $\widetilde{G}$ directly. Given two elements $a, b \in \widetilde{G}$, we have that $a \ll b$ if and only if $X \preceq Y$, where X is the equivalence class containing a and Y is the equivalence class containing b. A problem arises when we seek to find the supremum or infimum of two elements $a, b \in \widetilde{G}$ that are equivalent. The solution is to define the supremum (resp. infimum) to be the first operand, or the second operand, or some other equivalent element. In this sense the set $\widetilde{G}$ together with the relations $\ll$ and $\equiv$ defined on it, is a complete lattice (whose elements are not the elements of $\widetilde{G}$ but rather the equivalence classes of the partition of $\widetilde{G}$ induced by $\equiv$). In a similar fashion if we define the following pair of binary relations

$$\widetilde{f} \ll \widetilde{g} \Leftrightarrow \widetilde{f}(x) \ll \widetilde{g}(x) \ \forall x \in E, \text{ and}$$

$$\widetilde{f} \equiv \widetilde{g} \Leftrightarrow \widetilde{f}(x) \equiv \widetilde{g}(x) \ \forall x \in E,$$

where $\widetilde{f}, \widetilde{g} \in \mathcal{H}$, then the set $\mathcal{H}$ together with the relations $\ll$ and $\equiv$ defined on it, is a complete lattice. We call morphological operators defined on this complete lattice *fold-space morphological operators.*

4. Folding Induced Self-Dual Filters

Definition 3 (folding induced self-dual filter) *The product* $\sigma^{-1}\Gamma\sigma \in \mathcal{O}(\mathcal{F})$, *where* $\Gamma \in \mathcal{O}(\mathcal{H})$, *is called a* folding induced self-dual filter (FISF) *if for all* $f \in \mathcal{F}$, $\sigma^{-1}\Gamma\sigma(f) = \left[\sigma^{-1}\Gamma\sigma(f^*)\right]^*$.

Theorem 2 *Let* $\Gamma \in \mathcal{O}(\mathcal{H})$. *The product* $\sigma^{-1}\Gamma\sigma \in \mathcal{O}(\mathcal{F})$ *is a FISF if* $\Gamma\left(\widetilde{f}\right) = \left[\Gamma\left(\widetilde{f}^{\vee}\right)\right]^{\vee}$ *for all* $\widetilde{f} \in \mathcal{H}$ *where* $\widetilde{f}^{\vee} = (f_1, -f_2)$ *when* $\widetilde{f} = (f_1, f_2)$.

Proof: By definition $\sigma^{-1}\Gamma\sigma$ is a FISF if $\sigma^{-1}\Gamma\sigma(f) = \left[\sigma^{-1}\Gamma\sigma(f^*)\right]^*$ for all $f \in \mathcal{F}$. If we let $\widetilde{f} = (f_1, f_2) = \sigma(f)$ then we can write the LHS as $\sigma^{-1}\Gamma((f_1, f_2))$.

Using property 1 of Proposition 1 we can write the RHS as $\left[\sigma^{-1}\Gamma\left(\left(f_1, -f_2\right)\right)\right]^*$. If we let $\widetilde{g} = (g_1, g_2) = \Gamma\left(\left(f_1, f_2\right)\right)$ and $\widetilde{h} = (h_1, h_2) = \Gamma\left(\left(f_1, -f_2\right)\right)$ we can then write $\sigma^{-1}\left(\widetilde{g}\right) = \left[\sigma^{-1}\left(\widetilde{h}\right)\right]^*$. Using property 2 of Proposition 1 this can be written as $\sigma^{-1}\left(\left(g_1, g_2\right)\right) = \sigma^{-1}\left(\left(h_1, -h_2\right)\right)$. Hence if $\Gamma\left(\left(f_1, f_2\right)\right) = (g_1, g_2)$ then $\Gamma\left(\left(f_1, -f_2\right)\right) = (g_1, -g_2)$, i.e. $\Gamma\left(\widetilde{f}\right) = \left[\Gamma\left(\widetilde{f}^\vee\right)\right]^\vee$. □

This theorem in essence states that Γ must be an operator that is self-dual with respect to the second component of its argument.

4.1. Constructing FISFs

The ESJ folded closing can be written as the product $\sigma^{-1}\Gamma\sigma$ where $\Gamma\left(\widetilde{f}\right)(x) = \left(\phi_B\left(f_1\right)(x), f_2(x)\right)$, and $\phi_B \in \mathcal{O}\left(\mathcal{F}\right)$ is the closing $\phi_B = \varepsilon_B \delta_B$ for the dilation $\delta_B \in \mathcal{O}\left(\mathcal{F}\right)$ and the erosion $\varepsilon_B \in \mathcal{O}\left(\mathcal{F}\right)$ defined

$$\delta_B(f)(x) = \bigvee_{y \in B} \left\{f(x-y)\right\}, \text{ and } \varepsilon_B(f)(x) = \bigwedge_{y \in B} \left\{f(x+y)\right\},$$

where B is a compact set of E [10]. This filter acts only on the first component of $\widetilde{f}$. Hence according to Theorem 2 the ESJ folded closing is a FISF. We will call FISFs constructed in this manner *type 1 FISFs*.

It is precisely because the ESJ folded closing does not take into account the second component of $\widetilde{f}$ that it performs poorly as an impulse noise filter. Image values less than the fold point cannot be replaced by values greater than the fold point and vice versa. Consequently it "can not completely remove pepper noise from a light area or salt noise from a darker region" [3]. This then is our motivation for introducing the following definitions of the supremum and infimum, respectively, for the set $\widetilde{G}$:

$$a \vee b = \begin{cases} a, & \text{if } b_1 < a_1 \\ b, & \text{if } a_1 < b_1 \\ (a_1, 0), & \text{if } a_1 = b_1 \text{ and } a_2 \neq b_2 \\ a, & \text{if } a_1 = b_1 \text{ and } a_2 = b_2, \end{cases}$$

$$a \wedge b = \begin{cases} a, & \text{if } a_1 < b_1 \\ b, & \text{if } b_1 < a_1 \\ (a_1, 0), & \text{if } a_1 = b_1 \text{ and } a_2 \neq b_2 \\ a, & \text{if } a_1 = b_1 \text{ and } a_2 = b_2, \end{cases}$$

where $a, b \in \widetilde{G}$. These definitions dictate that the supremum or infimum of two distinct but equivalent elements is always the equivalent element with the second component equal to zero. A fold-space closing $\Gamma = E_B \Delta_B$ defined in terms of these definitions, where

$$\Delta_B\left(\widetilde{f}\right)(x) = \bigvee_{y \in B} \left\{\widetilde{f}(x-y)\right\}, \text{ and } E_B\left(\widetilde{f}\right)(x) = \bigwedge_{y \in B} \left\{\widetilde{f}(x+y)\right\},$$

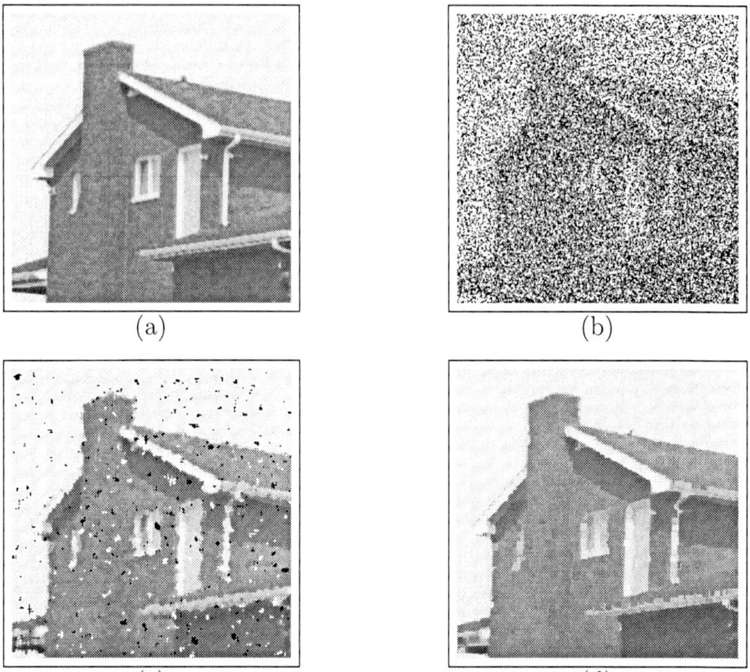

(a) (b) (c) (d)

Fig. 2. Demonstration of noise filtering of a natural scene. (a) Original 8-bit grey-scale image. (b) Image corrupted with 60% salt-and-pepper noise. (c) Result after applying a 5×5 median filter to the noisy image. (d) Result after applying a type 2 FISF, based on a fold-space closing with a 5×5 flat structuring element, to the noisy image.

satisfies Theorem 2 and can thus be used to construct a FISF. We will call FISFs constructed in this manner *type 2 FISFs*. Figure 2 demonstrates the effectiveness of this FISF in removing salt-and-pepper noise.

Unfortunately, as Figure 3 shows, both the ESJ folded closing and type 2 FISFs based on a fold-space closing perform poorly when large areas within the image are at either grey-value extreme. This then is the motivation for introducing the following definitions of the supremum and infimum, respectively,

$$\bigvee \{a, b, \ldots\} = \begin{cases} (\alpha, \mu) & \text{if } \mu \text{ is unique} \\ (\alpha, 0) & \text{otherwise,} \end{cases}$$

$$\bigwedge \{a, b, \ldots\} = \begin{cases} (\beta, \nu) & \text{if } \nu \text{ is unique} \\ (\beta, 0) & \text{otherwise,} \end{cases}$$

where $\alpha = \bigvee \{a_1, b_1, \ldots\}$, $\mu = \text{mode}\,\{x \mid (\alpha, x) \in \{a, b, \ldots\}\}$, $\beta = \bigwedge \{a_1, b_1, \ldots\}$, and $\nu = \text{mode}\,\{y \mid (\beta, y) \in \{a, b, \ldots\}\}$. A fold-space closing defined in terms of these definitions satisfies Theorem 2. The behaviour of this FISF is illustrated in Figure 3. We will call FISFs constructed in this manner *type 3 FISFs*.

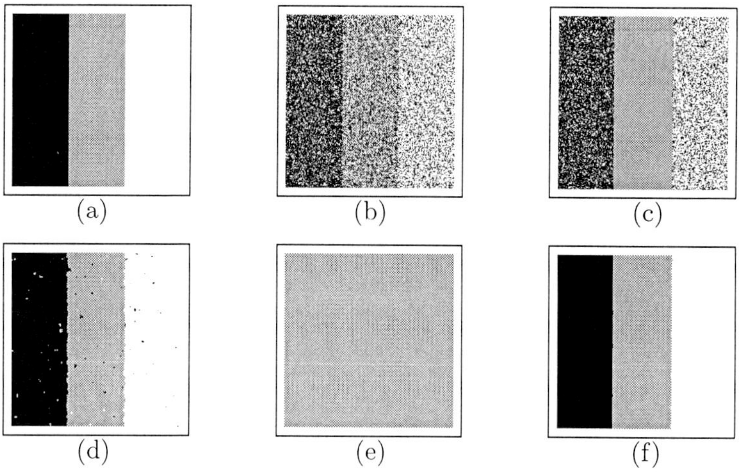

Fig. 3. Demonstration of noise filtering when significant parts of the image are at either extreme of the grey-value range. (a) Original 8-bit image with grey-value bands at 0, 128, and 255. (b) Original image corrupted with 50% salt-and-pepper noise. (c) Result after applying an ESJ folded closing with a 5×5 flat structuring element to the noisy image (type 1 FISF). (d) Result after applying a 5×5 median filter to the noisy image. (e) Result after applying a type 2 FISF, based on a morphological closing with a 5×5 flat structuring element, to the noisy image. (f) Result after applying a type 3 FISF, based on a morphological closing with a 5×5 flat structuring element, to the noisy image.

4.2. Implementation Issues

The implementation of fold-space morphological operators for digital images differs from the implementation of conventional grey-scale morphological operators only in that the supremum and infimum operations (either between two pixels or over a window of pixels) must propagate a template value in addition to a grey-value. Given that a digital image is typically represented using a finite number of grey-values $\{0, 1, \ldots, m\}$ where $m+1$ is a power of 2, the crease does not exist. However, the symmetric point with respect to the ordering does exist and so we can set $c = m/2$. Unfortunately c is not a representable grey-value. This presents a problem in the case of type 2 and type 3 FISFs when it comes to applying the inverse folding operator. One possible solution is to replace any pixel values that would map to the crease with the grey-value of the preceding or succeeding representable grey-value (though the resulting filter is now only approximately self-dual). Two other possible solutions, which preserve self-duality, are: (1) to use only an odd number of grey-values, and (2) at the unfolding step, to replace any pixel values that would map to the crease with the corresponding pixel produced by applying a median filter (or indeed any other self-dual filter) to the original image.

5. Discussion and Conclusion

We have proposed a novel method for the construction of self-dual operators, FISFs, from arbitrary morphological operators defined on what we call fold-space. We have shown that the folded closing proposed by Evans, Svalbe, and Jones [3] is a particular type of FISF. Whilst the ESJ folded closing performs poorly as a noise filter for impulse-type noise, we have shown that other types of FISF can be designed that are very effective noise filters. Further opportunities for research include an investigation of the possible applications for FISFs, and the theoretical characterisation of multiple folding and other extensions to the equivalence relation induced by folded ordering.

Users of MICROMORPH version 1.3 [1] can download an implementation of fold-space dilation and erosion (based on the supremum and infimum definitions for type 2 FISFs) from our ftp site.[2]

Acknowledgements

This research was funded by an Australian Research Council grant. The authors would also like to express their gratitude to the anonymous reviewers for their comments and suggestions.

References

1. V. Barnett. The ordering of multivariate data. *Journal Of The Royal Statistical Society Series A*, 139(3):318–354, 1976.
2. G. Birkhoff. *Lattice Theory*, volume 25 of *American Mathematical Society Colloquium Publications.* American Mathematical Society, New York, revised edition, 1948.
3. C. J. Evans, I. D. Svalbe, and R. Jones. Nonlinear image processing using alternative orderings. In B. I. Chaplin and W. H. Page, editors, *DICTA'97 and IVCNZ'97 Proceedings of the first joint Australia and New Zealand biennial conference on: Digital Image & Vision Computing—Techniques and Applications*, pages 177–182, Palmerston North, New Zealand, Dec. 1997. Massey University. ISBN:0-473-04947-3.
4. R. C. Gonzalez and R. E. Woods. *Digital Image Processing.* Addison-Wesley, Reading, Massachusetts, 1992.
5. H. Heijmans. Self-dual morphological operators and filters. *Journal of Mathematical Imaging and Vision*, 6:15–36, 1996.
6. H. Heijmans and C. Ronse. Annular filters for binary images. *IEEE Transactions on Image Processing*, 8(10):1330–1340, 1999.
7. H. J. Heijmans. *Morphological Image Operators.* Academic Press, London, 1994.
8. H. J. Heijmans. On the construction of morphological operators which are self-dual and activity-extensive. *Signal Processing*, 38:13–19, 1994.
9. R. Kresch. Extension of morphological operations to complete semilattices and its applications to image and video processing. In H. Heijmans and J. Roerdink, editors, *Mathematical Morphology and its Applications to Image and Signal Processing*, volume 12 of *Computational Imaging and Vision*, pages 35–42, Dordrecht, The Netherlands, 1998. Kluwer Academic Publishers.
10. J. Serra. *Image Analysis and Mathematical Morphology.* Academic Press, London, 1982.
11. J. Serra, editor. *Image Analysis and Mathematical Morphology. Volume 2: Theoretical Advances.* Academic Press, London, 1988.
12. J. Serra and L. Vincent. An overview of morphological filtering. *Circuits, Systems and Signal Processing*, 11(1):47–108, 1992.

---

[1] Software developed by the Centre for Mathematical Morphology, Paris School of Mines, and marketed by TRANSVALOR S.A., Paris, France.

[2] ftp://ftp.cssip.uq.edu.au/pub/other/fold.zip

FLEXIBLE LINEAR OPENINGS AND CLOSINGS

MICHAEL BUCKLEY and HUGUES TALBOT
CSIRO-MIS, Locked Bag 17, North Ryde NSW 1670, Australia
Email: {Michael.Buckley}{Hugues.Talbot}@cmis.CSIRO.AU
Tel: +61 2 9325-3224
Fax: +61 2 9325-3200

Abstract. We present here a new solution to a problem which is commonly encountered in applied image analysis: the problem of extracting *linear features* from an image. Examples of linear features are fibres, fractures and object boundaries. Such features, though linear, are usually not straight – in general they are curved.

A standard morphological tool for extraction of dark linear features is the minimum of straight-line closings. If the linear features are light in colour, the dual, a maximum of straight-line openings, may be used. This approach, however, is not robust to curvature in the linear features, especially if they are narrow.

In this paper we present an efficient algorithm inspired by local shortest-paths which is robust to curvature. This approach yields filters which are morphological closings or openings. Examples show the effect of the new method.

Key words: Morphological Filtering, Minimal Paths, Algorithms.

1. Introduction

This paper deals with the efficient implementation of algebraic openings and closings over specific sets of paths that can be used for image filtering purposes, typically as an intermediary step for segmenting thin linear features in images.

Our goal is to filter images where thin, elongated features in 2-D images are present. Figures 1(a) and 3(a) offer examples. On such images it is often necessary to filter an image so as to remove noise while at the same time not removing linear features which contain important information. Typically the noise which is to be removed consists of small spots or specks.

Such filtering is not easy because linear features are by definition thin. Many line detection algorithms have a low-pass filtering stage which can erase such features altogether, or make assumptions which are not necessarily verified. The typical mathematical morphology approach is to perform a minimum of closings or of openings with line structuring elements oriented in a large number of directions [7], assuming the features are uninterrupted and relatively straight (We refer to this filter as the SLC ("straight line closing") or the SLO ("straight line opening") filter), but when such features are not straight enough or are too thin, they get filtered out as well.

To account for non-strictly straight features, a morphological reconstruction by dilations or erosions can be performed after applying the SLC filter [9]. This approach can lead to other kinds of problems because the reconstruction stage is often hard to constrain properly [5]. In this case, reconstructed features are no longer necessarily narrow.

2. Optimal Path Approaches

For sufficiently well defined problems, a viable alternative to filtering the image to suppress unwanted noise prior to extracting the desired structure, is to search for features in the raw data.

2.1. Global paths

The *global optimal path problem* is the problem of finding paths of optimal cost from one extremity of an image to the other, for example from top to bottom or from left to right. Such paths will cross the whole image, hence the term "global".

When the problem is to find a dark path on a white background, one interesting path-cost criterion is the sum of grey levels along the path. While finding a path to minimise such a criterion may sound like a difficult problem, there is in fact an efficient algorithm that uses generalized distance transforms.

More precisely, let us suppose that we want to find a path P between two regions A and B of a discrete grey level image I. The graph connecting A and B can be the standard 4-connected or 8-connected grid or any given graph. Each vertex p of the graph is associated with a value $v_I(p) \in \mathbb{R}$. We want to find, amongst all paths P joining A and B, that for which

$$C_I(P) = \sum_{p \in P} v_I(p)$$

is minimal. The distance between sets A and B may now be defined as

$$d_I(A, B) = \min\{C_I(P) | P \text{ connects } A \text{ and } B\}.$$

It can be proved that p belongs to a minimal path P between A and B if and only if

$$d_I(A, \{p\}) + d_I(\{p\}, B) = d_I(A, B)$$

This suggests an efficient algorithm for finding minimal paths between A and B:

1. Compute $d_I(A, \{p\})$ for all p in I.
2. Compute $d_I(B, \{p\})$ for all p in I.
3. Sum the two results and find the points of lowest value: they define at least one minimal path between A and B.

Optimal paths are an interesting application of classic dynamic programming techniques to image analysis. Dijskstra [2] initially proposed an algorithm for finding minimal paths on graphs. More recently Rosen and Vincent [4] used a similar algorithm for images and applied it to feature extraction. Buckley and Yang [1], as well as Gruen and Li [3] introduced a series of regularity criteria to global paths.

2.2. Local paths

However, global optimal paths are useful tools only in a limited numbers of cases. The method is for example not generally applicable to the segmentation of randomly oriented short linear features in an image.

Recently, Vincent [10] introduced local optimal paths to overcome this sort of limitation. The idea is to consider, for a given positive value of L, paths of length L only. In global shortest-paths a cone $C_\theta^{\delta\theta} = [\theta - \delta\theta, \theta + \delta\theta]$ is notionally placed with its point at each pixel (x, y). The path-cost is minimized over all paths which (1) lie within $C_\theta^{\delta\theta}$, and (2) connect (x, y) with the image boundary.

For a given length L and $\delta\theta$, θ can be varied to produce a number of images I_θ. The minimum among the $I_\theta(x, y)$ provides an indication of the direction and value of the minimal path at pixel (x, y), although the minimal path itself is lost.

Vincent actually proposes an efficient parallel/recursive algorithm to implement this on 2-D images for arbitrary search cones, but we shall only here present the simpler case for $\theta = 0, \pi/2, \pi$ or $3\pi/2$, and $\delta\theta = \pi/4$. Vincent's algorithm is laid out in Algorithm 1.

Algorithm 1 (local minimal path)

For $\theta = 0, \pi/2, \pi$ *and* $3\pi/2$, *do:*
    $I_\theta \leftarrow I$
    repeat L *times*
        for i *in* $(-1, 0, +1)$
            $J_i \leftarrow I_\theta$ *shifted by one pixel in direction* $-\theta + i\pi/4$
        $I_\theta \leftarrow I_\theta + \min_i\{J_i\}$
Result: $J \leftarrow \min_\theta\{I_\theta\}$

In this algorithm, I is the original image, the I_θ and J_i are intermediate images, and J is an image holding the final result. In this case the search cones are just right-angled triangles of height L.

Figure 1 shows an example of the application of this algorithm to an image of pavement in which cracks are present. The cracks need to be detected for road maintenance reasons. Fig. 1(a) is the original image and Fig. 1(b) is the image filtered using Vincent's algorithm with $L = 20$. We can see how the algorithm filtered out a lot of the noise but also how it blurred the thin cracks. In the rest of the paper, we will show how to adapt Vincent's algorithm into a closing, which will not have this effect, while retaining some of this algorithm's properties, such as its ability to follow a non-straight path. The result of the new method, applied to the same image and with the same length parameter $L = 20$, is shown in Fig. 1(c).

For simplicity, in the rest of the paper we describe the closing only; the flexible linear opening can trivially be obtained by grey-level complementation.

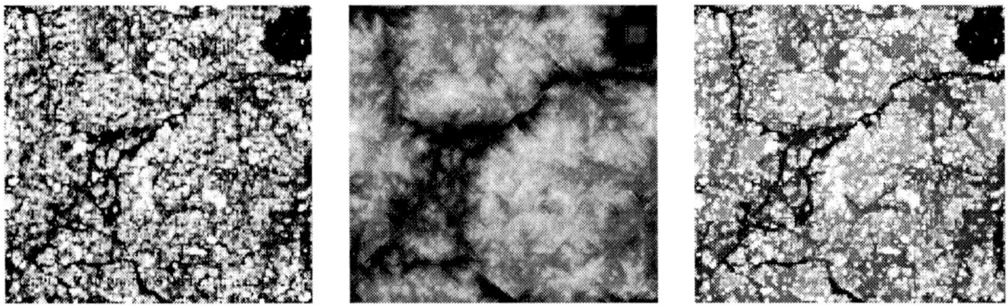

(a) (b) (c)

Fig. 1. An image of cracks in road pavement (a). Vincent's local minimal path algorithm (b), flexible linear closing (c).

3. Flexible Linear Closings

In this section we present the filtering that we want to achieve and we introduce an interesting decomposition that gives rise to an efficient algorithm.

3.1. Collections of structuring elements

The SLC (straight-line closing) filter uses straight-line structuring elements (SEs) such as that shown in Fig. 2(a). For a given length L there are effectively $2L$ such digital straight-lines at different angles [6]. The SLC filter computes closings with each of these SEs and returns the pixel-wise minimum of all of the closings.

Consider the flexible line segments shown in Fig. 2(b)-(e). Like Fig. 2(a), these objects are digital lines made up of $L = 10$ pixels. They are not *straight* lines, however. The idea of the *flexible linear closing* (FLC) filter is to compute the minimum of closing with a collection of SEs such as these. For a given value of the length parameter L, there are many more flexible linear segments than straight linear segments. Whereas as we have noted there are in practice only $2L$ different straight linear segments of length L, we will consider a collection of flexible linear segments of length L whose size is $O(3^L)$. By exploiting recursive structure in this collection of SEs, following the general approach of Vincent [10], we can compute the minimum of this very large collection of closings in the same order of computation time as that required by the SLC, namely $O(L)$. The FLC will preserve dark linear features if they are at least L pixels in length, even if they are narrow and curved.

However, it is possible that a collection of SEs may be *too* rich and therefore *too* flexible. Consider, for example, the SE shown in Fig. 2(f). It is, in some sense, too "curly". If SEs such as these were included, structures less than L pixels in diameter might be preserved. We define the collection of SEs used in the FLC filter as the union of four separate groups of SEs: the vertical group (V_L), the horizontal group (H_L), the forward diagonal group (D_L^+) and the backward diagonal group (D_L^-).

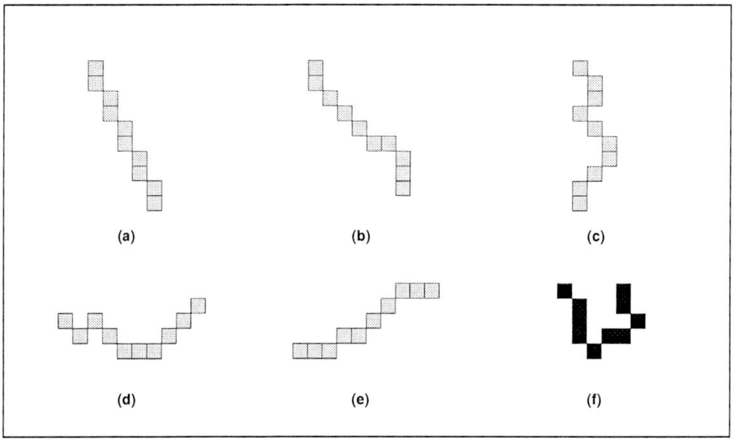

Fig. 2. Linear segments used as structuring elements (a-e). Unsuitable structuring element (f).

For a given length parameter L, all SEs considered have exactly L pixels and are 8-connected. To be included in group V_L a SE B must have one further property: the row indices of the points in B must all be distinct. Note that 8-connectivity implies that in this case the i-values must be a set of L contiguous integers. In Fig. 2, segment (c) is a member of V_{10}, as is the straight segment (a). Similarly group H_L contains SEs whose column indices j are distinct; SEs (d) and (e) are members of H_{10}.

The forward diagonal group D_L^+ is defined in terms of the difference, $i - j$, between the row and column indices. A SE B is a member of D_L^+ if the collection $\{i - j \mid (i,j) \in B\}$ contains no duplicates. (We assume that i is increasing downwards and j is increasing to the right.) SE (e) is a member of D_{10}^+. In similar fashion D_L^- is defined by the property that the values of $i + j$ for points $(i,j) \in B$ must be distinct. SEs (a) and (b) are in D_{10}^-.

Note that SE (f) is in none of the four groups, that the four groups are not disjoint, and that there is a relationship between these groups and Vincent's search cones: all the SEs belonging to V_L are contained in the union of two vertical cones of opening 90°and height L, one going up and the other going down, intersecting at their single-pixel extremity, when the origin of the SE is placed at the intersection of both cones. A similar result holds for horizontal and diagonal cones.

For simplicity, in the following, we only consider the discrete and finite case, and so we use the min/max notation rather than infimum/supremum ($\wedge/\vee$). We also denote the set of flexible linear structuring elements as:

$$\mathcal{P}_L = V_L \cup H_L \cup D_L^+ \cup D_L^- \quad (1)$$

and the flexible linear closing as:

$$\Phi_{\mathcal{P}} = \min_{B \in \mathcal{P}} \phi_B. \tag{2}$$

3.2. Decomposition of the problem

We begin by noting that, given the decomposition (1) of $\mathcal{P}_L$, we have

$$\Phi_{\mathcal{P}_L} = \min\{\Phi_{V_L}, \Phi_{H_L}, \Phi_{D_L^+}, \Phi_{D_L^-}\}. \tag{3}$$

We now describe a procedure which efficiently computes $\Phi_{V_L} = \min_{B \in V_L} \phi_B$. Only slight modifications are required for the computation of Φ_{H_L}, $\Phi_{D_L^+}$ and $\Phi_{D_L^-}$. These are then combined via pixel-wise minimum to obtain $\Phi_{\mathcal{P}_L}$.

Let us first introduce a general theorem on closing decomposition.

Theorem 1 (closing decomposition) *If B_x is any flat structuring element with origin x and f is a function of $\mathbb{Z}^n$ with values in a finite subset of $\mathbb{Z}$, then*

$$\phi_B(f)(z) = \min_{x \in B} \max_{y \in B} f(z + y - x) \tag{4}$$

The result follows directly from standard morphological equations. Combining Theorem 1 with (2) we have the following definition of the filter Φ_{V_L}, applied to an image f and evaluated at an arbitrary point z, in terms of maxima and minima:

$$\Phi_{V_L}(f)(z) = \min_{B \in V_L} \min_{x \in B} \max_{y \in B} f(z + y - x). \tag{5}$$

Let us define the cost $C_f(B) = \max_{x \in B} f(x)$ (Note that in Vincent's algorithm, it would be $\sum_{x \in B} f(x)$). Let $\mathcal{A}_L = \{P = B_x \mid B \in V_L \text{ and } x \in B\}$ be the set of all translations B_x which include the origin of members B of V_L. (Note that $0 \in B_x$ if and only if $x \in B$.) Equation (5) becomes

$$\Phi_{V_L}(f)(z) = \min_{P \in \mathcal{A}_L} C_f(P_{-z}). \tag{6}$$

That is, the value of the vertical flexible linear closing $\Phi_{V_L}(f)$ at point z is the minimum path cost over all paths in V_L which pass through z. The key step in the development of the algorithm for computation of the FLC is the following decomposition of the set $\mathcal{A}_L$ of paths. Essentially we partition this set according to the row positions of the uppermost points in each path. If P is any member of $\mathcal{A}_L$, then (1) P is a vertical path of length L (i.e. a member of V_L), and (2) the origin $(0,0)$ is an element of P.

Now let us break P into two pieces, with the break at the origin and the origin itself included in both parts. Let P^d be the upper part of P and P^u the lower part, so $P = P^d \cup P^u$.

The path P^d is a vertical path of length $l, 1 \leq l \leq L$ whose lowest point is the origin, that is, a length-l path proceeding upwards from the origin. Let Q_l be the set of all such paths, so $P^d \in Q_l$. Similarly let $\tilde{Q}_l$ be the set of all length-l paths proceeding downwards from the origin, so $P^u \in \tilde{Q}_{L-l+1}$. Our arbitrary path P is therefore a member of the collection

$$\mathcal{B}_L = \bigcup_{1\leq l\leq L} \{P^d \cup P^u \mid P^d \in Q_l \text{ and } P^u \in \tilde{Q}_{L-l+1}\} \tag{7}$$

so $\mathcal{A}_L \subset \mathcal{B}_L$. We do not have the room to expand the full proof here, but the converse is also true and $\mathcal{A}_L$ and $\mathcal{B}_L$ are in fact identical.

With this, (6) becomes

$$\Phi_{V_L}(f)(z) = \min_{P\in\mathcal{B}_L} C_f(P_{-z}) \tag{8}$$

which simplifies to

$$\Phi_{V_L}(f)(z) = \min_{1\leq l\leq L} \max\left\{D_f^l(z), \tilde{D}_f^{L-l+1}(z)\right\} \tag{9}$$

where $D_f^l(z) \equiv \min_{P^d\in Q_l} C_f(P_{-z}^d)$ is the minimal l-step upwards distance from z using a maximum-of-grey-values metric, and $\tilde{D}_f^l(z) \equiv \min_{P^u\in\tilde{Q}_l} C_f(P_{-z}^u)$ is the minimal l-step downwards distance. These terms may be computed for all points z in the image and all l, $1 \leq l \leq L$ using a simple modification of Vincent's local shortest-path algorithm. These are then combined via (9) to produce $\Phi_{V_L}(f)(z)$ for all points z in the image domain.

3.3. Algorithm

An algorithm for computing $\Phi_{V_L}(f)(z)$ is shown in Algorithm 2. The algorithms for computing Φ_{H_L}, $\Phi_{D_L^+}$ and $\Phi_{D_L^-}$ are similar, and must be combined by Eq. 3 to compute the FLC. The relationship to Vincent's algorithm is straightforward. This algorithm is only $O(L^2)$, but it can be made more efficient by remarking that the J_T and J_B are computed multiple times. By reordering the loops and taking advantage of the recursive implementation, a $O(L)$ algorithm that uses L times more memory can be derived. Due to lack of space we cannot present these details here.

4. Application

Figure 3 shows the result of a segmentation of a star field from the Hubble Space Telescope used for the detection of asteroids. Asteroids are close to earth as compared to distant stars and the parallax due to the motion of the HST around its orbit create curved paths on long exposures where asteroids are present. It is essential to distinguish curved trails from straight artifacts due to bright stars or cosmic rays. This is achieved using an FLO (flexible linear opening) to extract both curved and straight features, and a conventional SLO (straight line opening) followed by morphological reconstruction to extract the straight paths only. Note that this filter does reconstruct the asteroid trail as well (Fig 3(c)) but not as sharply as the FLO (Fig 3(b)). Thresholding the difference yields the desired result.

Algorithm 2 (vertical flexible linear closing)

For $0 \leq m \leq (L-1)$ *do:*
 For $\theta = \pi/2$, *do:*
  $I_\theta \leftarrow I$
  repeat m *times*
   for i *in* $(-1, 0, +1)$
    $J_i \leftarrow I_\theta$ *shifted by one pixel in direction* $-\theta + i\pi/4$
   $I_\theta \leftarrow \max(I_\theta, \min\{J_i\})$
 $J_T \leftarrow \min\{I_\theta\}$
 For $\theta = 3\pi/2$, *do:*
  $I_\theta \leftarrow I$
  repeat $L - m + 1$ *times*
   for i *in* $(-1, 0, +1)$
    $J_i \leftarrow I_\theta$ *shifted by one pixel in direction* $-\theta + i\pi/4$
   $I_\theta \leftarrow \max(I_\theta, \min\{J_i\})$
 $J_B \leftarrow \min\{I_\theta\}$
 $J^v_m = \max(J_T, J_B)$
Result: $J^v \leftarrow \min_m\{J^v_m\}$

5. Discussion and Future Work

As presented, the algorithm performs a similar task to the straight-line closing followed by morphological reconstruction, but is different enough to be of interest in some applications. A flexible linear closing will always remove less from the input image than the corresponding straight-line closing, but there is no obvious order relationship between the FLC and the SLC followed by morphological reconstruction (as illustrated in Fig 3(d)).

The algorithm presented is also translation invariant, whereas it is quite difficult to implement an efficient translation invariant SLC [8].

The algorithm as presented works only for paths which are included in 90°cones. For some applications, it might be interesting to restrict the set of paths to narrower cones (for example 45°or less), i.e. straighter paths. Vincent [10] actually presented a more complex algorithm for local path optimisation on such cones, but we have not yet adapted his algorithm to our problem, as the implementation is significantly harder and the benefits are not obvious.

It would also be interesting to apply the entire algorithm to better approximations of the circle than just squares. The paths would then all have approximately the same Euclidean length, which could be useful for some applications involving rotation invariance. It is possible to link this enhancement to the use of narrower cones in a relatively straightforward manner, but it is less obvious how to do so in the general case (of an arbitrary angular subset of an arbitrary regular polygon), without giving up the recursive nature of the algorithm, and therefore much of its efficiency.

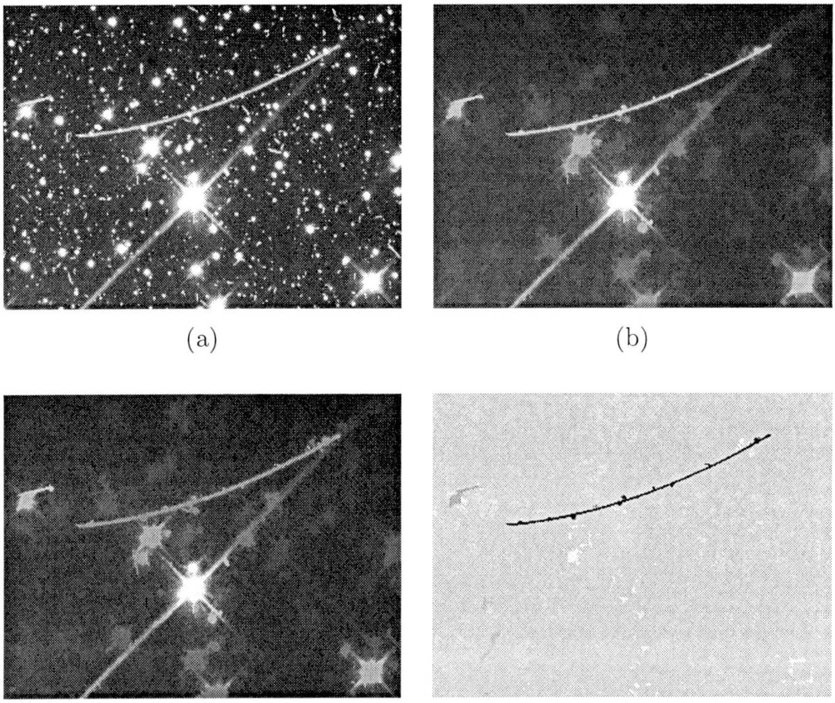

(a) (b) (c) (d)

Fig. 3. Application of flexible linear opening: a star field containing an asteroid trail (a). FLO of length 40 (b). SLO of length 10 followed by morphological reconstruction (c). (d) shows the difference (c)-(b) with the 0 level as mid-grey. Original image courtesy NASA/JPL/HST.

Another limitation is that there is no constraint on the smoothness (regularity) of the paths. One can imagine situations where limiting the proposed algorithm to smoother paths would be useful. Buckley and Yang [1] have done some work on regularised global shortest path, and it remains to be seen if this work can be adapted to ours.

This work can also be adapted to 3-D images with few difficulties other than practical implementation, but it hasn't been carried out yet.

6. Conclusion

We have presented an algorithm to perform a new series of transforms: the flexible linear closings and openings. The algorithm has polynomial complexity and is reasonably easy to implement. These transforms can be used to filter out noise in images while retaining thin, linear, but not necessarily perfectly straight features. These features can then for example be segmented more easily.

References

1. M. Buckley and J. Yang. Regularised shortest paths extraction. *Pattern Recognition Letters*, 18:621–629, 1997.
2. E. Dijkstra. A note on two problems in connexion with graph. *Numerische Mathematik*, 1:269–271, 1959.
3. A. Gruen and H. Li. Road extraction from aerial and satellite images by dynamic programming. *ISPRS Journal of Photogrammetry and Remote Sensing*, 50(4):11–20, 1995.
4. B. Rosen and L. Vincent. Morphological image processing techniques applied to detection of correlogram tracks. *U.S. Navy Journal of Underwater Acoustics*, 44(2):571–586, April 1994.
5. P. Salembier and J. Serra. Flat zone filtering, connected operators and filters by reconstruction. *IEEE Transactions on Image Processing*, 3(8):1153–1160, August 1995.
6. P. Soille. Grey scale convex hulls: definitions, implementations and applications. In *Mathematical Morphology and its Applications to Image and Signal Processing*, volume 12 of *Computational Imaging and Vision*, pages 83–90, Amsterdam, June 1998. Kluwer. Proceedings of ISMM'98.
7. P. Soille, E. Breen, and R. Jones. Recursive implementation of erosions and dilations along discrete lines at arbitrary angles. *IEEE Transactions on Pattern Analysis and Machine Intelligence*, 18(5):562–567, May 1996.
8. P. Soille and H. Talbot. Directional morphological filtering. Submitted.
9. L. Vincent. Morphological grayscale reconstruction in image analysis: Applications and efficient algorithms. *IEEE Transactions on Image Processing*, 22(2):176–201, 1993.
10. L. Vincent. Minimal path algorithms for the robust detection of linear features in images. In *Mathematical Morphology and its Applications to Image and Signal processing*, volume 12 of *Computational Imaging and Vision*, pages 331–338, Amsterdam, June 1998. Kluwer. Proceedings for ISMM'98.

SOME APPLICATIONS OF APERTURE FILTERS

R. HIRATA JR., EDWARD R. DOUGHERTY* and
JUNIOR BARRERA
Universidade de São Paulo, Departamento de Ciência da Computação
CP 66281 - CEP 05389-970 - São Paulo, SP, Brazil
email: hirata@ime.usp.br, jb@ime.usp.br

Abstract. Aperture filters is a class of non-linear filters that are translation invariant and locally defined in both space and range. The optimal design of a filter in this class is practically possible because one can impose the size of both the space and range window, and thereby greatly reduce the number of parameters to be estimated in filter design. This is accomplished without overly affecting the probability mass that would be used for full optimization. This paper provides several real-world applications of aperture filters, such as automatic extraction of markers, deblurring, and resolution enhancement.

Key words: Optimal Filter, Aperture Filter, Non-Linear Filter.

1. Introduction

There are two fundamental problems with designing nonlinear operators (filters) by computational learning or statistical optimization [1]. The first is the inordinate amount of data required for estimating conditional probabilities and the second is the algebraic problem involved in reducing the designed operator to a convenient representation. The present paper concerns aperture filters, which are grayscale operators that help overcome the first problem [2]. An aperture filter views the image through an *aperture*, which is the Cartesian product between a domain window and a range window, that is chosen according to the signal values in the domain window. Signal values above and below the range window are projected into the top and bottom of the aperture, respectively. This projection compresses the probability mass of the observed signal into a smaller set of variables in such a way as not to alter the mass of observations within the aperture (which carry the most mass) and minimally alter the mass of those outside the aperture. The aperture reduces greatly the dimensionality of the computational learning problem, while not substantially affecting the distribution of probability mass that would be used for full optimization.

This paper applies aperture filters to three basic image processing tasks: marker detection, deblurring, and resolution conversion. Application of nonlinear filters to deblurring has been difficult owing to the problem of estimating optimal nonincreasing nonlinear filters from image data. An exception has been in binary image processing for documents, where nonincreasing filters have

* Texas A&M University, Department of Electrical Engineering, College Station, TX 77843-3128, email: edward@ee.tamu.edu.

proven very effective. In particular, nonlinear filter design by computational learning has been very successful for resolution enhancement of binary images. Here, using aperture filters, we extend the approach to grayscale images.

2. Aperture Filters

Let E be a non empty set that is an Abelian group with respect to a binary operation denoted by $+$. Elements of E will be denoted by x and z. Let L and M be two subsets of $\mathbb{Z}$. Elements of L or M will be denoted by y. The operations of *maximum* and *minimum* on subsets of L or M will be denoted, respectively, $\vee$ and $\wedge$.

An *image* is a mapping from E to L or from E to M. Images will be denoted by f or g (Fig. 1a). The set of all images from E to L (resp., M) will be denoted $Fun[E,L]$ (resp., $Fun[E,M]$). For any $f \in Fun[E,L]$ and $x \in E$, the translation of f by x is the function f_x given by, for any $z \in E$, $f_x(z) = f(z-x)$ (Fig. 1b). An *image operator* is a mapping from $Fun[E,L]$ to $Fun[E,M]$. Image operators will be denoted by the capital Greek letter Ψ.

Let W be a finite subset of E. An *image window* is a mapping from W to L. Image windows will be denoted by u. The set of all image windows will be denoted $Fun[W,L]$. For any $f \in Fun[E,L]$, the *restriction* to W of f is the image window f/W given by, for any $z \in W$, $(f/W)(z) = f(z)$ (Fig. 1b). The *restriction class* of f to W, denoted $\mathcal{F}_{f/W}$, is the family of images whose restriction to W gives f/W, that is, $\mathcal{F}_{f/W} = \{g \in Fun[E,L] : f/W = g/W\}$ An image operator Ψ is called *spatially locally defined* in the window W if and only if (iff), for any $f \in Fun[E,L]$ and $x \in E$, $\Psi(f)(x) = \Psi(g)(x), \forall g \in \mathcal{F}_{f_{-x}/W}$. An image operator Ψ is called *spatially translation invariant* iff, for any $z \in E$, $\Psi(f_z) = \Psi(f)_z$. An operator that is both spatially locally defined in W and spatially translation invariant is called a *W-operator.*

A *characteristic function* is a mapping from $Fun[W,L]$ to M. Characteristic functions will be denoted by lower case Greek letters. When L and M are finite sets the characteristic functions will be also called *computational functions.* An important property of the W-operators is that they can be characterized uniquely by the characteristic functions, that is, there exists a bijection between the set of W-operators and the set of characteristic functions [3] [4]. The characterization of a W-operator Ψ by its characteristic function ψ is given by, for any $f \in Fun[E,L]$ and $x \in E$, $\Psi(f)(x) = \psi(f_{-x}/W)$. From now on, to simplify the presentation, we will consider just images with infinite range, that is, $L = M = \mathbb{Z}$.

Let $u \in Fun[W,\mathbb{Z}]$ and $y \in \mathbb{Z}$. The grayscale *translation* of u by y is given by, for any $z \in W$, $(u+y)(z) = u(z) + y$ (Fig. 1c). The subset $K \subset \mathbb{Z}$, $K = \{-k,\ldots,k\}$, where k is a positive integer, will be called a *grayscale*, or *range*, *window.* The *windowing* of u at y by the grayscale window K is the function $u/K_y \in Fun[W,K]$ given by, for any $z \in W$, $(u/K_y)(z) = \bigwedge\{\bigvee\{-k, u(z)-y\}, k\}$ (Fig. 1d, the shaded points are the projections of $u - u(o)$ into the range window K).

A characteristic function ψ from $Fun[W,\mathbb{Z}]$ to $\mathbb{Z}$ is called *locally defined* in

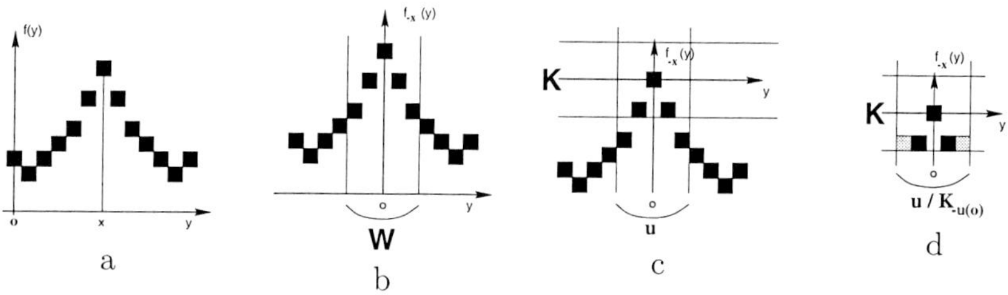

Fig. 1. (a) $f : Fun[E, \mathbb{Z}]$, (b) f_{-x}, (c) $u - u(o)$, (d) $u/K_{-u(o)}$

K iff, for any $u \in Fun[W, \mathbb{Z}]$, $\psi(u) = u(o) + \beta_{u(o)}(u/K_{u(o)})$, where $u(o)$ is the image of u at the origin and β_y, for any $y \in \mathbb{Z}$, is a computational function from $Fun[W, K]$ to K; ψ is called *translation invariant* iff, for any $y \in \mathbb{Z}$ and $u \in Fun[W, \mathbb{Z}]$, $\psi(u + y) = \psi(u) + y$. A characteristic function is called a *K-characteristic function* if it is both locally defined in K and translation invariant.

Theorem 1 *A K-characteristic function ψ is characterized by, for any $u \in Fun[W, \mathbb{Z}]$, $\psi(u) = u(o) + \beta_\psi(u/K_{u(o)})$, where β_ψ is a computational function from $Fun[W, K]$ to K.*

An operator Ψ characterized by a K-characteristic function ψ is called an *aperture filter* and is denoted $\Psi_{\mathcal{A}}$. The morphological representation of aperture filters have been treated previously (under the name of WK-operators) [5].

Definition 1 *An aperture filter $\Psi_{\mathcal{A}}$ is a mapping from $Fun[E, \mathbb{Z}]$ to $Fun[E, \mathbb{Z}]$, given by $\Psi_{\mathcal{A}}(f)(x) = u(o) + \beta_\psi(u/K_{u(o)})$, where $u = f_{-x}/W$.*

The definition above can be extended in two ways: one can use a computational function $\beta : Fun[W, K]$ to K', $K' \subset \mathbb{Z}$, $K' = \{-k', \ldots, k'\}$; and one can translate the function f by a function $s(u)$ (instead of $u(o)$), where $s(u)$ is a mapping from $Fun[W, \mathbb{Z}]$ to $\mathbb{Z}$ (which determines the positioning of the aperture).

The statistical design of aperture filters from image realizations requires estimation of the characteristic functions from sample image data (pairs of observed and ideal images) and their representation in an efficient computational form. The statistical design of aperture filters and its computational representation has been treated in a previous paper [2].

3. Markers Detector

The morphological segmentation method developed by Beucher and Meyer is useful to find the borders of specified objects [6]. The power of the method comes from the fact that it simplifies the segmentation process by reducing it

Fig. 2. (a) Observed, (b) ideal and (c,d) examples of application

to the problem of finding markers for the specified objects [7]. However, finding markers may not be easy. In this application, aperture filters are used in conjunction with the Beucher-Meyer paradigm to facilitate automatic segmentation.

Figure 2a shows part of an image of a face from which the objective is to segment the eyes. The images have about 500×500 points and the database has images of the same face at different angles of observation. In the experiment, several aperture filters have been trained with different window sizes and ranges using Figs. 2a and 2b as the observed and ideal images, respectively. The ideal image has been produced using the segmentation paradigm from human chosen markers. The result of the segmentation is a binary image labeled 255 for the eyes and 0 elsewhere. For this training pair, a good aperture is ($3\times5\times10$, 10) (width×height× k, k') positioned on the median (see [2] for details about the aperture positioning). Figures 2c and 2d show two results obtained by the proposed method (superposed to the original image). The results of the application of the designed operator have been filtered with morphological *area-open* filter to clean small statistical errors (small connected components). Nine images were used for testing. The result of the segmentation is correct for six images, partially correct for 2 images (missed one of the eyes) and one failure (missed both eyes).

An extension of this technique has been applied with success to segment objects in a sequence of images for video edition [8]. Figures 3a and 3b show one of the training pairs used to segment the racket and figures 3c and 3d show two examples of application of the method combined with the original image [aperture ($3\times3\times10$,10)].

4. Deblurring

Aperture filters have been shown to be a powerful tool for deblurring [2]. In those experiments, images of a random Boolean function model [9] whose primary function is pyramidal (Fig. 4a) were blurred by a (3×3) nonflat convolution kernel and digitized (Fig. 4b). The objective was to design an aperture filter to deblur images of that kind. Ten training and ten test images were used, each of size 256×256 points. Six apertures sizes were tested and we observed the decreasing of MAE (mean absolute error) and MSE (mean square

a b c d

Fig. 3. (a,b) Training pair and (c,d) two application images.

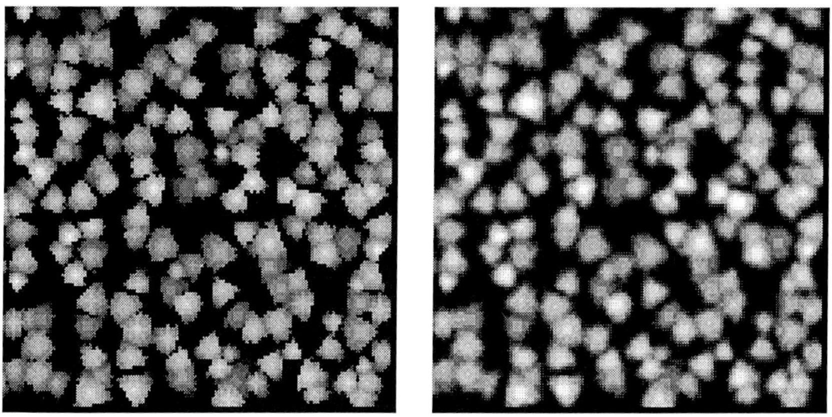

a b

Fig. 4. Part of a random Boolean function: (a) original; (b) blurred.

error) errors as the number of training examples increase. The increasing in the size of the aperture yields decreasing of MAE (Fig. 5) and MSE errors if a sufficient number of training examples are given. The designed aperture filters have been compared to optimal linear filters of up to 7×7 points (label 7×7 in Fig. 5). Figures 6a and 6b show part of a test image and the result from the application of the aperture. The main difference between the filters (linear and nonlinear) is that the aperture filters preserve the structure of the edges, while the linear filters do not.

There are clear theoretical reasons for aperture filters to outperform linear filters. For a given window W, the unconstrained optimal filter Ψ_{opt} will outperform the optimal aperture filter, Ψ_{aper}, and the optimal linear filter, Ψ_{lin}, because the later two represent constraints on optimality. The reason we do not use Ψ_{opt} is that the error of estimating Ψ_{opt} from data is too great. While the class of aperture filters does not include the class of linear filters, optimizing

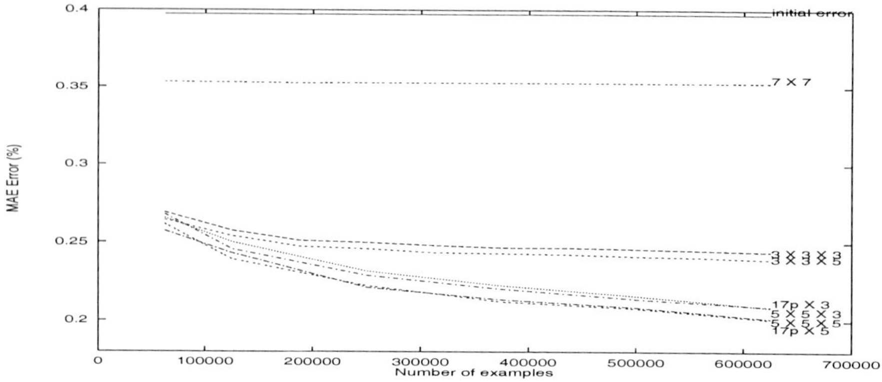

Fig. 5. MAE plots for deblurring random Boolean functions.

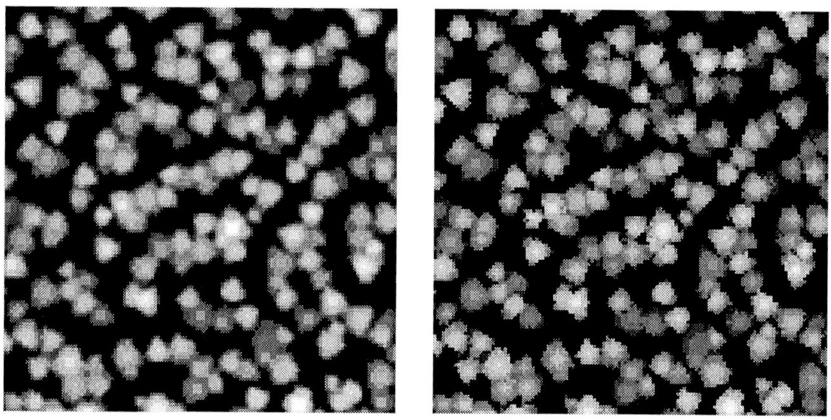

a b

Fig. 6. (a) Test image; (b) Result of the aperture applied on the left image.

over an aperture tends to be less constraining than requiring linearity, albeit, at the cost of increased estimation error for filter design. The great advantage of Ψ_{aper} over Ψ_{lin} is that, for observations strictly within the aperture, there is no constraint and therefore we get maximum structural restoration. This is similar to secondarily constrained binary filters [10]. Efficient structural constraint does not occur for linear filters, and this improved structural restoration for aperture filters can offset the training advantage of linear filters for sufficiently large data samples. A second reason aperture filters (and nonlinear filters, in general) are superior to linear filters is that digital signals are quantized, so that there really are no linear digital-signal filters, and the quantization of the optimal linear filter can have serious negative effects for insufficient quantization.

In another set of experiments, the objective is to deblur a 430 × 430 satel-

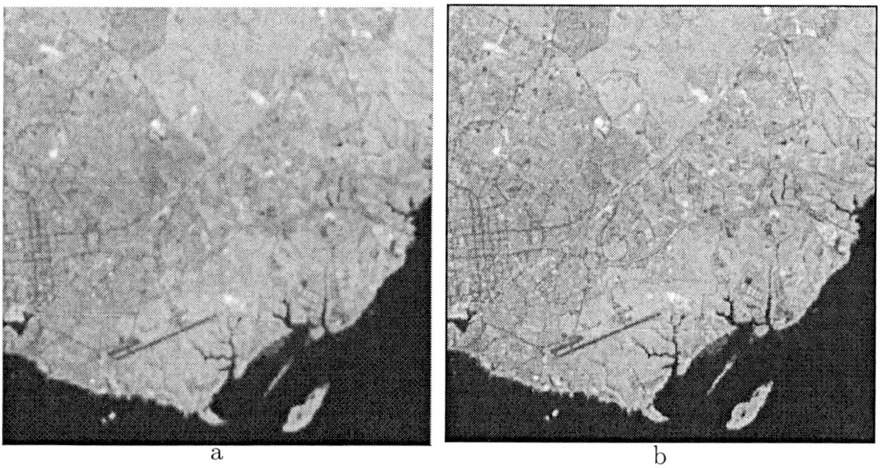

Fig. 7. (a) CBERS simulation and (b) resulting of the aperture filtering.

lite image (SPOT 4) blurred with a non-flat convolution kernel to simulate the image that would be seen by the new Chinese-Brazilian satellite (CBERS, Fig. 7a). A total of 35 experiments have been done using 7 different apertures $(W \times k,\, k')$: $(W \times 5, 20)$, $(W \times 5, 25)$, $(W \times 5, 30)$, $(W \times 10, 20)$, $(W \times 10, 25)$, $(W \times 10, 30)$, $(W \times 15, 15)$ and $(W \times 15, 30)$; where $(W \times k,\, k')$ gives the dimensions of the aperture (W being the five-point cross dilated by itself), and k and k' as defined in section 2. Since there is only one image available, the operators have been designed using 17629, 33566, 48130, 61101, and 72933 random points of the image. The whole image has been used for testing the designed operators. Figure 7b shows the resulting image for aperture size $(W \times 15, 30)$ using 72933 examples.

5. Integer Multiresolution Enhancement

Multiresolution enhancement is an important application for the printer industry because documents generated at a certain low resolution should not appear blocky in a high resolution printer [11]. The printer software has to enhance the image before printing. In this section, we show an experiment where the objective is to design an operator to enhance the resolution of an image digitized at 75 dpi (Fig. 8a) in order to estimate the image digitized at 150 dpi (as Fig. 8b). A simple way to shrink a 150 dpi image to 75 dpi is to choose, for each four points of the large image, one point and discard the other three. If this is done in a coherent form, then we obtain four different images (phases) of 75 dpi (which can be regrouped to compose the 150 dpi image again.) Figure 9 shows the whole process. The idea of multiresolution enhancement we applied is to design aperture filters to transform a low resolution image (i.e., the observed image is an image acquired at low resolution) to each of the phases of the same

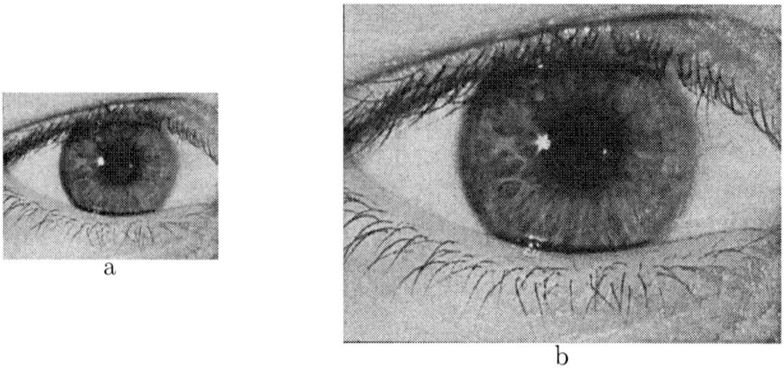

Fig. 8. (a) Eye image at 75 dpi and (b) Eye image at 150 dpi.

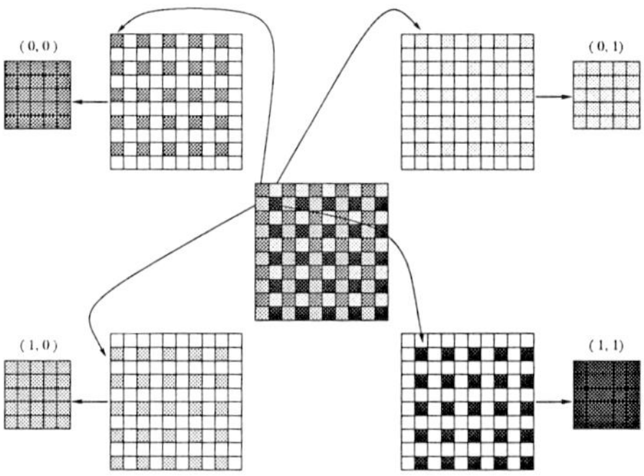

Fig. 9. Sampling an image in all its phases.

image acquired at high resolution (i.e., each sampled phase will be the ideal image used to estimate the aperture operator for that phase). When we apply the designed operators to images acquired at the same low resolution (similar to the ones we used for training), we end up with images of what is expected to be the phases of the ideal high resolution image. The process ends regrouping the images to get an enhanced image at a high resolution (following the arrows of Fig. 9 backwards). The result is expected to be good because the differences between observed and the phases of the ideal image are not large. Since the ratio of the resolution is an integer (in the example, 2, because of the 75 to 150 dpi conversion), the enhancement is said to be *integer-conversion.* A similar idea can be used to non-integer resolution conversion [11] (for instance, 80 to 120 dpi). Figure 10 shows the subsamplings resulting from sampling Fig. 8b. The images are labeled by the origin point where the sampling started. Four operators, Ψ_1, Ψ_2, Ψ_3 and Ψ_4, going to each of the four subsampled images, have been designed using the right half of the 75 dpi image (Fig. 8a) for train-

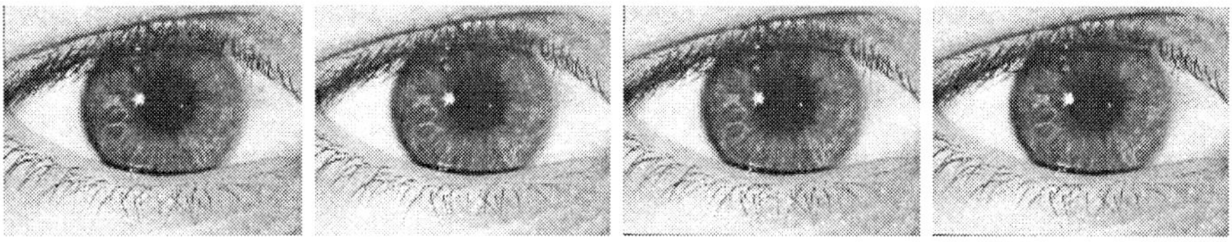

Fig. 10. Phases (0,0), (0,1), (1,0) and (1,1) of the 150 dpi image.

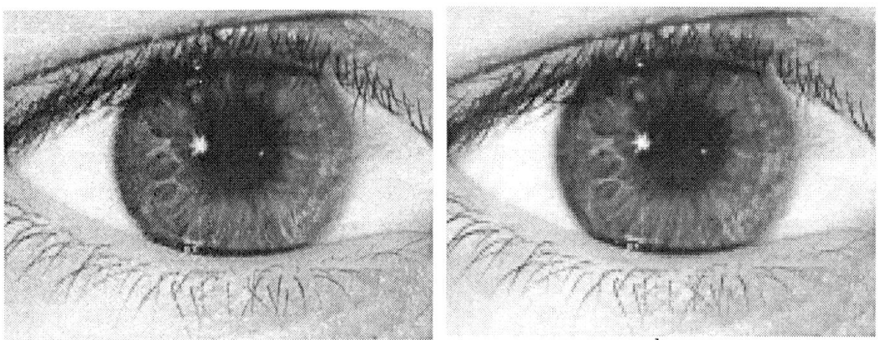

a b

Fig. 11. (a) Result for aperture (3×3×10, 30) and (b) Result for simple replication.

ing and the whole image for testing. Ten experiments have been tested with the following apertures ($W \times k$, k'): (3×3×10, 10), (3×3×10, 15), (3×3×10, 20), (3×3×10, 25), (3×3×10, 30), ($W \times 5$, 20), ($W \times 5$, 25), ($W \times 5$, 30), and ($W \times 10$, 30), where W is the cross-window defined previously.

The designed operators have been compared to the 150 dpi image and with two classical expanding algorithms (simple replication and averaging of nearest neighbor pixels.) For the aperture filters, the high resolution images have been put together by regrouping the resulting images from the application of the four operators. One regrouped result for (3×3×10, 25) aperture is shown by Fig. 11a. Figure 11b shows the result of simple replication method. The main differences between the methods are in the textures (they are better reconstructed by the aperture filter) and in the edges (classical methods tend to make them blurred or blocky.)

6. Conclusion

The ability of statistically designed aperture filters to solve some standard grayscale image filtering problems has been experimentally demonstrated. Naturally, the goodness of designed filters depends on the amount of training data and the aperture size needed for satisfactory estimation. Nonetheless, aperture filters can be used to extend the computational learning methods that have been successfully used in the design of nonincreasing binary filters. In the next future the authors plan to compare the aperture filters for multiresolution enhancement with another known linear approach as splines approximation [12] and other nonlinear approaches [13, 14].

Acknowledgements

The authors acknowledge Dr. Artyom M. Grigoryan for the images of the Boolean model and prof. Dr. G.J.F. Banon in conjunction to the National Institute of Spatial Research (INPE-Brazil) for the permission of use of the satellite images. The first author acknowledges support from CNPq.

References

1. J. Barrera, E. R. Dougherty, and N. S. Tomita. Automatic Programming of Binary Morphological Machines by Design of Statistically Optimal Operators in the Context of Computational Learning Theory. *Electronic Imaging*, 6(1):54-67, January 1997.
2. R. Hirata Jr., E. R. Dougherty, and J. Barrera. Aperture Filters. *Signal Processing*, 80(4), To appear.
3. E. R. Dougherty and D. Sinha. Computational Mathematical Morphology. *Signal Processing*, 38:21–29, 1994.
4. E. R. Dougherty and D. Sinha. Computational Gray-Scale Mathematical Morphology on Lattices (A Comparator-based Image Algebra) Part I: Architecture. *Real-Time Imaging*, 1:69–85, 1995.
5. J. Barrera and E. R. Dougherty. Representation of Gray-Scale Windowed Operators. In Henk J.A.M. Heijmans and Jos B.T.M. Roerdink, editors, *Mathematical Morphology and its Applications to Image and Signal Processing*, volume 12 of *Computational Imaging and Vision*, pages 19–26. Kluwer Academic Publishers, Dordrecht, May 1998.
6. F. Meyer and S. Beucher. Morphological Segmentation. *Journal of Visual Communication and Image Representation*, 1(1):21–46, September 1990.
7. R. Hirata Jr., J. Barrera, and R. A. Lotufo. A Tutorial on Image Segmentation via Mathematical Morphology with Khoros. In *Khoros´ 97 proceedings*, pages 192–206. Khoral Research, Khoral Research Inc., Editors, March 1997.
8. R. Hirata Jr., J. Barrera, F. C. Flores, and R. A. Lotufo. Automatic Design of Morphological Operators for Motion Segmentation. In J. Stolfi and C. L. Tozzi, editors, *Proc. of Sibgrapi'99*, pages 283–292, Campinas, SP, Brazil, 1999.
9. M. Schmitt and F. Preteux. *Image Analysis and Mathematical Morphology*, volume 2, chapter Boolean texture analysis and synthesis. Ed. J. Serra, Academic Press, New York, 1988.
10. O. V. Sarca, E. R. Dougherty, and J. Astola. Secondarily Constrained Boolean Filters. *Signal Processing*, 71(3):247–263, December 1998.
11. R. P. Loce and E. R. Dougherty. *Enhancement and Restoration of Digital Documents: Statistical Design of Nonlinear Algorithms.* SPIE - The International Society for Optical Engineering, Bellingham, 1997.
12. Andrew S. Glassner. *Principles of digital image synthesis.* Morgan Kaufmann, 1995.
13. F. Fekri, R. M. Mersereau, and R. W. Schafer. Enhancement of Text Images Using a Context Based Nonlinear Interpolative Vector Quantization Method. In *Proc. of ICIP'98*, 1998.
14. D. A. F. Florencio and R. W. Schafer. Post-Sampling Aliasing Control for Natural Images. In *Proceedings of ICASSP'95*, 1995.

GA OPTIMISATION OF MULTIDIMENSIONAL GREY-SCALE SOFT MORPHOLOGICAL FILTERS WITH APPLICATIONS IN ARCHIVE FILM RESTORATION

NEAL R. HARVEY
Space and Remote Sensing Sciences Group,
Nonproliferation and International Security Division,
Los Alamos National Laboratory
Los Alamos, NM 87544, USA

and

STEPHEN MARSHALL
Dept. of Electronic & Electrical Engineering
University of Strathclyde, Glasgow, G1 1XW, UK

Abstract. A technique using Genetic Algorithms (GAs) for the optimisation of multidimensional grey-scale soft morphological filters is described, which has applications in archive film restoration. It applies specifically to the problem of film-dirt removal in archive film sequences. Because of ageing, scratches and film dirt, much valuable film archive material is unusable. There is therefore a demand for automatic techniques which can remove these film artifacts whilst preserving image structure, fine detail and without introducing motion blur.

The optimisation criterion is based on mean absolute error (MAE). The optimisation is undertaken using genetic algorithms. An artificial training set is constructed from within the archive material by selecting relatively clean areas and transferring examples of noise and dirt to these frames.

Examples of applying the techniques described to a real film sequence, provided by the British Broadcasting Corporation (BBC), are shown.

Key words: Soft Morphology, Genetic Algorithms, Film Dirt, Archive Restoration.

1. Introduction

There has been a growing interest in recent years in the area of archive film restoration due, in part, to the emergence of digital television broadcasting, the growth in video sales and the expansion in the number of television channels. To satisfy the increasing demand for material to fill air time, it is becoming more attractive to offer archive material. Unfortunately, a lot of the available archive material has suffered some form of corruption and requires restoration in order to be made of a sufficient quality for resale or broadcast. Here we describe some of our recent research in which we investigated the application of grey-scale soft morphological filters in the removal of a specific type of corruption, known as *film dirt*, from archive film material.

2. The Film Dirt Problem

Film dirt is a common problem in archive film restoration. Film dirt occurs when foreign particles get caught in the film transport mechanism and damage the film, causing loss of information. This damage appears as "blotches" of random size, shape and intensity. These blotches are non-time correlated (temporally impulsive).

Fig. 1 shows an example of a region of an image extracted from a sequences which has been corrupted with film dirt.

Fig. 1. Region extracted from an image sequence corrupted with film dirt

3. Soft Morphological Filters

Soft morphological filters are a relatively recently introduced class of non-linear filters [1, 2]. Their original definition was related to the class of (standard/structural) morphological filters (discrete flat morphological filters), but they have since been extended to the grey-scale (function processing) case [3]. The idea behind soft morphological filters is to slightly relax the standard definitions of morphological filters in such a way as to achieve robustness whilst retaining most of their desirable properties. Whereas standard morphological filters are based on local maximum and minimum operations, in soft morphological filters these operations are replaced by more general weighted order statistics. The key idea of soft morphological operations is that the structuring element is divided into two parts: the *hard centre* which behaves like the standard structuring element and the *soft boundary*, where maximum and minimum are replaced by other order statistics. This makes the filters behave less rigidly in noisy conditions and makes them more tolerant to small variations in the shapes of the objects in the filtered image.

Just as the fundamental standard morphological operations are dilation and erosion, the fundamental soft morphological operations are *soft dilation* and *soft erosion*. In a manner similar to that of standard morphological operations, the secondary soft morphological operations of *soft opening* and *soft closing* and the tertiary soft morphological operations of *soft open-closing* and *soft close-opening* can be defined.

Before proceeding to the definitions of the soft morphological operations, some other concepts need to be defined:

The *Structuring System* $[b, a, r]$ consists of three parameters; functions a and b, having supports A and B, respectively ($A \subset B$) and a natural number, r, satisfying $1 \leq r \leq |B|$, where $|B|$ is the cardinality of B. Function b is called the *structuring function*, a its (hard) *centre* (A the support of its (hard) centre), $b \backslash a$ its (soft) *boundary* ($B \backslash A$, the support of its (soft) boundary) and r the *order index of its centre* which is also referred to as the *repetition parameter*.

3.1. Fundamental Grey-Scale Soft Morphological Operations

Grey-scale *soft dilation* of a signal f by the structuring system $[b, a, r]$ is denoted by $f \oplus [b, a, r]$ and is defined by:

$$f \oplus [b,a,r](x) = \text{the } r^{th} \text{largest value of the multiset}$$
$$\{r \diamond (f(x-\alpha) + a(\alpha)) | \alpha \in A\} \bigcup \{f(x-\beta) + b(\beta) | \beta \in B \backslash A\} \qquad (1)$$

Grey-scale *soft erosion* of a signal f by the structuring system $[b, a, r]$ is denoted by $f \ominus [b, a, r]$ and is defined by:

$$f \ominus [b,a,r](x) = \text{the } r^{th} \text{smallest value of the multiset}$$
$$\{r \diamond (f(x+\alpha) - a(\alpha)) | \alpha \in A\} \bigcup \{f(x+\beta) - b(\beta) | \beta \in B \backslash A\} \qquad (2)$$

The symbol $\diamond$ represents the repetition operator. That is: $r \diamond x = \overbrace{x, ..., x}^{r \text{ times}}$.

As an extreme case, grey-scale soft morphological operations by the structuring system $[b, a, r]$ reduce to the equivalent standard grey-scale morphological operations by the function b if $r = 1$, or, alternatively, if $A = B$. If $r > |B \backslash A|$, grey-scale soft morphological operations by the structuring system $[b, a, r]$ reduce to the equivalent grey-scale standard morphological operations by the structuring function a.

4. Optimisation of Soft Morphological Filters

Several methods have been described for the optimisation of soft morphological filters. Huttunen et al [4] and Kuosmanen et al [5] describe methods for the optimal choice of (2-D) flat (function-set processing) soft morphological structuring system. These methods do not, however, optimise the choice of soft morphological operation. Harvey [6, 7] described a method for the optimisation of (2-D/spatial) grey-scale soft morphological filters which is able to optimise not only the structuring system, but also the choice of soft morphological operation. In [8] these GA optimisation techniques have been applied to the restoration of archive film material.

In this paper we seek to illustrate the extension of the techniques to the optimisation of grey-scale soft morphological filters in the spatio-temporal domain and to illustrate their performance in the restoration of real corrupted image sequences, provided by the British Broadcasting Corporation.

5. Why extend to the Spatio-Temporal Domain?

Sequences that contain fast motion have always been a problem for archive film restoration methods. The reason that fast motion is a problem is that, temporally, fast motion is very similar to the film dirt. That is, if an object has fast motion it only appears briefly in a single frame, and not in the same position in adjacent frames. Film dirt has very similar temporal characteristics, i.e. it is non-time correlated (temporally impulsive). By extending the filtering from purely spatial to spatio-temporal it is anticipated that the resulting filters will make use of the temporal characteristics and in this way outperform their purely spatial counterparts. However, by careful coding of the filter parameters for genetic algorithm optimisation, the search space for spatio-temporal (3-D) filters will include the filters of lesser dimensions (i.e. 2-D and 1-D).

We do not provide a description of how these parameters are incorporated into a genetic algorithm optimisation strategy. The fundamentals of this technique have been described in a previous ISMM paper [6]. Instead we concentrate on the application of these methods to a "real-world" problem.

The GA will be capable of searching for any 3-D grey-level soft morphological filter which is a combination of four operations from the set { soft erode, soft dilate, do-nothing }, which will use a structuring function (hard centre and soft boundary) and repetition parameter chosen from all the possible variations within the overall region of support and maximum grey-level value, the bounds of which are pre-set. This search space encompasses (3-D) spatio-temporal, 2-D (purely spatial) and 1-D (purely temporal) soft morphological filters. In addition, the class of soft morphological filters encompasses several other classes of non-linear filters including standard morphological filters and rank-order filters.

6. Applying the GA Optimisation Method to the Film-Dirt Problem

In order to make use of a GA in the optimisation of filter parameters, there has to be some method of defining a *fitness value* to an individual chromosome representing a particular set of grey-scale soft morphological filter parameters. A *fitness function* has to be determined which provides some objective measure of the individual's *performance* in its *environment*. This fitness function is crucial to the successful implementation of the GA optimisation technique. The environment, in this case, is the image sequence which is to be filtered. Defining what is meant by performance, however, is a more complicated task. The nature of images is that they are to be viewed by humans. The general idea in the field of image restoration is that of improving the subjective quality of the images when viewed. Unfortunately, there does not exist a simple function which maps subjective image quality to some objective quality criterion. Criteria do exist which provide some objective measure of image quality. The majority of these criteria are based on a comparison with an ideal (uncorrupted) version of the image under consideration, and will contain some modification of signal-to-noise ratio (SNR) or the mean absolute error (MAE) [9].

Generally, in the case of film restoration, it is not possible to perform a comparison with an ideal image sequence, as such a thing does not exist. After all, if a non-corrupted version of the film exists, why bother trying to restore a corrupted version?

One method of addressing this problem is as follows. In most image sequences it is generally possible to find areas of the image which are uncorrupted. It is then possible to artificially *corrupt* this *ideal* image with particles of film dirt extracted from other similar, but corrupted, regions in the image sequence. In this way it is possible to produce the necessary *training set* which will allow the evaluation of a fitness value based on some measure of MAE and/or MSE.

6.1. GA Optimisation of Soft Morphological Filters using a Training Set

As mentioned above, it is generally not all that difficult to produce an artificial training set by finding areas of the image sequence (in a number of separate frames) which are uncorrupted and then artificially corrupting this ideal/reference image sequence with particles of film dirt extracted from other similar, but corrupted, regions in the image sequence. An example of just such a training set is shown below. Figure 2 shows a series of uncorrupted regions extracted from an image sequence and Fig. 3 shows the same sequence after having been artificially corrupted with film dirt. The sequence runs in a "raster scan": i.e. left to right, top to bottom.

6.1.1. *Fitness Function*

Having a training set, i.e. an ideal and corrupted version of the same image sequence, enables the fitness value of an individual (i.e. a particular set of filter parameters) to be based on a comparison of the filtered image sequence (processed with a filter having the parameters represented by the particular individual) with the ideal image sequence. The fitness of an individual is therefore determined as follows:

Let MAE_{max} be the maximum possible MAE for an image (for 8-bit greyscale images MAE_{max} would be 255). Let N be the number of images in the training sequence. Let mae_i by the MAE for the ith image in the filtered, corrupted sequence with respect to the ith image in the ideal sequence. Let $fitness_j$ be the overall fitness of the individual j.

Fig. 2. Uncorrupted regions extracted from image sequence

Fig. 3. Artificially corrupted regions extracted from image sequence

$$fitness_j = 100 \times ((\sum_{i=1}^{i=N}(1 - (mae_i \; / \; MAE_{max})))/N) \tag{3}$$

So, to put this in words, the "interim" fitness for an image in the sequence is a measure of how far away the filtered image is from the worse case (and hence how close it is to the ideal). The fitness value for an individual is then the average of all these interim fitness values over the whole image sequence, expressed as a percentage. A filter capable of *perfectly* restoring an image sequence would then have a fitness value of 100.

The actual "genetic algorithms" used in these techniques were based on what is often referred to as a *simple genetic algorithm*. There is much literature available on this topic, and in the interest of brevity we omit a description and instead refer the interested reader to [10, 11, 12].

7. Application to Real Image Sequences

The GA was run, using the same training set as illustrated above. The GA was set the task of optimising a soft morphological filter with an overall size of structuring function set at 5 × 5 × 3 (i.e. spatial dimensions of 5 × 5 and a temporal dimension of 3. The best filter found is shown in Fig. 4. This filter was then applied to an entire image sequence.

Hard Centre					Soft Boundary				
*	*	*	*	*	*	2	6	2	*
*	*	*	*	*	*	*	2	*	*
*	*	*	*	*	*	*	*	*	*
*	*	*	*	*	*	*	2	*	*
*	*	*	*	*	*	2	6	2	*

*	*	*	*	*	2	6	3	6	2
*	*	*	*	*	3	1	5	1	3
*	*	6	*	*	5	4	*	4	5
*	*	*	*	*	3	1	5	1	3
*	*	*	*	*	2	6	3	6	2

*	*	*	*	*	*	2	6	2	*
*	*	*	*	*	*	*	2	*	*
*	*	*	*	*	*	*	*	*	*
*	*	*	*	*	*	*	2	*	*
*	*	*	*	*	*	2	6	2	*

Repetition Parameter = 39

Filter Sequence: Soft Dilation -- Soft Erosion (i.e. Soft Closing)

Fig. 4. Best filter found using GA

Fig. 5 shows an examples of a region extracted from a sequence of images corrupted with film dirt, together with the same region after having been filtered with the grey-scale soft morphological filter found using the GA. The upper example of the pair is the corrupted version and the lower example is the filtered version.

Fig. 5. Region extracted from image corrupted with film dirt and the same region after filtering with the spatio-temporal grey-scale soft morphological filter found using the GA

8. Discussion

It can be seen that filter found has some aspects which would be expected of a suitable filter. For instance, the hard centre has a support of 1 pixel and that pixel is the origin of the structuring function (i.e. the pixel under consideration). So, the output of the filter is weighted towards the input pixel value. Also, the filter sequence found is *soft dilation - soft erosion* or *soft closing*.

This might be expected for the removal of "dark" artefacts within the image. However, due to the nature of soft morphological filters, the relationship between filter parameters and their effects on the image are not quite as intuitive as for "standard" morphological filters.

In general the results depicted here show that the filter found has excellent performance in attenuating/removing film-dirt from image sequences and has little, if any, effect on the image detail.

In spite of the filter being optimised for a small, artificially created training set, the filter still performs well when applied to the entire image sequence. The performance of the filter found can be further illustrated when we apply the same filter to totally different image sequences.

8.1. Application to Out-of-Training-Sample Data

Figs. 6 shows the results of applying the filter found using the GA and training set described above to entirely different image sequences. It can be seen that the filter is still able to perform extremely well with respect to its ability to remove film dirt and retain important image detail.

9. Conclusions

A technique for the optimisation of multi-dimensional grey-scale soft morphological filters has been developed which is able to optimise filters with respect to a criterion based on mean absolute error. This criterion necessitates the creation of an artificial training set. However, it has been shown that this is not an overly burdensome task. It has also been shown that the filter found during optimisation has excellent performance for data sets other than that used in the optimisation, provided the training set contains corruption which is sufficiently representative of the corruption in the other sequences to be restored.

References

1. L. Koskinen, J. Astola, and Y. Neuvo. Soft Morphological Filters. In *Proc. SPIE Symposium on Image Algebra and Morphological Image Processing II*, pages 262–270, San Diego, USA, July, 1992.
2. P. Kuosmanen. Soft Morphological Filtering. Ph.D. Thesis, Dept. of Mathematical Sciences, University of Tampere, Finland, 1993.
3. C. C. Pu and F. Y Shih. Soft Mathematical Morphology: Binary and Grey-Scale. In *Proc. International Workshop on Mathematical Morphology and its Application to Signal Processing*, pages 28–33, Barcelona, Spain, May 1993.
4. H. Huttunen, P. Kuosmanen, L. Koskinen, and J. Astola. Optimization of Soft Morphological Filters by Genetic Algorithms. In *Proc. of Image Algebra and Morphological Image Processing V*, San Diego, USA, July 1994.
5. P. Kuosmanen, P. Koivisto, H. Huttunen, and J. Astola. Optimization of Soft Morphological Filters under Shape Preservation Criteria. In *Proc. Image Algebra and Morphological Image Processing V*, San Diego, USA, July 1994.
6. N. R. Harvey and S. Marshall. Grey Scale Soft Morphological Filter Optimisation by Genetic Algorithms. In P. Maragos, P. Schafer, and R. Butt, editors, *Mathematical Morphology and its Applications to Image and Signal Processing*, pages 179–186, Kluwer Academic Publishers, 1996.

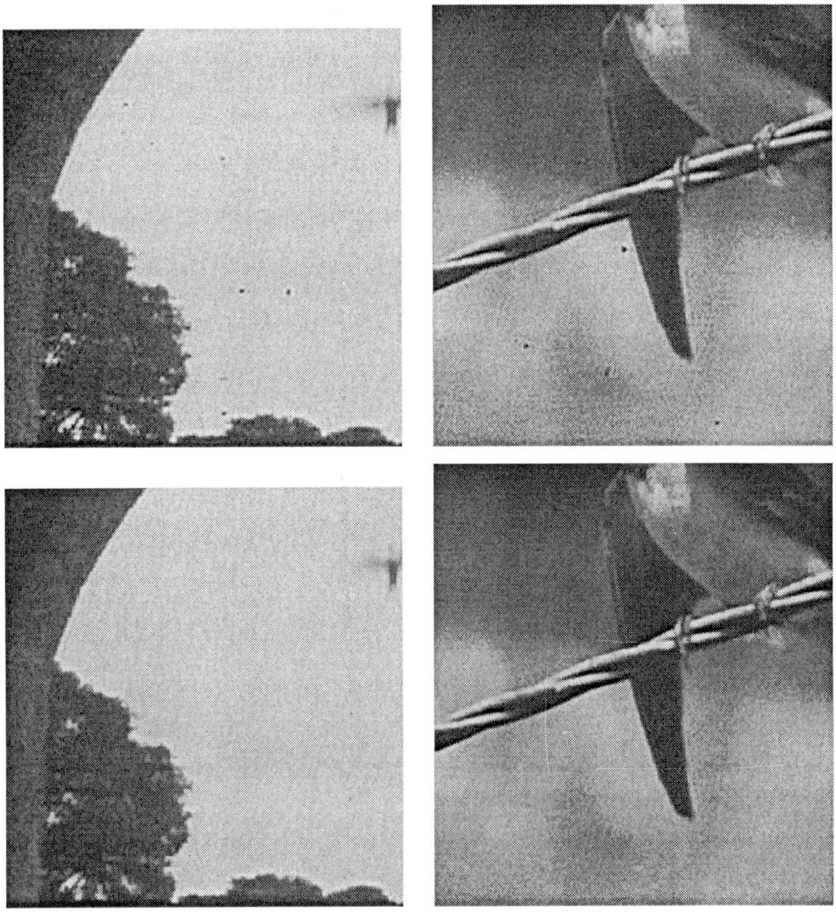

Fig. 6. Regions extracted from image sequences corrupted with film dirt and the same regions after filtering with the spatio-temporal grey-scale soft morphological filter found using the GA

7. N. R.Harvey. New Techniques for the Design of Morphological Filters using Genetic Algorithms. Ph.D. Thesis, Dept. of Electronic and Electrical Engineering, University of Strathclyde, UK, 1997.
8. N. R. Harvey and S. Marshall. Film Restoration Using Soft Morphological Filters. In *Proc. 6th International Conference on Image Processing and its Applications (IPIA'97)*, pages 279–282, Dublin, Ireland, July 1997.
9. R. E. Jacobson An Evaluation of Image Quality Metrics. *Journal of Photographic Science*, 43:7–16, 1995.
10. D. E. Goldberg. *Genetic Algorithms in Search, Optimization & Machine Learning.* Addison Wesley, Reading, Massachusetts, 1989.
11. M. Mitchell. *An Introduction to Genetic Algorithms.* MIT Press, Cambridge, Massachusetts, 1996.
12. M. D. Vose. *The Simple Genetic Algorithm: Foundations and Theory.* MIT Press, Cambridge, Massachusetts, 1999.

NEW INSIGHT ON DIGITAL TOPOLOGY

G. J. F. BANON
Instituto Nacional de Pesquisas Espaciais
CP 515
12 201-970, São José dos Campos, SP, Brazil

Abstract. The traditional approach of digital topology consists of using two different kinds of neighborhood for the black and white pixels of a binary image, and consequently two kinds of connectedness. In this paper, we are proposing to define connectedness in terms of a bounded subcollection of sets and to analyze the topological aspect of a binary image in an expanded domain in which it is sufficient to consider only one kind of connectedness. In the first part, we recall the definitions of neighborhood and connectedness of the traditional digital topology approach. In the second part, we define the notions of "bounded space", "connected bounded space" and of "connected subset of a bounded space". In the last part, we introduce two image operators (a dilation and an erosion) that produce expanded images whose connectedness is analyzed in relation to a bounded space obtained from the invariance domain of an opening. We show how the traditional two kinds of connectedness can be derived from this analysis.

Key words: Digital Topology, Adjacency, Bounded Space, Connectedness, Connected Class, Connected Subset, Connected Component, Connected Space, Dilation, Erosion, Opening, Connectivity Opening, Expanded Domain, Expansion Operators.

1. Introduction

Digital topology provides the theoretical foundations for image analysis and more specifically advanced image segmentation.

Traditionally, digital topology is based on the neighborhood concept. From it, other concepts, like connectedness and connected components, are derived.

When the pixels of a binary image are arranged along lines and columns, the need for using two kinds of neighborhood, one for the black pixels and one for the white, appears in the early studies on digital topology.

In this paper, we propose the notion of bounded space as first concept, instead of neighborhood. Then we can use the traditional topological approach (not digital) to define connectedness.

Our contribution consists of showing that it is possible to derive the traditional 4-connectedness and 8-connectedness from a unique connectedness defined on an expanded image domain.

This is obtained by using two expansion operators (one dilation and one erosion) and by studying connectedness in the expanded domain.

In the first section, we recall the definitions of neighborhood and connectedness of the traditional digital topology approach, and we show the need for two definitions.

In the second section, we introduce the definitions of "bounded space", "connected bounded space" and "connected subset of a bounded space".

In the last section, we show how the connectedness defined in terms of a bounded space obtained from the invariance domain of a morphological opening, and the traditional two kinds of connectedness are related.

2. Traditional Approach to Digital Topology

Let $\mathbf{Z}^2$ be the set of all ordered pairs of integers. The elements of $\mathbf{Z}^2$ will be called *points*. The set $\mathbf{Z}^2$ equipped with the usual addition is an Abelian group denoted $(\mathbf{Z}^2, +, o)$ where o is the null element of $+$ $(o = (0,0))$. From this group, we can build the notions of translate, transpose, symmetry, Minkowski addition, and translation invariant dilation (Serra, 1982).

Let u be a point of $\mathbf{Z}^2$, we denote by $B + u$ the *translate by* u of a subset B of $\mathbf{Z}^2$ and by B^{t} its transpose. If the subset B is equal to its transpose then it is said *symmetric*. We denote by δ_B the dilation by B defined by, for any subset Y of $\mathbf{Z}^2$,

$$\delta_B(Y) \triangleq Y \oplus B,$$

where $\oplus$ is the Minkowski addition.

By construction δ_B is a translation invariant (t.i.) operator.

Let H and V be the following subsets of $\mathbf{Z}^2$.

$$H \triangleq \{(0,-1),(0,0),(0,1)\} \text{ and } V \triangleq \{(-1,0),(0,0),(1,0)\}.$$

We denote by δ_4 and δ_8 the following dilations on $\mathbf{Z}^2$,

$$\delta_4 \triangleq \delta_{H \cup V} \text{ and } \delta_8 \triangleq \delta_{H \oplus V}.$$

Since H and V are symmetric, for n equals to 4 or 8 and for any points x_1 and x_2 in $\mathbf{Z}^2$,

$$x_1 \in \delta_n(\{x_2\}) \Leftrightarrow x_2 \in \delta_n(\{x_1\}) \tag{1}$$

Now we use these two dilations in the 4 and 8 neighborhood definitions.

Definition 1 *(neighborhood of a point) - Let x be a point of $\mathbf{Z}^2$, $\delta_4(\{x\})$ is its* 4-neighborhood *and $\delta_8(\{x\})$ is its* 8-neighborhood.

From neighborhood we can define adjacency.

Definition 2 *(adjacency between two points) - Two points x_1 and x_2 of $\mathbf{Z}^2$ are said to be n-*adjacent *if x_1 belongs to the n-neighborhood of x_2.*

Because of (1), the order of the points in Definition 2 is irrelevant. Furthermore, since $H \cup V \subset H \oplus V$, the 4-adjacency implies the 8-adjacency.

The adjacency between points can be extended to subsets.

Definition 3 *(adjacency between two subsets) - Two subsets X_1 and X_2 of* $\mathbf{Z}^2$ *are said to be* n-adjacent *if there exists a point x_1 in X_1 and a point x_2 in X_2 such that x_1 and x_2 are n-adjacent.*

We are now ready to recall the definitions of n-connected subset and points.

Definition 4 *(connected subset) - A subset X of* $\mathbf{Z}^2$ *is* n-connected *if it cannot be partitioned into two subsets which are not n-adjacent to each other.*

Definition 5 *(connected points) - Let X be a subset of* $\mathbf{Z}^2$. *Two points x_1 and x_2 of* $\mathbf{Z}^2$ *are said to be* n-connected in X *if there exists in X an n-connected subset which contains x_1 and x_2.*

If x_1 and x_2 are n-connected in X, we say that x_1 is n-connected in X to x_2. The binary relation "is n-connected in X to" is an equivalence relation. This observation leads to the notion of connected components of a subset.

Definition 6 *(connected components of a subset) - Let X be a subset of* $\mathbf{Z}^2$. *The equivalence classes of the equivalence relation "is n-connected in X to" are called the* n-components of X.

The subset X representing the original image of Figure 1 has four 4-connected components and just one 8-connected component.

Now, if we apply the above definitions to its complement $X^c \triangleq \mathbf{Z}^2 - X$, we come up with a problem because X^c has two 4-connected components and just one 8-connected component.

If we assume, for example, that X should be connected and should separate its complement into two components, then, in this case, the topological analysis could **not** be made by using the same connectedness for X and its complement.

In early studies on digital topology (see for example Rosenfeld's paper [5]), it was suggested using different connectedness for a set and its complement to get round the problem. In the above example one should use the 8-connectedness for X and the 4-connectedness for its complement X^c.

In the next section, we introduce a new approach that makes the topological analysis of a binary image more coherent because it is based on a unique definition of connectedness.

3. Connected Bounded Space

In topology, the connectedness of a subset of a set E is defined in relation to a subcollection of subsets of E called *topology* for which the unique open and closed sets are the empty set and the proper set E. In mathematical morphology, Serra [7] has introduced the notion of *connected class* without an explicit reference to a subcollection of subsets of E.

In this section, we will show a relationship between both concepts.

Based on the observation that the connectedness notion used in topology can be defined in terms of a subcollection of subsets of E which do **not** need to be a topology, we introduce the definition of bounded space.

Definition 7 *(bounded space) - Let $\mathcal{P}(E)$ be the collection of all subsets of E. A* bounded space *is a pair $(E, \mathcal{B})$, where $\mathcal{B}$ is any subcollection of $\mathcal{P}(E)$ containing at least the empty set $\emptyset$ and the set E.*

We use the expression "bounded" because $\mathcal{B}$ must contain the universal bounds of the poset $(\mathcal{P}(E), \subset)$, which are $\emptyset$ and E.

We can now introduce the notion of connectedness for a bounded space.

Definition 8 *(connected bounded space) - A bounded space $(E, \mathcal{B})$ is connected (or E is connected w.r.t. $\mathcal{B}$) if $\emptyset$ and E are the unique sets X in $\mathcal{B}$ such that X and X^c (the complement of X in E) are both in $\mathcal{B}$.*

Like in topology ([4], p. 84), we have the following characterization of a connected bounded space.

Proposition 1 *(characterization of a connected bounded space) - A bounded space $(E, \mathcal{B})$ is connected if and only if E cannot be partitioned into two sets in $\mathcal{B}$.*

Let $\mathcal{S}$ be a subcollection of $\mathcal{P}(E)$ and let X be a subset of E, then we denote by $\mathcal{S}_X$ the subcollection $\{S' \in \mathcal{P}(E) : \exists S \in \mathcal{S}, S' = S \cap X\}$.

If $\mathcal{B}$ contains the universal bounds of $\mathcal{P}(E)$, then $\mathcal{B}_X$ contains the universal bounds of $\mathcal{P}(X)$, hence we can state the next definition.

Definition 9 *(bounded subspace) - Let $(E, \mathcal{B})$ be a bounded space and let X be a subset of E, then $(X, \mathcal{B}_X)$ is called a* bounded subspace *of $(E, \mathcal{B})$.*

We are now ready to define the connectedness of a subset.

Definition 10 *(connected subset) - A subset X of E is a connected subset of $(E, \mathcal{B})$ (or a connected subset w.r.t. $\mathcal{B}$), if the bounded subspace $(X, \mathcal{B}_X)$ is connected.*

As we see, this connectedness definition is based on the notion of connected bounded space, contrasting with Definition 4 based on adjacency.

The subcollection $\mathcal{C}$ of the connected subsets of $(E, \mathcal{B})$ satisfies:

(i) the empty subset of E, and the singletons of E belong to $\mathcal{C}$;

(ii) let $\mathcal{S}$ be a subcollection of $\mathcal{C}$, if $\cap \mathcal{S}$ is nonempty, then $\cup \mathcal{S}$ belongs to $\mathcal{C}$.

Property (ii) is similar to Proposition 30 of Lima's book on topology [4].

Actually, a subcollection of $\mathcal{P}(E)$ which satisfies Properties (i) and (ii) is called a *connected class* by Serra ([7], p. 51) or a *connectivity class* by Heijmans ([3], p. 317). Hence we can state the following proposition.

Proposition 2 *(bounded space and connected class) - The connected subsets of any bounded space form a connected class.*

In other words, to each bounded space, we can associate a connected class, but the converse is not true (the subcollections of 4-connected and 8-connected subsets are counterexamples).

Proposition 2 is interesting because Serra ([7], Theorem 2.8, p. 52) has proved that each connected class is characterized by a family of openings, called *connectivity openings* by Heijmans ([3], p. 317). Hence to each bounded space we can associate a family of connectivity openings.

The practical problem is to find useful bounded spaces. In the next section we will see an important example of bounded space.

4. Digital Connectedness on $\mathbf{Z}^2$

In this section, we introduce a special bounded space for the topology analysis of the binary images defined on $\mathbf{Z}^2$.. This bounded space will be obtained from the invariance domain of a morphological opening.

Let a and b be the mappings from $\mathbf{Z}^2$ to $\mathcal{P}(\mathbf{Z}^2)$ defined by, for any y in $\mathbf{Z}^2$,

$$a(y) \triangleq \delta_8(\{2y\}) \quad \text{and} \quad b(y) \triangleq \delta_8(\{2y\})^{\mathrm{c}},$$

and let δ_a and ${}_b\varepsilon$ be the two mappings from $\mathcal{P}(\mathbf{Z}^2)$ to $\mathcal{P}(\mathbf{Z}^2)$ defined by, for any Y in $\mathcal{P}(\mathbf{Z}^2)$,

$$\delta_a(Y) \triangleq \bigcup_{y \in Y} a(y) \quad \text{and} \quad {}_b\varepsilon(Y) \triangleq \bigcup_{y \in Y^{\mathrm{c}}} b(y).$$

We know (see [1]) that δ_a and ${}_b\varepsilon$ are, respectively, a dilation and an erosion, a and b being their respective structuring functions. When applying these operators to an image we get two expanded images as shown in Figure 1.

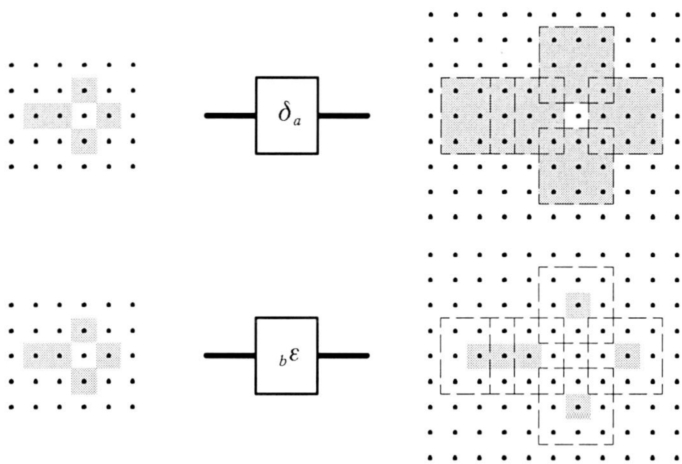

Fig. 1. Original and expanded images.

We use the expression "expanded image" because the distance between two components of the output image is twice this distance in the input image.

The operators δ_a and ${}_b\varepsilon$ are *mutually negative* in the sense that

$$ {}_b\varepsilon(Y^{\mathrm{c}}) = \delta_a(Y)^{\mathrm{c}} \text{ or equivalently } {}_b\varepsilon(Y)^{\mathrm{c}} = \delta_a(Y^{\mathrm{c}}) \tag{2} $$

From (2), we can prove $\Leftarrow$ in the following equivalence. For any subsets Y_1 and Y_2 of $\mathbf{Z}^2$,

$$ Y_1 \subset Y_2 \Leftrightarrow {}_b\varepsilon(Y_1) \subset \delta_a(Y_2) \tag{3} $$

With respect to ${}_b\varepsilon$, we have the following two properties.

Let y_1 and y_2 be 8-adjacent but not 4-adjacent points in a subset Y, then

$$ y_1 + y_2 \in {}_b\varepsilon(Y) \Leftrightarrow \exists y_3 \text{ and } y_4 \in Y, y_3 \text{ and } y_4 \in \delta_4(\{y_1\}) \cap \delta_4(\{y_2\}). \tag{4} $$

When Y reduces to two 4-adjacent points y_1 and y_2 in $\mathbf{Z}^2$, we have,

$$ {}_b\varepsilon(\{y_1, y_2\}) = \{2y_1, y_1 + y_2, 2y_2\} \tag{5} $$

Let ε_a and ${}_b\delta$ be the erosion and the dilation forming, respectively with δ_a and ${}_b\varepsilon$ two Galois connexions $(\varepsilon_a, \delta_a)$ and $({}_b\delta, {}_b\varepsilon)$, the first one being a Galois connexion between $(\mathcal{P}(\mathbf{Z}^2), \supset)$ and $(\mathcal{P}(\mathbf{Z}^2), \subset)$, and the second one between $(\mathcal{P}(\mathbf{Z}^2), \subset)$ and $(\mathcal{P}(\mathbf{Z}^2), \supset)$.

We observe that $\varepsilon_a \circ \delta_a$ and ${}_b\delta \circ {}_b\varepsilon$ are two identity operators. In this sense, the two expansions δ_a and ${}_b\varepsilon$ transform an image into an expanded image without loss of information.

By construction $\delta_a \circ \varepsilon_a$ is a morphological opening on $\mathbf{Z}^2$ [6]. Its invariance domain is the image of $\mathcal{P}(\mathbf{Z}^2)$ through δ_a, i.e., $\delta_a(\mathcal{P}(\mathbf{Z}^2))$ ([6], p. 53).

Since $\delta_a(\mathcal{P}(\mathbf{Z}^2))$ contains $\emptyset$ and $\mathbf{Z}^2$, the pair $(\mathbf{Z}^2, \delta_a(\mathcal{P}(\mathbf{Z}^2)))$ forms a bounded space.

It is interesting to note that $\delta_a(\mathcal{P}(\mathbf{Z}^2))$ is a subset of the *Khalimskii topology* ([3], p.222), nevertheless, the connected classes obtained from both subcollections are the same.

Hence, the connectedness of the binary images defined on $\mathbf{Z}^2$ can be analyzed indifferently using the Khalimskii topology space or the bounded space $(\mathbf{Z}^2, \delta_a(\mathcal{P}(\mathbf{Z}^2)))$. That is, we can define connectedness from the Khalimskii topological open sets or equivalently from the morphological open sets of $\delta_a \circ \varepsilon_a$. Since the latter are simpler we prefer to work with them.

Associated with the bounded space $(\mathbf{Z}^2, \delta_a(\mathcal{P}(\mathbf{Z}^2)))$, we have the adjacency graph or grid of Figure 2. This grid is obtained by drawing an edge between two points forming a connected subset w.r.t. $\delta_a(\mathcal{P}(\mathbf{Z}^2))$.

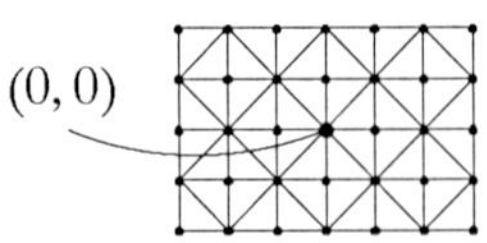

Fig. 2. Adjacency graph.

From now on, we suggest analyzing the digital topology of the binary images on $\mathbf{Z}^2$ by considering their expanded versions, through δ_a or ${}_b\varepsilon$, as subsets of the bounded space $(\mathbf{Z}^2, \delta_a(\mathcal{P}(\mathbf{Z}^2)))$.

The interesting result is that the 4-connectedness and 8-connectedness can now be characterized from this unique bounded space. To show this, we present first, without proof, two properties of the translations on $\mathbf{Z}^2$.

Proposition 3 *(translation properties) - For any subsets B, B_1 and B_2 of $\mathbf{Z}^2$, and any points x, x_1, x_2 and x_3 in $\mathbf{Z}^2$,*
(i) if $x_1 \in B_1 + x_2$ and $x_2 \in B_2 + x_3$, then $x_1 \in (B_1 \oplus B_2) + x_3$,
(ii) if $2x \in B^2 \Rightarrow x \in B$ then $2x_1 \in B^2 + 2x_2 \Rightarrow x_1 + x_2 \in B + 2x_2$.

These properties will be used to prove the next proposition. We will use, in particular, the fact that $H \cup V$ and $H \oplus V$ satisfy the condition $2x \in B^2 \Rightarrow x \in B$ of property (ii).

Proposition 4 *(adjacency characterization) - The following four statements are equivalent. For n equals 4 or 8, and for any points y_1 and y_2 in $\mathbf{Z}^2$,*
(i) y_1 and y_2 are n-adjacent,
(ii) $y_1 + y_2 \in \delta_n(\{2y_1\})$,
(iii) $y_1 + y_2 \in \delta_n(\{2y_2\})$,
(iv) $\delta_n(\{2y_1\}) \cap \delta_n(\{2y_2\}) \neq \emptyset$.

Proof - For any y_1 and y_2 in $\mathbf{Z}^2$,

$$\begin{aligned}
y_1 + y_2 \in \delta_n(\{2y_1\}) &\Leftrightarrow \delta_n(\{y_1\}) + y_1 && (\delta_n \text{ is t.i.})\\
&\Leftrightarrow y_2 \in \delta_n(\{y_1\}) && (\text{translation definition})\\
&\Leftrightarrow y_1 \text{ and } y_2 \text{ are } n\text{-adjacent} && (\text{adjacency definition})\\
&\Leftrightarrow y_1 + y_2 \in \delta_n(\{2y_2\}). && (\text{same steps as above})
\end{aligned}$$

This prove the equivalence between (i), (ii) and (iii). Furthermore, (ii) and (iii) implies (iv). Let us prove that (iv) implies (iii). For any points y, y_1 and y_2 in $\mathbf{Z}^2$,

$$\begin{aligned}
y \in \delta_n(\{2y_1\}) \cap \delta_n(\{2y_2\}) &\Leftrightarrow y \in \delta_n(\{2y_1\}) \text{ and } y \in \delta_n(\{2y_2\}) && (\cap \text{ definition})\\
&\Leftrightarrow 2y_1 \in \delta_n(\{y\}) \text{ and } y \in \delta_n(\{2y_2\}) && (\text{Property (1)})\\
&\Rightarrow y_1 \in \delta_n(\delta_n(\{2y_2\})) && ((\text{i}) \text{ of Proposition 3})\\
&\Rightarrow y_1 + y_2 \in \delta_n(\{2y_2\}). && ((\text{ii}) \text{ of Proposition 3})
\end{aligned}$$

That is, (iii) and (iv) are equivalent. □

Before extending the characterization adjacency to the subsets of $\mathbf{Z}^2$, we need one more proposition.

Proposition 5 *(expansion operator property) - For any distinct points y_1 and y_2 in $\mathbf{Z}^2$ ($y_1 \neq y_2$), we have,*

$$\delta_4(\{2y_1\}) \cap \delta_4(\{2y_2\}) = a(y_1) \cap a(y_2) \cap {}_b\varepsilon(\{y_1, y_2\}).$$

Proof - An exhaustive study with respect to each point u in $H \cup V - \{o\}$ shows that the number of points of $(H \cup V) \cap (H \cup V + 2u)$ is always one. Consequently, when y_1 and y_2 are 4-adjacent, the number of points of $\delta_4(\{2y_1\}) \cap \delta_4(\{2y_2\})$ is always one. From Proposition 4, this point is $y_1 + y_2$. That is, for any distinct points y_1 and y_2 in $\mathbf{Z}^2$,

$$\delta_4(\{2y_1\}) \cap \delta_4(\{2y_2\}) = \begin{cases} \{y_1 + y_2\} & \text{if } y_1 \text{ and } y_2 \text{ are 4-adjacent} \\ \emptyset & \text{otherwise.} \end{cases}$$

From Proposition 4, if y_1 and y_2 are 4-adjacent then $\{y_1+y_2\} \subset a(y_1)\cap a(y_2)$. Observing that $u \in H \cup V - \{o\}$ implies that $2u \notin H \cup V$, we have, for any two 4-adjacent distinct points y_1 and y_2 in $\mathbf{Z}^2$, $2y_1 \notin a(y_2)$ and $2y_2 \notin a(y_1)$. Therefore, using (5) and Proposition 4, we have, for any distinct points y_1 and y_2 in $\mathbf{Z}^2$,

$$a(y_1) \cap a(y_2) \cap {}_b\varepsilon(\{y_1, y_2\}) = \begin{cases} \{y_1 + y_2\} & \text{if } y_1 \text{ and } y_2 \text{ are 4-adjacent} \\ \emptyset & \text{otherwise.} \end{cases}$$

□

Using the previous proposition we can state the following result.

Proposition 6 *(subset adjacency characterization) - For any disjoint subsets Y_1 and Y_2 of $\mathbf{Z}^2$,*
(i) Y_1 and Y_2 are 4-adjacent $\Leftrightarrow \delta_a(Y_1) \cap \delta_a(Y_2) \cap {}_b\varepsilon(Y_1 \cup Y_2) \neq \emptyset$,
(ii) Y_1 and Y_2 are 8-adjacent $\Leftrightarrow \delta_a(Y_1) \cap \delta_a(Y_2) \cap \delta_a(Y_1 \cup Y_2) \neq \emptyset$.

Proof - Let us prove (i). For any disjoint subsets Y_1 and Y_2 of $\mathbf{Z}^2$,

$$\begin{array}{llr} Y_1 \text{ and } Y_2 \text{ are 4-adjacent} & \Leftrightarrow \begin{matrix}\exists y_1 \in Y_1 \\ \exists y_2 \in Y_2\end{matrix}, \; y_1 \text{ and } y_2 \text{ are 4-adjacent} & \text{(Def. 3)} \\ & \Leftrightarrow \begin{matrix}\exists y_1 \in Y_1 \\ \exists y_2 \in Y_2\end{matrix}, \; \delta_4(\{2y_1\}) \cap \delta_4(\{2y_2\}) \neq \emptyset & \text{(Prop. 4)} \\ & \Leftrightarrow \begin{matrix}\exists y_1 \in Y_1 \\ \exists y_2 \in Y_2\end{matrix}, \; a(y_1) \cap a(y_2) \cap {}_b\varepsilon(\{y_1, y_2\}) \neq \emptyset & \text{(Prop. 5)} \\ & \Leftrightarrow \delta_a(Y_1) \cap \delta_a(Y_2) \cap {}_b\varepsilon(Y_1 \cup Y_2) \neq \emptyset. & \end{array}$$

The $\Rightarrow$ of the last equivalence follows from the increasing property of δ_a and ${}_b\varepsilon$. Let us prove $\Leftarrow$. The non empty intersection implies that there exist y_1 in Y_1, y_2 in Y_2 and x in ${}_b\varepsilon(Y_1 \cup Y_2)$ such that $x \in a(y_1) \cap a(y_2)$ (the intersection of two distinct 3 by 3 squares). By Proposition 4, y_1 and y_2 are 4-adjacent or 8-adjacent. If they are 4-adjacent, then, by Propositions 4 and 5, $a(y_1) \cap a(y_2) \cap {}_b\varepsilon(\{y_1, y_2\}) \neq \emptyset$. If they are 8-adjacent but not 4-adjacent, then, the number of points of $a(y_1) \cap a(y_2)$ is one and, by Proposition 4, $x = y_1 + y_2$. Therefore, by (4), there exists y_3 in Y_1 (or Y_2) 4-adjacent to y_2 (or y_1) and we can apply again the previous 4-adjacency analysis. This proves (i).

The proof of (ii) is much easier since we can use the property that a dilation commutes with union, that is, $\delta_a(Y_1 \cup Y_2) = \delta_a(Y_1) \cup \delta_a(Y_2)$ and consequently the equality $\delta_a(Y_1) \cap \delta_a(Y_2) = \delta_a(Y_1) \cap \delta_a(Y_2) \cap \delta_a(Y_1 \cup Y_2)$. □

From the previous proposition, we can state our final result.

Proposition 7 *(characterization of digital connectedness) - For any subset Y of $\mathbf{Z}^2$,*
(i) Y is 4-connected if and only if ${}_b\varepsilon(Y)$ is a connected subset of $(\mathbf{Z}^2, \delta_a(\mathcal{P}(\mathbf{Z}^2)))$,
(ii) Y is 8-connected if and only if $\delta_a(Y)$ is a connected subset of $(\mathbf{Z}^2, \delta_a(\mathcal{P}(\mathbf{Z}^2)))$.

Proof - Let us prove (i). Let $\mathcal{B}$ be the subcollection $\delta_a(\mathcal{P}(\mathbf{Z}^2))$. For any Y of $\mathbf{Z}^2$,

$$Y \text{ is not 4-connected} \Leftrightarrow \begin{cases} \exists Y_1, Y_2 \subset \mathbf{Z}^2, \\ Y_1 \cup Y_2 = Y \text{ and } Y_1 \cap Y_2 = \emptyset \text{ and} \\ Y_1 \text{ and } Y_2 \text{ are not 4-adjacent} \end{cases} \quad \text{(Def. 4)}$$

$$\Leftrightarrow \begin{cases} \exists Y_1, Y_2 \subset \mathbf{Z}^2, \\ Y_1 \cup Y_2 = Y \text{ and } Y_1 \cap Y_2 = \emptyset \text{ and} \\ \delta_a(Y_1) \cap \delta_a(Y_2) \cap {}_b\varepsilon(Y_1 \cup Y_2) = \emptyset \end{cases} \quad \text{(Prop. 6)}$$

$$\Leftrightarrow \begin{cases} \exists X_1, X_2 \in \mathcal{B}_{{}_b\varepsilon(Y)}, \\ Y_1 \cup Y_2 = {}_b\varepsilon(Y) \text{ and} \\ X_1 \cap X_2) = \emptyset \end{cases} \quad (*)$$

$$\Leftrightarrow {}_b\varepsilon(Y) \text{ is not a connected subset of } (\mathbf{Z}^2, \mathcal{B}). \quad \text{(Def. 10)}$$

* The $\Rightarrow$ of the equivalence follows by choosing $X_1 = \delta_a(Y_1) \cap {}_b\varepsilon(Y)$ and $X_2 = \delta_a(Y_2) \cap {}_b\varepsilon(Y)$, and by applying property (3) and the property that a dilation commutes with union. The implication $\Leftarrow$ follows by observing that for any subset Y of $\mathbf{Z}^2$, $\mathcal{B}_{{}_b\varepsilon(Y)} = \{X \in \mathcal{P}(E) : \exists Y' \subset Y, X = \delta_a(Y') \cap {}_b\varepsilon(Y)\}$ and ${}_b\varepsilon(Y) \subset \delta_a(Y_1 \cup Y_2)$.

The last inclusion is used in conjunction with (3) to prove that $Y_1 \cup Y_2 = Y$.

The proof of (ii) is similar and simpler since Proposition 6 is not needed. □

The operators δ_a and ${}_b\varepsilon$ being mutually negative, we see that analyzing the 4-connectedness of the foreground of Y and the 8-connectedness of its background (Y^c), is (from Proposition 7) the same as analyzing the connectedness of the foreground of ${}_b\varepsilon$(Y) and its background (${}_b\varepsilon(Y)^c$ which is $\delta_a(Y^c)$) w.r.t. $\delta_a(\mathcal{P}(\mathbf{Z}^2))$.

In the same way, we see that analyzing the 8-connectedness of the foreground of Y and the 4-connectedness of its background (Y^c), is (from Proposition 7) the same as analyzing the connectedness of the foreground of δ_a(Y) and its background ($\delta_a(Y)^c$ which is ${}_b\varepsilon(Y^c)$) w.r.t. $\delta_a(\mathcal{P}(\mathbf{Z}^2))$.

The important fact, is that in all situations we analyze the connectedness with respect to the same bounded space $(\mathbf{Z}^2, \delta_a(\mathcal{P}(\mathbf{Z}^2)))$.

In the case of the bounded space $(\mathbf{Z}^2, \delta_a(\mathcal{P}(\mathbf{Z}^2)))$, we could verify that the connectivity openings γ_x (see Section 3) associated with this space are given by, for any point x in $\mathbf{Z}^2$ and any subset X of $\mathbf{Z}^2$,

$$\gamma_x(X) \triangleq \bigcup_{n \geq 1} \delta_c{}^n(\{x\}|X),$$

where $\delta_c(\cdot|X)$ is the conditional dilation on $\mathbf{Z}^2$ given by, for any subsets X and X' of $\mathbf{Z}^2$,

$$\delta_c(X'|X) \triangleq \bigcup_{x \in X'} c(x) \cap X,$$

and where c is a mapping from $\mathbf{Z}^2$ to $\mathcal{P}(\mathbf{Z}^2)$ given by, for any point x in $\mathbf{Z}^2$,

$$c(x) \triangleq \begin{cases} H \oplus V & \text{if } \exists y \in \mathbf{Z}^2, x = 2y \text{ or } x = 2y + (1,1) \\ H \cup V & \text{otherwise.} \end{cases}$$

Hence for testing if a subset X of $\mathbf{Z}^2$ is connected w.r.t. $\delta_a(\mathcal{P}(\mathbf{Z}^2))$, we choose a point x in X and we compute $\gamma_x(X)$. If $\gamma_x(X) = X$ then X is connected, otherwise it is not.

It is interesting to observe that applying the same approach to the hexagonal lattice, the expanded image by the dilation and the one by the erosion have the same connectedness which corresponds to the usual digital 6-connectedness.

5. Conclusion

In this work we introduced an alternative method for studying connectedness of a binary image. This was done based on two observations.

The first observation was that we can define connectedness simply on bounded spaces and, for our purpose, we did not need to consider a much elaborated structure like a topology space for example. An interesting result was that behind each bounded space there is a connected class as defined by Serra. An open problem is how to express the related connectivity opening family in terms of a bounded space.

The second observation was that by expanding the image domain it was possible to get two interesting results. The first one was the definition of a useful bounded space to study the connectedness of a binary image. This bounded space was obtained from the invariance domain of a morphological opening. The second result was that the foreground and background in the expanded domain can be studied coherently by using the same connectedness. Probably, other interesting definitions, like a border definition, could be found in the expanded image domain.

Finally, we would like to point out that the suggested bounded space for connectedness analysis was obtained from a subcollection of morphological open sets. This subcollection which is a subset of the Khalimskii topology appeared to be sufficient for our purpose. This shows the importance of the morphological open sets and could justify a notion of "morphological space" which would be a bounded space $(E, \mathcal{M})$ where $\mathcal{M}$ is a $\cup$-closed subcollection of subsets of E.

References

1. G. J. F. Banon and J. Barrera. *Set operator decomposition and conditionally translation invariant elementary operators*, in *Mathematical Morphology and its Applications to Image Processing*, Eds: Serra, Jean and Soille, Pierre, pages 5–12. Kluwer Academic, Dordrecht 1994. UR*Lib* repository: `<dpi.inpe.br/banon/1995/09.01.10.00>`.
2. J. Dugundji. *Topology*. Allyn and Bacon Inc., Boston, 1967.
3. H. J. A. M. Heijmans. *Morphological Image Operators*. Academic Press, Boston, 1994.
4. E. L. Lima. *Elementos de Toplogia Geral*. Livros Ténicos e Científicos Editora S. A., Rio de Janeiro, 1976.
5. A. Rosenfeld. Connectivity in digital pictures. *Journal of the ACM*, 17(1):146–160, January 1970.
6. J. Serra. *Image Analysis and Mathematical Morphology*. Academic Press, London, 1982.
7. J. Serra. *Image Analysis and Mathematical Morphology, Volume 2: Theoretical Advances*. Academic Press, London, 1988.

APPROXIMATE CONNECTIVITY AND MATHEMATICAL MORPHOLOGY

ANTONY T. POPOV
St. Kliment University of Sofia, Faculty of Mathematics and Informatics, 5, James Bourchier Blvd., BG-1164 Sofia, BULGARIA
e-mail: atpopov@fmi.uni-sofia.bg, fax (+359 2) 687 180
http://www-it.fmi.uni-sofia.bg/atpopov/

Abstract. As known from the works of Serra, Ronse, and Haralick and Shapiro, the connectivity relations are found to be useful in filtering binary images. But it can be used also to find roadmaps in robot motion planning, i.e. to build discrete networks of simple paths connecting points in the robot's configuration space capturing the connectivity of this space. This paper generalises and puts together the notion of a connectivity class and the notion of a separation relation. This gives an opportunity to introduce approximate epsilon-connectivity, and thus we show the relation between our approach and the Epsilon Geometry introduced by Guibas, Salesin and Stolfi. Ronse and Serra have defined connectivity analogues on complete lattices with certain properties. As a particular case of their work we consider the connectivity of fuzzy compact sets, which is a natural way to study the connectivity of grey - scale images. This idea can be transferred also in planning robot trajectories in the presence of uncertainties. Since based on fuzzy sets theory, our approach is intuitively closer to the classical set oriented approach, used for binary images and robot path planning in known environment with obstacles.

Key words: Connectivity, Complete Lattice, Fuzzy Set, Morphological Operations, ε-Geometry.

1. Imprecise Computations

In many practical tasks we operate with imprecise or uncertain data, especially when this data comes as an output from some measuring instrument. This problem appears in image processing because of the distortion effect, and often because of non-precise calibration of the camera. Therefore lots of work have been done for the development of algorithms which give reliable results operating with imprecise data. A good framework for imprecise computations in robust geometrical algorithms is the *Epsilon Geometry*, which uses inaccurate primitives. It is based on a general model of imprecise computations, which includes rounded-integer and floating-point arithmetic as special cases [4]. Another more general approach for imprecise geometrical reasoning is the *Fuzzy Geometry* approach. It generalises in some sense the Epsilon Geometry, and moreover gives the opportunity to transfer geometric ideas directly from black-and-white image processing techniques to the analysis of grey-scale images [14]. Our approach to fuzzy connectivity is more general than this in [2] and [14] since it is based on the choice of a connectivity class.

2. Mathematical Morphology – Background

Serra [16] and Heijmans [7] have shown that morphological operations can be formulated on any complete lattice. A set $\mathcal{L}$ with a partial ordering "$\leq$" is called a *complete lattice* if every subset $\mathcal{H} \subseteq \mathcal{L}$ has a supremum $\bigvee \mathcal{H} \in \mathcal{L}$ (least upper bound) and infimum (greatest lower bound) $\bigwedge \mathcal{H} \in \mathcal{L}$.

An operator $\varphi : \mathcal{L} \mapsto \mathcal{M}$, where $\mathcal{L}$ and $\mathcal{M}$ are complete lattices, is called dilation if it distributes over arbitrary suprema: $\varphi(\bigvee_{i\in I} X_i) = \bigvee_{i\in I} \varphi(X_i)$, and erosion if it distributes over arbitrary infima. Erosions and dilations are increasing operations [7]. An operator $\psi \in \mathcal{L}^*$ is called a closing if it is increasing, idempotent ($\psi^2 = \psi$) and extensive ($\psi(X) \geq X$). An operator $\psi \in \mathcal{L}^*$ is called an opening if it is increasing, idempotent and anti-extensive ($\psi(X) \leq X$) [7]. A pair of operators (ε, δ), is called an adjunction, if for every two elements $X, Y \in \mathcal{L}$ it follows that $\delta(X) \leq Y \iff X \leq \varepsilon(Y)$. In [7] it is proved that if (ε, δ) is an adjunction then ε is erosion and δ is dilation. If (ε, δ) is an adjunction, then the composition $\varepsilon\delta$ is a closing, and $\delta\varepsilon$ is an opening. As an example, let us consider the lattice $\mathcal{L}$ with elements the subsets of a linear space E, i.e. $\mathcal{L} = \mathcal{P}(E)$. Here and henceforth, by $\mathcal{P}(X)$ we denote the power set of X, i.e. the family of all subsets of the set X . Then every translation-invariant dilation is represented by the standard Minkowski addition: $\delta_A(X) = A \oplus X = X \oplus A$, and its adjoint erosion is given by Minkowski subtraction: $\varepsilon_A(X) = X \ominus A$ [7]. Then closing and opening of A by B are defined as $A \bullet B = (A \oplus B) \ominus B$, $A \circ B = (A \ominus B) \oplus B$. These operations are referred to as *classical* or *binary morphological operations*.

3. Connectivity of Binary Images

In mathematics, the notion of connectivity is formalized in the topological framework in two different ways. First, a set is called to be connected when it cannot be partitioned as a union of two open, or two closed sets. In practice, it is more suitable to work with the so - called *arcwise connectivity*. A set X is said to be arcwise-connected when for every two distinct points a and b from X there exists a continuous curve joining a and b and lying entirely in X. Arcwise connectivity is more restrictive than the general one. It is not difficult to show that any arcwise-connected set in $\mathbf{R}^n$ is connected. The opposite is not true in general, but for open sets in $\mathbf{R}^n$ the topological connectedness is equivalent to arcwise connectedness. Arcwise connectivity is widely used in robot motion planning. The motion planning task is to find a path, i.e. a continuous sequence of collision-free configurations of the robot (or any moving agent referred as a robot), connecting two arbitrary input configurations - the start configuration q_b and the final configuration q_e, whenever such a path exists, or indicate that no such path exists. The negative result means that the query points q_b and q_e lie in different connected components of the free configuration space. In this case the profit of studying approximate connectivity is evident, because if there are uncertainties in the robot metrics and control parameters, the robot may collide with the obstacles when moving through narrow passages in the workspace [8]. If the geometric models of the robot and the workspace are imprecise, the approximate connectivity approach could be useful in practice, especially when the path planner captures the connectivity of the robots configuration space by building a probabilistic roadmap [8], a network of simple paths connecting points picked in random in this space. In image analysis, several notions of digital connectivity have been introduced. Usually, they exploit the definition of arcwise connectivity in a discrete way – depending on the regarded neighbourhood relation (4-square, 8-square, hexagonal, etc.) [17]. Following the works of Serra [15, 16, 17] and Haralick and Shapiro [6], an abstract connectivity framework, suited mainly for analysis and processing of binary images has been developed. It is strongly related with the mathematical morphology concepts [15]. The base concept is the *connectivity class*:

Definition 1 *Let E be an arbitrary set. A family $\mathcal{C}$ of the subsets of E is called a connectivity class if the following properties hold:*

1. $\emptyset \in \mathcal{C}$ *and* $\{x\} \in \mathcal{C}$ *for every* $x \in E$ *;*
2. *If* $E_i \in \mathcal{C}$ *for* $i \in I$ *and* $\bigcap_{i \in I} E_i \neq \emptyset$ *, then* $\bigcup_{i \in I} E_i \in \mathcal{C}$.
3. *If* $E \supseteq A \in \mathcal{C}$, *then for every* $x \in E$ $\tau_x(A) \subset \mathcal{C}$.

The third axiom can be referred to as translation -invariance of the connectivity. It hasn't been imposed in the previous works [17, 6, 13] since it doesn't give any topological impact to the notion of connectivity. However it is essential from morphological point of view since the translation-invariance is one of the main bases of mathematical morphology. This condition can be replaced by more general one, namely *affine invariance of connectivity*, or in the case of arbitrary complete lattice by T-*invariance* with respect to a given Abelian group of automorphisms T [7]. In our work it is sufficient to work only with translation-invariant operators.

Given a connectivity class in a universal set E we can define the maximal connected component of a set $A \subseteq E$ containing a particular point x:

$$\gamma_x(A) = \bigcup \{C \in \mathcal{C} \,|\, x \in C \text{ and } C \subseteq A\}$$

Then it can be proved easily (see [16] or [7]) that :

- For every $x \in E$ γ_x is an algebraic opening in E;
- $\gamma_x(\{x\}) = \{x\}$;
- either $\gamma_x(A) = \gamma_y(A)$ or $\gamma_x(A) \cap \gamma_y(A) = \emptyset$;
- $x \notin A$ implies that $\gamma_x(A) = \emptyset$;
- $\bigcup_{x \in E} \gamma_x(A) = A$.

It is easy to demonstrate that $X \in \mathcal{C}$ if and only if for every two points $x, y \in X$ it follows that $\gamma_x(X) = \gamma_y(X)$.

Theorem 1 *If X and A are connected with respect to the connectivity class $\mathcal{C}$, then $X \oplus A$ is connected with respect to $\mathcal{C}$ as well.*

This theorem generalises the result of Theorem 9.59 from [7], where only arcwise connectivity is considered. It follows directly from more general results in [17], [12] and [13], but for completeness we present the proof.

Proof: Because of translation invariance it is sufficient to work only with structuring elements A containing the zero point. Then $X \subseteq X \oplus A$ Then because of the monotonicity of openings, for every $x \in X$ we have that

$$\emptyset \neq X = \gamma_x(X) \subseteq \gamma_x(X \oplus A).$$

Therefore, if $y \in X$ then $\gamma_y(X \oplus A) = \gamma_x(X \oplus A)$ because both sets contain the set $\gamma_x(X)$ which is the same as $\gamma_y(X)$. Since $0 \in A$, then $x \in \tau_x(A) \subset \mathcal{C}$. On the other hand $\tau_x(A) \subseteq X \oplus A$. Then from the third property of the maximal component, we have that for every $a \in A$ $\gamma_x(X \oplus A) = \gamma_{x+a}(X \oplus A)$ since both sets contain $\tau_x(A)$ which is nonempty. Analogically, for every $y \in X$ and every $b \in A$ we have $\gamma_y(X \oplus A) = \gamma_{y+b}(X \oplus A)$. Then $\gamma_{x+a}(X \oplus A) = \gamma_{y+b}(X \oplus A)$, which automatically proves the theorem because of the arbitrary choice of x, y, a and b. □

Let $\mathcal{S}$ be a binary relation between the subsets of a universal set E, i.e. $\mathcal{S} \subseteq \mathcal{P}(E) \times \mathcal{P}(E)$.

Definition 2 *We say that $\mathcal{S}$ is a separation if it satisfies the following conditions [6] :*

– *$\mathcal{S}$ is symmetric, which means that $(X, Y) \in \mathcal{S}$ if and only if $(Y, X) \in \mathcal{S}$.*
– *$\mathcal{S}$ is exclusive, which means that $(X, Y) \in \mathcal{S}$ implies $X \cap Y = \emptyset$.*
– *$\mathcal{S}$ is hereditary, which means that $(X, Y) \in \mathcal{S}$ implies $(X', Y') \in \mathcal{S}$ for each $X' \subseteq X$ and $Y' \subseteq Y$.*

We have omitted the last axiom of Haralick and Shapiro as too restrictive for our studies.

Following Ronse [13], we may consider two more axioms characterizing the separation relation:

– If $\bigcap_{i \in I} A_i \neq \emptyset$ and for each $i \in I$ $(A_i, E \setminus A_i) \in \mathcal{S}$, then $(\bigcap_{i \in I} A_i, E \setminus \bigcup_{i \in I} A_i) \in \mathcal{S}$.
– Given three nonvoid pairwise disjoint sets A, B and C from E, such that $(A, B) \in \mathcal{S}$ and $(A \cup C, B) \in \mathcal{S}$, then we must have $(A, B \cup C) \in \mathcal{S}$.

Every connectivity class defines a separation relation and vice versa. If $\mathcal{C}$ is a connectivity class, then we can define the separation relation $\mathcal{S}$ in the following way: say that $(X, Y) \in \mathcal{S}$ when $\gamma_x(X \cup Y) \cap \gamma_y(X \cup Y) = \emptyset$. Then it is easy to demonstrate that the conditions for $\mathcal{S}$ to be a separation relation are satisfied. Then if $\mathcal{S}$ is a separation defined for the subsets of the universal set E, we can define a connectivity class as $\mathcal{C}$ as any subset of $\mathcal{P}(E)$ with the properties:

1. $\emptyset \in \mathcal{C}$ and $\{x\} \in \mathcal{C}$ for every $x \in E$;
2. A set $\emptyset \neq A \subseteq E$ belongs to $\mathcal{C}$ if for every its two subsets X and Y such that $X \cap Y = \emptyset$, $X \cup Y = A$ it follows that $(X, Y) \notin \mathcal{S}$. Such a set will be called indecomposable.

To show that $\mathcal{C}$ is a connectivity class we must demonstrate that given the indecomposable sets $A_i \subseteq E$ for $i \in I$ and $\bigcap_{i \in I} A_i \neq \emptyset$, then the set $A = \bigcup_{i \in I} A_i$ is indecomposable as well. The last is proved in [13].

If K is a connected structuring element then $X \circ K = \bigcup_i X_i \oplus K$, where X_i are the maximal connected components of X (see for instance theorem 5.2 in

[6]). Then if A is separated from B and the structuring element K is connected it simply follows that $(A \cup B) \circ K = (A \circ K) \cup (B \circ K)$. Then suppose that the random dropout noise N removes points from the ideal image A to produce the observed image B. If the noise N is separated from the complement of A^c, the complement of A, then $A = B \bullet K$. Therefore the connectivity relations give us the possibility of perfect reconstruction of the original image.

4. Epsilon-Geometry Approach

The Epsilon Geometry framework defines the notion of an *epsilon predicate* as a means for creating approximate tests. Let $\mathcal{O}$ be a set of objects in a space supplied with a metric d . Let $\mathtt{P}$ be a predicate defined on $\mathcal{O}$. Then for any $X \in \mathcal{O}$ and any $\varepsilon > 0$, let us define an epsilon version of $\mathtt{P}$ [3, 4]. Say that $\varepsilon - \mathtt{P}(X) = \mathtt{true}$ if and only if $\mathtt{P}(X') = \mathtt{true}$ for **some** X', $d(X, X') \leq \varepsilon$. $(-\varepsilon) - \mathtt{P}(X) = \mathtt{true}$ if and only if $\mathtt{P}(X') = \mathtt{true}$ for **all** X', $d(X, X') \leq \varepsilon$. For instance a polygon is said to be $(-\varepsilon)$-convex if it remains convex under any perturbation to its vertices in disks with radius $\varepsilon > 0$.

Let γ_x be the connectivity openings in E associated with the path-connectivity (the usual arcwise connectivity in the continuous case $E = \mathbf{R}^n$, or its discrete analogs as mentioned above). Let $K \subseteq E$ be a connected structuring element which contains the origin. Then we can define another family of openings : $\gamma'_x(A) = A \cap \gamma_x(A \oplus K)$ for every point $x \in A$. These openings are connectivity openings with respect to the connectivity class

$$\mathcal{C}' = \{D \subseteq \mathbf{R}^n \mid \text{there exists C } \in \mathcal{C} \text{ such that } D \subseteq C \subseteq D \oplus K\}$$

as shown in Example 9.62 from [7].

Then having as a base the usual path connectivity in its continuous or discrete versions, we can introduce the notion of *epsilon connectivity* .

Definition 3 *A set $X \subseteq E$ is called ε-connected if it is connected with respect the connectivity class $\mathcal{C}'$ when $K = B_\varepsilon(0)$.*

Here and henceforth $B_r(x)$ denotes the closed ball (disk in two-dimensional space) with centre x and radius r. An example of this notion is given on Fig. 1. For completeness we can refer to the connected sets as 0-connected.

If P and Q are non-empty compacts in $\mathbf{R}^n$, then

$$\text{dist}\,(P, Q) \;=\; \inf\{\varepsilon \,|\, Q \subseteq P \oplus B_\varepsilon(0)\,,\, P \subseteq Q \oplus B_\varepsilon(0)\}. \tag{1}$$

is known as *Hausdorff distance* between P and Q. In practice, in all image processing and robot control tasks we work with compact sets. Therefore our definition of epsilon - connectivity is correct, since for every set X' from $\mathcal{C}'$ there exists a set X from $\mathcal{C}$ such that dist $(X, X') \leq \varepsilon$. Namely, the set X plays the role of C in definition 3 Here, the notion of $(-\varepsilon)$-connectivity is not useful, since we should have a connectivity class, which elements must have the property, that all sets at a distance not greater than ε should be path - connected for $\varepsilon > 0$.

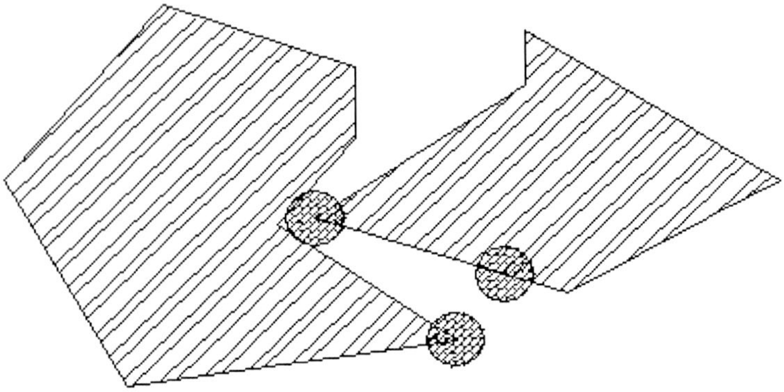

Fig. 1. The hatched region becomes ε-path-connected adding the disk, with ε equal to the radius of the disk

5. Fuzzy Morphological Operations

Consider the set E called the universal set. A fuzzy subset A of the universal set E can be considered as a function $\mu_A : E \mapsto [0,1]$, called the membership function of A. $\mu_A(x)$ is called the degree of membership of the point x to the set A. The ordinary subsets of E, sometimes called 'crisp sets', can be considered as a particular case of a fuzzy set with membership function taking only the values 0 and 1.

Let $0 < \alpha \leq 1$. An α-cut of the set X (denoted by $[X]_\alpha$) is the set of points x, for which $\mu_X(x) \geq \alpha$.

The usual set-theoretical operations can be defined naturally on fuzzy sets: Union and intersection of a collection of fuzzy sets is defined as supremum, resp. infimum of their membership functions. Also, we say that $A \subseteq B$ if $\mu_A(x) \leq \mu_B(x)$ for all $x \in E$. The complement of A is the set A^c with membership function $\mu_{A^c}(x) = 1 - \mu_A(x)$ for all $x \in E$. If the universal set E is linear, like the d-dimensional Euclidean vector space $\mathbf{R}^d$ or the space of integer vectors with length d, then any geometrical transformation arising from a point mapping can be generalised from sets to fuzzy sets by taking the formula of this transformation for graphs of numerical functions, i.e. for any transformation ψ like scaling, translation, rotation etc. we have that $\psi(\mu_A(x)) = \mu_A(\psi^{-1}(x))$. Therefore we can transform fuzzy sets by transforming their $\alpha-$cuts like ordinary sets.

Let us now consider the dilation and erosion as defined by Werman and Peleg [19].

$$\mu_{\mathcal{D}(B,A)}(x) = \sup_{b \in E} \min(\mu_A(b), \mu_B(x-b)), \tag{2}$$

$$\mu_{\mathcal{E}(B,A)}(x) = \inf_{b \in E} \max(\mu_B(b), 1 - \mu_A(b-x)), \tag{3}$$

It is easy to find an example showing that they do not form an adjunction [10]. So, let us now consider only the Werman-Peleg dilation. It can be checked directly that the adjoint erosion of δ_B, denoted by ε_B, takes the form $\mu_{\varepsilon_B(A)}(x) = \inf_{y\in\mathcal{U}}\left(h\left(\mu_A(y) - \mu_B(y-x)\right)(1-\mu_A(y)) + \mu_A(y)\right)$, where $h(x) = 1$ when $x \geq 0$ and is zero otherwise. In this case we shall denote $\delta_A(B) = \delta_B(A) = A \oplus B$, and $\varepsilon_B(A) = A \ominus B$.

Definition 4 *A function $F : \mathbf{R}^n \mapsto \mathbf{R}$ is called upper semicontinuous (u.s.c.) if for any $t \in \mathbf{R}$ such that $t > F(x)$ there exists a neighbourhood $B_r(x)$ such that $t > F(y)$ for every point $y \in B_r(x)$.*

It is proved in [11] that if A and B are fuzzy sets from $\mathbf{R}^d$ whose membership functions are upper semicontinuous with bounded support, then so is $A \oplus B$, and $[A \oplus B]_\alpha = [A]_\alpha \oplus [B]_\alpha$.

Definition 5 *Following [14], let us say that the points $x, y \in E$ are connected in the fuzzy set A if there exists a path Γ from x to y such that*

$$\inf_{z\in\Gamma} \mu_A(z) \geq \min[\mu_A(x), \mu_A(y)].$$

Then we may call the fuzzy set A $\alpha-$path-connected if every two points x and y, such that $\mu_A(x) \geq 1-\alpha$ and $\mu_A(y) \geq 1-\alpha$, are connected in A. Let consider further a fuzzy set whose membership function is u.s.c. with bounded support. Consider we are given a connectivity class $\mathcal{C}$ in the universal set E (either $\mathbf{R}^d$ or $\mathbf{Z}^d$). Then we may generalise the upper definition not only for path-connectivity saying that a fuzzy set A is $\alpha-$connected for any $\alpha \in [0, 1)$ if $[A]_\beta \subset \mathcal{C}$ for every positive β less than or equal to $1-\alpha$. For completeness we may call a set A to be 1-connected if its support is in the connectivity class, which means $\bigcup_\alpha [A]_\alpha \subset \mathcal{C}$. We made our definitions consonant with the Epsilon Geometry framework: in the fuzzy sense an image is 0-connected if and only if its essential parts are connected.

Applying directly Theorem 1 we show that if for a given $\alpha \in [0, 1]$ the fuzzy sets X and A are $\alpha-$connected, then $X \oplus A$ is also $\alpha-$connected. So we define the *degree of connectivity* of the fuzzy set A in the following way:

$$dc(A) = 1 - \inf\{\alpha \,|\, A \text{ is } \alpha - \text{connected}\}.$$

It is straightforward to show that $dc(A \oplus B) \leq \min[dc(A), dc(B)]$.

This is a general way to define degree of connectivity, which can be used as a feature in different pattern recognition tasks depending on the basic connectivity class. On Fig. 2 one can see an example of X-ray mammogram. The degree of connectivity of the outer darker part is 0.33, the light internal region has degree of connectivity 0.78. The degree of connectivity of the whole image is 0.24 disregarding the background. On the lower part of the image one can see two calcifications (the small light spots at the bottom). Therefore for any small region containing the calcifications, the degree of connectivity will be approximately 0.24. Any of the two calcifications themselves have high degree of connectivity ≈ 0.8. Therefore we can use the ratio between the degree of connectedness and the variance between the maximal and the minimal grey levels of the region as a feature of the region. If the value of this feature is small one could suppose the existence of abnormalities in the region.

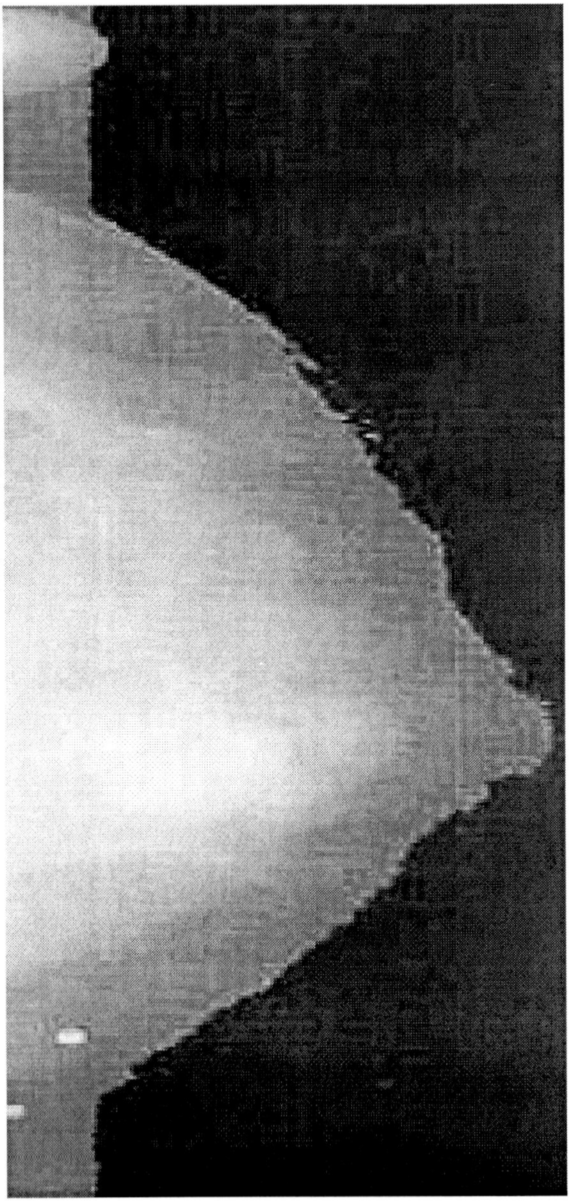

Fig. 2. An example of a mammogram with cancerous calcifications

6. Conclusions and Future Research

We have shown that we can deal with ε-connected sets when imprecise description of the objects is available. This approach gives a safer solution in the field of robotics and vehicle guidance. Also, a general definition of connectivity of fuzzy sets have been established, suited for direct implementation to grey-scale images. Since we work with adjoint fuzzy morphological operations, we have real opening and closing filters (idempotent, increasing, [anti]-extensive), which gives us the opportunity to apply fuzzy closing operation directly to the grey-scale image to remove sparse noise. Such noise may appear with the old CCD cameras when some pixels are dead, but also in the process of photocopying where some ink drops may damage the copy. Also, this noise can be considered as the most severe case of random noise. Therefore, if we demonstrate that a connectivity based filtering works well to such dropout noise, it is most likely that it can apply to any random noise. Since in the crisp case every convex set is connected, it should be interesting to find a relation between the values of the degree of convexity defined in [11], and the degree of connectivity of a given fuzzy set.

References

1. I. Bloch and H. Maître, Fuzzy mathematical morphologies: A comparative study, *Pattern Recognition*, Vol. 28(9), pp. 1341–1387, 1995.
2. I. Bloch, Fuzzy connectivity and mathematical morphology, *Pattern Recognition Letters*, Vol.14, 483–488, 1993.
3. L. Guibas, D. Salesin, J. Stolfi, Epsilon geometry: building robust algorithms from imprecise computations. *Proc. 5th Annual ACM Symposium on Computational Geometry*, pp. 100–111, 1983.
4. L. Guibas, D. Salesin, J. Stolfi, Constructing strongly convex approximate hulls with inaccurate primitives. *Algorithmica*, Vol. 9, pp. 534–560, 1993.
5. P. Diamond, P. Kloeden, Metric spaces of fuzzy sets, *Fuzzy sets and systems*, Vol. 35, pp. 241 – 249, 1990.
6. Haralick, R.M. and L. G. Shapiro, *Computer and robot vision, volume 1.* Addison-Wesley, 1992.
7. H. J. A. M. Heijmans. *Morphological image operators,* Academic Press, Boston 1994.
8. J. -C. Latombe, *Robot motion planning.* Kluwer Academic Publishers, Boston, 1991.
9. G. Matheron. *Random sets and integral geometry*, Wiley, New York 1975.
10. A. T. Popov, Morphological operations on fuzzy sets, *Proc. 5th Int. Conf. on Image Processing and its Applications*, Edinburgh, pp. 837 – 840, 1995.
11. A. T. Popov, Convexity indicators based on fuzzy morphology, *Pattern Recognition Letters*, 18 (3), pp. 259 –267, 1997.
12. C. Ronse, Openings: main properties, and how to construct them. *Unpublished,* 1990.
13. C. Ronse, Set theoretical algebraic approach to connectivity in continuous or digital spaces. *Journal of Mathematical Imaging and Vision*, 8, pp. 41 - 58, 1998.
14. A. Rosenfeld, The fuzzy geometry of image subsets. *Pattern Recognition Letters*, 2, pp. 311–317, 1984.
15. J. Serra. *Image analysis and mathematical morphology*, Academic Press, London 1982.
16. J. Serra. Mathematical morphology for complete lattices. In J. Serra, editor, *Image analysis and mathematical morphology*, vol. 2, Academic Press, London 1988.
17. J. Serra, Connectivity on complete lattices. *Journal of Mathematical Imaging and Vision*, 9, pp. 231 - 251, 1998.

18. D. Sinha and E. R. Dougherty, Fuzzification of Set Inclusion, Theory and Applications. *Fuzzy Sets and Systems*, 55, pp. 15–42, 1993.
19. Werman M. and Peleg S., *Min - max operators in texture analysis,* IEEE Trans. on Pattern Analysis and Machine Intelligence, 7(2), pp. 730–733, 1985.
20. L. Zadeh, *Fuzzy sets and their applications to cognitive processes.* Academic Press, London, 1975.

MULTIRESOLUTION CONNECTIVITY: AN AXIOMATIC APPROACH*

ULISSES M. BRAGA-NETO and JOHN GOUTSIAS
Center for Imaging Science
and Department of Electrical and Computer Engineering
The Johns Hopkins University
Baltimore, MD 21218, USA

Abstract. In this paper, we introduce an axiomatic framework for the notion of multiresolution connectivity on complete lattices. This framework extends the notion of connectivity classes, introduced by Serra in the late eighties. We introduce multiresolution connectivities by means of two equivalent notions: connectivity measures and connectivity pyramids. We present examples of multiresolution connectivities based on pyramids of dilations and of morphological sampling operators. We study the application of multiresolution connectivity to various image analysis tasks, such as pyramid decompositions, hierarchical segmentations, and multiresolution features.

Key words: Connectivity, Mathematical Morphology, Multiresolution Image Analysis.

1. Introduction

In this paper, we present an axiomatic framework for the notion of multiresolution connectivity. This approach extends the concept of a connectivity class, introduced by Serra in the late eighties [7]. In our framework, a connectivity class corresponds to a single-resolution connectivity.

Multiresolution connectivities are introduced by means of axiomatizations for the equivalent notions of connectivity measures and connectivity pyramids. A connectivity measure is a nonnegative function on the lattice of interest that quantifies the idea of a varying degree of connectivity. The equivalent notion of a connectivity pyramid consists of an upper semi-continuous family of nested connectivities, which extends the notion of connectivity classes to a multiresolution setting.

We have organized this paper as follows. In Section 2, we introduce notation and provide a brief overview of single-resolution connectivity. In Section 3, we deal with the axiomatic definition of multiresolution connectivity. In Section 4, we present examples of multiresolution connectivities based on pyramids of dilations and morphological sampling operators. In Section 5, we describe the relationship of multiresolution connectivity with multiresolution image analysis tools, such as pyramid decompositions, hierarchical segmentations and multiresolution features. Finally, in Section 6, we present our conclusions.

* This work was supported by the Office of Naval Research, Mathematical, Computer, and Information Sciences Division, under ONR Grant N00014-90-1345. The first author was also supported by the CNPq Scholarship 200725196-3.

2. Mathematical Preliminaries

In this section, we introduce notation and present a brief overview of concepts needed in the sequel. For more details, the reader is referred to [1, 3].

A *complete lattice* $\mathcal{L}$ is a partially ordered set (poset) in which every family of elements has an infimum and a supremum, denoted by $\bigwedge$ and $\bigvee$, respectively. If the set is totally ordered, it is called a *complete chain*. Following Serra [8], whenever we use the terms "lattice" and "chain" we mean "complete lattice" and "complete chain," respectively. By definition, every lattice $\mathcal{L}$ must posses a *least element* O and a *greatest element* I, given by $O = \bigwedge \mathcal{L}$ and $I = \bigvee \mathcal{L}$, respectively. A subset $\mathcal{S}$ of a lattice $\mathcal{L}$ is called a *sup-generating family* for $\mathcal{L}$ if every element of $\mathcal{L}$ can be written as the supremum of elements in $\mathcal{S}$. An element of the sup-generating family $\mathcal{S}$ is called a *sup-generator*. It is assumed that O is not a sup-generator; i.e., $O \notin \mathcal{S}$ (of course, O is sup-generated voidly by any family $\mathcal{S} \subseteq \mathcal{L}$). In this paper, subsets of $\mathcal{L}$ will be denoted by script letters, such as $\mathcal{C}, \mathcal{F}, \mathcal{G}, \mathcal{H}$. The elements of $\mathcal{L}$ are generally denoted by uppercase letters, such as A, B, C. In order to distinguish the elements of the sup-generating family $\mathcal{S}$, we denote them by lowercase letters, such as x, y, z. Given an element $A \in \mathcal{L}$, we define $\mathcal{S}(A) = \{x \in \mathcal{S} \mid x \leq A\}$ as the family of all sup-generators majorated by A.

Two lattices $\mathcal{L}$ and $\mathcal{L}'$ are *isomorphic* if there is a bijection $\psi : \mathcal{L} \to \mathcal{L}'$ that preserves the ordering, i.e., $A \leq B \iff \psi(A) \leq \psi(B)$, for $A, B \in \mathcal{L}$. The bijection ψ is said to be an *isomorphism* between the lattices $\mathcal{L}$ and $\mathcal{L}'$.

Given a lattice $\mathcal{L}$, with sup-generating family $\mathcal{S}$, a family $\mathcal{C} \subseteq \mathcal{L}$ is called a *connectivity class* on $\mathcal{L}$ if the following conditions are satisfied:

(*i*) $O \in \mathcal{C}$

(*ii*) $\mathcal{S} \subseteq \mathcal{C}$

(*iii*) if $C_i \in \mathcal{C}$ and $\bigwedge C_i \neq O$, then $\bigvee C_i \in \mathcal{C}$.

The elements of the connectivity class $\mathcal{C}$ are the *connected elements* of $\mathcal{L}$. It can be shown that a connectivity class $\mathcal{C}$ is uniquely determined by its family of *connectivity openings* $\{\gamma_x \mid x \in \mathcal{S}\}$, where $\gamma_x(A) = \bigvee\{C \in \mathcal{C} \mid x \leq C \leq A\}$.

A *partition*, or *segmentation*, of an element $A \in \mathcal{L}$ is a mapping P_A: $\mathcal{S}(A) \to \mathcal{L}$, such that:

(*i*) $x \leq P_A(x) \leq A$, for every $x \in \mathcal{S}(A)$

(*ii*) $P_A(x) = P_A(y)$ or $P_A(x) \wedge P_A(y) = O$, for every $x, y \in \mathcal{S}(A)$.

Each $P_A(x)$ is called a *zone* of the partition P_A of A. The partition is said to be *connected*, with respect to a given connectivity class $\mathcal{C}$, if $P_A(x) \in \mathcal{C}$, for all $x \in \mathcal{S}(A)$. For any two partitions P_A and $P_A{}'$, we say that P_A is *finer* than $P_A{}'$, which we denote by $P_A \sqsubseteq P_A{}'$, if $P_A(x) \leq P_A{}'(x)$, for each $x \in \mathcal{S}(A)$ (in this case, we may also say that $P_A{}'$ is *coarser* than P_A). It is easy to see that $\sqsubseteq$ defines a partial order on the set $\mathcal{T}_A$ of all partitions of A. As a matter of fact, the poset $\mathcal{T}_A$ is a complete lattice: the infimum is simply $(\sqcap P_A^i)(x) = \bigwedge P_A^i(x)$, while the supremum is given by $\sqcup P_A^i = \sqcap\{P_A \mid P_A \sqsupseteq P_A^i,\ i \in I\}$.

3. Multiresolution Connectivities

We now provide equivalent axiomatizations for the multiresolution notions of connectivity measures and connectivity pyramids. All proofs, as well as additional results and examples, can be found in [1].

Definition 1 *Let $\mathcal{L}$ be a lattice with sup-generating family $\mathcal{S}$. A function ϕ: $\mathcal{L} \to \overline{\mathbb{R}}$ is said to be a* connectivity measure *on $\mathcal{L}$ if:*
(i) $0 \leq \phi(A) \leq \infty$, *for all* $A \in \mathcal{L}$
(ii) $\phi(O) = \phi(x) = \sup\{\phi(A) \mid A \in \mathcal{L}\}$, *for* $x \in \mathcal{S}$
(iii) $\phi(\bigvee A_i) \geq \inf\{\phi(A_i)\}$, *for all* $A_i \in \mathcal{L}$ *such that* $\bigwedge A_i \neq O$.

The value $\phi(A)$ indicates the degree of connectivity of an element $A \in \mathcal{L}$. If $\phi(A) = 0$, A is said to be *disconnected*, whereas, if $\phi(A) = \sup\{\phi(A') \mid A' \in \mathcal{L}\}$, A is said to be *fully connected*. The zero element and the sup-generators are always fully connected. Condition (*iii*) requires that the degree of connectivity of the supremum of overlapping elements of $\mathcal{L}$ must not become smaller than the smallest connectivity measure of the individual elements. We say that ϕ is *discrete* if the range $\phi(\mathcal{L})$ does not have accumulation points in $\overline{\mathbb{R}}$.

A simple example of a discrete connectivity measure, with $\mathcal{L} = \mathcal{P}(\mathbb{Z}^2)$ and sup-generating family $\mathcal{S} = \{\{v\} \mid v \in \mathbb{Z}^2\}$, is given by $\phi(A) = 1$, if A is 4-connected, $\phi(A) = 0.5$, if A is 8- but not 4-connected, and $\phi(A) = 0$, otherwise.

Given a connectivity class $\mathcal{C}$ on $\mathcal{L}$, we define a simple binary connectivity measure by taking $\phi(A) = 1$, if $A \in \mathcal{C}$, and $\phi(A) = 0$, otherwise. Hence, standard connectivities may be viewed as *single-resolution* connectivities, where the degree of connectivity is all-or-nothing. We mention that a connectivity measure can be interpreted, with minor modifications, as the membership function of a *fuzzy connectivity class*, although we do not pursue this interpretation here.

In the sequel, unless stated otherwise, $J \subseteq \overline{\mathbb{R}}_+$ will be a closed index set, with $0 \in J$. We say that α is an *upper accumulation point* of J, if the interval $(\alpha - \epsilon, \alpha]$, for any $\epsilon > 0$, contains infinitely many points of J.

Definition 2 *Let $\mathcal{L}$ be a lattice with sup-generating family $\mathcal{S}$. A function* $\mathbf{C}$: $J \to \mathcal{P}(\mathcal{L})$ *is a* connectivity pyramid *on $\mathcal{L}$ if:*
(i) $\mathbf{C}(0) = \mathcal{L}$
(ii) $\mathbf{C}(\alpha)$ *is a connectivity class for each* $\alpha \in J$
(iii) $\mathbf{C}(\alpha) \subseteq \mathbf{C}(\beta)$, *if* $\alpha \geq \beta$
(iv) $\mathbf{C}(\alpha) = \bigcap_{\beta<\alpha} \mathbf{C}(\beta)$, *if α is an upper accumulation point of J.*

At times, it will be convenient to denote the connectivity classes $\mathbf{C}(\alpha)$ by $\mathcal{C}_\alpha$, and write $\mathbf{C} = \{\mathcal{C}_\alpha\}_{\alpha \in J}$. Condition (*iii*) requires the connectivity classes $\mathcal{C}_\alpha$ to be nested, so that the criterion for connectivity becomes stricter as α increases. On the other hand, the upper semi-continuity condition (*iv*) provides a smoothness constraint "from below" on the pyramid. If J does not contain any accumulation points, $\mathbf{C}$ is said to be a *discrete* connectivity pyramid, in which case condition (*iv*) becomes void.

Examples of connectivity pyramids, based on families of decreasingly connected graphs and families of increasingly finer topologies, are given in [1].

Given a connectivity pyramid $\mathbf{C} = \{\mathcal{C}_\alpha\}_{\alpha \in J}$ on $\mathcal{L}$, we associate a family of *α-connectivity openings* on $\mathcal{L}$, given by $\gamma_{\alpha,x}(A) = \bigvee\{C \in \mathcal{C}_\alpha \mid x \leq C \leq A\}$, for $\alpha \in J$, $x \in \mathcal{S}$. It can be shown that these α-connectivity openings satisfy nesting and upper semi-continuity properties similar to conditions (*iii*) and (*iv*) of connectivity pyramids. Moreover, a connectivity pyramid is uniquely determined by its family of α-connectivity openings.

The set of connectivity pyramids defined on J is a complete lattice $\mathcal{P}(\mathcal{L}, J)$, with partial order defined by $\mathbf{C} \leq \mathbf{C}' \iff \mathbf{C}(\alpha) \subseteq \mathbf{C}'(\alpha)$, $\alpha \in J$. This is implied [7] by the fact that $\mathcal{P}(\mathcal{L}, J)$ is an inf semi-lattice, where the infimum of a family $\mathbf{C}_i$ of connectivity pyramids is given by $\mathbf{C}(\alpha) = \bigcap \mathbf{C}_i(\alpha)$, $\alpha \in J$, with universal element $\overline{\mathbf{C}}(\alpha) = \mathcal{L}$, $\alpha \in J$. Similarly, the set of connectivity measures that take on values in J is a complete lattice $\mathcal{M}(\mathcal{L}, J)$, with the partial order defined by $\phi \leq \phi' \iff \phi(A) \leq \phi(A)$, $A \in \mathcal{L}$, since $\mathcal{M}(\mathcal{L}, J)$ is an inf semi-lattice, where the infimum of a family ϕ_i of connectivity measures is given by $\phi(A) = \inf\{\phi_i(A)\}$, $A \in \mathcal{L}$, with universal element $\overline{\phi}(A) = \sup J$, $A \in \mathcal{L}$.

In similar fashion to the "second generation" connectivity classes of [8], we have the following result.

Proposition 1 *Let $\mathbf{C} = \{\mathcal{C}_\alpha\}_{\alpha \in J}$ be a connectivity pyramid on a lattice $\mathcal{L}$ with sup-generating family $\mathcal{S}$, and let δ be an extensive dilation on $\mathcal{L}$ such that $\delta(x) \in \mathcal{C}_\alpha$, for every $x \in \mathcal{S}$ and $\alpha \in J$. Then,*

$$\mathbf{C}^\delta = \{\delta^{-1}(\mathcal{C}_\alpha)\}_{\alpha \in J}$$

defines a new connectivity pyramid on $\mathcal{L}$, which is less strict than the original connectivity pyramid; i.e., $\mathbf{C} \leq \mathbf{C}^\delta$.

The following result states the equivalence between the class of connectivity pyramids defined on J and the class of connectivity measures that take on values in J.

Theorem 1 *Let $\mathcal{L}$ be a lattice with sup-generating family $\mathcal{S}$. The lattice of connectivity measures $\mathcal{M}(\mathcal{L}, J)$ that take on values in J is isomorphic to the lattice $\mathcal{P}(\mathcal{L}, J)$ of connectivity pyramids defined on J. Moreover, the isomorphism X: $\mathcal{M}(\mathcal{L}, J) \to \mathcal{P}(\mathcal{L}, J)$ is given by*

$$X(\phi)(\alpha) = \{A \in \mathcal{L} \mid \phi(A) \geq \alpha\}, \quad \alpha \in J, \tag{1}$$

with inverse Y: $\mathcal{P}(\mathcal{L}, J) \to \mathcal{M}(\mathcal{L}, J)$ given by

$$Y(\mathbf{C})(A) = \sup\{\alpha \in J \mid A \in \mathbf{C}(\alpha)\}, \quad A \in \mathcal{L}. \tag{2}$$

A multiresolution connectivity assumes either a connectivity measure ϕ, or a connectivity pyramid $\mathbf{C}$, where the latter can also be specified by a family of α-connectivity openings $\{\gamma_{\alpha,x} \mid \alpha \in J, x \in \mathcal{S}\}$. Given any one of these three *equivalent* ways to specify a multiresolution connectivity, one can refer liberally to the other two. We say that an element $A \in \mathcal{L}$ is *α-connected* if one of the three equivalent conditions are satisfied: (a) $\phi(A) \geq \alpha$, or (b) $A \in \mathcal{C}_\alpha$, or (c) $\gamma_{\alpha,x}(A) = A$, for $x \in \mathcal{S}(A)$. An *α-connected component*, or *α-grain*, of $A \in \mathcal{L}$ is an α-connected element $C \in \mathcal{L}$ such that $C \leq A$ and there is no α-connected element $C' \in \mathcal{L}$ with $C \leq C' \leq A$. It can be shown that the family of α-grains of A is given by $\mathcal{C}_\alpha(A) = \{\gamma_{\alpha,x}(A) \mid x \in \mathcal{S}(A)\}$.

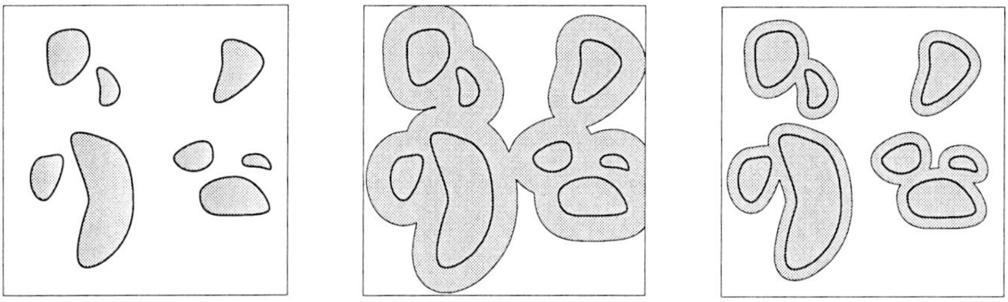

A $\quad$ $A \oplus (m-\alpha)B$ $\quad$ $A \oplus (m-\beta)B$

Fig. 1. Dilation-based multiresolution connectivity. Since $A \oplus (m-\alpha)B$ is connected, but $A \oplus (m-\beta)B$ is not, it follows that $\alpha \le \phi(A) < \beta$.

4. Multiresolution Connectivities based on Pyramids of Operators

A family $\{\psi_\alpha \mid \alpha \in J\}$ of operators on a lattice $\mathcal{L}$, where J is an arbitrary index set of real numbers, is said to be a *bottom-up* (resp. *top-down*) *pyramid of operators* if there exists a $\gamma \in J$ such that $\psi_\alpha = \psi_\gamma \psi_\beta$ (resp. $\psi_\beta = \psi_\gamma \psi_\alpha$), for all $\alpha \ge \beta$ [6]. The next two subsections present useful multiresolution connectivities based on top-down pyramids of morphological operators.

4.1. Dilation-Based Multiresolution Connectivity

Consider the dilations $\delta_\alpha(A) = A \oplus (m-\alpha)B$, for $\alpha \in J$, defined on $\mathcal{K}(\mathbb{R}^d)$ (the lattice of all compact subsets of $\mathbb{R}^d$ in the usual Euclidean topology), where B is the closed unit ball and $J = [0, m]$. It is easy to see that $\{\delta_\alpha \mid \alpha \in J\}$ defines a top-down pyramid of dilations.

Proposition 2 *Let $\mathcal{L} = \mathcal{K}(\mathbb{R}^d)$ with sup-generating family $\mathcal{S} = \{\{v\} \mid v \in \mathbb{R}^d\}$, furnished with the usual Euclidean topological connectivity. Let $\{\delta_\alpha \mid \alpha \in J\}$ be the top-down pyramid of dilations defined above. Then, $\mathbf{C} = \{\mathcal{C}_\alpha\}_{\alpha \in J}$, where*

$$\mathcal{C}_0 = \mathcal{L}; \qquad \mathcal{C}_\alpha = \{A \in \mathcal{L} \mid \delta_\alpha(A) \in \mathcal{C}\}, \quad \alpha \subset J \setminus \{0\},$$

defines a connectivity pyramid on $\mathcal{K}(\mathbb{R}^d)$.

The associated connectivity measure is clearly given by

$$\begin{aligned} \phi(A) &= \sup \{\alpha \in J \mid A \oplus (m-\alpha)B \text{ is connected}\} \\ &= m - \inf \{\alpha \in J \mid A \oplus \alpha B \text{ is connected}\}. \end{aligned}$$

Figure 1 provides an illustration of this multiresolution framework. Note that 2-D digital versions of the dilation-based multiresolution connectivity discussed here can be defined as well.

4.2. Multiresolution Connectivity Based on Morphological Sampling

Another interesting example of multiresolution connectivity, based on morphological sampling [4], was suggested by Henk Heijmans.[1] It turns out that the

[1] Personal communication.

resulting multiresolution connectivity scheme is based on a top-down pyramid of closings. By necessity, the discussion in this subsection will be very brief. We refer the reader to [1, 3, 4] for more details.

Consider, as the domain of definition of images, the lattice $\mathcal{F}(\mathbb{R}^d)$ of all closed subsets of $\mathbb{R}^d$. Let the *sampling grid* $S \subset \mathbb{R}^d$ be given by $S = \{k_1 u_1 + \cdots + k_d u_d \mid k_i \in \mathbb{Z}\}$, where u_i are linearly independent vectors in $\mathbb{R}^d$. Let the *sampling element* $C \subset \mathbb{R}^d$ be an open set such that $0 \in C$, $C \cap S = \{0\}$ and $S \oplus C = \mathbb{R}^d$. This last condition is known as the *covering assumption*. For instance, we may take $C = \{x_1 u_1 + \cdots + x_d u_d \mid -a < x_i < a\}$, where $a > 1/2$. Given the above conditions, one can show that $\sigma(A) = \{s \in S \mid C_s \cap A \neq \emptyset\}$ defines a dilation from $\mathcal{F}(\mathbb{R}^d)$ into $\mathcal{P}(S)$, called the *sampling operator*. It can also be shown that the adjoint erosion from $\mathcal{P}(S)$ into $\mathcal{F}(\mathbb{R}^d)$, known as the *reconstruction operator*, is given by $\rho(V) = \{v \in \mathbb{R}^d \mid \check{C}_v \cap S \subseteq V\}$, where $\check{C} = \{-v \mid v \in C\}$ is the reflection of C. Thus, the composition $\pi = \rho\sigma$ defines a closing from $\mathcal{F}(\mathbb{R}^d)$ into itself. This is known as the *approximation operator*.

Proposition 3 *Let $\mathcal{C}$ be a connectivity class on $\mathcal{F}(\mathbb{R}^d)$ and let C be a sampling element such that, for every $s \in S$,*

$$C_s \setminus \bigcup_{s' \in S'} C_{s'} \in \mathcal{C}, \quad \forall S' \subseteq S \setminus \{s\}. \tag{3}$$

Then, the approximation operator π is connectivity-preserving; i.e., $\pi(\mathcal{C}) \subseteq \mathcal{C}$.

We remark that condition (3) is actually easy to check in practical situations (it usually involves a small finite number of tests).

In order to extend the above theory to a multiresolution setting, we consider the morphological sampling approach known as *covering discretization* [3]. Consider the sampling grid S, defined previously, and let the sampling element be $C = \{x_1 u_1 + \cdots + x_d u_d \mid -1 < x_i < 1\}$. Define

$$S_n = \frac{1}{2^{n-1}} S, \qquad C_n = \frac{1}{2^{n-1}} C, \qquad n \geq 1. \tag{4}$$

It can be easily checked that the sampling elements C_n satisfy condition (3), for each $n \geq 1$, with the usual topological connectivity in $\mathbb{R}^d$. In addition, $S_n \oplus C_n = \mathbb{R}^d$, so that the covering assumption is satisfied for each $n \geq 1$. This leads to a family of approximation operators $\{\pi_n \mid n \geq 1\}$, which can be shown to define an anti-granulometry on $\mathcal{F}(\mathbb{R}^d)$; i.e., a top-down pyramid of closings. We now arrive at the main result of this subsection.

Proposition 4 *(Multiresolution connectivity based on morphological sampling). Let $\mathcal{L} = \mathcal{F}(\mathbb{R}^d)$ with sup-generating family $\mathcal{S} = \{\{v\} \mid v \in \mathbb{R}^d\}$, furnished with a connectivity class $\mathcal{C}$. Let C_n be the sampling elements discussed in connection with (4), and let $\{\pi_n \mid n \geq 1\}$ be the associated family of approximation operators. If condition (3) is satisfied for the underlying connectivity class $\mathcal{C}$, then $\mathbf{C} = \{\mathcal{C}_n\}_{0 \leq n \leq \infty}$, where*

$$\mathcal{C}_0 = \mathcal{F}(\mathbb{R}^d); \quad \mathcal{C}_n = \pi_n^{-1}(\mathcal{C}) = \{A \in \mathcal{L} \mid \pi_n(A) \in \mathcal{C}\},\ 0 < n < \infty; \quad \mathcal{C}_\infty = \mathcal{C},$$

defines a discrete connectivity pyramid on $\mathcal{F}(\mathbb{R}^d)$.

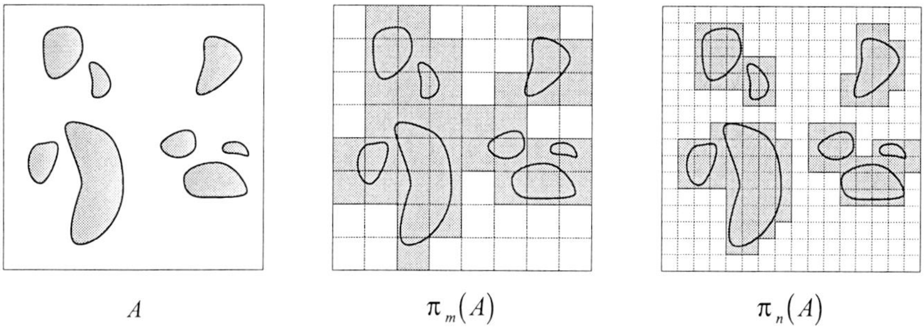

Fig. 2. Multiresolution connectivity based on morphological sampling. Since $\pi_m(A)$ is connected, but $\pi_n(A)$ is not, it follows that $m \leq \phi(A) < n$.

Figure 2 illustrates this multiresolution connectivity scheme.

5. Multiresolution Tools

In this section, we briefly discuss the application of multiresolution connectivities to a number of useful multiresolution image analysis tasks.

5.1. Pyramid Decompositions

In the case of a discrete index set J (e.g., $J = \{0, 1, \ldots, m\}$), a multiresolution connectivity can be used to derive a pyramid representation scheme (we refer the reader to [2] for more details on pyramid decompositions).

Given a lattice $\mathcal{L}$ with sup-generating family $\mathcal{S}$, we define the *root marker set* $\mathcal{R}$ to be a non-empty subset of $\mathcal{S}$; i.e., $\emptyset \neq \mathcal{R} \subseteq \mathcal{S}$. The analysis and synthesis operators of the pyramid scheme are given by

$$\psi_j^{\uparrow} = \bigvee_{x \in \mathcal{R}} \gamma_{j+1,x} \quad \text{(Analysis)}, \tag{5}$$

$$\psi_j^{\downarrow} = \mathbf{id} \qquad \text{(Synthesis)}, \tag{6}$$

respectively, where $\{\gamma_{j,x} \mid j \in J, x \in \mathcal{S}\}$ is the family of α-connectivity openings associated with the multiresolution connectivity.

Assume that there exist addition and subtraction operations $+, -$ defined on $\mathcal{L}$, such that $\psi_j^{\uparrow}(A) + (A - \psi_j^{\uparrow}(A)) = A$, for every given $A \in \mathcal{L}$ (e.g., in the binary case, we may take $+$ to be set union and $-$ to be set difference). The analysis operators then give rise to a pyramid decomposition: $A \to \{D_0, A_1\} \to \{D_0, D_1, A_2\} \to \cdots \to \{D_0, D_1, \ldots, D_{m-1}, A_m\}$, where

$$\begin{cases} A_0 &= A \\ A_{j+1} &= \psi_j^{\uparrow}(A_j) = \bigvee_{x \in \mathcal{R}} \gamma_{j+1,x}(A_j) \\ D_j &= A_j - A_{j+1}, \end{cases} \tag{7}$$

such that A can be recovered from the *approximation signal* A_m and the *detail signals* $\{D_j\}$ via a simple summation, $A = A_m + \sum_{j=0}^{m-1} D_j$. Note that the above

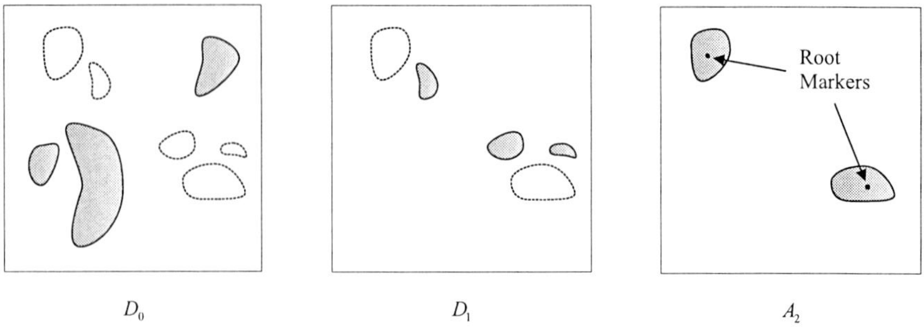

Fig. 3. Pyramid decomposition of the image A in Fig. 2, assuming dilation-based multiresolution connectivity. The root marker set $\mathcal{R}$ is indicated on the approximation image A_2. Note that the sum of D_0, D_1 and A_2 recovers the original image A.

decomposition scheme depends on the particular root marker set $\mathcal{R}$. The name "root marker" comes from the fact that the m-connected components marked by one of the elements of $\mathcal{R}$ are present in all approximation signals A_j.

Figure 3 depicts a binary example that shows three levels of a pyramid decomposition, based on the above scheme, where dilation-based multiresolution connectivity is assumed.

5.2. Hierarchical Segmentation

The concept of hierarchical segmentation is of fundamental importance in the framework of multiresolution-based applications, such as adaptive bit-rate object-oriented coding [5]. Below, we formalize this notion in the context of multiresolution connectivities.

Definition 3 *Let $\mathcal{L}$ be a lattice with sup-generating family $\mathcal{S}$. A* hierarchical partition *or* hierarchical segmentation *of A is a family $\mathcal{P}_A = \{P_A^\alpha \mid \alpha \in J\}$, where P_A^α are partitions of A such that:*
(i) $P_A^\alpha \sqsubseteq P_A^\beta$, if $\alpha \geq \beta$
(ii) $P_A^\alpha = \sqcap_{\beta<\alpha} P_A^\beta$, if α is an upper accumulation point of J.
Each $P_A^\alpha(x)$ is called an α-zone of the hierarchical partition. Moreover, we say that $\mathcal{P}_A$ is connected, *with respect to a given multiresolution connectivity $\mathbf{C} = \{\mathcal{C}_\alpha\}_{\alpha\in J}$, if $P_A^\alpha(x) \in \mathcal{C}_\alpha$, for all $x \in \mathcal{S}(A)$ and $\alpha \in J$.*

Condition (i) above says that a hierarchical partition of A is a family of increasingly finer partitions of A. On the other hand, condition (ii) imposes a smoothness requirement on this family.

Proposition 5 *Let $\mathcal{L}$ be a lattice with sup-generating family $\mathcal{S}$, furnished with a multiresolution connectivity, given by the family of α-connectivity openings $\{\gamma_{\alpha,x} \mid x \in \mathcal{S}, \alpha \in J\}$. Given an element $A \in \mathcal{L}$, the mappings $C_A^\alpha : \mathcal{S}(A) \to \mathcal{L}$ defined by*

$$C_A^\alpha(x) = \gamma_{\alpha,x}(A), \quad x \in \mathcal{S}(A), \, \alpha \in J, \tag{8}$$

provide a connected hierarchical partition $\mathcal{C}_A = \{C_A^\alpha \mid \alpha \in J\}$ of A.

We refer to the hierarchical partition $\mathcal{C}_A$ as the *hierarchical partition of connected components* of A. Note that, since in a multiresolution connectivity we have that $\mathcal{C}_0 = \mathcal{L}$, the bottom level of a hierarchical partition of connected components corresponds to the original object.

5.3. Multiresolution Connectivity-Based Image Features

Given a lattice $\mathcal{L}$ and an element $A \in \mathcal{L}$, a *partition feature* is a *monotone* mapping ν: $\mathcal{T}_A \to \overline{\mathbb{R}}$, where $\mathcal{T}_A$ is the lattice of partitions of A. By a monotone mapping, it is meant that either ν is increasing, i.e., $P_A \sqsubseteq P_A' \Rightarrow \nu(P_A) \leq \nu(P_A')$, or that ν is decreasing, i.e., $P_A \sqsubseteq P_A' \Rightarrow \nu(P_A) \geq \nu(P_A')$. An increasing (resp. decreasing) partition feature ν is *upper semi-continuous* if, for any chain $\{P_A^i \mid i \in I\}$ of increasingly finer partitions in $\mathcal{T}_A$, we have that $\nu(\sqcap P_A^i) = \inf\{\nu(P_A^i)\}$ (resp. $\sup\{\nu(P_A^i)\}$).

A useful partition feature is the *counting feature* μ, given by $\mu(P_A)$ = number of zones of P_A, if this number is finite, or $\mu(P_A) = \infty$, otherwise. This feature is clearly decreasing. In addition, we have the following result.

Proposition 6 *The counting feature μ is upper semi-continuous on $\mathcal{T}_A$.*

Now, consider a multiresolution connectivity $\mathbf{C} = \{\mathcal{C}_\alpha\}_{\alpha \in J}$ on $\mathcal{L}$, where $J = [0, m]$ is an interval of real or integer numbers (we may have $m = \infty$). Let $A \in \mathcal{L}$ and let ν be a partition feature on $\mathcal{T}_A$. We define the *clustering curve* X_A^ν of A with respect to ν to be the function X_A^ν: $J \to \overline{\mathbb{R}}$ given by:

$$X_A^\nu(\alpha) = \nu(C_A^\alpha), \quad \alpha \in J, \tag{9}$$

where C_A^α is level α of the hierarchical partition of connected components of A, defined in (8).

Proposition 7 *Let $J = [0, m]$ be a continuous interval. We have that:*
(i) The clustering curve $X_A^\nu : J \to \overline{\mathbb{R}}$ is a monotone, continuous function on J, except for a countable number of jump discontinuities.
(ii) If ν is upper semi-continuous, then X_A^ν is left-continuous at the jumps.

The variation in X_A^ν indicates how the connectivity feature changes as clusters of components break apart or merge. For the counting feature μ, X_A^μ gives the variation of the number of α-grains of A. From the previous discussion, it is clear that X_A^μ is an increasing, piecewise constant, left-continuous function.

We define the *clustering spectrum* Y_A^ν of A as the derivative of X_A^ν:

$$Y_A^\nu(\alpha) = \begin{cases} \frac{d}{d\alpha} X_A^\nu(\alpha), \ \alpha \in [0, m], & \text{if } J \text{ is continuous} \\ X_A^\nu(\alpha) - X_A^\nu(\alpha - 1), \ \alpha = 1, 2, \ldots, m, & \text{if } J \text{ is discrete} \end{cases}$$

It is understood that, at points α where X_A^ν has jump discontinuities, $Y_A^\nu(\alpha)$ will be assigned an impulse of magnitude equal to the jump. Figure 4 illustrates the clustering curve and the clustering spectrum of a binary image.

We remark that, in the case of a binary lattice (e.g., when $\mathcal{L} = \mathcal{P}(\mathbb{R}^2)$), the counting feature can be applied to derive other interesting topological multiresolution features. For instance, the curve $X_{A^c}^\mu$ gives the variation of the number of *pores* of A, while the curve $X_A^\mu - X_{A^c}^\mu$, gives the variation of the *genus* of A.

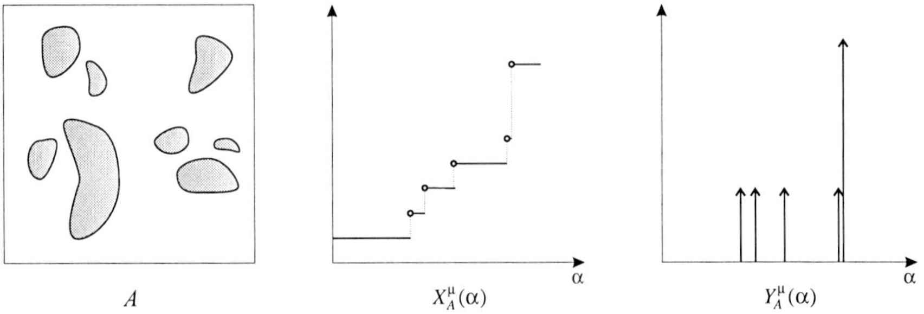

Fig. 4. Clustering curve X_A^μ and clustering spectrum Y_A^μ associated with the image on the left, in the case of a dilation-based multiresolution connectivity, with the counting feature μ (the vertical axes of the plots are not drawn to the same scale).

6. Concluding Remarks

In this paper, we have presented an axiomatic formulation of multiresolution connectivity, which extends the single-resolution notion of connectivity classes. We have presented multiresolution connectivity examples based on pyramids of dilations and of morphological sampling operators. We have also demonstrated the use of this framework in some useful multiresolution image analysis tasks.

An interesting issue that was not discussed in this paper is the relationship between connected operators [6] and multiresolution connectivities. This problem is studied in [1]. In this reference, we present an additional example of multiresolution connectivity based on a pyramid of operators, namely, a pyramid of openings. Moreover, we investigate the concept of hyperconnectivity [8], which provides a better framework for connectivity on graylevel and multispectral images, and extend this notion to the multiresolution case.

References

1. U. M. Braga-Neto and J. Goutsias, "A multiresolution approach to connectivity on complete lattices," Tech. Rep. JHU/ECE, Department of Electrical and Computer Engineering, The Johns Hopkins University, Baltimore, MD, Under Preparation.
2. J. Goutsias and H. J. A. M. Heijmans, "Multiresolution signal decomposition schemes. Part 1: Linear and morphological pyramids," Technical Report PNA–R9810, CWI, Amsterdam, The Netherlands, October 1998.
3. H. J. A. M. Heijmans, *Morphological Image Operators.* Boston, Massachusetts: Academic Press, 1994.
4. H. J. A. M. Heijmans and A. Toet, "Morphological sampling," *Computer Vision, Graphics, and Image Processing: Image Understanding*, vol. 54, pp. 384–400, 1991.
5. P. Salembier, "Morphological multiscale segmentation for image coding," *Signal Processing*, vol. 38, pp. 359–386, 1994.
6. P. Salembier and J. Serra, "Flat zones filtering, connected operators, and filters by reconstruction," *IEEE Transactions on Image Processing*, vol. 4, pp. 1153–1160, 1995.
7. J. Serra, ed., *Image Analysis and Mathematical Morphology. Volume 2: Theoretical Advances.* London, England: Academic Press, 1988.
8. J. Serra, "Connectivity on complete lattices," *Journal of Mathematical Imaging and Vision*, vol. 9, pp. 231–251, 1998.

CONNECTED OPERATORS BASED ON REGION-TREE PRUNING

PHILIPPE SALEMBIER and LUIS GARRIDO
Universitat Politècnica de Catalunya, Barcelona, SPAIN
philippe@gps.tsc.upc.es, oster@gps.tsc.upc.es

Abstract. This paper discusses region-based representations useful to create connected operators. The filtering approach involves three steps: first, a region tree representation of the input image is constructed. Second, the simplification is obtained by pruning the tree and third, and output image is constructed from the pruned tree. The paper focuses in particular on several pruning strategies that can be used on tree representation.

Key words: Connected Operator, Max-Tree, Min-Tree, Binary Partition Tree, Viterbi.

1. Introduction

Filtering techniques commonly used in image processing are defined by an input/output relationship that relies on a specific signal $h(x)$ called impulse response, window or structuring element. The three classical cases are:

Linear convolution and impulse response: the output of a linear translation-invariant system is given by: $\psi_h(f)(x) = \sum_{k=-\infty}^{\infty} h(k)f(x-k)$. The impulse response, $h(x)$, defines the filter properties. For image processing, the main drawback of linear filters is the blurring they introduce. The blurring characteristics is directly related to the extension and shape of the impulse response.

Median filter and window: considering a window W, the output of a median filter is defined by: $\psi_W(f)(x) = Median_{k \in W}\{f(x-k)\}$. Here also, the basic properties of the filter are defined by its window. The major drawback of this filtering strategy is that every region tends to be round after filtering with most commonly used windows (circles, squares, etc.).

Morphological erosion/dilation and structuring elements: dilation by a structuring element $h(x)$ is defined in a way similar to the convolution: $\delta_h(f)(x) = \bigvee_{k=-\infty}^{\infty}(h(k)+f(x-k))$, where $\bigvee$ denotes the supremum. The erosion is given by $\epsilon_h(f)(x) = \bigwedge_{k=-\infty}^{\infty}(h(k) - f(x+k))$, where $\bigwedge$ denotes the infimum. Based on these two primitives, morphological opening: $\gamma_h(f) = \delta_h(\epsilon_h(f))$, and closing: $\varphi_h(f) = \epsilon_h(\delta_h(f))$, can be constructed. These operators also introduce severe distortions due to the shape of the structuring element.

Most people would say that the heart of the filter design is to appropriately select the impulse response, the window or the structuring element. However, for image processing, this selection implies some drawbacks. Since $h(x)$ (or W) is not related at all with the input signal, its shape introduces severe distortions in the output. Many connected operators used in practice choose a completely different approach: the filtering is done without using any specific signal such

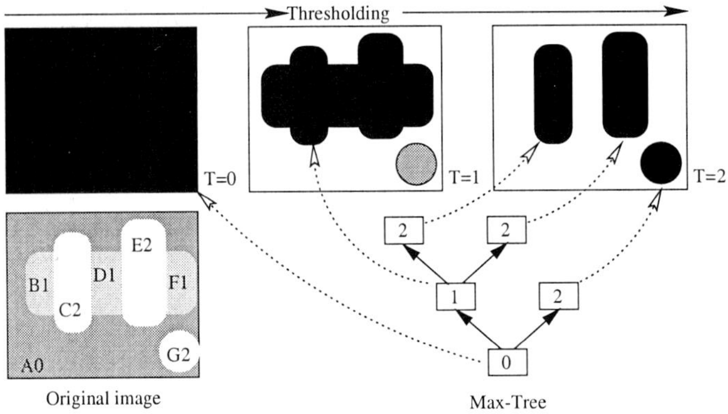

Fig. 1. Max-tree representation of images.

as an impulse response, a window or a structuring element. As a result, no distortion related to a priori selected signals is introduced in the output. Gray level connected operators [6] act by merging of elementary regions called flat zones. They cannot create new contours and, as a result, they cannot introduce in the output a structure that is not present in the input. Furthermore, they cannot modify the position of existing boundaries between regions and, therefore, have very good contour preservation properties. Several approaches can be used to create connected operators. One of the most popular approach consists in using the classical pixel-based representation of the image and a reconstruction process [7, 2]. An alternative approach relies on the definition of a region-based representation of the image and the definition of a region merging process [5, 4]. The goal of this paper is to discuss this second approach assuming that the region-based representation is a tree. The organization of this paper is as follows. Section 2 defines two region tree representations: the Max-tree (or Min-tree) and the Binary Partition Tree. The filtering strategies are discussed in section 3. Conclusions are reported in section 4.

2. Region Tree Representations

2.1. MAX-TREE AND MIN-TREE

The first tree representation is called a Max-tree [5]. This representation enhances the maxima of the signal. Each node $\mathcal{N}_k$ in the tree represents a connected component of the space that is extracted by the following thresholding process: for a given threshold T, consider the set of pixels X of gray level value larger than T and the set of pixels Y of gray level value equal to T:

$$X = \{x \text{ , such that } f(x) \geq T\} \text{ and } Y = \{x \text{ , such that } f(x) = T\} \quad (1)$$

The tree nodes $\mathcal{N}_k$ represent the connected components of X such that $Y \neq \emptyset$. An example of Max-tree is shown in Fig. 1. The original image is made of 7

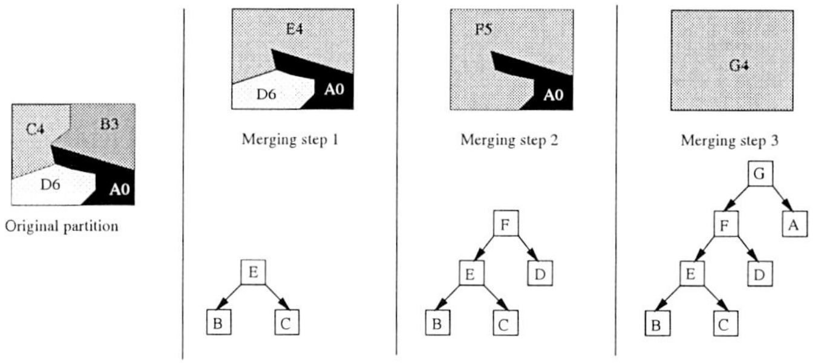

Fig. 2. Example of Binary Partition Tree creation with a region merging algorithm.

flat zones: {A,...,G}. The number following each letter defines the gray level value of the flat zones. The binary images, X, resulting from the thresholding with $0 \leq T \leq 2$ are shown in the center of the figure. Finally, the Max-tree is given in the right side. It is composed of 5 nodes that represent the connected components shown in black. The number inside each square represents the threshold value where the component was extracted. Finally, the links in the tree represent the inclusion relationships among the connected components following the threshold values. Note that when the threshold is set to $T = 1$, the circular component does not create a connected component that is represented in the tree because none of its pixels has a gray level value equal to 1. However, the circle itself is obtained when $T = 2$. The regional maxima are represented by three leaves and the tree root represents the entire image support.

2.2. Binary Partition Tree

The second example of region-based representation is the Binary Partition Tree [4]. It represents a set of regions that can be obtained from the partition of flat zones. The leaves of the tree represent the flat zones of the original signal. The remaining nodes represent regions that are obtained by merging the regions represented by the children. As in the cases of the Max-tree and Min-tree, the root node represents the entire image support. This representation should be considered as a compromise between representation accuracy and processing efficiency. Indeed, all possible merging of flat zones are not represented in the tree. Only the most "useful" ones are represented. However, as will be seen in the sequel, the main advantage of the tree representation is that it allows the fast implementation of sophisticated processing techniques.

The Binary Partition Tree should be created in such a way that the most "useful" regions are represented. This issue can be application dependent. However, a possible solution, suitable for a large number of cases, is to create the tree by keeping track of the merging steps performed by a segmentation algorithm based on region merging (see [3, 1]). In the following, this information is called the *merging sequence*. Starting from the partition of flat zones, the algorithm merges neighboring regions following a homogeneity criterion until a single region is obtained. An example is shown in Fig. 2. The original partition involves four regions. The regions are indicated by a letter and the number indicates the grey level value of the flat zone. The algorithm merges the four regions in three steps. In the first step, the pair of most similar regions, B and C, are merged to create region E. Then, region E is merged with region D to create region F. Finally, region F is merged with region A and this creates region G corresponding to the region of support of the whole image. In this example, the merging sequence is: $(B, C)|(E, D)|(F, A)$. This merging sequence progressively defines the Binary Partition Tree as shown in Fig. 2.

To create the Binary Partition Trees used in this paper, the merging algorithm following the color homogeneity criterion described in [1] has been used. It should be noticed however that the homogeneity criterion has not to be restricted to color. For example, if the image for which we create the Binary Partition Tree belongs to a sequence of images, motion information should also be used to generate the tree: in a first stage, regions are merged using a color homogeneity criterion, whereas a motion homogeneity criterion is used in the second stage. Furthermore, additional information of previous processing or detection algorithms can also be used to generate the tree in a more robust way. For instance, an object mask can be used to impose constraints on the merging algorithm in such a way that the object itself is represented as a single node in the tree. Typical examples of such algorithms are face, skin, character or foreground object detection. If the functions used to create the tree are self-dual, the tree itself is a self-dual representation appropriate to derive self-dual connected operators. By contrast, the Max-tree (Min-tree) is adequate for anti-extensive (extensive) connected operators. Note that in all cases, the trees are hierarchical region-based representations. They encode a large set of regions and partitions that can be derived from the flat zones partition of the original image without adding new contours.

3. Filtering Strategy

Once the tree representation has been created, the filtering strategy consists in *pruning* the tree and in reconstructing an image from the pruned tree. The simplification is performed by pruning because the idea is to eliminate the image components that are represented by the leaves and branches of the tree. The nature of these components depends on the tree. In the case of Max-trees (Min-trees), the components that may be eliminated are regional maxima (minima) whereas the elements that may be simplified in the case of Binary Partition Trees are unions of the most similar flat zones. The simplification itself is governed by a criterion which may involve simple notions such as size, contrast or more complex ones such as texture, motion or even semantic criteria.

3.1. Increasing criteria

One of the interests of the tree representations is that the set of possible merging steps is fixed (defined by the tree branches). As a result, a large number of simplification (pruning) strategies may be designed. A typical example deals with non-increasing simplification criteria. A criterion $\mathcal{C}$ assessed on a region

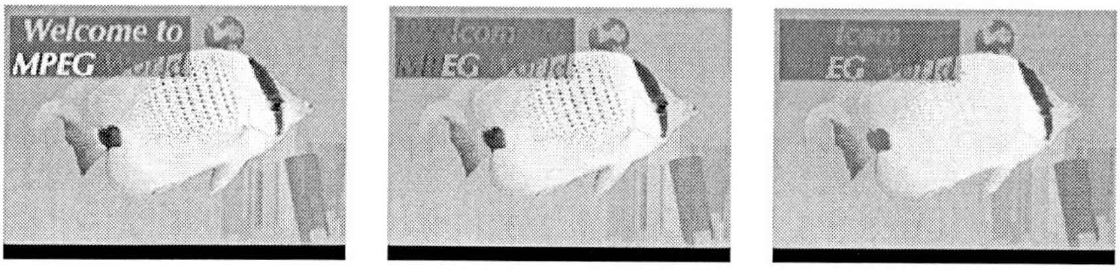

Fig. 3. Area filtering: original (left), area opening (center), area open-close (right).

R is increasing iff: $\forall R_1 \subseteq R_2 \Rightarrow \mathcal{C}(R_1) \leq \mathcal{C}(R_2)$. Assume that nodes where the criterion value is lower than a given threshold should be removed by merging. If the criterion is increasing, the pruning strategy is straightforward because if a node has to be removed all its descendants have also to be removed. A typical example is the area opening [8]. One of its possible implementation consists in creating a Max-tree and in measuring the area (the number of pixels) $\mathcal{A}_k$ contained in each node $\mathcal{N}_k$. If the area $\mathcal{A}_k$ is smaller than a threshold, $\mathcal{T}_\mathcal{A}$, the node is removed. The simplification effect of the area opening is illustrated in Fig. 3. The operator removes small bright components. If the simplified image is processed by the dual operator, the area closing, small dark components are also removed. Using the same strategy, a large number of connected operators can be obtained.

3.2. Non-increasing criteria

If the criterion is not increasing, the pruning strategy is not trivial since the descendants of a node to be removed have not necessarily to be removed. In practice, the non-increasingness of the criterion implies a lack of robustness of the operator [5]. For example, similar images may produce quite different results or small modifications of the criterion threshold involve drastic changes on the output. A possible solution consists in applying a transformation on the set of decisions. The transformation should create a set of increasing decisions while preserving as much as possible the decisions defined by the criterion. This problem may be viewed as dynamic programming issue that can be efficiently solved with the Viterbi algorithm.

The dynamic programming algorithm will be explained and illustrated in the sequel on a binary tree (see Fig. 4). The extension to N-ary trees is straightforward. The trellis on which the Viterbi algorithm is applied has the same structure as the region tree except that two trellis states, *preserve* $\mathcal{N}_k^P$ and *remove* $\mathcal{N}_k^R$, correspond to each node $\mathcal{N}_k$ of the tree. The two states of each child node are connected to the two states of its parent. However, to avoid non-increasing decisions, the *preserve* state of a child is not connected to the *remove* state of its parent. As a result, the trellis structure guarantees that if a node has to be removed its children have also to be removed. The cost associated to each state is used to compute the number of modifications the algorithm has to do to create an increasing set of decisions. If the criterion value states that the node of the tree has to be removed, the cost associated to

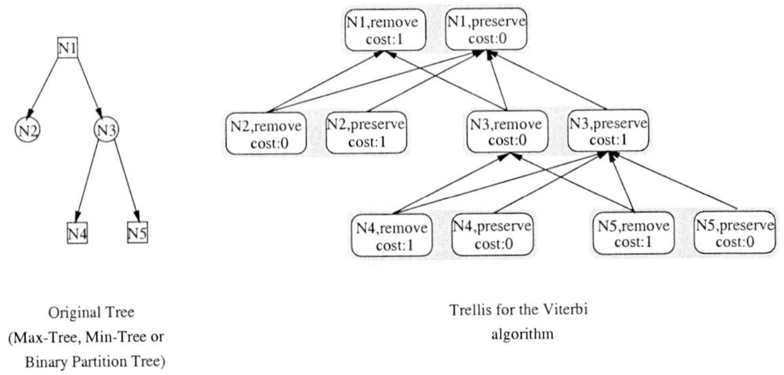

Fig. 4. Creation of the trellis for the Viterbi algorithm. A circular (square) node on the Tree indicates that the criterion value states that the node has to be removed (preserved).

the *remove* state is equal to zero (no modification) and the cost associated to the *preserve* state is equal to one (one modification). Similarly, if the criterion value states that the node has to be preserved, the cost of the *remove* state is equal to one and the cost of the *preserve* state is equal to zero. The cost values appearing in Fig. 4 assume that nodes $\mathcal{N}_1$, $\mathcal{N}_4$ and $\mathcal{N}_5$ should be preserved and that $\mathcal{N}_2$ and $\mathcal{N}_3$ should be removed. The goal of the Viterbi algorithm is to define the set of increasing decisions such that $\sum_k \text{Cost}(\mathcal{N}_k)$ is minimized.

To find the optimum set of decisions, a set of paths going from all leaf nodes to the root node is created. For each node, the path can go through either the *preserve* or the *remove* state of the trellis. The Viterbi algorithm is used to find the paths that minimize the global cost at the root node. The optimization is achieved in a bottom-up iterative fashion. For each node, it is possible to define the optimum paths ending at the *preserve* state and at the *remove* state:

Let us consider a node $\mathcal{N}_k$ and its *preserve* state $\mathcal{N}_k^P$. A path $Path_k$ is a continuous set of transitions between nodes $(\mathcal{N}_\alpha \rightarrow \mathcal{N}_\beta)$ defined in the trellis: $Path_k = (\mathcal{N}_\alpha \rightarrow \mathcal{N}_\beta) \cup (\mathcal{N}_\beta \rightarrow \mathcal{N}_\gamma) \cup \ldots \cup (\mathcal{N}_\psi \rightarrow \mathcal{N}_k)$. The path $Path_k^P$ starting from a leaf node and ending at that state is composed of two sub-paths[1]: the first one, $Path_k^{P,Left}$, comes from the left child and the second one, $Path_k^{P,Right}$, from the right child (see Fig. 5). In both cases, the path can emerge either from the *preserve* or from the *remove* state of the child nodes. If $\mathcal{N}_{k_1}$ and $\mathcal{N}_{k_2}$ are respectively the left and the right child nodes of $\mathcal{N}_k$, we have:

$$\begin{aligned} Path_k^{P,Left} &= Path_{k_1}^R \bigcup (\mathcal{N}_{k_1}^R \rightarrow \mathcal{N}_k^P) \quad \text{or } Path_{k_1}^P \bigcup (\mathcal{N}_{k_1}^P \rightarrow \mathcal{N}_k^P) \\ Path_k^{P,Right} &= Path_{k_2}^R \bigcup (\mathcal{N}_{k_2}^R \rightarrow \mathcal{N}_k^P) \quad \text{or } Path_{k_2}^P \bigcup (\mathcal{N}_{k_2}^P \rightarrow \mathcal{N}_k^P) \\ Path_k^P &= Path_k^{P,Left} \bigcup Path_k^{P,Right} \end{aligned} \tag{2}$$

The path cost is equal to the sum of the costs of its individual state transi-

[1] In the general case of an N-ary tree, the number of incoming paths may be arbitrary.

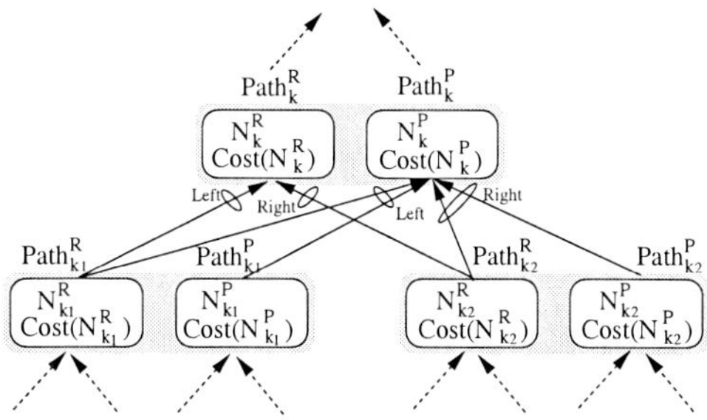

Fig. 5. Definition of *Path* and cost for the Viterbi algorithm (see Eqs. 2, 3 and 4).

tions. Therefore, the path of lower cost for each child can be easily selected.

$$\begin{array}{llll}
\text{If } Cost(Path_{k_1}^R) & < Cost(Path_{k_1}^P) & & \\
\quad\text{then } \{ & Path_k^{P,Left} & = & Path_{k_1}^R \bigcup (\mathcal{N}_{k_1}^R \rightarrow \mathcal{N}_k^P); \\
 & Cost(Path_k^{P,Left}) & = & Cost(Path_{k_1}^R); \} \\
\quad\text{else } \{ & Path_k^{P,Left} & = & Path_{k_1}^P \bigcup (\mathcal{N}_{k_1}^P \rightarrow \mathcal{N}_k^P); \\
 & Cost(Path_k^{P,Left}) & = & Cost(Path_{k_1}^P); \} \\
\text{If } Cost(Path_{k_2}^R) & < Cost(Path_{k_2}^P) & & \\
\quad\text{then } \{ & Path_k^{P,Right} & = & Path_{k_2}^R \bigcup (\mathcal{N}_{k_2}^R \rightarrow \mathcal{N}_k^P); \\
 & Cost(Path_k^{P,Right}) & = & Cost(Path_{k_2}^R); \} \\
\quad\text{else } \{ & Path_k^{P,Right} & = & Path_{k_2}^P \bigcup (\mathcal{N}_{k_2}^P \rightarrow \mathcal{N}_k^P); \\
 & Cost(Path_k^{P,Right}) & = & Cost(Path_{k_2}^P); \} \\
Cost(Path_k^P) \;=\; & Cost(Path_k^{P,Left}) & + & Cost(Path_k^{P,Right}) + Cost(\mathcal{N}_k^P);
\end{array} \tag{3}$$

In the case of the *remove* state, $\mathcal{N}_k^R$, the two sub-paths can only come from the *remove* states of the children. So, no selection has to be done. The path and its cost are constructed as follows:

$$\begin{array}{lll}
Path_k^{R,Left} & = & Path_{k_1}^R \bigcup (\mathcal{N}_{k_1}^R \rightarrow \mathcal{N}_k^R); \\
Path_k^{R,Right} & = & Path_{k_2}^R \bigcup (\mathcal{N}_{k_2}^R \rightarrow \mathcal{N}_k^R); \\
Path_k^R & = & Path_k^{R,Left} \bigcup Path_k^{R,Right}; \\
Cost(Path_k^R) & = & Cost(Path_{k_1}^R) + Cost(Path_{k_2}^R) + Cost(\mathcal{N}_k^R);
\end{array} \tag{4}$$

This procedure is iterated in a bottom-up fashion until the root node is reached. One path of minimum cost ends at the *preserve* state of the root node and another path ends at the *remove* state. Among these two paths, the one of minimum cost is selected. This path connects the root node to all leaves and the states it goes through define the final decisions. By construction, these decisions are increasing and as close as possible to the original decisions.

An example of motion filtering is shown in Fig. 6. The objective of the operator is to remove all moving objects. The criterion is the mean displaced frame difference estimated on each node (non-increasing criterion). In this sequence, all objects are still except the ballerina behind the two speakers and

Fig. 6. Example of motion connected operator preserving fixed objects: original frame (left), motion connected operator (center), residue (right).

the speaker on the left side. The connected operator with the Viterbi algorithm removes all moving components.

3.3. Global optimization under constraint

In this section, we illustrate a more complex pruning strategy involving a global optimization under constraint. Let us denote by $\mathcal{C}$ the criterion to optimize (for example, minimize) and by $\mathcal{K}$ the constraint. Moreover, assume that the criterion and the constraint are additive over the regions $\mathcal{N}_k$: $\mathcal{C} = \sum_{\mathcal{N}_k} \mathcal{C}(\mathcal{N}_k)$ and $\mathcal{K} = \sum_{\mathcal{N}_k} \mathcal{K}(\mathcal{N}_k)$. The problem is therefore to define a pruning strategy such that the resulting partition is composed of nodes $\mathcal{N}_i$ such that:

$$\text{Min} \sum_{\mathcal{N}_i} \mathcal{C}(\mathcal{N}_i) \text{ , with } \sum_{\mathcal{N}_i} \mathcal{K}(\mathcal{N}_i) \leq T_{\mathcal{K}} \tag{5}$$

This problem is equivalent to the minimization of the Lagrangian: $\mathcal{L} = \mathcal{C} + \lambda\mathcal{K}$ where λ is the Lagrange parameter. Both problems have the same solution if we find λ^* such that $\mathcal{K}$ is equal (or very close) to the constraint threshold $T_{\mathcal{K}}$. Therefore, the problem consists in using the tree to find by pruning a set of nodes creating a partition such that:

$$\text{Min} \left(\sum_{\mathcal{N}_i} \mathcal{C}(\mathcal{N}_i) + \lambda^* \sum_{\mathcal{N}_i} \mathcal{K}(\mathcal{N}_i) \right) \tag{6}$$

Assume, in a first step, that the optimum λ^* is known. In this case, the pruning is done by a bottom-up analysis of the tree. If the Lagrangian value corresponding to a given node $\mathcal{N}_0$ is smaller than the sum of the Lagrangians of the children nodes $\mathcal{N}_i$, then the children are pruned:

$$\text{If } \mathcal{C}(\mathcal{N}_0) + \lambda^* \mathcal{K}(\mathcal{N}_0) < \sum_{\mathcal{N}_i} \mathcal{C}(\mathcal{N}_i) + \lambda^* \sum_{\mathcal{N}_i} \mathcal{K}(\mathcal{N}_i) \text{, prune } \mathcal{N}_i. \tag{7}$$

This procedure is iterated up to the root node. In practice of course, the optimum λ^* parameter is not known and the previous bottom-up analysis of the tree is embedded in a loop that searches for the best λ parameter. The computation of the optimum λ parameter can be done with a gradient search

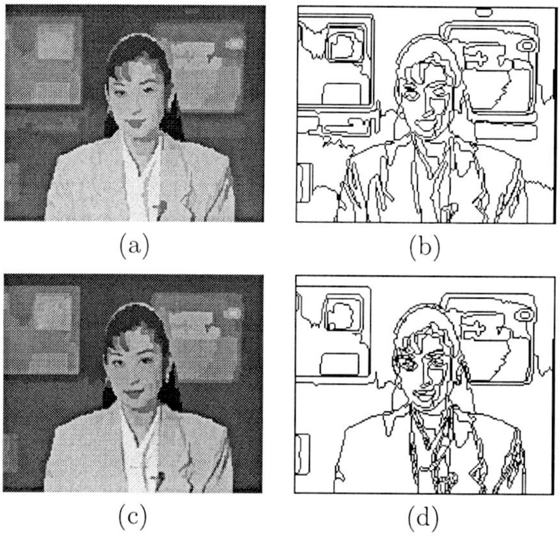

Fig. 7. Example of optimization strategies under a squared error constraint of 31 dB. (a) Minimization of the number of the flat zones, (b) contours of the flat zones of Figure 7(a) (87 flat zones, perimeter length: 4491), (c) Minimization of the total perimeter length, (d) contours of the flat zones of Figure 7(c) (219 flat zones, perimeter length: 3684).

algorithm. The bottom-up analysis itself is not expensive in terms of computation since the algorithm has simply to perform a comparison of Lagrangians for all nodes of the tree. The part of the algorithm that might be expensive is the computation of the criterion and the constraint values associated to the regions. Note, however, that this computation has to be done once.

This type of pruning strategy is illustrated by two examples relying on a Binary Partition Tree representation. In the first example, the goal of the connected operator is to simplify the input image by minimizing the number of flat zones of the output image: $\mathcal{C}_1 = \sum_{\mathcal{N}_k} 1$. In the second example, the criterion is to minimize the total length of the contours of the flat zones: $\mathcal{C}_2 = \sum_{\mathcal{N}_k} Perimeter(\mathcal{N}_k)$. In both cases, the criterion has no meaning if there is no constraint because the algorithm would prune all nodes. The constraint we use is to force the output image to be a faithful approximation of the input image: the squared error between the input and the output images $\mathcal{K} = \sum_{\mathcal{N}_k} \sum_{x \in \mathcal{N}_k} (\psi(f)(x) - f(x))^2$ is constrained to be below a given quality threshold. In the examples shown in Figure 7, the squared error is constrained to be of at least 31 dB. Figure 7(a) shows the output image when the criterion is the number of flat zones. The image is visually a good approximation of the original image but it involves a much lower number of flat zones: the original image is composed of 14335 flat zones whereas only 87 flat zones are present in the filtered image. The second criterion is illustrated in Figure 7(c). The approximation provided by this image is of the same quality as the previous one. However, the characteristics of its flat zones are quite different. The total length of the perimeter of its flat zones is equal to 3684 pixels whereas the example of Figure 7(a) involves a total perimeter length of 4491 pixels. The reduction of perimeter length is obtained at the expense of a drastic increase of the number of flat zones: 219 instead of 87. Figures 7(b) and 7(d) show the flat zone contours which are more complex in the first example but the number of flat zones is higher in the second one.

This kind of strategy can be applied for a large number of criteria and constraints. Note that without defining a tree structure such as a Max-tree, a Min-tree or a Binary Partition Tree, it would be extremely difficult to implement this kind of connected operators.

4. Conclusions

This paper has discussed two region-based representations useful to create connected operators: Max-tree (Min-tree) and Binary Partition Tree. The filtering approach involves three steps: first, a region-based representation of the input image is constructed. Second, the simplification is obtained by pruning the tree and third, and output image is constructed from the pruned tree. The tree creation defines the set of regions that the pruning strategy can use to create the final partition. It represents a compromise between flexibility and efficiency: on the one hand side, not all possible merging of flat zones are represented in the tree, but on the other hand side, once the tree has been defined complex pruning strategies can be defined. In particular, it is possible to deal with non-increasing criteria using dynamic programming approach such as the Viterbi algorithm or to involve constrained optimization criterion.

References

1. L. Garrido, P. Salembier, and D. Garcia. Extensive operators in partition lattices for image sequence analysis. *Signal Processing*, 66(2):157–180, April 1998.
2. F. Meyer. The levelings. In *Fourth Int. Symposium on Mathematical Morphology*, pages 199–206, Amsterdam, The Netherlands, June 1998. Kluwer.
3. O. Morris, M. Lee, and A. Constantinides. Graph theory for image analysis: an approach based on the shortest spanning tree. *IEE Proceedings, F*, 133(2):146–152, April 1986.
4. P. Salembier, and L. Garrido. Binary partition tree as an efficient representation for image processing, segmentation and information retrieval. *IEEE Trans. on Image Processing*, April, 2000.
5. P. Salembier, A. Oliveras, and L. Garrido. Anti-extensive connected operators for image and sequence processing. *IEEE Trans. on Image Processing*, 7(4):555–570, April 1998.
6. J. Serra and P. Salembier. Connected operators and pyramids. In SPIE *Image Algebra and Mathematical Morphology*, Vol. 2030, pp. 65–76, San Diego (CA), USA, July 1993.
7. L. Vincent. Morphological gray scale reconstruction in image analysis: Applications and efficient algorithms. *IEEE Trans. on Image Processing*, 2(2):176–201, April 1993.
8. L. Vincent. Grayscale area openings and closings, their efficient implementation and applications, *First Workshop on Mathematical Morphology and its Applications to Signal Processing*, pp.22-27, Barcelona, Spain, May 1993.

IMAGE SEGMENTATION BASED ON THE DERIVATIVE OF THE MORPHOLOGICAL PROFILE

MARTINO PESARESI
European Commission, Space Applications Institute
Environmental and Geo-Information Unit TP 441
21020 Ispra, (Varese) Italy.

and

JON ATLI BENEDIKTSSON
Department of Electrical and Computer Engineering
University of Iceland
Hjardarhaga 2-6
107 Reykjavik
Iceland.

Abstract. A new segmentation method based on the morphological characteristic of connected components in images is proposed. The formalisation of the morphological characteristic is based on a composition of the residuals of morphological opening and closing transforms by reconstruction. In case of multi-scale segmentation, this concept is generalised through the derivative of the morphological profile. Multi-scale segmentation is particularly well suited for complex image scenes such as aerial or fine-resolution satellite images, where very thin, enveloped and/or nested regions have to be retained. The proposed method performs well in the presence of both low radiometric contrast and relative low spatial resolution, which may produce textural and border effects and ambiguity in the object/background distinction. Examples of the proposed segmentation approach applied on satellite images are given.

Key words: Mathematical Morphology, Morphological Segmentation, Levelling, High Resolution Satellite Imagery.

1. Introduction

In this paper, a new segmentation method is proposed. The proposed method is based on the residuals of morphological opening and closing transforms with a geodesic metric. It may be considered analogous to a region growing technique. However, in contrast to using statistical local properties, like in region growing approaches, the proposed method uses a pixel similarity rule based on the morphological characteristic of the connected components in the image. Morphological residuals between the original grey-level function and the composition of a granulometry and an anti-granulometry by reconstruction are used to build a so-called *morphological profile function.* This function is interpreted as a fuzzy membership function related to a set of morphological characteristics of the connected components in the image. Then, the labeling phase is formalised as a decision rule based on the greatest value of the derivative of the morphological profile function. The proposed method can be applied with both single-scale and multi-scale approaches. Since standard morphological segmentation approaches are based on an edge-detection phase (watershed line extraction on a gradient image), the original contribution of this work is the definition of a morphological segmentation method, which avoids the gradient calculation, and can be applied either to single-scale or multi-scale image processing problems.

Watershed line detection [1] is the main tool of mathematical morphology for image segmentation. Watershed segmentation was introduced in image analysis by Beucher and Lantujoul [2] and defined mathematically by both Meyer [3] and Najman and Schmitt [4]. However, except for a few simple cases where the target object is brighter than the background or vice versa, watershed segmentation cannot be applied directly. Another well-known problem of the standard approach in watershed segmentation is its severe over-segmentation, which is difficult to overcome. The standard non-linear solution to the over-segmentation problem has been introduced by Meyer and Beucher [1] as a strategy which involves a marker selection followed by flooding of the relief formed by the gradient obtained from these markers. The marker detection is the main problem for this approach. If there is no external information available, the marker detection problem is generally solved by morphological filtering (usually geodesic closing) of the gradient image, followed by a thresholding of the filtered gradient. Since closing is defined by morphological dilation followed by erosion, this variation of the watershed-plus-marker approach assumes that the local minima of the gradient, which are smaller (thinner) than the structuring element (SE), are not relevant. The same applies to grey level edges with values less than a given threshold.

All the above-mentioned approaches assume that the region of interest for detection is large and homogenous, relative to the spatial and spectral resolution of the sensor. Consequently, these approaches are very hard to apply in segmentation of textured or very complex scenes, and they often lead to results that are unreliable.

2. The Proposed Approach

It is well known that there are two fundamentally different strategies for image segmentation, i.e., edge detection and region growing. Even though the standard approach to morphological segmentation is dependent on edge-detection, it is possible to consider a different morphological approach to the segmentation problem. The idea here is to try to characterise image features by their morphological intrinsic characteristics, instead of using their boundary. In a hypothetical approach that uses a morphological region growing technique, the border of a detected feature can be of size zero, thus avoiding the above-mentioned surface-of-edges problems in segmentation of complex imagery. Thus, a feature or an "object" in the image could be defined as a connected component (region of pixels) with the same characteristics, measured by some kind of a morphological operator.

It is a common practice to use the opening and closing transforms in order to isolate bright (opening) and dark (closing) structures in images, where bright/dark means brighter/darker than the surrounding features in the images. In order to isolate features with a thinner support than a given SE, a widely used technique is to take the residuals of the opening, closing and original images, by a morphological transformation called *top-hat* and *inverse top-hat* (or bot-hat) [5].

The chosen approach for the opening and closing calculation uses a non-Euclidean metric known as filtering by reconstruction [6,7]. The reason for using the reconstruction approach is that this family of morphological filters has proven to have a better shape preservation than classical morphological filters. In fact reconstruction filters introduce less shape noise since an interaction between the shape of the structures present in the image and the shape of the structuring element is used in the filtering.

The simpler and intuitive taxonomy of morphological characteristics could, for a given spatial domain, be the set T = {"flat", "concave", "convex"}, local curvature of the grey level function surface, where the spatial domain is determined by a given SE. As described above, such segmentation requires the precise definition of the spatial domain where the method is applied in terms of a SE size. Some structures may have a high response for a given SE size, and a lower response for other SE sizes, depending on the interaction between the SE size and the size of the structure. Sometimes we know ex-ante the size of the structures that we want to detect. However, that is often not possible, and then a single-SE-size approach appears to be too simplistic. For these reasons, in exploratory or more complex cases, it can be a good idea to use a range of different SE sizes in order to explore a range of different hypothetical spatial domains, and to use the best response of the structures in the image for segmentation. Given the above-proposed notion of morphological characteristic, it is straightforward to extend the same concept to multi-scale processing, by introducing the concepts of *morphological profile* and of the *derivative of the morphological profile* (DMP).

2.1. Definition

Let $\Pi\gamma(x)$ be the *opening profile* (a vector) at the point x in the image I defined as:

$$\Pi\gamma(x) = \{\Pi\gamma_\lambda : \Pi\gamma_\lambda{}^*(x), \forall\lambda \in [0, \ldots, n]\} \tag{1}$$

and let $\Pi\phi(x)$ be the *closing profile* (a vector) at the point x of the image I defined as:

$$\Pi\phi(x) = \{\Pi\phi_\lambda : \Pi\phi_\lambda{}^*(x), \forall\lambda \in [0, \ldots, n]\} \tag{2}$$

with $\gamma_0{}^*(x) = \phi_0{}^*(x) = I(x) \rightarrow \Pi\gamma_0(x) = \Pi\phi_0(x) = I(x)$ for $\lambda = 0$ by definition of opening and closing by reconstruction. Given (1) and (2), the opening profile can also be defined as a granulometry made with opening by reconstruction, while the closing profile can be defined as anti-granulometry made with closing by dual reconstruction. The derivative of the morphological profile is defined as a vector where the measure of the slope of the opening-closing profile is stored for every step of an increasing SE series.

The *derivative of the opening profile* $\Delta\gamma(x)$ is defined as the vector:

$$\Delta\gamma(x) = \{\Delta\gamma_\lambda : \Delta\gamma_\lambda = \mid \Phi\gamma_\lambda - \Phi\gamma_{\lambda-1} \mid, \forall\lambda \in [1, \ldots, n]\} \quad (3)$$

By duality, the *derivative of the closing profile* $\Delta\phi(x)$ is the vector:

$$\Delta\phi(x) = \{\Delta\phi_\lambda : \Delta\phi_\lambda = \mid \Phi\phi_\lambda - \Phi\phi_{\lambda-1} \mid, \forall\lambda \in [1, \ldots, n]\} \quad (4)$$

Generally, the *derivative of the morphological profile* $\Delta(x)$ or the DMP can be written as the vector:

$$\Delta(x) = \begin{cases} \Delta_{c+\lambda} : \Delta\gamma_\lambda, \forall \in [1, \ldots, n] \\ \Delta_{c-\lambda+1} : \Delta\varphi_\lambda, \forall \in [1, \ldots, n] \end{cases} \quad (5)$$

for an arbitrary integer c with n equal to the total number of iterations.

Given all the above, the morphological multi-scale characteristic Φ of the image I at the point x can be defined as the SE size (iteration number) with the greatest associate value in the Δ function:

$$\Phi(x) = \{\varepsilon : \Delta_\varepsilon(x) = \vee\Delta(x), \forall\varepsilon \in [c - n + 1, \ldots, c + n]\} \quad (6)$$

Equation (6) can be rewritten in order to maintain information about the type of structure that is detected. In order to do that, the *multi-scale-opening characteristic* $\Phi\Delta(x)$ of the image I at the point x can be defined as:

$$\Phi\gamma(x) = \{\lambda : \Delta\gamma_\lambda = \vee\Delta\gamma(x)\} \quad (7)$$

while the *multi-scale-closing characteristic* can be defined as:

$$\Phi\varphi(x) = \{\lambda : \Delta\varphi_\lambda = \vee\Delta\varphi(x)\} \quad (8)$$

With these definitions, an algorithm for multi-scale segmentation of the image I, based on its characteristic, can be written as:

$$\Phi(x) = \begin{cases} \hat{\kappa}_\lambda = \Phi\gamma(x) : \vee\Delta\gamma(x) > \vee\Delta\varphi(x) \\ \check{\kappa}_\lambda = \Phi\varphi(x) : \vee\Delta\gamma(x) < \vee\Delta\varphi(x) \\ \bar{\kappa} = 0 : \vee\Delta\gamma(x) = \vee\Delta\varphi(x) \end{cases} \quad (9)$$

where the label of the morphological characteristic is the iteration code of the opening or closing series that correspond to the greatest value of the derivative. If this greatest derivative value is strictly equal for both the opening and closing series, the "flat" label $\bar{\kappa}$ is applied. In this sense, an image feature is a set of connected pixels or a connected component with the same value of Φ. The function Φ takes values in the range $\Phi(x) \in [\hat{\kappa}_1, \ldots, \hat{\kappa}_n]$ in case of prevalently "convex" regions, values in the range $\Phi(x) \in [\check{\kappa}_1, \ldots, \check{\kappa}_n]$ in case of prevalently "concave" regions, and the value $\Phi(x) = \bar{\kappa} = 0$ in case of prevalently "flat" or morphologically "indifferent" regions for all the used sizes of $SE \in [1, \ldots, n]$.

In case of uncertainty or ambiguity of the distinction between scene foreground and background, it is also possible to soften the conditions of the morphological characteristic by rewriting (9) as:

$$\Phi^\sigma(x) = \begin{cases} \hat{\kappa}_\lambda = \Phi\gamma(x) : \vee\Delta\gamma(x) > \vee\Delta\varphi(x) > \sigma \\ \check{\kappa}_\lambda = \Phi\varphi(x) : \vee\Delta\gamma(x) < \vee\Delta\varphi(x) > \sigma \\ \bar{\kappa} = 0 : \vee\Delta\gamma(x) = \vee\Delta\varphi(x) \leq \sigma \end{cases} \quad (10)$$

for a given level of contrast $\sigma \geq 0$. With $\sigma = 0$, (9) and (10) give the same result, while by increasing the value of $\sigma > 0$ the level of necessary contrast is increased in order to avoid the labeling of the pixels with the "flat" label. Thus, the level σ could be interpreted as a threshold used in distinguishing between image foreground and background.

2.2. Discussion

The intuitive idea of a multi-scale morphological profile can theoretically be interpreted as a variation of the notion of a morphological spectrum. It can be defined as an extension of the *opening spectrum* studied by Haralick*et al.* [9] or the *pattern spectrum* defined in Maragos [10]. Both are based on the definition of some kind of granulometry [11], as the opening spectrum, i.e., the image sequence created by computing the differences between successive images in a granulometry generated by a flat SE family with an integral index set. Applications of variations of the morphological spectrum have been proposed for image noise reduction [12] and pseudo bandpass image decomposition [13].

While the above-mentioned approaches do not require a particular metric for the morphological transforms, the DMP approach requires the use of granulometry and anti-granulometry made by opening and closing by reconstruction, using a geodesic metric. In addition, the above-mentioned approaches are mainly used for filtering purposes, while the definitions of the derivative of the morphological profile in (6) and the morphological multi-scale characteristic in (7) and (9), allow us to reach the segmentation phase, avoiding problems related to the calculus of a gradient function.

Theoretically, the proposed method and watershed segmentation share the same geodesic approach, but there is an important difference. In contrast to watershed segmentation, which is intrinsically non self-dual (and in fact is implemented with basin flooding algorithms), the proposed approach is fully self-dual and then treats darker or brighter connected components exactly in the same way. This is an important positive characteristic of the proposed method, since it allows us to avoid the calculus of the gradient function in the segmentation process.

Moreover, in watershed segmentation, we typically only take into account the *absolute contrast* (any kind of grey-level difference) between adjacent connected components in the region merging process. However, in the proposed approach we introduce the concept of *relative contrast*, where relative means relative to a given spatial domain of application. This way, contrast is not only defined trivially as the grey-level difference between adjacent connected components, but it also includes a dimension describing the scale with which the image is observed where different levels of simplification may produce different relative contrasts. In fact, to use the derivative of the morphological profile as in (6), corresponds to detecting the spatial domain having the greatest contrast (in darker or in brighter structures) with respect to the surrounding connected components. Thus, the contrast is always measured relative to a given spatial domain. An important consequence of this characteristic is that the proposed method is expected to dramatically reduce the over-segmentation problem of the classical watershed approach, by introducing some sort of an intrinsic hierarchy.

In a recent paper, Crespo *et al.* [14] give similar observations, to what has been presented here, about problems related to the use of the classical watershed segmentation procedure in cases where it is also necessary to retain very thin structures. As here, the adopted solution in [14] involves filtering by reconstruction. However, the overall strategy there is completely different from our approach. The procedure proposed in [14] requires

1. a morphological filtering phase for image simplification;
2. the extraction of a set of masks by thresholding of some criteria (five such criteria are used in [14]);
3. the extraction of a final mask using a set union of the preceding criteria;
4. a region merging phase based on a measure of region dissimilarity, calculated as the absolute difference between the grey-level average of adjacent regions.

Differences between the method proposed here, which is based on the derivative of the morphological profile (DMP), and the method proposed in [14], concern both the self-duality and the complexity of the structure of parameters that are responsible for the final segmentation output.

In contrast to the DMP approach, the morphological filtering phase of the "flat" approach in [14] is intrinsically non self-dual because it uses alternate filtering in order to detect flat areas. Therefore, the operator has to decide in advance what kinds of structures (brighter or darker) are more important to retain. However, that is often arbitrary and may provoke undesired effects on the final output. Moreover, the DMP approach requires the definition of only two parameters (the step of the granulometry/anti-granulometry and the maximal SE size used, or the maximal number of iteration), while the "flat" approach in [14] requires a more complex structure of parameters. Those parameters may be less manageable and can provoke instability or robustness problems on the output. An example is the use of a gradient-to-area ratio criterion that often produces instability when it is needed to manage at the same time regions of very different areas. These regions can be of various sizes, from one pixel to many hundreds of pixels. The solution in [14], i.e., to exclude regions of area less than a given number of pixels (5 pixels in the cited paper), is questionable. It eliminates most of the small regions and it further complicates the general parameter structure by introducing another threshold. These shortcomings are overcome by the approach proposed here.

3. Applications

3.1. Previous experiments

The idea to use a composition of opening transforms for a morphological segmentation of satellite data was proposed some time ago for the detection of different urban structures [15,16]. In the experiments in [15,16], segmentation labels were obtained after the arithmetic summing of an opening series with an increasing SE. The method is only applicable to Boolean maps (binary or 2-

grey-level images) and it does not use geodesic metric. More recently, Pesaresi and Kanellopoulos [17] used a composition of geodesic opening and closing operations of different sizes in order to build a morphological profile. Then, they used a neural network approach for the classification of features. The difference from their method to the method proposed here, is that in [17] the absolute residual between the original image and the opened or closed one was used as a morphological characteristic function. Therefore, the method in [17] cannot be used for multi-scale segmentation since it limits the explored spatial domain by a restraint range of SEs. The method proposed here, which is based on the derivative of the morphological profile, is both more general and more robust than all the methods above, as will be demonstrated by the following examples.

3.2. Examples

Example of the application of the proposed method is now given for the segmentation of a satellite high-resolution image. The original image used in this experiment was recorded by the IRS-1C panchromatic sensor, which has ground spatial resolution of 5x5 meters, and covers a surface of about 15.000 x 15.000 pixels (75 km x 75 km). The subsample showed here is taken from an agricultural are connoted by scattered settlement (N-E of Athens, Greece), and covers a surface of 665 x 966 pixels (3.3 km x 4.8 km). Figure 1 (left) shows the subsampled image with a min-max histogram stretching. It is easy to note the co-presence in the scene of objects/regions of different sizes. As a consequence, in this experiment the multi-scale approach was applied as defined in (9). Figure 1 (right) shows the segmentation results. The spatial domain explored in this experiment was given by a range of 10 increasing SEs with an octagonal shape and a diameter ranging from 7 pixels (45 meters) up to 61 pixels (305 meters). The step from iteration λ to iteration $\lambda + 1$ was then equal to 6 pixels (30 m.). Consequently, the final number of labels was n=10+10+1=21, counting also the "flat" label. The function $\phi(x)$ assumes values in the range $\Phi(x) \in [\hat{\kappa}_1, \ldots, \hat{\kappa}_n]$ in case of prevalently "convex" regions, values in the range $\Phi(x) \in [\check{\kappa}_1, \ldots, \check{\kappa}_n]$ in case of prevalently "concave" regions, and the value $\phi(x) = \bar{\kappa} = 0$ in case of prevalently "flat" or morphologically "indifferent" regions for all the used sizes of $SE \in [1, \ldots, 10]$. Great regions as well as smaller ones are retained, without undesired loss of details in the segmented image. That is also the case for nested, thin and, complex regions. It is interesting to note that the proposed multi-scale approach seems to have a hierarchical effect. Large regions appear to have the same label. Also, no over-segmentation effect is detected due to the presence of non-relevant local minima and local maxima which is usual in classical segmentation by watershed approach.

Figure 2 shows a comparison between the proposed approach and classical watershed segmentation for a 100 x 100 pixel area of the Athens data set. The subsample is placed in the centre of Athens and is taken over a compact urban area with an internal vegetated area (park). Subimage 1 shows the original radiometric data enhanced with min-max histogram stretching for visualisation purposes. The classical morphological approach requires the detection

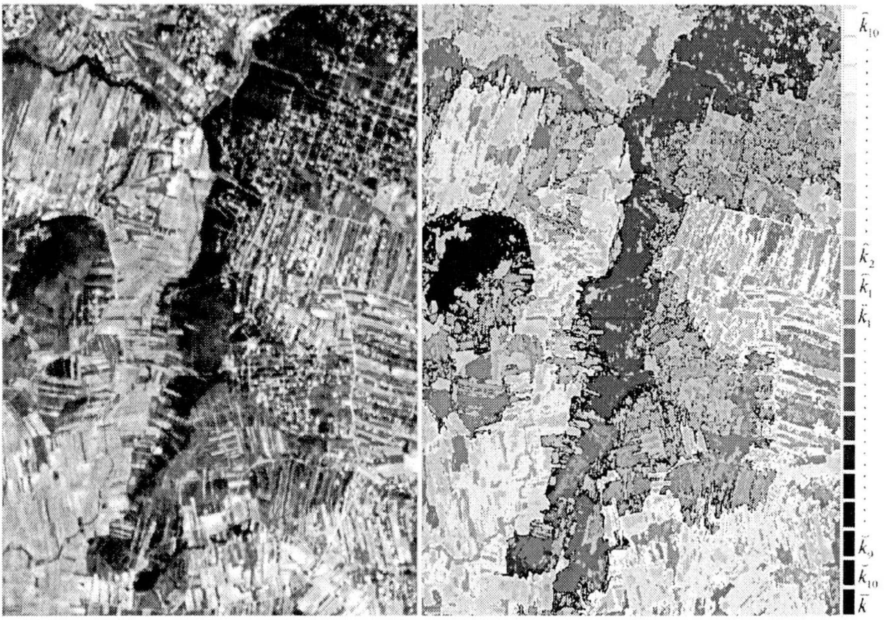

Fig. 1. Left: portion of IRS-1C panchromatic scene over an agricultural and scattered settlement area N-E of Athens, Greece, 1998. The image covers an area of 3.3 x 4.8 square kilometers with a resolution of 5x5 meters (665 x 966 pixels). The image has been enhanced for visualisation by min-max histogram stretching. Right: multi-scale segmentation obtained by (9). The explored spatial domain ranges from an octagon of 7 pixels (45 meters) to an octagon of 61 pixels (305 meters), with 10 steps of 6 pixels (30 meters) each.

of the "border" of the regions, and subimage 2 is the direct application of a morphological gradient transform (defined as the difference between dilation and erosion) to the original data. In this complex context, it is possible to observe that attempting to start from edges of regions leads to the production of "surfaces of edges" where most of the pixels are connoted as "border pixels." Another problem with the classical approach, which is evident here, is the over-segmentation generated by non-relevant local minima of the gradient function. Subimage 3 is the gradient of the filtered data where a morphological filter was applied, defined as the opening of the closed image with a flat SE equal to 3x3 pixels. Consequently, the situation in subimage 3 appears to be simpler than in subimage 2. Subimage 4 shows the results of watershed segmentation using the gradient image in subimage 3. Subimage 5 shows the output of the multi-scale morphological segmentation defined in (9). We can note that the proposed approach retains a better description of the original structural information, introducing less shape noise than the classical watershed segmentation approach. Another positive characteristic of the proposed method is the intrinsic hierarchy that reduces dramatically the over-segmentation effect. This can be detected in the case of the green area that is labeled as only one region by the proposed method but a set of non-homogenous regions by the classical approach.

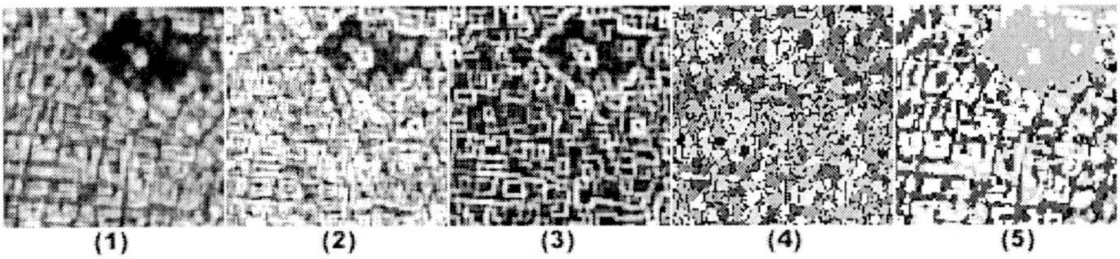

Fig. 2. Comparison of the proposed morphological segmentation approach and classical watershed segmentation. 1. Original radiometric information (IRS-1C panchromatic sensor) after linear histogram stretching; 2. gradient of the original data; 3. gradient of the filtered data; 4. image obtained by watershed segmentation; 5. image obtained by multi-scale segmentation as defined in (9).

4. Conclusions

Morphological segmentation by the derivative of the morphological profile was proposed. The proposed method is based on the use of residuals from opening and closing by reconstruction. In experiments, the proposed method demonstrated excellent performance even where the classical morphological approach had problems. In particular, the proposed approach gives a better shape description than the classical approach. It also retains significant small regions in images, and has an effect of intrinsic hierarchy that reduces dramatically the over-segmentation problem of the classical approach.

The proposed method is particularly well suited for the segmentation of complex image scenes such as aerial or fine-resolution satellite images where very thin, enveloped and/or nested regions may have to be retained, and where the gradient calculation has a major drawback. The method performs well in presence of both low radiometric contrast and relative low spatial resolution which may produce a textural effect, a border effect, and ambiguity in object/background distinction. All these factors are critical and lead to an instability effect if segmentation methods based on an edge-detection approach are applied.

The drawback of the proposed method concerns the necessity of looking at a range of increasing opening and closing by reconstruction operations, which may cause a heavy computational burden. As a consequence, for images with very large and homogeneous regions, it is possible that a gradient-plus-watershed approach may be more efficient, since it does not need to explore a very wide range of different SE sizes. For the above reasons, the method presented here is particularly well suited for segmentation of complex image scenes such as aerial or satellite images where very thin, enveloped and/or nested regions have to be retained. It is also well suited for images with low radiometric contrast and relatively low spatial resolution, which produce textural effects, border effects, and ambiguity in the object/background distinction. All these factors are critical and can lead to instability effects if segmentation methods based on the edge-detection approach are used.

References

1. F. Meyer and S. Beucher. Morphological Segmentation. *J. of Visual Communication and Image Represent.*, 11:21–46, 1990.
2. S. Beucher and C. Lantuejoul. Use of watersheds in contour detection. Int. Workshop on Image Processing, real-time edge and motion detection/estimation, CCETT/IRISA, Rennes, France, Sept. 79.
3. F. Meyer. Integrals, gradients and watershed lines. In *Proc. of Mathematical Morphology and its applications to Signal Processing*, pages 70–75, Barcelona, May 1993.
4. L. Najman and M. Schmitt. Definition and some properties of the watershed of a continuous function. In *Proc. of Mathematical Morphology and its applications to Signal Processing*, pages 75–81, Barcelona, May 1993.
5. J. Serra. *Image Analysis and Mathematical Morphology.* Academic Press, London, 1982.
6. J. Serra and P. Salembier. Connected operators and pyramids. In *Proc. SPIE, Non-Linear Algebra and Morphological Image Processing*, vol. 2030, pages 65–76, San Diego, 1993.
7. J. Crespo, J. Serra, and R. Schafer. Theoretical aspects of morphological filters by reconstruction. *Signal Processing*, 47:201–225, 1995.
8. F. Meyer. From connected operators to levelings. In Henk J.A.M. Heijmans and B.T.M. Roerdink, editors, *Mathematical Morphology and its Applications to Image and Signal Processing*, pages 191–198, Kluwer Academic Publisher, Dordrecht, Netherlands, 1998.
9. R.M. Haralick, E.R. Dougherty, and P.L.Katz. Model-based morphology: the opening spectrum. In Y. Mahavieh and R.C. Gonzales, editors, *Advances in Image Analysis*, SPIE Optical Engineering Press, Bellingham, WA, USA, 1992.
10. P. Maragos. Pattern spectrum and multiscale shape representation. *IEEE Trans. Pattern Anal. Machine Intell.*, 11:701–716, 1989.
11. E.J. Kraus, H.J. Heijmans, and E.R. Dougherty. Greyscale granulometries compatible with spatial scaling. *Signal Processing*, 34:1–17, 1993.
12. R.A. Peters. A new algorithm for image noise reduction using mathematical morphology. *IEEE Transaction on Image Processing*, 4:554-568, 1995.
13. R.A. Peters. Morphological Pseudo Bandpass Image Decomposition. *Journal of Electronic Imaging*, 5:198–213, 1995.
14. J. Crespo, R.W.Schafer, J.Serra, C.Gratin, and F.Meyer. The flat zone approach: A general low-level region merging segmentation method. *Signal Processing*, 62:37–60, 1997.
15. M. Pesaresi. Analisi numerica dello spazio edificato nella citt diffusa. Technical Report IUAV DAEST, Venice, Italy, 1993.
16. A. Bianchin and M. Pesaresi. Outils de morphologie mathmatique applique aux images satellite pour l'analyse de l'urbanization diffuse. In *Proc. EGIS-MARI 94*, pages 2085–2094, Paris, France, March 29 - April 1994.
17. M. Pesaresi. and I. Kannellopulos. Detection of Urban Features Using Morphological Based Segmentation and Very High Resolution Remotely Sensed Data. In I. Kanellopoulos, G.G. Wilkinson, and T. Moons, editors, *Machine Vision and Advanced Image Processing in Remote Sensing*, pages 271–284, Springer-Verlag, Berlin, Germany, 1999.

FLOODING AND SEGMENTATION

FERNAND MEYER
Centre de Morphologie Mathématique - Ecole des Mines de Paris
35, rue Saint-Honoré - 77305 Fontainebleau (FRANCE)
Email: meyer@cmm.ensmp.fr

Abstract. Flooding is at the heart of morphological segmentation. The properties of floodings, defined as upper levelings are first studied. If g is a flooding of f, its catchment basins are unions of the catchment basins of f. For this reason, it is possible to construct multiscale segmentations associated to families of increasing floodings of a given reference function. Various useful families of floodings are presented, together with ways to combine them. We then show how traditional segmentation techniques belong to this framework if one floods with a uniform level of water and with markers. Replacing this uniform flooding by others permits to favour the segmentation of regions having some characteristics such as large size or contrast. Finally the introduction of fuzzy markers establishes a continuum between traditional segmentation with markers and pure multiscale segmentation.

Key words: Flooding, Hierarchy, Minimum Spanning Tree, Ultrametric Distance, Multiscale Segmentation, Fuzzy Markers.

1. Introduction

Flooding seems to be at the heart of morphological segmentation: both the watershed as the swamping of a topographic surface imposing a set of markers as regional minima are based on flooding. Floodings are upper levelings, i.e. closings by reconstruction. Their properties are studied in a first part.

The catchment basins associated to a series of increasing floodings form a hierarchy, which may be summarized as an ultrametric distance between the catchment basins of the finest partition. A lexicographic ultrametric distance is introduced, allowing to combine the hierarchies associated to two families of increasing floodings. It is then shown how to use this distance for deriving segmentations in various modes: unsupervised, with markers or interactively.

In order to enlarge the palette of available tools we present several particular modes of flooding: as a matter of fact, the characteristics of the results will depend upon the criteria governing the flooding. Synchronous flooding where all lakes share some common size, such as altitude, depth, area or volume. Finally, we establish a continuum between multiscale segmentation and segmentation with markers by using fuzzy markers, where sources are placed, whose flood is slowed down by a factor associated to each marker.

2. Flooding a Topographic Surface

2.1. Definition of a flooding

Levelings have been presented in [5]. The simplest class of them are defined as follows.

Definition 1 *An image g is a leveling of the image f if and only if, for any couple of neighboring pixels p, q, $g_p > g_q \Rightarrow f_p \geq g_p > g_q \geq f_q$.*

As shown by this definition, the levelings are neither extensive nor anti-extensive. In this paper we will concentrate on the subclass of extensive levelings, which we call floodings. From the definition of levelings we immediately derive the definition of floodings:

Definition 2 *A function g is a flooding of a function f if and only if $g \geq f$ and for any couple of neighboring pixels (p, q): $g_p > g_q \Rightarrow g_p = f_p$.*

The following criterion is easily derived from the definition of floodings if one interprets the implication $A \Rightarrow B$ of the definition as [*nonA or B*].

Criterion 1 Flood: *A function g is a flooding of a function f if and only if $g = f \vee \varepsilon g$.*

2.2. Properties of floodings

Creation of lakes
A flooding g is obtained from a function f, by creating a number of lakes on the topographic surface of f. All connected components where $g > f$ are flat if g is a flooding of f:

$$\text{for any couple of neighboring pixels}(p, q): \left| \begin{array}{l} g_q > f_q \\ g_p > f_p \end{array} \right| \Rightarrow g_p = g_q.$$

We will call lake of a flooding g any flat zone of g containing at least a pixel p for which $f_p > g_p$. Let us consider a lake L of a flooding g of a reference function f. If all neighbors of L have a higher altitude, then L is a regional minimum. If L has a lower neighbor, there exists a couple of pixels (p, q), p belonging to L and $g_p > g_q$. This implies that $g_p = f_p$, meaning that the level of the flooding g and the level of the ground f are the same at pixel p : the lake cannot build a wall of water without solid ground to hold the water. If the lake L also has higher neighbors, this means that from the pixel p to the pixel q the level of g decreases; from p inwards towards the center of the lake, the altitude of g also decreases; whereas, if one follows the outside boundary of the lake, in both directions the altitude will increase. Hence p is a saddle point of the function f.

Algebraic properties
It is easy to check using their definition that:

- If g and h are two floodings of f, then $g \vee h$ and $g \wedge h$ also are floodings of f.

- If g and h are floodings of f and $g \geq h$ then g is a flooding of h.
- The relation $\{g \text{ is a flooding of } f\}$ is reflexive, antisymmetric and transitive: it is an order relation. In particular, if f and h are two functions such that $f \leq h$, then the family of floodings (g_i) of f verifying $g_i \leq h$ form a complete lattice for this order relation. The smallest element is f itself and the largest is obtained by repeating the geodesic erosion of h above f: $h^{n+1} = f \vee \varepsilon h^n$ until stability, that is when $h^{n+1} = h^n$ (this operation is known in the literature as reconstruction closing of f from h). The criterion "Flood" given above shows that the result at convergence effectively is a flooding of f. Convergence may be obtained faster when using a recursive or a data driven implementation of the algorithm using hierarchical queues. This operation also is known as reconstruction closing of f using h as marker.
- These properties permit various constructions of increasing families of floodings (g_i): it is necessary and sufficient that g_j is a flooding of g_{j-1}.

The simplest flooding is uniform and obtained by threshold $f_\lambda = \begin{matrix} f \text{ if } f > \lambda \\ \lambda \text{ if } f \leq \lambda \end{matrix}$. h-closings ([1]), area closings ([8]) or volumic closings ([7]) are floodings associated to the size of the catchment basins. Reconstruction closings with markers permit a manifold of floodings, associated to marker functions. Swamping with markers is a particular case ([6]).

2.3. Watershed and floodings: absorption of catchment basins during flooding

If g is a flooding of a reference function f, we may imagine a progressive flooding of f producing g. During this flooding process, the *catchment basins* (CB) of f merge according to two mechanisms: a) the level of the water reaches a saddle point within a CB X but has a lower level in the neighboring catchment basin Y: X is absorbed by Y; b) two previously disconnected lakes merge: the corresponding CB also merge; the merging also occurs at the localisation of a saddle point. As a result, each CB of g is either identical with a CB of f or equal to the union of several CB of f.

2.4. Flooding spanning tree

The neighborhood graph $G = (X, U)$ of a topographic surface f is defined as follows: the nodes X are the catchment basins, the edges U link neighboring catchment basins. The edge u between the nodes a and b is weighted by the altitude of the saddle point between a and b. However, during increasing floodings, the catchment basins merge only along some edges of the neighborhood graph: this set of edges is the minimum spanning tree $\Theta = (X, T)$ of G (unique if the weights of the edges of U are all different). Hence Θ summarizes all useful information associated to floodings of f. If g is a flooding of f, a number of catchment basins of f are covered by a lake. If we assign to each of the corresponding nodes of X the level of their lake, we are able to derive the catchment basins of g. An edge u of T will be considered as flooded if the level of water at one of the adjacent nodes is higher or equal to the weight of u. The binary tree obtained from Θ by cutting all edges which are not flooded is a forest where each tree represents a catchment basin of g.

3. Hierarchy Associated to an Ordered Series of Floodings

3.1. Hierarchy of the catchment basins

Let us now consider a family F of increasing floodings $(g_i)_{i\leq N}$ of a reference function f, verifying $g_i \leq g_j$ for $i < j$, and $g_0 = f$. The set of catchment basins of g_i is called A_i; $A = \bigcup A_i$ is the set of all catchment basins. Each A_i is a partition of E. The partitions are nested: each element a of A_i is the union of CB of A_{i-1}.

3.2. Hierarchy and ultrametric structure

The hierarchy of the catchment basins A associated to the family of flooding F is equivalent to the definition of an ultrametric distance. The distance $Fd(a,b)$ between two basins a and b of A is defined as the index of the lowest flooding of F for which a CB will contain them both. Such a basin always exists if the highest flooding g_N has only one minimum. In the opposite case, it may not exist: the distance between a and b is then infinite: $Fd(a,b) = \inf\{i \mid \exists\ c \in A_i\ ;\ a \cup b \subset c\}$.

The restriction of Fd to A_0 is an ultrametric distance, as it verifies:

* reflexivity : $Fd(x,x) = 0$

* symmetry: $Fd(x,y) = Fd(y,x)$

* ultrametric inequality: for all x,y,z: $Fd(x,y) \leq \max\{Fd(x,z), Fd(z,y)\}$

The first two axioms are obviously verified. The last one may be interpreted as follows: the minimum index of a flooding of F for which a CB contains both catchment basins x and y is lower than or equal to the index of the lowest flooding for which a CB contains all three catchment basins x and y and z. The level of this last flooding is precisely $\max\{Fd(x,z), Fd(z,y)\}$.

An ultrametric distance is a distance, as the ultrametric inequality is stronger than the triangular inequality. A closed ball for the ultrametric distance with centre a and radius n is the set of all regions b for which $Fd(a,b) \leq n$. The balls associated to an ultrametric distance have two unique features, which will be useful in segmentation. The radius of a ball is equal to its diameter, i.e. to the largest distance between two elements in the ball. Each element of a ball is the centre of this ball. The union of all closed balls of radius i precisely constitute the set A_i.

3.3. Useful families of floodings

If a hierarchy may be associated to each family of increasing floodings, it is now time to indicate useful families of floodings from which to derive multiscale segmentations. As a matter of fact, the quality of segmentation will depend to a great extent on the family of floodings on which it is build.

3.3.1. *Uniform flooding*
A flooding of a function f is uniform if the level of all lakes is at the same level λ. The family $\{f_\lambda\}_{\lambda \geq 0}$ used with markers is at the base of all traditional morphological segmentation: placing sources at the position of the markers and flood the topographic surface with uniform flooding. The resulting catchment basins are the expected segmentation.

3.3.2. *Size oriented flooding*
Size oriented flooding is produced by placing sources at each minimum and flooding the surface in such a way that all lakes share some common measure (height, volume or area of the surface). As the flooding proceeds, the level of some lakes cannot grow any further, as the level of the lowest path point has been reached. In fig. 1, flooding starts from all minima in such a way that all lakes always have uniform depth. The resulting hierarchy is called dynamics in case of depth driven flooding and has first been introduced by M. Grimaud. Area and volume criteria have been studied by C. Vachier ([7]). Size oriented flooding permits to produce hierarchical segmentation with good psychovisual properties. The depth criterion ranks the region according to their contrast, the area according to their size and the volume offers a nice balance between size and contrast.

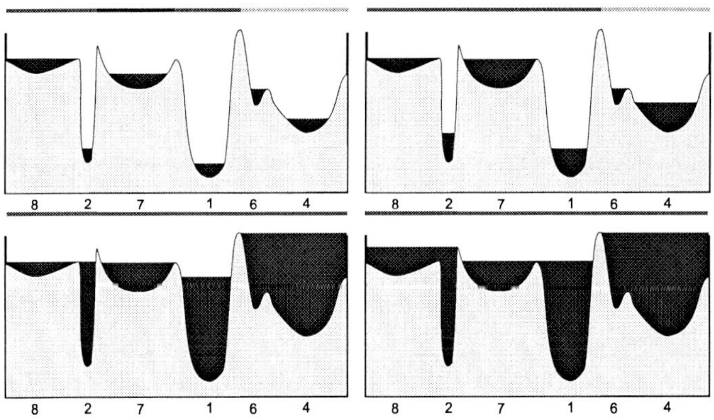

Fig. 1. Example of a height synchronous flooding. Four levels of flooding are illustrated; each of them is topped by a figuration of the corresponding catchment basins.

Remark 1 *Any stage of synchronous flooding represents a closing of the topographic surface. For the height, it will be h-closing, for the area, area closings and for the volume, volumic closings. If the criterion governing the growth of the lakes is the height, the associated flooding becomes uniform flooding.*

3.3.3. *Tailored flooding for favouring some types of regions.*
In some cases, while using one of the size criteria, it appears desirable to favour some regions. As an example: in many cases, the topographic surface to be flooded is the gradient image ∂h of an image h. The catchment basins of ∂h correspond to flat zones in h, which may be regional minima, maxima or step zones. However minima and maxima of h are perceptually more important than transition flat zones. For this reason, it may be worthwhile to push minima and maxima of h higher in the hierarchy.

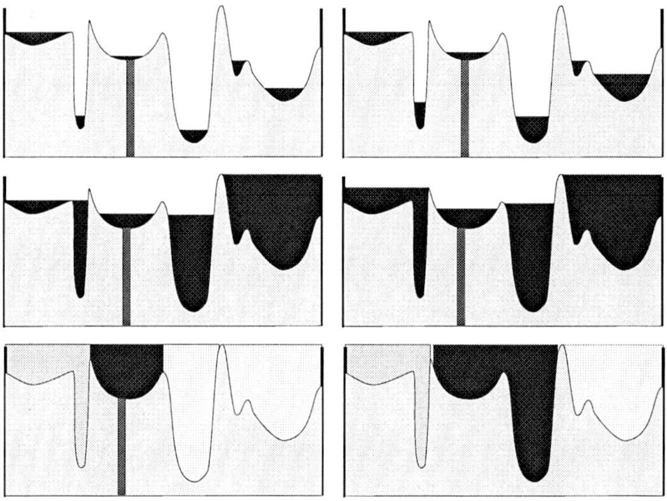

Fig. 2. 4 levels of tailored synchronous flooding, where the minimum marked red is slowed down by a factor 5. As a result we show the corresponding segmentation into 3 regions compared to the segmentation in 3 regions if no source is slowed down.

It is easy to obtain this result during synchronous flooding: the flooding of the minima we wish to favour is slowed down by some factor. In fig. 2 we have a case where depth synchronous flooding is performed. However the depth of the minimum above a black bar grows five times slower than the depth in the other catchment basins. For this reason, the minimum above a black bar survives much longer any absorption.

3.3.4. *Swamping: flooding speed=0 at the localisation of the markers*
In presence of markers, the corresponding minimum has no source at all. Or in other terms the flooding of this basin is infinitely slow. Hence the corresponding minima stay minima for ever, and catch their neighbouring basins, the final resulting flooding is the traditional swamping. If there are N minima, cutting the $N-1$ highest edges of the MST yields the segmentation based on markers. The advantage of this approach over the traditional approach of segmentation with markers, is the possibility to cut more than $N-1$ edges, yielding a segmentation with more regions than the number N of markers. Inversely, by adding a second criterion (see the lexicographic combination of criteria below)

it is possible to break the ties between the markers and obtain segmentations with less than N regions.

3.3.5. *Fuzzy markers for a maximal versatility*
Grouping both notions together, tailored flooding and flooding with markers, leads to the notion of fuzzy markers: each minimum is considered as a fuzzy marker and assigned a fuzzy level (1 means a hard marker, where no source is placed; 0 means no marker at all, and the source is not slowed down; λ means a fuzzy marker, and the corresponding source is slowed down by a factor of λ). Fuzzy markers permit to establish a continuum between traditional multiscale segmentation and segmentation with markers.

3.3.6. *Implementation of fuzzy markers through uniform flooding*
Synchronous flooding is a nice vehicle for presenting the concepts, but not really easy to implement. On the other hand, uniform flooding is extremely efficiently implemented with the help of hierarchical queues [3]. And it is possible to produce the same weights for the edges of the MST during uniform flooding as those produced by synchronous flooding, as we will see now. Let us illustrate the procedure with the depth criterion; furthermore, the source at minimum m_i is slowed down by the factor λ_i. A different label is assigned to each minimum of f and the topographic surface is flooded. Each time two lakes i and j with different labels merge, their depths d_i and d_j are measured. The weighted measures $\frac{d_i}{\lambda_i}$ and $\frac{d_j}{\lambda_j}$ (the measure is infinite for $\lambda = 0$, corresponding to a hard marker, which absorbs all its neighbors) are compared: the lake with the smallest measure is considered to be absorbed by the other and takes its label. The two lakes are necessarily neighbors on the flooding tree: the weighted size of the smallest lake is assigned to the edge of the flooding tree which connects them. When the uniform flooding is completed, all edges of the flooding tree will be assigned weights and the ultrametric distance can be computed.

3.4. Combining two different hierarchies

We have presented a number of mechanisms for constructing a hierarchy. We may still construct more by combining two of the preceding ones by one of the following mechanisms. The ultrametric distance Fd between any couple of elements of A_0 is easily computed as soon as we know it for all couples which are linked by an edge of the flooding tree: so it is sufficient to define a weight $Fd(x, y)$ for each edge (x, y) of the MST. The ultrametric distance Fd between two catchment basins a and b of A_0 is then simply obtained: $Fd(a, b)$ is equal to the highest weight on the unique path on the tree FT between a and b. The most natural ways to combine two hierarchies are the following ones.

3.4.1. *Maximum and minimum of two flooding hierarchies*
If F_1d and F_2d are the distances associated to two distinct flooding families, we define the maximum and the minimum of these distances. For the maximum we create a flooding tree $(F_1 \vee F_2)T$ where each edge (x, y) has the weight

$F_1d(x,y) \vee F_2d(x,y)$. The resulting ultrametric distance is equal to $F_1d \vee F_2d$. Replacing $\vee$ by $\wedge$ produces the infimum $(F_1 \wedge F_2)d$ which is the greatest ultrametric distance smaller then F_1d and F_2d: it is equal to $F_1d \wedge F_2d$ on each edge of the MST and higher or equal between nodes which are not neighbors on the MST.

3.4.2. *A lexicographic ultrametric distance*
If the number of different floodings within a family F_1 is reduced, it does not offer a rich multiscale representation of the reference function f: from one flooding g_i to the next g_{i+1} the number of catchment basins will be strongly reduced, each catchment basin of g_{i+1} being the union of several catchment basins of g_i. In some cases, it may be desirable to have a richer hierarchy with more levels and more regions. However, we desire that the major stratification of F_1 be respected. This is the case for instance if one uses markers, as explained above: all regions with markers are equivalent and appear at the highest level of the hierarchy. Introducing a second criterion will make it possible to tie the breaks between regions with markers and get additional levels of the hierarchy with less regions. The solution is to combine F_1 with another hierarchy F_2 into a lexicographic ultrametric distance. We define the lexicographic distance $F_1F_2d(a,b) = \{F_1d(a,b), F_2d(a,b)\}$. The order relation is classically defined as: $F_1F_2d(a,b) > F_1F_2d(c,d) \Leftrightarrow \{F_1d(a,b) > F_1d(c,d)\}$ or $\{F_1d(a,b) = F_1d(c,d)$ and $F_2d(a,b) > F_{21}d(c,d)\}$. Let us now take the flooding spanning tree and assign to each of its edges adjacent to two nodes a and b the distance $F_1F_2d(a,b)$. The lexicographic distance $L_{12}d(a,b)$ between any two nodes a and b will be defined as the maximum taken by the distance F_1F_2d along the unique path on the flooding tree between a and b. This yields again an ultrametric distance, which leads to a hierarchy which is finer than the hierarchies induced by the family F_1.

4. Application to Segmentation

We are now able to summarize a number of various multiscale criteria in form of a weighted spanning tree. This flooding spanning tree may now be used in each of the three classical segmentation modes.

4.1. Segmentation with markers

In many situations one has a seed for the objects to segment. It may be the segmentation produced in the preceding frame when one has to track an object in a sequence. It may also be some markers produced either by hand or automatically. As a result, some nodes of the flooding tree may be identified as markers. The resulting segmentation associated to these markers will then still be a minimum spanning forest, but constrained in that each tree is rooted in a marker. Several algorithms exist for constructing the minimum spanning forest, closely related to the classical algorithms for constructing the MST of a graph (see [4], for more details). The simplest conceptually is the following: take any couple of markers and cut the highest edge on the unique path between them. Repeat this operation until a forest is created where each tree contains one and only one marker. We obtain the classical segmentation with markers if we chose the ultrametric distance associated to uniform flooding, i.e. the tree Θ itself. Using a different set of weights, associated to another mode of flooding will permit to produce another balance between regions: for instance if the weights are based on area synchronous flooding, the same set of markers will tend to detect large regions, whereas a set of weights based on depth will stress contrasted regions.

4.2. Unsupervised segmentation

The primary aim of a hierarchical approach is to be able to easily produce segmentations with an arbitrary number of regions. The best segmentation into n regions should be a partition in n regions such that each region is as homogeneous as possible and two distinct regions as different as possible. The approach based on ultrametric distances defines the homogeneity of a region as its diameter: it is the largest distance between two nodes in this region. We may then define the diameter of a partition as the largest diameter of a region in the partition. A partition P_1 in n regions will be better than another partition P_2 also in n regions, if the diameter of P_1 is smaller than the diameter of P_2. It is easy to verify that the best partition into n regions, i.e. the partition with the smallest diameter is then obtained by cutting the $n-1$ highest edges in this tree, producing $n-1$ subtrees, each of them representing a region of the desired segmentation. This is illustrated on the top row of fig. 3, where two levels of a hierarchy are illustrated, with a lower and a higher number of regions. The quality of the obtained segmentation relies entirely upon the particular choice of hierarchy, or more precisely upon the family on floodings underlying this hierarchy.

4.3. Interactive segmentation

Besides the traditional segmentation technique based on markers, new interactive segmentation techniques may also be developed. A toolbox for interactive editing is currently under development ([2]), based on a hierarchy of segmentations. A mouse position is defined by its $x-y$ coordinates in the image but also by its depth z in the segmentation tree. If the mouse is active, the whole tile containing the cursor is activated and added or suppressed from the segmentation mask. For the same $x-y$ position, a mouse displacement towards lower levels of the hierarchy will result in a resegmentation of the region, whereas a displacement towards higher levels represents a fusion of adjacent regions. This is illustrated in fig. 3, where two levels of a hierarchy are illustrated on the top row, with a lower and a higher number of regions. None of them gives completely satisfying results. For this reason, some regions in the background will be merged by replacing them with their oversets in higher levels of the hierarchy (balls of the ultrametric distance with a larger radius) , while other regions in the person will be resegmented by moving lower in the hierarchy (balls of the ultrametric distance with a smaller radius).

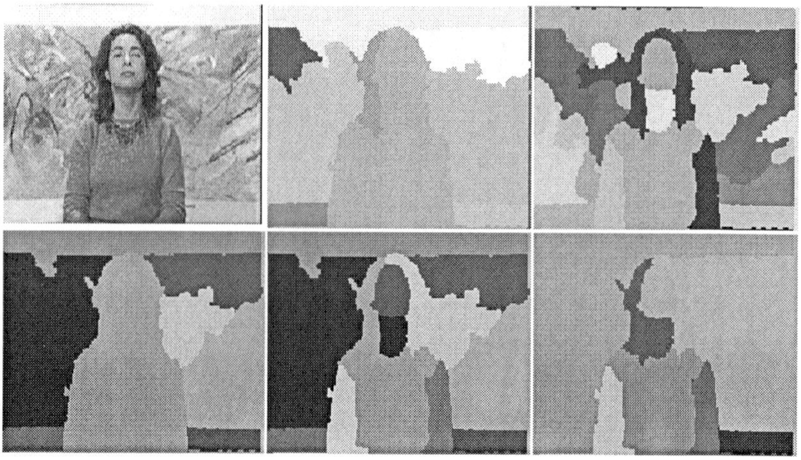

Fig. 3. Top row: a) initial image; b) and c) two partitions with a decreasing diameter associated to volumic driven floodings. Bottom row: local variations of the radius of the balls permit to resegment some regions and merge others.

5. Conclusion

We have defined a unified framework for hierarchical morphological segmentation based on families of increasing floodings of a gradient image. Traditional morphological segmentation appears as a particular case. It is possible to use this framework for unsupervised, marker driven or interactive segmentation. Size oriented flooding permits to favour size or contrast or a balance between both. Using fuzzy markers permits to obtain a continuum between marker driven segmentation and unsupervised multiscale segmentation. Finally, the use of lexicographic ultrametrics permits to combine various criteria and obtain fine grained hierarchical segmentations.

References

1. M. Grimaud. New measure of contrast : dynamics. *Image Algebra and Morphological Processing III, San Diego CA, Proc. SPIE*, 1992.
2. F. M. M. F. Zanoguera, B. Marcotegui. An interactive colour image segmentation system. In *Wiamis'99 : Workshop on Image Analysis for Multimedia Interactive Services*, pages 137–141. Heinrich-Hertz Institut Berlin, 1999.
3. F. Meyer. Algorithmes à base de files d'attente hiérarchique. Technical Report NT-46/90/MM, Sept. 1990.
4. F. Meyer. Minimal spanning forests for morphological segmentation. *ISMM94 : Mathematical Morphology and its applications to Signal Processing*, pages 77–84, 1994.
5. F. Meyer. The levelings. In H. Heijmans and J. Roerdink, editors, *Mathematical Morphology and Its Applications to Image Processing*, pages 199–207. Kluwer, 1998.
6. F. Meyer and S. Beucher. Morphological segmentation. 1(1):21–46, Sept. 1990.
7. C. Vachier. *Extraction de Caractéristiques, Segmentation d'Image et Morphologie Mathématique*. PhD thesis, E.N.S. des Mines de Paris, 1995.
8. L. Vincent. Grayscale area openings and closings, their efficient implementation. In J. Serra and P. Salembier, editors, *First Workshop on Mathematical Morphology and its Applications to Signal Processing*, pages 22–27, Barcelona, Spain, May 1993.

A MORPHOLOGICAL MULTI-SCALE GRADIENT FOR COLOR IMAGE SEGMENTATION

M. C. D'ORNELLAS and R. VAN DEN BOOMGAARD
Intelligent Sensory Information Systems
University of Amsterdam, Faculty WINS
Kruislaan, 403 - 1098 SJ Amsterdam
E-mail: {ornellas,rein}@wins.uva.nl
FAX: (31)(20) 525-74-90

Abstract. Segmentation and edge detection are key points in image analysis. Mathematical morphology employs the watershed transform to obtain the edges of the objects in an image. Usually, the watershed is significantly influenced by the morphological gradient. Furthermore, the direct segmentation of the gradient by the watershed transform results in an extreme oversegmentation. In this paper, we propose a morphological approach to compute the multi-scale gradient applied to color images. The main property of this technique, established on color morphology, is that it does not split the color channels in contrast to other methods in the literature. The experiments have shown that the suggested technique enhances the segmentation results generating more precise watershed lines.

Key words: Color Morphology, Edge Detection, Color Multi-Scale Gradient, Morphological Segmentation, Watershed.

1. Introduction

One of the main goals of image analysis is to isolate regions that are likely to come from a single object in order to analyze and recognize geometrical properties and the structure of the objects. The geometrical analysis of the objects must be quantitative, since only such an analysis and description of the objects can provide a coherent mathematical framework for describing the spatial organization. The quantitative description of geometrical structures is the objective of mathematical morphology.

So far, the use of such framework has allowed the development of a class of morphological algorithms to deal with binary and grayscale images. Multi-channel processing has been of growing interest recently, especially in color image processing. By comparison to grayscale images, color images contain three times as much data. Thereby, their use allows getting a much more robust segmentation toward lighting conditions and a better accuracy concerning the extracted regions.

The segmentation task this problem presents to mathematical morphology is far from easy. Indeed, the conventional morphological segmentation technique is the watershed transform. However, watershed is intrinsically a grayscale transformation, with its applicability depending on the existence of an order relation on pixel values. By defining an ordering relation that induces a lattice structure on the image, one can extend mathematical morphology to color images and therefore, to the watershed segmentation.

The purpose of this work is to introduce a new morphological multi-scale gradient for color images and stress that by including this operator in the morphological segmentation process, one can improve the watershed results.

The paper is organized as follows. Section 2 refers to color morphology and explains how morphological operators can be obtained by defining a vectorial ordering relation on the image. Further on, the proposed definition of the color multi-scale gradient is presented in section 3. Section 4 discuss about color watershed techniques and demonstrates the way color segmentation based on watersheds can be achieved and improved by using the multi-scale gradient. The experimental results are shown in section 5. Finally, conclusions are drawn in section 6.

2. Color Morphology

Every morphological operator we apply to the grayscale images can be applied to the color channels separately, because it commutes with infimum and supremum respectively. This somewhat marginal processing is equivalent to the vectorial approach defined by the canonic lattice structure when only supremum and infimum operators and their compositions are involved and induces a totally ordered lattice presented in [5] and expressed by $X \leq Y \iff X(i) \leq Y(i), \forall i \in 1, \ldots, N$.

With these relations, the supremum of a family $\{X_j\}$ is the vector $\vee X$ where each component $\vee X(i)$ is the supremum of the $\{X_j(i)\}$. Respectively, the infimum of a family $\{X_j\}$ is the vector $\wedge X$ where each component $\wedge X(i)$ is the infimum of the $\{X_j(i)\}$. Using this procedure, new colors not contained in the input image will appear even for flat structuring functions. Despite of this major drawback of introducing new colors into the image, there is an inverse effect of color reduction suggested in [2].

The essential point to extend mathematical morphology to color images is to define a vectorial ordering relation that induces a lattice structure on the data. Recently, in [1] and [10] an extension from grayscale to color morphology have been introduced, following the idea of adjunctions and h-adjunctions. Firstly, the input image is coded. Secondly, a mapping $h : R^N \rightarrow R$ is used to rank the vectors that means that each vector pixel is represented by a single scalar value. The main idea is that a lexicographical ordering[1] is used, inducing a total ordering and determines clearly the infimum and supremum of each set of vectors.

With this approach, it is possible to perform any grayscale morphological operator on the coded image, and to decode the result afterwards. Thus, the vectorial ordering relation can be considered as an extension to a total order. It follows that, the output vector of any morphological operator is necessarily one of the input vectors. Nevertheless, it is sensitive to the choice of the vectorial ordering relation and to the color model in use. The suggested method in [1]

[1] An ordered pair (i, j) is lexicographically earlier than (i', j') if either $i \leq i'$ or $i = i'$ and $j \leq j'$. It is lexicographic since it corresponds to the dictionary ordering of letter words.

is slightly modified to obtain a morphological multi-scale gradient for color images that will drive the watershed segmentation.

3. Morphological Edge Detection

3.1. Edge Detection

Many applications in computer vision require edge detection. Edges are of extreme importance, as they convey essential information in a picture. Accurate edge detection is necessary for a number of image analysis and recognition techniques including watershed segmentation. Ideally, edges are steep transitions between smoothly varying luminance areas. In real scenes, transitions are not so steep, which means that their extraction requires careful processing.

Several edge detection techniques exist in the literature [6]. Usually, these techniques use first or second derivatives of image luminance and adopt some adaptive filtering techniques to control the effects of noise. Although this is a sure way to reduce noise effects, it leads, most likely, to discontinuities in the detected edges and in contour misplacement.

3.2. Color Multi-scale Gradient

In morphological image processing, several edge detectors have been developed. However, we will concentrate on the classical morphological gradient defined as:

$$\nabla(f) = \delta_g(f) - \varepsilon_g(f) \tag{1}$$

It is clear that the gradient magnitude obtained by subtracting the erosion $\varepsilon_g(f)$ from a dilation $\delta_g(f)$ is influenced by the size and shape of the structuring function g. Since the gradient is used at the basis of the general morphological approach to segmentation, it is of primary importance to compute it properly. However, this gradient is inappropriate because when large structuring functions are employed, edges that are close to each other may get merged by smoothing and may be detected as only one single edge. On the other hand, small structuring functions result in too many noise points due to insufficient averaging and low intensity edges. Unfortunately, this problem cannot be solved by using gradients by erosion or dilation because they produce biased results depending on the brightness of the region.

In order to solve these problems, we propose a multi-scale approach to the morphological gradient to work on color images. Some mathematical equations must be provided. Equation 2 is a morphological filter $\psi(f)_{i+1}$. The filter chosen was an opening ψ of the color image f scaled by g_i. Clearly, $\psi(f)_1$ corresponds to the color image f. By adding the morphological gradient of the filtered color image in equation 3 to the top-hat by opening of its own gradient in equation 4 (i.e. white top-hat) and taking the average based on scale we arrive at the color multi-scale gradient in equation 5.

$$\psi(f)_{i+1} = \varphi_{g_i}(\psi(f)_i) \tag{2}$$

$$\nabla(\psi(f)_{i+1}) = \delta_{g_i}(\psi(f)_{i+1}) - \varepsilon_{g_i}(\psi(f)_{i+1}) \tag{3}$$

$$\Gamma(\nabla(\psi(f)_{i+1})) = \nabla(\psi(f)_{i+1}) - \varphi_{g_i}(\nabla(\psi(f)_{i+1})) \tag{4}$$

$$G(f) = \frac{1}{scale} \sum_{i=0,scale} \nabla(\psi(f)_{i+1}) + \Gamma(\nabla(\psi(f)_{i+1})) \tag{5}$$

The color multi-scale gradient allows enhancing the real boundaries, which exist in each scale, compared to boundaries that are present in one scale only. Moreover, due to the averaging operation, it is more robust to noise caused by scaling. The top-hat transform in equation 4 is a very good contrast detector suitable for enhancing the bright and narrow objects in the image [9].

It is interesting to compare the results obtained by conventional methods against the proposed color multi-scale gradient. Figure 1 shows the original RGB color image and three versions of the gradient. For the sake of comparison, all gradients images are shown in grayscale. Figure 1(b) shows the result obtained by applying the morphological gradient (see equation 1) to the color image. Figure 1(c) is achieved by taking the maximum of the gradients from the color channels separately and figure 1(d) shows the results obtained by the color multi-scale gradient when scale is set to 5.

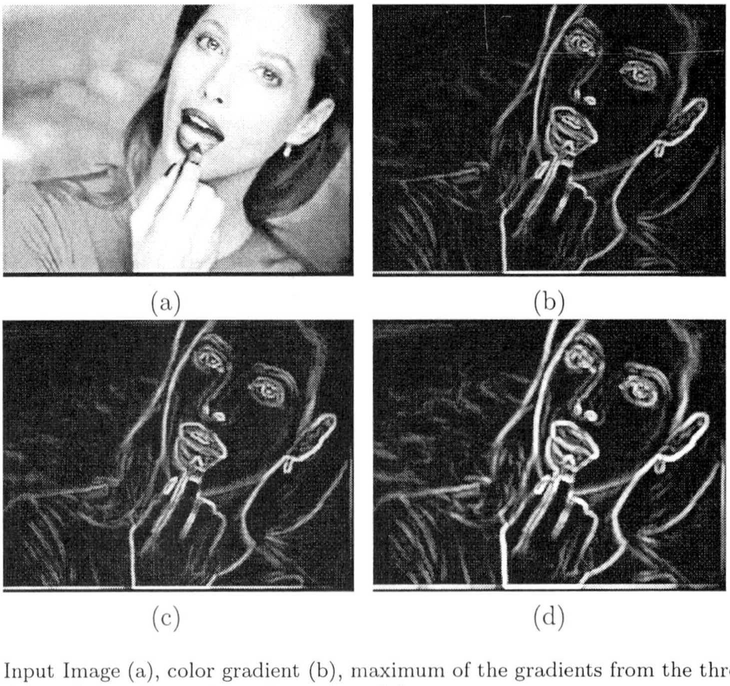

Fig. 1. Input Image (a), color gradient (b), maximum of the gradients from the three-color channels (c), and the color multi-scale gradient (d).

In this paper, we use an interval of Euclidean disks centered at origin and parameterized by i. The interval is generated from a single small convex symmetrical structuring function g by letting $g_0 = g$ and g_i for $(i \geq 1)$ given by i Minkowski additions. Euclidean disks reduce the anisotropy that is present in other shapes like a line, a square, or a diamond.

4. Color Image Segmentation with Watersheds

The idea of watershed is drawn by considering an image as a topographical surface. Suppose we pierce holes at every regional minimum and dip the image surface into water, then water will flood areas adjacent to regional minima. A regional minimum is a connected plateau from which it is impossible to reach a point of lower gray level by an always-descending path. As the image surface is immerged, some of the flood areas (catchment basins) will tend to merge. When two or more different flood areas are touched, watershed lines (i.e. dams) are constructed between them. When finished, the resulting networks of dams define the watershed of the image.

4.1. Color Watershed Techniques

The watershed method is meaningful only for grayscale image analysis. It is based on the existence of a total ordering relation embedded in the complete lattice. Since such a relation does not exist in color space, the watershed transform is not applicable immediately. One should consider experimenting particular approaches in order to understand its advantages and drawbacks. There are so many techniques that could be employed to exploit the idea of watersheds to color image segmentation. Different techniques may produce different watershed lines and catchment basins. In this paper, we will restrict ourselves to the ones shown in figure 2.

The first two techniques shown in figure 2(a) and (b) start through the splitting of the color channels followed by an image filtering. Every channel is filtered to make easier the image segmentation. Morphological filters are used for this task. These filters remove regions that are smaller than a given size but preserve the contours of the remaining objects [8]. Subsequently, depending on the order in which the combination of the channels is performed, the gradient of the filtered image is approximated by the use of a morphological gradient operator. After this step, the gradient is used as an input to the marker selection for the watershed algorithm to partition an image into homogeneous regions.

The third technique in figure 2(c) follows almost the same steps described in the previous approaches (i.e, image filtering, morphological gradient, marker selection, and watersheds). The novelty of this technique is that, despite of splitting color channels, it makes use of color morphology for image filtering and for the multi-scale gradient.

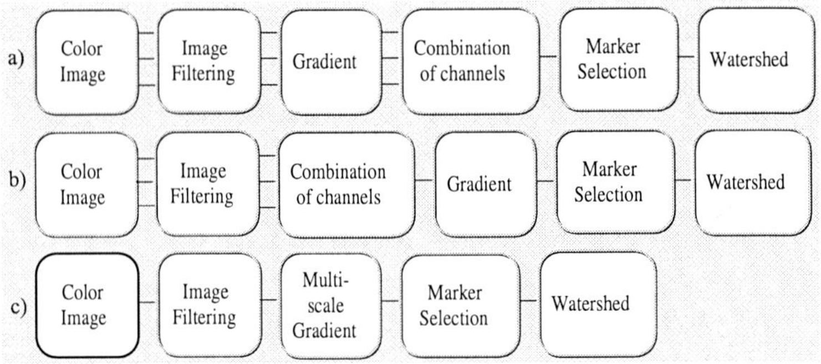

Fig. 2. Color Watershed Approaches.

4.2. Marker Selection

The procedure to extract the extrema of a function is called geodesic reconstruction [11]. Lets us consider two functions f and r and suppose that we want to impose some regional minima R to f. We construct $r(x,y) = +\infty \; if \; (x,y) \notin R; r = 0 \; if \; (x,y) \in R$. Function f is called the mask image and r is the marker. Denote by n the elementary structuring function. The reconstruction of f by r is obtained by iterating the following operation until stability is reached:

$$r^{k+1}(x,y) = \wedge(f(x,y), \varepsilon_n r^k(x,y)) \tag{6}$$

where k is the level of iteration performed.

Among various applications of this useful transformation, valley removal is of interest for us now. To extract valleys with contrast greater than a height value h of an image I, it suffices to reconstruct I from $I + h$. By algebraic difference between I and the reconstructed function, one gets the desired image without valleys. To extract the peaks, one can use a dual reconstruction of I from $I - 1$. It should be noted that the image produced by the valley removal step is subjected to the combination of channels before a regional minimum detection is applied.

4.3. Watershed Algorithm

The watershed algorithm was introduced for the purpose of segmentation in [7] and [12]. The generic marker-based watershed algorithm used in this paper, recently introduced in [3], is established on the wave front propagation interpretation. It consists of a discrete wavefront initialization and data driven propagation. The initial wavefront (i.e., the set of markers) is considered as sources of propagation. Amid waves arriving at a point, the first one is added to the discrete front. This stage is repeated for every point on the wavefront. The envelope of all infinitesimal wavefronts forms the macroscopic wavefront. The algorithm can be applied to different image types and makes use of a priority queue to speed up the propagation step.

5. Experimental Results

The experimental results described in this section have been partially obtained using `mmorph`, a novel morphological library for `Matlab` [2]. Without loss of generality, the color model chosen was the classical RGB despite of color features being highly correlated. Nevertheless, other color models for color segmentation and recognition could be used [4].

The evaluation criterion chosen was to preserve the majority of all important edges and reduce the number of remaining segments. Figure 3(a) and (b) show the watershed results on top of the original image when the watershed schemes (a) and (b) from figure 2 are applied. The height h was set to 20 for the marker selection. Figure 3(c) and (d) show the watershed results on top of the original image when the watershed scheme (c) from figure 2 is applied. The scale s and height h were set to [4,20] and [5,20] respectively.[3]

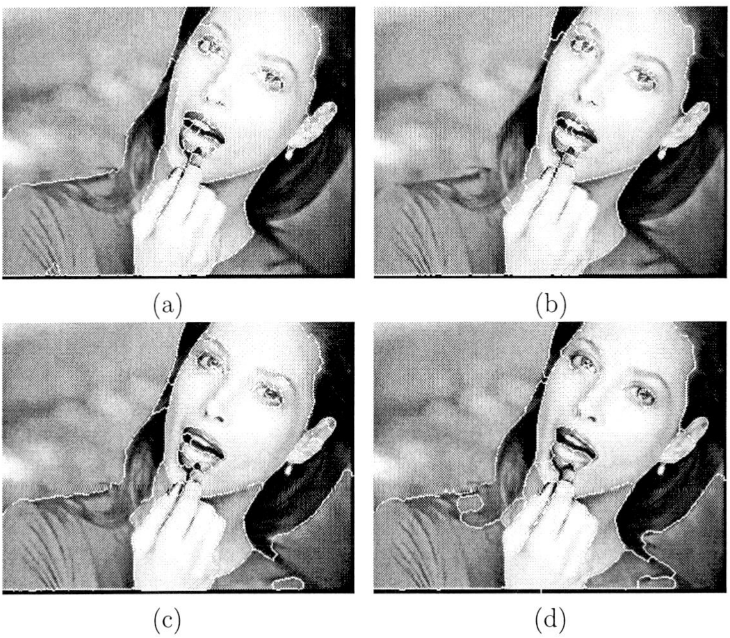

Fig. 3. Input Image and Watershed Results.

6. Conclusions

The performance of morphological techniques based on the watershed approach is strongly influenced by the gradient algorithm used as a starting step for segmentation. Using a conventional gradient operator, the watershed approach

[2] Additional information about this library can be obtained at http://www.mmorph.com

[3] Due to the costs of color printing and the inherent distortions associated with the size reduction and the printing process, the corresponding color images will be made available through http://www.wins.uva.nl/~ornellas/images/ismm2000

leads to an oversegmented image with many irrelevant regions. In this paper, we have presented a color multi-scale gradient and a new method for the watershed segmentation using color morphology without splitting the image in its own channels.

The reconstruction algorithm embedded in the marker selection step can efficiently remove irrelevant minima caused by noise and quantization error in the resulting gradient images. Hence, watershed transformation based on the color multi-scale gradient results in meaningful segmentation with more precise watershed lines.

Acknowledgements

The work reported here was accomplished partially under the support of CAPES Foundation grant BEX 2780/950. We gratefully acknowledge the ideas of Hieu Tat Nguyen and Luisa Faella, which have contributed to this work.

References

1. J. Chanussot and P. Lambert. Total ordering based on space filling curves for multi-valued morphology. In H. J. A. M. Heijmans and J. B. T. M. Roerdink, editors, *Mathematical Morphology and Its Applications to Image Processing*, pages 51–58. Kluwer Academic Publishers, The Netherlands, 1998.
2. C.-H. Demarty and S. Beucher. Color segmentation algorithm using an hls transformation. In H. J. A. M. Heijmans and J. B. T. M. Roerdink, editors, *Mathematical Morphology and Its Applications to Image Processing*, pages 231–238. Kluwer Academic Publishers, The Netherlands, 1998.
3. M. C. d'Ornellas and R. v.d. Boomgaard. Generic algorithms for morphological image operators - a case study using watersheds. In H. J. A. M. Heijmans and J. B. T. M. Roerdink, editors, *Mathematical Morphology and Its Applications to Image Processing*, pages 323–330. Kluwer Academic Publishers, The Netherlands, 1998.
4. T. Gevers and A. W. M. Smeulders. Color-based object recognition. *Pattern Recognition*, 32:453–464, 1999.
5. J. Goutsias, H. J. A. M. Heijmans, and K. Sivakumar. Morphological operators for image sequences. *Computer Vision and Image Understanding*, 62:326–346, 1995.
6. R. M. Haralick and L. G. Shapiro. *Computer and Robot Vision - Volume I.* Addison Wesley, New York, 1993.
7. F. Meyer and S. Beucher. Morphological segmentation. *Journal of Visual Communications and Image Processing*, 1(1):21–46, 1990.
8. J. Serra and L. Vincent. An overview of morphological filtering. *IEEE Transactions on Circuits, Systems and Signal Processing*, 11:47–108, 1992.
9. P. Soille. *Morphological Image Analysis.* Springer-Verlag, Barcelona, 1999.
10. H. Talbot, C. Evans, and R. Jones. Complete ordering and multivariate mathematical morphology. In H. J. A. M. Heijmans and J. B. T. M. Roerdink, editors, *Mathematical Morphology and Its Applications to Image Processing*, pages 27–34. Kluwer Academic Publishers, The Netherlands, 1998.
11. L. Vincent. Morphological grayscale reconstruction in image analysis: Applications and efficient algorithms. *IEEE Transactions on Image Processing*, 2:176–201, 1993.
12. L. Vincent and P. Soille. Watersheds in digital spaces: An efficient algorithm based on immersion simulations. *IEEE Transactions on Pattern Analysis and Machine Intelligence*, 13(6):583–598, 1991.

AUTOMATIC WATERSHED SEGMENTATION OF COLOR IMAGES

I. VANHAMEL and H. SAHLI
VUB - ETRO, Pleinlaan 2, B-1050 Brussels - Belgium.
Tel: +32-2-629.2916, Fax: +32-2-629.2883

and

I. PRATIKAKIS
IRISA - ViSTA, Campus de Beaulieu, 35042 Rennes, France.
Tel: +33-2-99.84.75.16, Fax: +33-2-99.84.71.71

Abstract. This paper presents a fully automatic watershed color segmentation scheme which is an extension to color images of a previously reported approach dedicated to segmentation of scalar images. The importance of this extension lies mainly on its ability to automatically select an optimum result out of a hierarchical stack. This achievement is realized through the introduction of new evaluation methods for the segmentation quality of each level of the hierarchy which considers a tradeoff between the preservation of details and the suppression of heterogeneity. The first method estimates the local color error of the regions and combines it with the amount of regions. The second evaluates the contrast of the segmented image by combining a region uniformity with an inter-region contrast measure for all regions. These two methods are compared with respect to an existing one. Experimental results demonstrate the improvement which has been achieved by using the new evaluation criteria.

Key words: Hierarchical Watershed Segmentation, Color, Evaluation Criteria.

1. Introduction

Color image segmentation refers to the partitioning of a multi-valued image into meaningful objects. The additional information provided by color along with the continuously increasing number of applications which deal with analysis tasks of color images, advocate its prominent position among the interests of the image processing community.

In this paper, we present an automatic color image segmentation algorithm based on the principles of hierarchical segmentation using the watershed transformation. It is an extension of an earlier hierarchical watershed segmentation of scalar images [7].

Similar to the grey level case, the segmentation of color images using the watershed transformation can be translated as the elimination of its main drawback, namely over-segmentation. In the literature, treatment of the over-segmentation problem in color images can be divided into three main categories. The first one deals with the selection of an appropriate set of *markers* which will be used as a guide for an exclusive selection of regions. The marking procedure is a very difficult task and sometimes impossible to solve. This approach has been followed by Meyer [6] who used color differences between pixels to establish a stopping criterion for a region growing process which propagated the labels of selected markers. The second category is linked to the *flat zone* concept. Crespo and Schafer [2] propose a scheme which establishes a flat zone inclusion relationship, affecting only those flat zones that belong to an extracted feature computed from all color bands. This feature is computed by employing a gradient-to-area criterion. Recently, a tendency exists towards a third approach which attributes a *hierarchy* among the gradient watersheds. This additionally implies that the computation of the gradient magnitude of a selected color scheme has to be considered. Shafarenko et al. [8] apply the *waterfall* hierarchical segmentation algorithm [1] on a LUV gradient image and employ a termination criterion for the merging process based on the topology of randomly textured images. Finally, Demarty and Beucher [3] propose two merging schemes for gradient watersheds which are computed on an HLS gradient magnitude image. Either they apply the waterfall algorithm or they merge the basins according to the color of their minima.

In our approach, we extend an earlier hierarchical watershed segmentation scheme [7] for color images, experimenting with RGB and $L^*a^*b^*$ color space. The gradient computation for these color spaces is based on the *di Zenzo multi-band gradient* [9]. We select the optimum segmented level out of a hierarchical stack by evaluating a criterion which takes into account the trade off between preservation of details and suppression of heterogeneity. Three different evaluation criteria were tested. Experimental results are given for all combinations between color spaces and evaluation criteria.

The organization of this paper is as follows: In section 2 we describe the proposed hierarchical segmentation scheme. The evaluation criteria that are introduced for automatically extracting the best segmentation level out of the produced hierarchical stack are given in section 3. Section 4 is dedicated to the discussion of our experimental results and conclusions.

2. The Hierarchical Segmentation Scheme

The hierarchical segmentation allows to obtain iteratively less detailed image partitioning. In our approach, a *hierarchical stack* $\{HL_k\}$, $k = 1, 2, \cdots, l_m$ is obtained. Each *hierarchical level* HL_k is defined as a partitioning $\mathbf{P}_k = \{R_{k,1}, R_{k,2}, \cdots, R_{k,n_k}\}$ of the original color image $\mathbf{I}$, which preserves the inclusion relationship $\mathbf{P}_k \supseteq \mathbf{P}_{k-1}$. In other words, each region $R_{k,i}$ of $\mathbf{P}_k$ is a assumed to be the union of a unique set of $\mathbf{P}_{k-1}$ regions. $\mathbf{P}_k$ is a mosaic color image consisting of a patchwork of regions of constant color, estimated as the mean color vector of the region. Our approach, called *constrained dynamics of contours* is detailed in a previous paper [7] dealing with scalar images. Our extension of the original algorithm [7] for the segmentation of color images concerns the computation of the gradient magnitude in multi-valued images and the modification of the hypothesis test. The adapted scheme which deals with color images is described in the flowchart of Figure 1 and is detailed in the sections which follow.

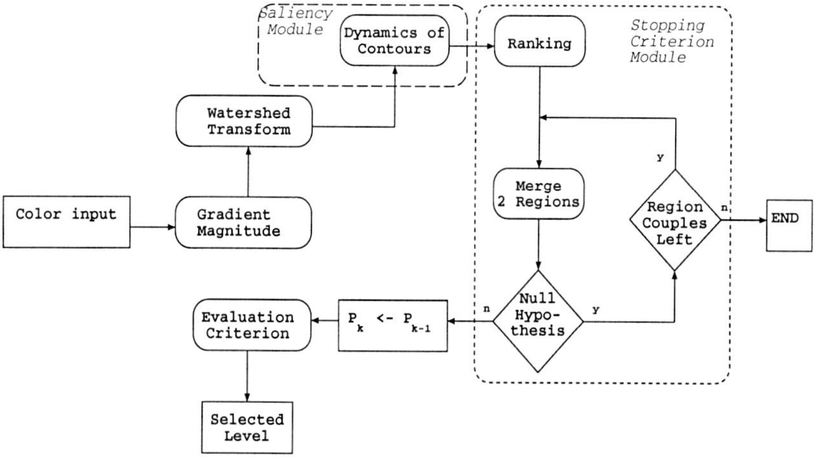

Fig. 1. Flowchart of the *constrained dynamics of contours* for color images

2.1. Color gradient estimation

Several approaches for color gradient estimation were proposed in the literature. An estimation of the color gradient as the sum over the gradient of each color component has been reported by Demarty and Beucher [3]. Another approach is to take the RMS of the component gradient magnitudes as the magnitude of the resultant gradient [9]. Rather than combining the different gradient components, the color scalar gradient can be estimated as the largest color distance between neighboring pixels. In case of an uniform color space, the Euclidean distance is used [8].

The above mentioned approaches make use of color components; but however, do not consider that color images are *vector-valued* functions for the multiband gradient estimation. An explicit formula for multi-band gradient estimation has been reported by S. di Zenzo [9]:

Let a m-band image be represented by a function $f : R^2 \to R^m$, that maps a point $P(x_1, x_2)$ in the image plane to a m-vector $\mathbf{f} = [f_1(x_1, x_2), \ldots, f_m(x_1, x_2)]^t$. For color images we have $m = 3$. Since we are looking for variations in the image, we consider the difference at two nearby image points P and Q: $\Delta\mathbf{f}(P, Q) = \mathbf{f}(P) - \mathbf{f}(Q)$. Considering an infinitesimal displacement dP, the difference becomes the differential $df^2 = \sum_{i=1}^{2} \frac{\partial \mathbf{f}}{\partial x_i} dx_i$ and its squared norm:

$$df^2 = \sum_{i-1}^{2} \sum_{j-1}^{2} \frac{\partial \mathbf{f}}{\partial x_i} \cdot \frac{\partial \mathbf{f}}{\partial x_j} dx_i dx_j \tag{1}$$

where the dot denotes inner product. Note that, speaking about the inner product of vectors implies that some kind of metric has been defined in R^m.

Using the Euclidean metric we get:

$$df^2 = \sum_{i=1}^{2}\sum_{j=1}^{2} g_{ij}dx_i dy_j \tag{2}$$

where the g_{ij}'s are given by $g_{ij} = \sum_{k=1}^{m} \frac{\partial f_k}{\partial x_i}\frac{\partial f_k}{\partial x_j}$.

The extrema of the quadratic form defined by Eq 2 are obtained in the direction of the eigenvectors of the matrix $[g_{ij}]$, given by $\mathbf{n}_{\pm} = (\cos\theta_{\pm}, \sin\theta_{\pm})$; $\theta_{+} = \frac{1}{2}\arctan\frac{2g_{12}}{g_{11}-g_{22}}$, $\theta_{-} = \theta_{+} + \pi/2$; and the values at these locations correspond to the eigenvalues $\lambda_{\pm} = (g_{11} + g_{22} \pm \sqrt{(g_{11} - g_{22})^2 + 4g_{12}})/2$. The multi-band gradient is then defined by:

$$\nabla\mathbf{f} = \sqrt{\lambda_{+} - \lambda_{-}} \tag{3}$$

In this paper we use, for both the RGB and $L^*a^*b^*$ color space, the *di Zenzo multi-band gradient* as input for the watershed segmentation.

2.2. Hierarchical stack of gradient color watersheds

After the application of the watershed transformation on the gradient magnitude image we construct a tree structure following a model that consists of two modules. The first module is dedicated to valuate the saliency of each contour arc of the oversegmented image using the *dynamics of contours* [5], while the second module identifies the different hierarchical levels using a *stopping criterion*. When the contour valuation of adjacent region couples has terminated, a ranking of the values providing the priority of merging is applied. The hierarchical segmentation algorithm will be completed after the application of the merging stopping criterion phase which retrieves the different hierarchical levels HL_k. For this purpose, a statistical decision is employed through a *hypothesis test*, leading to the creation of a new hierarchical level in the case that the homogeneity constraint imposed in the regions is violated during the region merging process. The hypothesis set is defined as follows:

- H_0^k: Two adjacent regions belong to the same label at level k
- H_1^k: Two adjacent regions belong to different labels at level k

This stopping criterion phase has been extended in the case of multi-valued images defined in RGB and $L^*a^*b^*$ color space. The generalization of the *chi-square* test approach for multivariate analysis is the *Hotteling's* T^2 test. The application of this test to the merged region $R_i \cup R_j$ is given by:

$$H_0^k : \tilde{F}^2_{(ij)_k} < F_{(3,n_{(R_i\cup R_j)_k}-4)}(\alpha) \Rightarrow P(i,j) = true \tag{4}$$

$$H_1^k : \tilde{F}^2_{(ij)_k} \geq F_{(3,n_{(R_i\cup R_j)_k}-4)}(\alpha) \Rightarrow P(i,j) = false \tag{5}$$

where $\tilde{F}^2_{(ij)_k} = \frac{1}{3(n_{(R_i\cup R_j)_k}+2)}\frac{(\mu_i-\mu_j)^t\Sigma^{-1}_{(ij)}(\mu_i-\mu_j)}{\frac{1}{n_{R_{k,i}}}+\frac{1}{n_{R_{k,j}}}}$, μ_i and μ_j are the mean color vector of regions $R_{k,i}$ and $R_{k,j}$ respectively, $\Sigma_{(ij)}$ is the covariance matrix approximated as a diagonal matrix under the assumption that the components in RGB and $L^*a^*b^*$ color space are independent, and $F_{(m,l)}$ denotes a random variable with a *F-distribution* which has m and l degrees of freedom. The confidence interval α is set to 0.05 throughout all our experiments.

3. Criteria for Automatic Hierarchical Level Selection

In most hierarchical segmentation schemes, the desired degree (level) of segmentation remains interactive. In other words, the user may select visually and qualitatively, the hierarchical level at which the resulting segmentation is acceptable. In this paper, we introduce an automatic hierarchical level selection based on a quantification of the segmentation quality for each level of the hierarchical tree via an evaluation criteria. Various empirical evaluation methods for image segmentation have been proposed in the literature [10]. They rely on segmentation quality measures such as (i) intra-region uniformity, (ii) inter-region contrast and (iii) region shape (boundaries should be smooth and accurate).

Liu and Yang [4] suggested a segmentation evaluation function $\mathcal{LY}(.)$ which expresses the trade-off between the suppression of heterogeneity and preservation of details. Using the same notation as in Section 2.2, $\mathcal{LY}(.)$ is defined as:

$$\mathcal{LY}(\mathbf{I},\mathbf{P}_k) = \sqrt{n_k} \sum_{R_{k,i}\in\mathbf{P}_k} \frac{\sum_{x_j\in R_{k,i}} {d_E}^2(\mathbf{I}(x_j),\mathbf{P}_k(x_j))}{\sqrt{n_{R_{k,i}}}} \tag{6}$$

where $\mathbf{I}$ is the original color image, $\mathbf{P}_k$ is the partitioning at level k, n_k denotes the number of regions at level k and $d_E(.,.)$ is the Euclidean distance. The term $\sqrt{n_k}$ is a global measure which penalizes the segmentation with too many regions. The term $\sum {d_E}^2/\sqrt{n_{R_{k,i}}}$ is a local measure which penalizes small regions or regions with a large color error. The smaller the value of $\mathcal{LY}(\mathbf{I},\mathbf{P}_k)$, the better the segmentation result is.

When taking into consideration different metrics $d(.,.)$, for color distance estimation, in any selected color space a generalization of Liu and Yang evaluation function is proposed:

$$\mathcal{GLY}(\mathbf{I},\mathbf{P}_k) = \sqrt{n_k} \sum_{R_{k,i}\in\mathbf{P}_k} \sum_{x_j\in R_{k,i}} d^2(\mathbf{I}(x_j),\mathbf{P}_k(x_j)) \tag{7}$$

In this formulation we ensure that significant small regions can be also considered (not penalized).

The segmentation evaluation functions $\mathcal{LY}(\mathbf{I},\mathbf{P}_k)$ and $\mathcal{GLY}(\mathbf{I},\mathbf{P}_k)$ do not incorporate directly the quality measures (ii) and (iii). In view of this, we propose an evaluation function $\mathcal{CH}(\mathbf{I},\mathbf{P}_k)$ which combines a region homogeneity measure $H_{k,i}$ (local color error) with an inter-region contrast measure $C_{k,i}$ (color difference between adjacent regions). It is denoted as :

$$\mathcal{CH}(\mathbf{I},\mathbf{P}_k) = \frac{1}{\sum_{R_{k,i}\in\mathbf{P}_k} n_{R_{k,i}}} \sum_{R_{k,i}\in\mathbf{P}_k} n_{R_{k,i}} CH_{k,i} \tag{8}$$

where for a given region $R_{k,i}$, the term $CH_{k,i}$ is defined as:

$$CH_{k,i} = \begin{cases} 1 - \frac{H_{k,i}}{C_{k,i}}, & if\ H_{k,i} < C_{k,i} \\ 0 & otherwise \end{cases} \tag{9}$$

We estimate the region homogeneity $H_{k,i}$ and the border contrast $C_{k,i}$ as follows:

$$H_{k,i} = \frac{1}{n_{R_{k,i}}} \sum_{x_j \in R_{k,i}} d(\mathbf{I}(x_j), \mathbf{P}_k(x_j)) \tag{10}$$

$$C_{k,i} = \frac{1}{n_{B(R_{k,i})}} \sum_{x_j \in B(R_{k,i})} \max_{x_l \in N(x_j)} (d(\mathbf{I}(x_j), P(x_l))) \tag{11}$$

where $d(.,.)$ is the metric defined for the selected color space, $B(R_{k,i})$ is the set of border pixels of region $R_{k,i}$, and $N(x_j)$ denotes the neighborhood for pixel x_j. $H_{k,i}$ is very similar to the definition of a uniformity criterion, and $C_{k,i}$ expresses a trade-off between the border accuracy of a region and the difference between the region and its neighbors. $CH_{k,i}$ penalizes those regions with large color error and low border contrast. If the color error inside the region is low and the adjacent regions are significantly different (high border contrast), its value is high. This conforms to the quality measures (i) and (ii). In the proposed segmentation evaluation criterion $\mathcal{CH}(\mathbf{I}, \mathbf{P}_k)$, the higher the value is, the better the segmentation result is.

4. Experimental Results - Conclusions

As already mentioned in Section 1, the fundamental problem in watershed color segmentation is over-segmentation. This can be seen in Figures 2(a)-(d). In this paper, we tried to overcome the over-segmentation problem by an automatic color segmentation scheme based on the construction of a stack with hierarchical levels evaluated by functions which describe the desired optimum segmentation. We have tested three different evaluation functions which were analytically described in Section 3. The behavior of our proposed algorithm was demonstrated in *RGB* (Fig. 3) and $L^*a^*b^*$ (Fig. 4) color space. For the sake of clarity, a subset of successive hierarchical levels for each color space is presented, including the selected optimum segmentation according to each of the evaluation functions that have been considered. Additionally, in Fig. 5 the behavior of the evaluation functions for all the levels in the hierarchical stack, is shown. We recall that in the case of $\mathcal{CH}(.)$ we search for the global maximum, while in the case of $\mathcal{LY}(.)$ and $\mathcal{GLY}(.)$ we search for the global minimum. Note that for the global optimum the last level of the hierarchical stack is not considered since this attributes one segment to the whole image. After our experiments with various images - only two are presented herein due to lack of space - $\mathcal{GLY}(.)$ and $\mathcal{CH}(.)$ are improvements of $\mathcal{LY}(.)$ evaluation function. Consistent to its definition $\mathcal{GLY}(.)$ penalizes segmentation levels with a high number of regions when the color error term is low. This causes the suppression of significant image details (small regions). In the case of $\mathcal{CH}(.)$, significant image details are preserved.

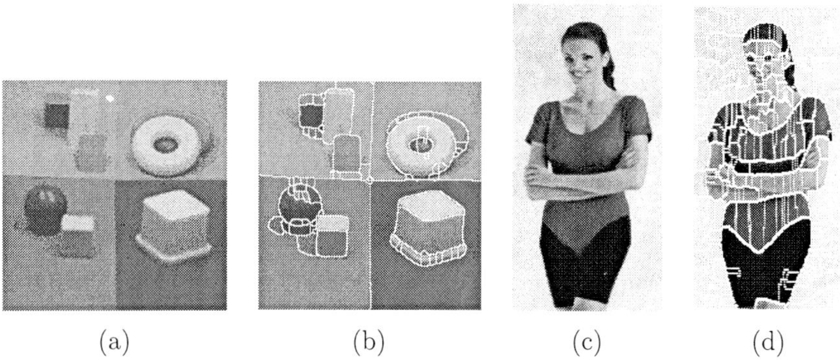

Fig. 2. Original images (a)Toys, (c) Woman, and their oversegmented results (b),(d)

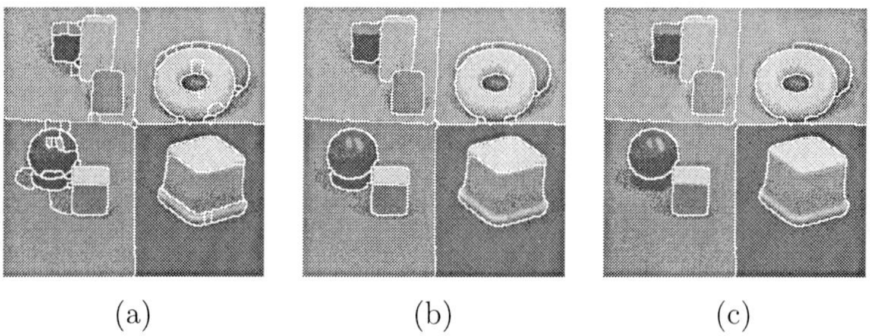

Fig. 3. (a)-(c) Three successive hierarchical levels ($L^*a^*b^*$ case) : (b) Automatic level selection by $\mathcal{GLY}(\mathbf{I},\mathbf{P}_k)$ and $\mathcal{CH}(\mathbf{I},\mathbf{P}_k)$, (c) Automatic level selection by $\mathcal{LY}(\mathbf{I},\mathbf{P}_k)$

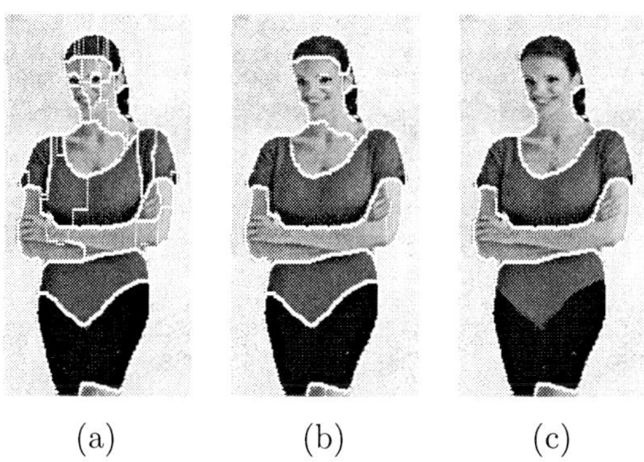

Fig. 4. (a)-(c) Three successive hierarchical levels (RGB case) : (b) Automatic level selection by $\mathcal{CH}(\mathbf{I},\mathbf{P}_k)$, (c) Automatic level selection by $\mathcal{LY}(\mathbf{I},\mathbf{P}_k)$ and $\mathcal{GLY}(\mathbf{I},\mathbf{P}_k)$

In this paper, no panacea solution is provided, but our experimental results encourage further investigation guided by the proposed scheme within the field of automatic hierarchical level selection out of a hierarchical stack in color image segmentation.

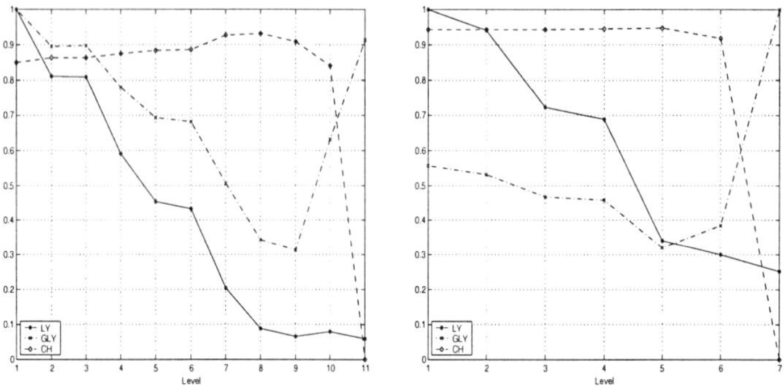

(a) (b)

Fig. 5. The evaluation functions $\mathcal{LY}(\mathbf{I}, \mathbf{P}_k)$, $\mathcal{GLY}(\mathbf{I}, \mathbf{P}_k)$ and $\mathcal{CH}(\mathbf{I}, \mathbf{P}_k)$ for each hierarchical level, in the case of (a) Toys and (b) Woman image

References

1. S. Beucher. *Watershed, hierarchical segmentation and waterfall algorithm.* In J. Serra and P. Soille, editors, *Mathematical Morphology and its applications on Image and Signal processing (ISMM'94)*, pages 69–76. Kluwer Academic Press, 1994.
2. J. Crespo and R. Schafer. *The flat zone approach and color images.* In J. Serra and P. Soille, editors, *Mathematical Morphology and its applications on Image and Signal processing (ISMM'94)*, pages 85–92. Kluwer Academic Press, 1994.
3. C.H. Demarty and S. Beucher. *Color segmentation algorithm using an HLS transformation.* In H. Heijmans and J. Roerdink, editors, *Mathematical Morphology and its applications on Image and Signal processing (ISMM'98)*, pages 231–238. Kluwer Academic Press, 1998.
4. J. Liu and Y.H. Yang. *Multiresolution color image segmentation. IEEE Transactions on Pattern Analysis and Machine Intelligence*, 16(7):689–700, 1994.
5. L.Najman and M. Schmitt. *Geodesic saliency of watershed contours and hierarchical segmentation. IEEE Transactions on Pattern Analysis and Machine Intelligence*, 18(12):1163–1173, 1996.
6. F. Meyer. *Color image segmentation.* In *4th IEE Conference on Image Processing and applications*, volume 354, pages 53–56, 1992.
7. I. Pratikakis, H. Sahli, and J. Cornelis. *Low level image partitioning guided by the gradient watershed hierarchy. Signal Processing*, 75(2):173–195, 1999.
8. L. Shafarenko, M. Petrou, and J. Kittler. *Automatic watershed segmentation of randomly textured color images. IEEE Transactions on Image Processing*, 6(11):1530–1544, 1997.
9. S. Di Zenzo. *A note on gradient of a multi-image. Computer Vision, Graphics and Image Processing*, 33:116–125, 1986.
10. Y.J. Zhang. *A survey on evaluation methods for image segmentation. Pattern Recognition*, 29(8):1335–1346, 1996.

MOTION SEGMENTATION USING SEEDED REGION GROWING

RICHARD BEARE and HUGUES TALBOT
CSIRO Mathematical and Information Sciences
Locked Bag 17, North Ryde, Australia, 1670
Richard.Beare{Hugues.Talbot}@cmis.csiro.au

Abstract. Motion based segmentation techniques usually begin with an estimate of the velocity flow field and utilise clustering techniques to segment the flow field. This paper investigates the practicality of performing segmentation using the information typically applied to the velocity estimation stage. A region growing process is described that can reliably segment rigid objects observed by moving cameras without requiring shape or motion models.

Key words: Motion Segmentation, Region Growing.

1. Introduction

Motion information is a powerful cue that can assist in segmenting image sequences. In some situations motion information may be the most significant cue, possibly far more significant than colour or texture. Segmentation of image sequences is important in many applications, including robotics, video coding, multimedia and security, and a wide variety of approaches have been investigated.

Motion based segmentation has been under investigation for many years. Typical approaches involve two steps – estimation of image velocities followed by clustering of some kind. A wide variety of approaches to both steps has been developed.

This paper explores the merits of applying a region-based segmentation technique – seeded region growing – to a form of information that is usually used for velocity estimation.

2. Background

2.1. Velocity Estimation

Velocity estimation techniques can be broadly described as either feature based tracking schemes or spatio-temporal filtering schemes. The first type of scheme includes edge, corner and block based schemes [1, 2] which typically produce sparse flow fields, while the second type includes gradient based optical flow schemes [3] and energy models [4, 5, 6] which produce dense flow fields.

2.2. Region-based segmentation techniques

In the past region-based segmentation techniques have been mostly used to segment grey-level or colour still images [7, 8, 9]. These schemes assume that it is possible to define subsets of the regions of interest (markers), some form of region descriptor (control field) and a measure of compatibility between a region's control field and border pixels. The markers are grown by individually considering pixels on the outer edge of each marker. A decision on whether to add an individual pixel to a growing region is based on this compatibility measure.

Region-based segmentation techniques – using the mean colour of the regions as the measure on the control field – have been proposed for the segmentation of time series of images, with applications to sequence coding [10]. These techniques are attractive due to the high level nature of the features and the robustness of region growing techniques. The high level nature of the region based tokens can simplify the matching process.

2.3. Other motion segmentation techniques

Velocity based segmentation techniques may generally be considered as local or global – local schemes attempt to find discontinuities in the velocity field while global schemes attempt to find regions of consistency. Local schemes may be susceptible to noise but are often attractive from the implementation point of view. Global schemes commonly attempt to fit regions of the flow field to analytic functions [11, 12]. Hybrid schemes that formulate the problem as a global optimization that can be solved using an iterative local computation have also been investigated [16].

Shape and motion models of non-rigid objects, like humans, have been incorporated into robust contour tracking frameworks by [13, 14]. These techniques do not rely on tracking of feature points and can maintain a segmentation in cluttered environments.

2.4. Our Scheme

The segmentation scheme described in this paper does not operate on velocity fields. A region growing technique is applied to the information from which velocity fields are typically derived. The aim of the scheme is to demonstrate that motion based segmentation may be performed without using velocity estimates.

3. Fundamental Motion Descriptors

The velocity at each location in a dense flow field is derived from a set of possible solutions (henceforth termed a *motion distribution*). The process involved in making this decision (or *collapsing the distribution*) may be purely local, or employ more global information.

This paper describes a segmentation technique that operates directly on a motion distribution image, therefore avoiding the problem of collapsing the motion distributions to produce a flow field. In this example the simplest and least robust measure - single pixel color differences - is used to produce the motion distribution. This distribution may be computed quickly - it is simply the absolute difference between a pixel in one frame and each pixel in a region of interest in the subsequent frame. The size and shape of the region of interest (motion descriptor kernel) is defined by the expected range of motions. This is equivalent to defining a search space for a feature based scheme. Automated techniques for deciding the region size have not been investigated.

4. Using Motion Descriptors in Seeded Region Growing

Seeded region growing is a well known, and intuitive technique, that is normally used to segment scenes using greyscale or colour information. The basic principle of operation is to place pixels from the borders of the regions onto a priority queue according to a measure of similarity between a control field describing the region and candidate pixels. For instance, when segmenting colour or multispectral images, each pixel is represented by a vector, and each region can be described by the multispectral average of all of the pixels within the region. The similarity between any pixel and a region can be computed by the magnitude of a suitable vector difference.

The concept can be extended to motion analysis. The control field can be a “motion image” in which each pixel is described by a motion descriptor. The problem is now to define a similarity measure. Each region may also be described by a motion descriptor, but the similarity between two motion descriptors is not well described by a vector or element-wise difference.

4.1. Region Representation

To solve the problem of the similarity measure, it is convenient to represent a region in a form that is similar to the pixels to which it is going to be compared.

The simplest way to represent a region is to use a motion descriptor that is simply the element-wise sum of all motion descriptors associated with pixels belonging to the region. This control field can be easily updated every time a pixel is added to a region and is therefore efficient to implement. Another option would be to convert the motion descriptor to a probability distribution function. This representation may be useful in many circumstances, but is more computationally intensive.

4.2. Determining pixel priority

The order in which pixels are assigned to regions is critical to region growing techniques, and is dependent on the pixel priority. The process of determining the pixel priority based on motion information is different to the processes used for spectral information for a number of reasons:

- The two dimensional nature of the motion descriptor means that simple vector differences aren't meaningful.
- A large proportion of the motion descriptors may not be relevant. The quality of match in some parts of the descriptor kernel may be very bad (e.g. in the case of a small feature and significantly different background covering a significant area of the motion descriptor), so using all of the

kernel to determine the priority is not sensible.

A simple scheme has been developed to evaluate the priority of a motion descriptor. (A one dimensional version of this is shown in Fig. 1.) A region motion descriptor based on color differences indicates the relative likelihoods of different motion vectors. The most likely motion corresponds to the point of minimum pixel color difference. Therefore the most important characteristic of the motion descriptor describing a region is the location (not the value) of the minimum (or maximum if a correlation kernel is used) (i in Fig. 1). This location may be an unreliable estimate of region velocity if the region is small, but can be expected to become more reliable as the region grows. A measure of similarity between a region motion descriptor and a pixel motion descriptor may be simply defined as follows. Suppose the minimum error in the region motion descriptor is at location (i, j). The value of the error in the pixel motion descriptor at the same location may be taken as a measure of compatibility between the two kernels and used to determine a priority. If the error is large then the priority is low because the two descriptors are unlikely to be representing similar velocities.

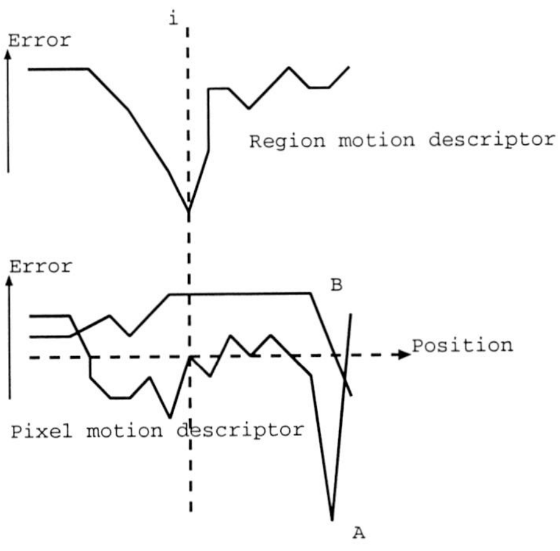

Fig. 1. Priority measure for motion descriptor – 1 dimensional version shown for simplicity. The vertical axis shows error magnitude. In this case pixel A will have a higher priority than pixel B because the error magnitude at location i is lower.

4.3. Enforcing temporal continuity

Seeded region growing techniques require that seeds be selected. In our scheme, the initial seed selection is manual – one seed is assigned to the border of the image being investigated (the "background" seed), and another is assigned to each object. In order to enforce spatial continuity, the eroded [15] results of the segmentation at time n are used to derive the seeds for the segmentation at time $n + 1$. This part of the process is somewhat haphazard at present –

seeded region growing does not allow regions to shrink, so it is important that the erosion structuring element produces an updated seed that is contained entirely within the object of interest. Automated ways of achieving this have not yet been investigated. One option includes using motion information to update the position of the seed rather than relying on overlap between frames.

5. Results

This section shows the segmentation results for a number of scenes using the technique just described.

5.1. Randomly textured objects

Fig. 2 shows a scene in which a randomly textured object moves randomly in front of a background with a statistically identical random texture and motion. This is a variant of common psychological experiments that demonstrate the power of motion information as a segmentation cue. (Note that it is impossible to appreciate this from the still images.) This scene can only be segmented using motion information – it is impossible to define a filter to find a square in individual frames. This example is also difficult for feature based schemes because the motion is random, so it is hard to generate reliable velocity estimates. The region growing segmentation is able to function successfully without knowledge of the shape of the object or any complex motion models.

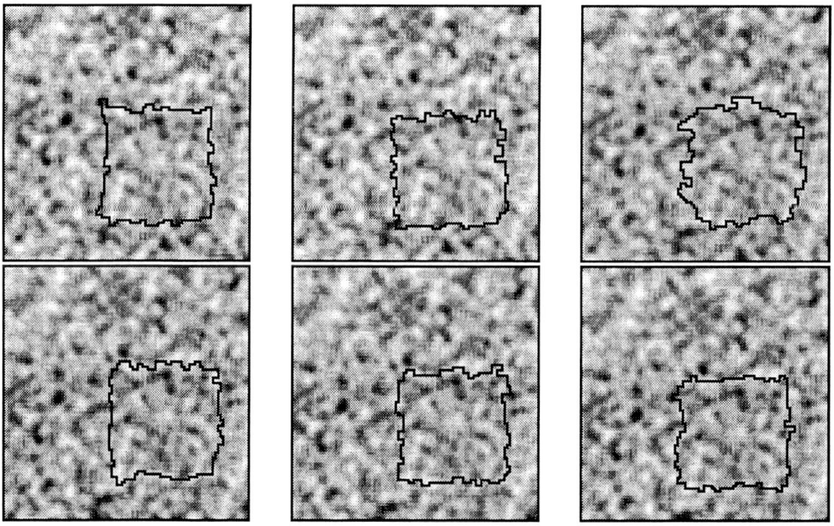

Fig. 2. Random textures segmented using only motion information. Results show every second frame.

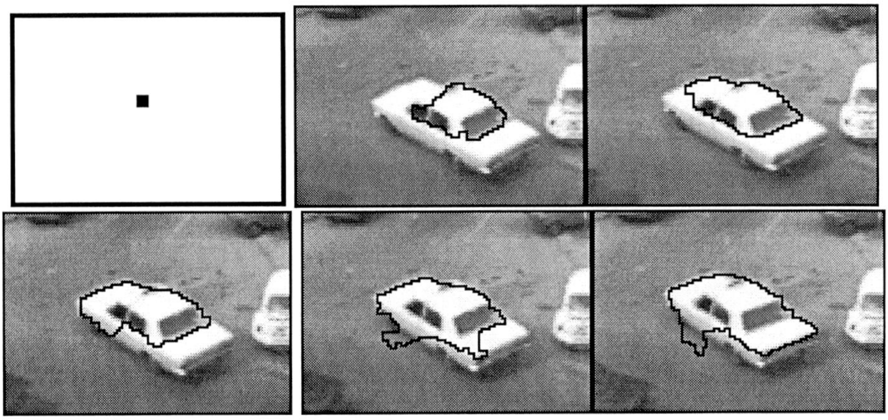

Fig. 3. A subset from the Hamburg taxi sequence. Segmented using only motion information. First image shows the initial seeds. Results show every second frame.

Fig. 4. A subset from a highway scene. Segmentation using only motion information. First image shows the initial seeds. Results show every second frame.

5.2. Hamburg Taxi

Fig. 3 shows the results of segmenting a small subset of the Hamburg taxi sequence. The initial result does not encompass the entire car, although the results do improve over time. The segmentation produced is reasonable, but takes some time to evolve, because there are large flat regions where the motion information is not very strong. The borders of the segmentation are not particularly stable.

5.3. Highway scene

Fig. 4 is a relatively complex scene with a moving background and an object that is shrinking in size. The background also contains some independently moving objects. The motion only segmentation produces quite good results, although the borders are not particularly stable. The shadow of the car is not segmented. This is understandable because it overlaps the border of the image, which is the second seed. Some of the problems with border stability in this scene could be due to the simple minded update of seeds between frames in the presence of high speed motion. The high displacement of objects makes it necessary to use a large erosion to force the seed to be entirely contained within the object of interest. This means that the constraints on temporal border smoothness are not very strong.

6. Discussion and Further Work

The techniques described in this paper work well on a variety of scenes, without relying on complex scene models, by exploiting the simplest forms of motion information - single pixel differences. Experiments with more reliable forms of motion information, such as correlation based measures, show some improvements in segmentation but suffer significant increase in computational cost. These examples show that motion segmentation techniques operating on motion descriptors, rather than flow fields, are an interesting avenue of research.

We haven't tackled yet the problem of the initial seeds, which is an absolute necessity in many real world applications. Also, we think that a simple erosion of results between frames to provide temporal continuity is only adequate if the object being tracked does not move too far between frames. A more sophisticated approach may improve robustness. Speed is also an issue for real time applications. The speed of the process is largely dependent on the nature of the motion descriptors.

The technique described here can be applied to any form of motion descriptor, including correlation based approaches and spatio-temporal filter-based approaches. Some experiments have been done using sum of absolute difference based measures, however no filter-based measures have been used yet - this may be an interesting area for future investigation.

Preliminary investigations into the use of colour information in regions where motion cues are unreliable have suggested that the costs of including the extra information are not justified by improvements in results.

Investigations into the extension of the region growing process to use 3 dimensional data have begun, but the results so far are inconclusive.

We also plan to experiment with replacing the simple erosion of the marker between frames with something that can take advantage of the motion information - i.e. change the position of the seed rather than just the size.

7. Conclusions

This paper has discussed a motion based segmentation technique that uses the type of information usually applied to velocity estimation. The technique avoids the problems associated with estimating a reliable flow field. The technique is able to use an initial set of seeds to establish a segmentation that is maintained over time.

References

1. S. M. Smith and J. M. Brady ASSET-2: Real-time motion segmentation and shape tracking IEEE transactions on pattern analysis and machine intelligence, 18(8):814–820, 1995.
2. R. Deriche. and O. D. Faugeras Tracking Line Segments Image and Vision Computing, 8(4):261–270, 1990.
3. B. K. P Horn and B. G. Schunck Determining Optical Flow Artificial Intelligence, 17, 185-203, 1981,
4. D. J. Fleet and A. D. Jepson Hierarchical Construction of Orientation and Velocity Sensitive Filters IEEE Transactions on pattern analysis and machine intelligence, 11(3): 315–325, 1989.
5. E. H. Adelson and J. R. Bergen Spatiotemporal energy models for the perception of motion Journal of the Optical Society of America, 2(2):284–299, 1985.
6. A. B. Watson and A. J. Ahumada. Model of human visual-motion sensing J. Opt Soc Am, 2(2):322–342, 1985.
7. F. Meyer and S. Beucher Morphological segmentation Journal of Visual Communication and Image Representation, 1(1):21–46, 1990.
8. T. Sandor and D. Metcalf and Y.J. Kim Segmentation of brain CT images using the concept of region growing International Journal Biomedical Computing, 29:133–147, 1991.
9. R. Adams and L. Bischof Seeded region growing IEEE Transactions on Pattern Analysis and Machine Intelligence,16:641–647, 1994.
10. Ph. Salembier and L. Torres and F. Meyer and C. Gu Region-based Video Coding using Mathematical Morphology Proceeding of the IEEE (Invited paper), 83(6):843–857, 1995.
11. G. Adiv Determining Three-Dimensional Motion and Structure from Optical Flow Generated by Several Moving Objects IEEE Transactions on pattern analysis and machine intelligence, 4:384–401, 1985.
12. A. Bab-Hadiashar and D. Suter Motion Segmentation: A Robust Approach IEEE International Workshop on the Interpretation of Visual Motion, Santa Barbara, California:3–9, 1998.
13. A. Blake and R. Curwen and A. Zisserman A framework for spatio-temporal control in the tracking of visual contours International Journal of Computer Vision, 11(2):127–145, 1993.
14. A. Blake and M. Isard and D. Reynard Learning to track the visual motion of contours Artificial Intelligence, 78:101–103, 1995.
15. J. Serra *Image analysis and mathematical morphology* Academic Press, 1982.
16. D. W. Murray and B. F. Buxton Scene segmentation from visual motion using global optimization IEEE Transactions on Pattern Analysis and Machine Intelligence, 9(2):220–228, 1987.

A SEGMENTATION PYRAMID FOR THE INTERACTIVE SEGMENTATION OF 3-D IMAGES AND VIDEO SEQUENCES

F. ZANOGUERA, B. MARCOTEGUI and F. MEYER
Centre de Morphologie Mathématique - Ecole des Mines de Paris
35, rue Saint-Honoré, 77305 Fontainebleau Cedex, France
email: zanoguera@cmm.ensmp.fr

Abstract. This paper presents a technique for the generation of a segmentation pyramid for a video sequence, designed for use in an interactive segmentation context. The pyramid is represented as a minimum spanning tree, which allows an efficient access to the information and avoids unnecessary recalculations. Several ways to introduce user interaction are also proposed.

Key words: Segmentation Pyramid, Minimum Spanning Tree, Interactive Segmentation, 3-D Image Segmentation, Video Sequence Segmentation, Region Hierarchy.

1. Introduction

Image segmentation is the first step in many computer vision systems, and it has been used for a long time in industrial and medical applications. Recently, along with the growth of the Internet and the availability of powerful computers to the general public, new standards like MPEG-4 and MPEG-7 are appearing. They address the so-called content-based applications, where images and video-sequences are treated not at the pixel level, but at the object level. These applications rely on a successful segmentation of the objects present in the scene.

In multimedia applications, the treated images can be very diverse and automatic segmentation becomes problematic as no *a priori* knowledge on the contents of the images is available. In this context, interactive segmentation offers an attractive solution by automating the task of finding homogeneous regions while leaving object definition to the user. Segmentation pyramids are very well suited to interactivity, as they offer a high degree of flexibility. The user can navigate between the different resolution levels, choosing the regions that form the object of interest.

Several types of bottom-up segmentation pyramids have been proposed in the literature [13], [12], [14], [5], [1], [2], [15]. In these approaches, regions are progressively merged into larger regions until a single region is obtained or a stop condition is satisfied. The hierarchy is then represented in the form of a graph. Two main issues arise when examining the existing literature:
1. The merging criterion is in most cases based on a *local* dissimilarity measure, which does not take into account global measures such as size of the regions. This produces good results at the first merging steps, but at coarser levels of the hierarchy semantic aspects start to play a role and local merging criteria

are no longer valid.
2. Although some of the techniques proposed can directly be applied to the segmentation of 3D images, no proposals have been made for the generation of a pyramid which is *coherently preserved* throughout a complete video sequence shot. In most cases, moving video sequences are treated in a tracking-based manner, where an initial mask is interactively defined for the first frame of the sequence, to be tracked in the remaining frames [7], [3]. Whenever a hierarchy is used, it is for the analysis of individual images and it is not preserved from one frame to the next. It would however be very desirable to produce a hierarchy which is coherently preserved throughout the video sequence shot. In this way, when the user interacts to re-partition a certain region or to merge it with a neighbour, this interaction affects all the images of the sequence and the process of video object generation is much faster.

The first issue has been addressed in [17], [6], in the context of interactive segmentation of still images. In [17], a segmentation pyramid is produced by establishing a hierarchy between the regions of the watershed. The merging order defining the hierarchy is based on the volume extinction values [16], a global measure which is well adapted to the characteristics of human perception, due to the trade-off between size and contrast it makes. As the complete hierarchy can be obtained in a single watershed of the image, it results in very fast calculation. The hierarchy is represented in the form of a Minimum Spanning Tree, which allows for very fast manipulation.

This paper deals with the issue of video sequence processing by proposing an extension of the techniques described in [17] to the segmentation of video sequences and 3-D images. We consider video sequences as a particular case of 3-D images with two spatial dimensions and one temporal dimension. Video sequences being too long to process them as a single 3D block, a time recursive approach is proposed.

A segmentation pyramid for the whole sequence (or 3-D image) is made available to the user, who can navigate between the different resolution levels to create or modify the object partition. The actions carried out by the user affect the whole sequence.

2. General Description

To produce a segmentation pyramid for a 3-D image or a video sequence, we use a 3-D approach based on flat-zone detection followed by a watershed, both using 3-D connectivity. The 3-D approach has the advantage that the same algorithms can be applied to both 3-D images and video sequences. In the sequel we will focus on video sequences, although the technique is also valid for 3-D image segmentation.

As a video sequence can be very long and the computer memory is limited, the sequence must be divided into blocks of images for processing. In order to be able to preserve the region hierarchy between consecutive blocks, our algorithm divides the video sequence into *overlapping blocks* (one image overlap).

An intra-block algorithm produces the segmentation pyramid for each block.

In order to ensure a coherence of the regions and their hierarchy between blocks, a single global hierarchy is recursively produced, using a procedure that we will call *pyramid projection.*

The segmentation pyramid for the video sequence is represented by a minimum spanning tree (MST). This tree is progressively updated at each projection step using the information contained in each block of the sequence.

Section 3 explains the generation of the segmentation pyramid for an individual block. Section 4 describes the pyramid projection between two consecutive blocks. Finally, Section 5 proposes some mechanisms to interact with the segmentation pyramid.

3. Intra-Block Segmentation

This section describes how the segmentation pyramid is produced for a block of images and how it can be represented as a tree. The computation is carried out in two steps. Firstly, a fine partition of the block is produced and represented by a neighbourhood graph. Then, from the neighbourhood graph a minimum spanning tree representing the segmentation pyramid is derived.

3.1. Creation of a fine partition through flat-zone detection and gradient flooding

In our previous work with still images in [17], a fine partition is created by flooding a morphological gradient from all its minima. This produces a partition with as many regions as there are minima on the gradient image. This technique can be directly extended to still 3-D images by using a 3-D gradient and flooding using 3-D connectivity. However, for moving video sequences, using a 3-D gradient will produce erroneous spatial contours around the moving areas, as these areas will show high gradient values due to the image-to-image changes. Using 2-D connectivity to calculate the gradient is not a good approach either, as it will produce leaks in the propagation: when two regions with low 2-D gradient values become temporally connected due to motion, temporal propagation will produce the label of the first region to be propagated inside the second region from one image to the next.

Creating the fine partition by flat-zone detection on the initial image rather than detecting the catchment basins of a gradient image restricts the possibility of temporal leaks as the propagation is made only across regions having the same colour. However, the contours of the flat zones are in general very irregular. To take advantage of the better contours produced by the watershed without increasing the risk of leaks, a hybrid approach is used. Large 3-D flat zones are detected and taken as markers for a 3-D watershed propagation inside a gradient image produced using only 2-D connectivity. Small flat zones are discarded, considering that they belong to a transition area, where the gradient will very likely have a high value (and thus no leaks will take place). To take advantage of colour information, three separate 3-D watersheds for the three colour components are calculated, taking flat zones or pseudo-flat zones [11] as markers. The intersection of the three partitions is then calculated. In a last step, very small regions are removed and a last watershed is computed to find the final contours. The partition intersection further minimises the risk of leaks (spatial and temporal), as there will only be a leak if it exists in the three partitions.

As an important pre-processing step, filtering by levelings [10] is applied to the three normalised colour components. This allows the enlargement of the flat zones without modifying the regions' contours.

3.2. The neighbourhood graph

Any partition can be represented by a neighbourhood graph. Each region of the partition is represented by a node of the graph. Two nodes are connected if the regions they represent are neighbours on the image. The graph edges can be weighted to express a local dissimilarity measure between regions. Possible dissimilarity measures that can be used to weight the graph edges are the lowest pass point along the border separating the two regions on the gradient image, the average gradient value along the border or a distance measure based on the colour components inside the region.

The neighbourhood graph provides a simplified representation of the image, and it can be flooded using a watershed algorithm where the edge weights represent the height of the borders the water must cross [9].

3.3. Minimum spanning tree and segmentation pyramids

In the watershed transformation, the lakes corresponding to two catchment basins A and B always meet through the path of lowest sup-section (sup-section: highest value along the path) between A and B. If we consider all possible pairs of catchment basins, we obtain a set of paths of minimal sup-section. This set of paths forms the minimum spanning tree of the neighbourhood graph [8]. Hence, the minimum spanning tree of the neighbourhood graph contains all the necessary information for flooding purposes. It can be created during the flooding of the neighbourhood graph by adding the edges to the MST in the order they are flooded on the neighbourhood graph, as long as they do not introduce a loop. When an edge introduces a loop on the MST, it is not added. At the end of the flooding, the MST has recorded the paths followed by the water to merge the different lakes. This algorithm corresponds to Bohuslav's algorithm for the calculation of the Minimum Spanning Tree.

On the MST, there is a unique path between any pair of nodes. Suppressing an edge of this tree produces a forest with two trees, which corresponds to a partition of the image into two regions. In the same way, suppressing $n-1$ edges of the tree partitions the image into n regions. Removing a variable number of edges of the tree produces a segmentation pyramid. It is then a question of knowing which edges to remove to obtain meaningful segmentations. The edge weights representing a dissimilarity measure between neighbouring regions, it seems reasonable to suppress them by decreasing weights. In this way, the most different regions will be separated first in the hierarchy.

The selection of the edge weights is a crucial issue to obtain meaningful segmentations. The approach used here consists of weighting the edges of the neighbourhood graph with a local dissimilarity measure based on a colour distance. The MST of the graph is then computed and its edges are simultaneously re-weighted based on the volume extinction value [16], which can be seen as a global dissimilarity measure. When an image is flooded by placing sources on all its minima, every time that two lakes meet an absorption takes place. The lake with smaller volume is considered to be absorbed by the lake with larger volume. When a lake is absorbed by a larger lake, its volume at the moment of the absorption is called the volume extinction value of the corresponding catchment basin. At the end of the flooding each catchment basin except one has been assigned an extinction value. The same procedure can be used during the graph flooding, where the volume of a region can be approximated as its surface multiplied by the current flooding level. The volume extinction values rate the regions in a way which is close to human perception, as they take into account both the size and contrast of the regions. Figure 1 shows a comparison between the 15 best regions found using a dynamics criterion [4] and the volume extinction values.

(a) Original image (b) Dynamics (c) Volume extinction values

Fig. 1. Best 15 regions as found using the dynamics and the volume extinction values.

Both the MST and the extinction values can be calculated by storing some extra information during the graph flooding.

Figure 2 shows three different resolution levels of the segmentation pyramid for four consecutive images belonging to the same block (for a block of 5 images).

4. Pyramid Projection

As treating a whole video sequence into one single block is not possible nor convenient, a projection step becomes necessary to put past and present information into correspondence.

For this purpose, the sequence is divided into overlapping blocks (one image overlap), and two continuity conditions are imposed on the common image between two blocks:

- The fine partition must be exactly the same on the common image for the two blocks.

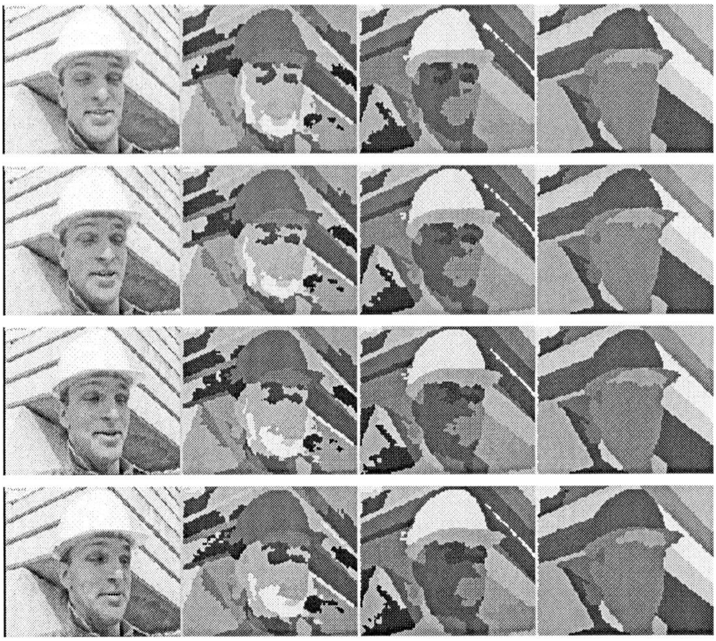

Fig. 2. Three levels of the segmentation pyramid for four consecutive images belonging to the same block.

– The hierarchy of the regions that exist in the two blocks should be preserved. New regions will be accommodated into this hierarchy.

In order to fulfill the first condition, the marker image used to produce the fine partition is modified. Section 3.1 described how 3-D flat or pseudo-flat zones were detected and used as markers for constructing the watershed of a gradient image. The marker image is therefore a 3-D image containing the detected flat zones. Now the first frame of the block containing the flat zones is replaced by the last frame of the fine partition corresponding to the previous block. The flat zones of the remaining images of the block are then relabeled to establish a correspondence with the regions of the imposed partition. Flat zones that do not obtain a label from the partition of the previous block are considered as new regions. The resulting 3-D image is then taken as marker for the watershed in the current block.

To satisfy the second condition, a recursive approach is used to dynamically update the MST to represent the sequence up to the current block.

Consider that the block N is currently being processed, and that an updated tree ϕ_{N-1} is available representing the segmentation pyramid up to the block $N-1$. The new tree ϕ_N is obtained by incorporating into the existing tree ϕ_{N-1} all the regions newly appeared in block N. In order to accommodate the new regions into the hierarchy, their volume extinction values must be calculated.

The nodes corresponding to the new regions will then be plugged into ϕ_{N-1} with edges weighted with this measure.

For the calculation of the extinction values for the new regions, a complete neighbourhood graph G_N is computed for the current block. Also, a new graph ϕ_N is obtained from ϕ_{N-1} by adding a node to ϕ_{N-1} for each new region, but with no link. To determine the links between the new nodes, and between the new and the old nodes, a *partial flooding* of G_N is carried out. In the partial flooding, only the nodes of the graph corresponding to new regions are flooded. Flooding the existing nodes is not necessary, as they are already linked in ϕ_N. When two new regions are joined by the flooding, one of them extinguishes as explained in Section 3.3. An edge is then added to ϕ_N between these two nodes, weighted with the volume of the extinguished region. When a new node meets with an old node, the new node is considered as extinguished (without volume comparison) and an edge is added to ϕ_N, weighted with the volume of this region. At the end of this procedure, a tree ϕ_N has been obtained. The paths between regions existing in ϕ_{N-1} are preserved with the same weights. New regions have been plugged into the tree, weighted with their volume extinction values.

At each projection step the tree ϕ_N grows by adding the new regions into the hierarchy while retaining previous information.

Figure 3 shows the results of the pyramid projection for images 1, 40 and 100 of the *Mother and Daughter* sequence.

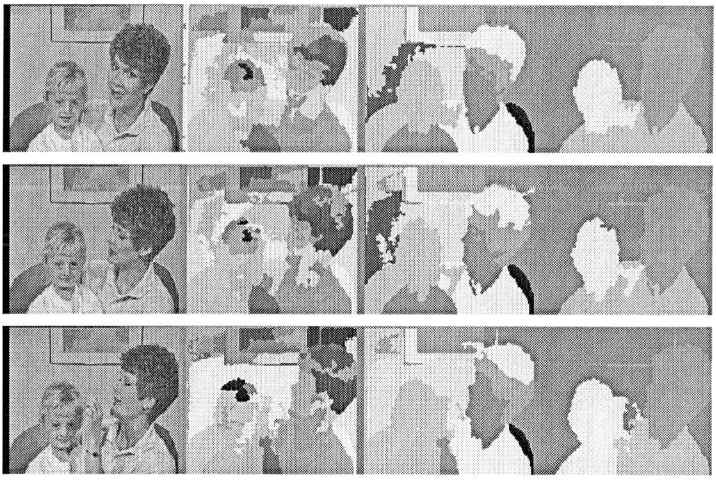

Fig. 3. Three different levels of the segmentation pyramid for images 1, 40 and 100 of the Mother and Daughter sequence. The segmentation pyramid has been projected block-to-block to span the whole sequence.

5. User Interaction

This section describes some possibilities for user interaction with the segmentation pyramid. From the user point of view, the interaction simply consists of re-segmentation and merging of regions, and marker drawing. The tree representation is not visible to the user.

Three types of interactions are proposed: automatic segmentation into a certain number of regions, local re-segmentation/merging of regions and segmentation from markers.

5.1. Selection of the total number of regions

A starting point in the segmentation process could be for the user to ask for the image to be segmented into a number n of regions. This corresponds to a request for an automatic segmentation of the image into n regions. Such a request may easily be satisfied by suppressing the $n-1$ edges of the tree with highest weight. The interaction can be presented in a very intuitive way by means of a sliding bar which slides up and down to have more or less resolution.

5.2. Local interactions

The interaction type described in the previous section treats the image as a whole, finding the n best regions. However, the user may be interested in having some regions segmented with more detail than others. In this case, the user must be offered the possibility to refine a certain area or to coarsen it by merging it with neighbouring regions. This is done by locally suppressing/adding edges from/to the tree. Two operations allow the user to locally navigate up and down the pyramid.

In the *refine* operation, the user clicks on a certain area with the mouse. At the same time, the number of regions in which the selected region must be subdivided may be specified. If it is not, a default value is used. The $n-1$ (n being the parameter specified by the user) edges of highest weight are suppressed, but this time only the edges inside the selected region are considered.

In the *coarsen* operation, the user selects a region with a mouse click, and again a parameter n may be specified. The $n-1$ most similar neighbouring regions are merged to the selected one. Among the previously eliminated edges, the $n-1$ of lowest weight that link a node belonging to the selected region with an external one are re-inserted.

Figure 4 shows an example of local actions carried out by the user. As a starting step, an automatic segmentation into 10 regions is obtained. The user then clicks on the helmet and requests a re-segmentation of the region into two regions. The helmet is then separated from the background on all the images of the sequence where the helmet is present. The user then requests a re-segmentation of the shoulder region, in order to separate it as well from the background. In this way, finer or coarser segmentations can be obtained on certain areas by simple mouse-clicking.

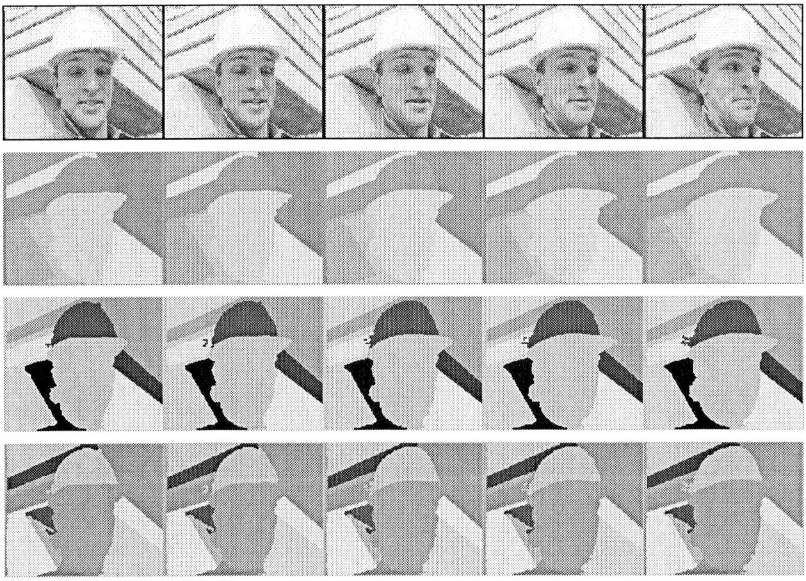

Fig. 4. Sequence of interactions, shown for one block of the video sequence. Top to bottom: original image, automatic segmentation into 10 regions, re-segmentation of the helmet area, re-segmentation of the shoulder.

6. Segmentation from Markers

Finally, another type of interaction is the classical marker drawing. The user is asked to roughly mark the objects of interest, including the background. The markers are imposed on the MST and the resulting segmentation is obtained for the whole sequence. An algorithm can be found in [8]. As a segmentation pyramid is available, further re-segmentation/merging are possible if the regions obtained are not completely satisfactory.

7. Conclusions

We have presented a technique to create a segmentation pyramid corresponding to a video sequence. Instead of a single partition, a segmentation pyramid is available. This approach is very well suited to interactivity as it allows the user to build the desired video object by taking regions from different resolution levels and frames. The correspondence between regions is made *at all resolution levels* for all the images of the sequence. In this way, the actions carried out by the user automatically affect all the images of the video sequence.

References

1. J. Cichosz and F. Meyer. Morphological multiscale image segmentation. *Workshop on Image Analysis for Multimedia Interactive Services, WIAMIS'97*, 1997.
2. L. Garrido, P. Salembier and D. Garcia. Extensive operators in partition lattices for image sequence analysis. *IEEE Transactions on Image Processing*, April 1998.
3. C. Gomila and F. Meyer. Tracking of video objects for videophony applications. *Proc. Workshop on Image Analysis for Multimedia Interactive Services, WIAMIS'99*, Berlin, May-June 1999.
4. M. Grimaud. New measure of contrast: dynamics. *Proc. Image Algebra and Morphological Processing III, SPIE*, San Diego CA, 1992.
5. B. Marcotegui and F. Meyer. Bottom-up segmentation of image sequences for coding. *Annales des Télécommunications*, 52(7–8):397–407, 1997.
6. B. Marcotegui, P. Correia, F. Marqués, R. Mech, R. Rosa, M. Wollborn and F. Zanoguera. A video object generation tool allowing friendly user interaction. *Proc. IEEE International Conference on Image Processing, ICIP'99*, Kobe, Oct. 1999.
7. F. Marqués and J. Llach. Tracking of generic objects for video object generation. *Proc. IEEE International Conference on Image Processing, ICIP'98*, Chicago, 1998.
8. F. Meyer. Minimum spanning forests for morphological segmentation. *Mathematical Morphology and its Applications to Image Processing, ISMM'94*, pages 77–84, 1994. Kluwer Academic Publishers.
9. F. Meyer. Morphological segmentation on a neighborhood graph. *Acta Stereologica*, 16(3):175–182, 1997.
10. F. Meyer. From connected operators to levelings. *Mathematical Morphology and its Applications to Image Processing, ISMM'98*, pages 191–198, June 1998.
11. F. Meyer. The levelings. *Mathematical Morphology and its Applications to Image Processing, ISMM'98*, pages 199–206, June 1998.
12. A. Montanvert, P. Meer and A. Rosenfeld. Hierarchical image analysis using irregular tessellations. *IEEE Transactions on Pattern Analysis and Machine Intelligence*, 13:307–316, April 1991.
13. O. Morris, M. Lee and A. Constantinides. Graph theory for images analysis: an approach based on the shortest spanning tree. *Proc. IEE*, 133(2):146–152, April 1986.
14. P.F.M. Nacken. Image segmentation by connectivity preserving relinking in hierarchical graph structures. *Pattern Recognition*, 28(6):907–920, 1995.
15. X. Shen, M. Spann and P. Nacken. Segmentation of 2D and 3D images through a hierarchical clustering based on region modeling. *Pattern Recognition*, 31(9):1295–1309, 1998.
16. C. Vachier. Extraction de charactéristiques, segmentation d'image et morphologie mathématique. PhD thesis, Ecole Nationale Supérieure des Mines de Paris, 1995.
17. F. Zanoguera, B. Marcotegui and F. Meyer. A toolbox for interactive segmentation based on nested partitions. *Proc. IEEE International Conference on Image Processing*, Kobe, Oct. 1999.

PARTITION LATTICE OPERATORS FOR EXTRACTION OF SEMANTIC VIDEO OBJECTS

DANIEL GATICA-PEREZ and MING-TING SUN
Department of Electrical Engineering, University of Washington. Box 352142. Seattle, WA, 981095, USA.

and

CHUANG GU
Microsoft Corporation. One Microsoft Way. Redmond, WA 98052, USA.

Abstract. We conceive the problem of multiple semantic video object (SVO) extraction as an issue of designing operators on a complete lattice of partitions. In this paper, we propose a framework based on accurate spatial partition generation and application of optimal extraction operators on the generated partitions. Under the framework, we introduce a spatio-temporal regional maximum likelihood operator for extraction purposes. Some theoretical properties of the operators are established. Experimental results show that our scheme is capable of successfully handling multiple SVOs in a variety of scenarios.

Key words: Semantic Video Object Extraction, Partition Lattice Operators.

1. Introduction

SVO extraction can be considered as a process of segmenting and tracking *arbitrary* collections of image regions (that correspond to objects in the real world) with *pixel-wise accuracy.* The task, crucial for the next generation of multimedia standards MPEG-4 and MPEG-7 [12], is formidable because SVOs are *human abstractions* that are not invariant, either in spatial features or in motion. Several approaches, based on different object representations (contours, regions, active meshes) have been recently proposed [2], [7], [11], [10].

When 2D regions are selected to represent an object, two factors define the quality of the extraction result: the precision of the spatial partition, and the selected tracking technique. On one hand, a good segmentation technique should preserve the contours of the scene objects, as human perception is sensitive to artifacts in borders. On the other hand, the tracking process is responsible for keeping an accurate SVO representation along time. Several methods consist of the computation of an initial object partition and its tracking by a *prediction-adjustment* process, in which the original partition is updated to generate the partitions at each time [3], [6], [10]. However, this process is not trivial if pixel-wise accuracy is required: noisy motion information, which constitutes the key factor of the procedure, often introduces inaccuracy and ambiguity in defining the boundaries of the video objects [7], [11]. Furthermore, some of these techniques unfortunately rely on heuristics, and/or cannot handle the case of several objects.

This paper presents a different approach. We propose a general framework for 2D region-based SVO extraction based on spatial partition generation and the application of one or more extensive operators in the lattice of partitions. We are interested in developing a systematic approach in which such operators can be defined, their properties can be analyzed, and that allows for the extraction of multiple SVOs from natural video sequences.

The rest of the paper is organized as follows. Section 2 presents an overview of our methodology. Section 3 introduces the partition lattice operators for SVO extraction. Section 4 shows results for several MPEG-4 test sequences. Section 5 provides some concluding remarks.

2. Proposed Approach

A complete lattice of partitions is an appropriate morphological framework to analyze segmentation problems [15], [14]. We start by reviewing such a notion. Given a space $\mathcal{E}$ and its power set $\mathcal{P}(\mathcal{E})$, a *partition* of $\mathcal{E}$ is a mapping $P : \mathcal{E} \rightarrow \mathcal{P}(\mathcal{E})$ such that $\forall x, y \in \mathcal{E}$, (i) $x \in P(x)$, and (ii) $P(x) = P(y)$ or $P(x) \cap P(y) = \emptyset$. $P(x)$ is called the *zone* or *region* of P that contains x. It can be proved that the set of all partitions of $\mathcal{E}$ constitutes a complete lattice, denoted by Π, where the partial ordering relationship is defined as $P_i \leq P_j \iff P_i(x) \subseteq P_j(x), \forall x \in \mathcal{E}, P_i, P_j \in \Pi$. In this case, P_i is said to be *finer* than P_j. The infimum of a set of partitions $\{P_i, i \in \mathcal{I}\}$ is defined as $(\bigwedge_i P_i)(x) = \cap_i P_i(x)$ $\forall x \in \mathcal{E}$, i.e., it corresponds to the partition made of the intersections of all the regions in the original set of partitions. Additionally, the supremum of a set $\{P_i\}$ is given by $(\bigvee_i P_i)(x) = \cap\{B : B = \cup_i \cup_{y \in B} P_i(y), x \in B, B \in \mathcal{P}(\mathcal{E})\}$ which is the finest partition that is larger than each of the individual P_i. For the two-partition case, $(P_i \vee P_j)(x) = (P_i \vee P_j)(y)$ if $P_i(x) = P_i(y)$ or $P_j(x) = P_j(y)$. Finally, the least and greatest elements of Π correspond to the finest partition P_O and the coarsest partition P_I, such that $P_O(x) = x$ and $P_I(x) = \mathcal{E}$ for all $x \in \mathcal{E}$.

For purposes of indexing of the zones of a partition, it is convenient to use the following notation: $P = \{R_i, i \in \mathcal{I}\}$, where $R_i = \cup\, x \in \mathcal{E}$ such that $P(x) = R_i$.

The extraction of the SVOs of a scene corresponds to one special case of partition of the image support. This can be defined as follows.[1]

Definition 1 *Let* $\mathbf{I} = \{\mathbf{I}^t \mid t \in \mathsf{Z}\}$ *be a multivalued image sequence, with domain* $\mathcal{E} = \mathcal{D}(\mathbf{I}^t) \subset \mathsf{Z}^2$. *Let* $P^t = \{R_i^t, i \in \{1, ..., N\}\}$ *denote a partition of* $\mathcal{E}$ *at time* t. *The j-th Semantic Video Object of the scene depicted in* $\mathbf{I}$ *(consisting of* M *objects) is defined by* $SVO_j = \{SVO_j^t\}$, *where*

$$SVO_j^t = \bigcup_{i=1}^{N_j^t} R_i^t \tag{1}$$

[1] Multivalued images and random variables are both denoted by bold letters. The meaning should be clear from the context.

In MPEG-4 terminology, SVO_j^t represents the j-th Video Object Plane (VOP) at time t, each composed of N_j^t regions of P^t ($\sum_j N_j^t = N$). Eq. 1 naturally allows for the definition of multiple SVOs. The associated partition of SVOs at time t, denoted by P_{SVO}^t, is the collection

$$P_{SVO}^t = \{SVO_j^t, j \in \{1, ..., M\}\} \tag{2}$$

We propose to achieve SVO extraction by (1) generating, at each time instant, spatial partitions P^t that do not depend on inaccurate motion information and that preserve the true object contours, so that the frontiers between objects can be discerned even though they are of similar color, and (2) finding optimal homomorphisms between the generated partitions and the set of SVOs in the scene. Therefore, we claim that SVO tracking can be formalized as a process of applying operators $\{\psi^t()\}$ in the lattice of partitions [14], [15]. These operators can be designed by specifying spatio-temporal statistical criteria.

In addition, we consider the interactive introduction of semantics as essential, such that one or more of the following functions can be implemented by a user: initial definition of the SVOs, creation of a multiview representation, correction of the automatic results, or specification of context. Some typical works in this area include [2] and [7].

SVO extraction is then defined as a *combined process of spatial partition generation and subsequent application of partition lattice operators, with user intervention at (possibly) different instants.*

For the partition generation stage, we have previously developed a four-band (color+intensity edges) morphological multivalued spatial segmentation method that improves the contour localization properties of the traditional watershed techniques [5]. The following section concentrates on the formulation of the partition operators.

3. Partition Lattice Operators for Semantic Video Object Extraction

Some basic partition operators for segmentation were originally proposed in [14]. More recently, a work that pointed out the connection between region merging algorithms and connected operators was presented in [4].

In our approach, once a partition P^t is generated at each time t, the problem becomes the *construction* of P_{SVO}^t from P^t by introducing temporal information that allows the implementation of the tracking function. This information is represented by a *temporal reference SVO partition set*, denoted by TR, composed of the SVO partitions of the scene at different time instants, i.e., partitions that correspond to different scene views. The partition reference set is then expressed as $TR = \{P_{SVO}^{t_k},\ k \in \{1, ..., K\}\}$. For example, if $K = 1$, and $t_K = t - 1$, then $TR = \{P_{SVO}^{t-1}\}$, which means that the generation of the current SVO partition will depend on information provided by the previous one. If $K > 1$, the decision for the construction of the current SVO partition will include information from multiple instances. Note that the partitions in TR can be computed either by off-line user interaction or automatically as part of the extraction process. We now define a SVO extraction operator.

Definition 2 *Let* Π *be the complete lattice of partitions of* $\mathcal{E} = \mathcal{D}(\mathbf{I}^t)$. *Let*

$P^t \in \Pi$, and $TR = \{P_{SVO}^{t_1}, ..., P_{SVO}^{t_K}\}$. An SVO extractor operator is a mapping $\psi_{TR}^t : \Pi \to \Pi$, so that

$$P_{SVO}^t = \psi_{TR}^t(P^t) \tag{3}$$

The explicit dependence on TR will be usually omitted as this set is used as a reference. With this formulation, several tracking schemes can be formulated: monoview ($k = 1$) or multiview ($k > 1$); causal ($t_k < t$) or non-causal ($t_k > t$). The superscript in the operators notation means that they can be time-dependent.

As we mentioned, we have assigned the accurate extraction of region boundaries to the partition generation phase. As a result, the extraction operators $\{\psi^t\}$ can be thought of as a classification mechanism, that assigns each region $R_i^t \in P^t$ to the appropriate SVO, so that no new spatial contours are introduced in the partition P_{SVO}^t. This concept is illustrated in Fig. 1. In fact, all possible operators that can be designed with this idea in mind satisfy the following property.

Property 1 *The operators $\{\psi^t\}$ are extensive.*

Proof. It directly follows from Eqs. 1 and 2.

This property implies a relation between this class of operators and connected operators [4], [9]. In addition, we would like the operators $\{\psi^t\}$ not to be injective. From the practical point of view, this would represent some robustness in the extraction operation, by allowing several partitions $\{P^t\}$, all comparable by the ordering relation, to be mapped to the same object partition P_{SVO}^t. Finally, idempotence represents another desired property, as it would imply that the extraction would be done in one single step.

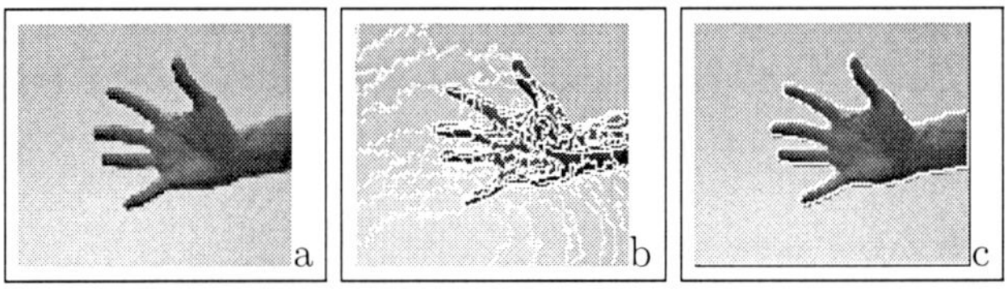

Fig. 1. SVO extraction operator. *Hand* sequence. (a) Original image $\mathbf{I}^1$. (b) Partition P^1 (segmentation model superimposed); (c) SVO partition $P_{SVO}^1 = \psi(P^1)$ (original image superimposed).

In the following section, we present one operator for the case in which the reference set is given by $TR = \{P_{SVO}^{t-1}\}$, and ψ is non-time-adaptive.

3.1. Partition Operator Based on Regional Maximum Likelihood.

The design of the partition operators can be formulated in terms of an optimality criterion to be satisfied. Note that, from the statistical point of view, SVO extraction (as has been formulated here) represents a process of assigning the regions of P^t to a given class, namely the objects in the scene; from the algebraic point of view, it corresponds to the design of extensive partition operators. We initially propose a partition operator $\psi_j(P^t)$ to construct *each* SVO_j^t

from the partition P^t using the previous SVO partition (P^{t-1}_{SVO}) as reference,

$$P^t_{SVO_j} = \psi_j(P^t)$$

where $P^t_{SVO_j}$ is the partition that divides the image support into the j-th SVO and the rest of the scene. The generation of P^t_{SVO} is then straightforward.

Let $V_{\mathbf{I}^t,\mathbf{I}^{t-1}} : \mathcal{P}(\mathcal{E}) \rightarrow \mathsf{Z}^2$ be the mapping that computes a region motion vector, assuming a pure translational model, using the image sequence $\mathbf{I}^t$ at times t and $t-1$, so that $V^t_i = V_{\mathbf{I}^t,\mathbf{I}^{t-1}}(R^t_i)$ denotes the motion vector computed for the region $R^t_i \in P^t$. Additionally, let $[X]_h$ represent the translated version of $X \subset \mathsf{Z}^2$ by $h \in \mathsf{Z}^2$: $[X]_h = \{x + h \mid x \in X\}$. The region attribute that will be used to construct the SVO partition at each time t, using the temporal reference $P^{t-1}_{SVO} = \{SVO^{t-1}_j, j \in \{1,...,M\}\}$ is defined as follows.

Definition 3 *Given a partition P^t, and the SVO partition P^{t-1}_{SVO}, the normalized overlapped area between the i-th region $R^t_i \in P^t$ and the j-th SVO is given by:*

$$noa^t_{ij} = \frac{card([R^t_i]_{V^t_i} \bigcap SVO^{t-1}_j)}{card(R^t_i)} \tag{4}$$

This measure takes values between zero (no overlapping) and one ($R^t_i \subseteq SVO^{t-1}_j$), and will decide for the assignment of each region in P^t to the corresponding SVO. Obviously, each $R^t_i \in P^t$ belongs either to the j-th SVO or to any other SVO in the scene depicted in the image sequence. In hypothesis testing terms,

$$H_0 : R^t_i \subseteq SVO^t_j \quad ; \quad H_1 : H^c_0 \tag{5}$$

The normalized overlapped area can be modeled as a continuous random variable **noa**, taking values noa in [0,1] (we drop the index t in what follows to simplify the notation). Let $svo_j, j = 1,...,M$ represent the j-th possible class (i.e. the j-th SVO), with prior probabilities $P(svo_j)$, and let svo^c_j denote the set of all classes except the j th one, which implies $P(svo^c_j) = 1 - P(svo_j)$. With this setting, $P(svo_j/noa)$ and $P(svo^c_j/noa)$ represent the a posteriori conditional probabilities that correspond to H_0 and H_1, respectively. We use the Maximum a Posteriori (MAP) criterion to map each region to an SVO [13]:

$$P(svo_j/noa) \underset{H_0}{\overset{H_1}{\lessgtr}} P(svo^c_j/noa) \tag{6}$$

such that the hypothesis H_x that is chosen is the one that has a larger a posteriori probability. Applying Bayes theorem on both sides of the expression and rearranging terms,

$$\frac{p(noa/svo_j)}{p(noa/svo^c_j)} \underset{H_0}{\overset{H_1}{\lessgtr}} \frac{P(svo^c_j)}{P(svo_j)} \tag{7}$$

where $p(noa/svo_j)$ represents the class-conditional probability density function. For the two-object case, we can assume equal priors ($P(svo_j) = P(svo^c_j)$), as foreground and background video objects may have any size and shape, and the expression reduces to the maximum likelihood criterion

$$L(noa) \equiv \frac{p(noa/svo_j)}{p(noa/svo_j^c)} \underset{H_0}{\overset{H_1}{\lessgtr}} 1 \tag{8}$$

For the cases of larger number of objects, however, the exact expression is Eq. 7. Let k^t denote the ratio $P(svo_j^c)/P(svo_j)$. We propose to model the class-conditional probability density functions by exponential distributions:

$$p(noa/svo_j^c) = \lambda_1 e^{-\lambda_1 noa} u(noa) \quad ; \quad p(noa/svo_j) = \lambda_2 e^{-\lambda_2(1-noa)} u(1-noa)$$

where $u(x)$ designates the step function. These distributions approximately model the real data: due to segmentation errors, $p(noa/svo_j)$ should be highly concentrated around $noa = 1$, and rapidly decay as $noa \to 0$. The dual situation holds for $p(noa/svo_j^c)$. In addition, the parameter values λ_i should make the conditional probabilities outside the interval $[0,1]$ negligible. The problem has been reduced to finding an optimal threshold for noa,

$$noa \underset{H_0}{\overset{H_1}{\lessgtr}} \frac{\lambda_2 - \ln(\lambda_2/k^t\lambda_1)}{\lambda_1 + \lambda_2} = T_{noa} \tag{9}$$

We can now write an expression for the proposed partition operator:

$$P_{SVO_j}^t = \psi_j(P^t) = \{SVO_j^t, \mathcal{E}\backslash SVO_j^t\} \tag{10}$$

where $A\backslash B$ denotes set difference and

$$SVO_j^t = \bigcup_i R_i^t \text{ such that } noa_{ij}^t \geq T_{noa} \tag{11}$$

The parameters λ_i and k^t can be estimated from the actual data. However, if we assume symmetry between the exponential distributions ($\lambda_1 = \lambda_2$), and $\lambda_i \gg k^t$, the expression for the optimal threshold can be further simplified and approximated as:

$$T_{noa} = \frac{\lambda_2 - \ln(\lambda_2/\lambda_1)}{\lambda_1 + \lambda_2} + \frac{ln\ k^t}{\lambda_1 + \lambda_2} \approx \frac{1}{2} \tag{12}$$

This analysis shows that ψ_j, under the described assumptions, is equivalent to a tracking algorithm recently reported in [7] for the two-SVO case.

To extract the M SVOs present in the scene, ψ_j should be applied $M-1$ times (the $M-th$ SVO is always selected as the scene background). Finally, P_{SVO}^t can be directly generated from the set of partitions $\{\psi_j(P^t)\}_{j=1}^{M-1}$, by defining a partition operator $\psi_{RML}()$ for *regional maximum likelihood*:

$$P_{SVO}^t = \psi_{RML}(P^t) = \bigwedge_{j=1}^{M-1} \psi_j(P^t) = \bigwedge_{j=1}^{M-1} P_{SVO_j}^t \tag{13}$$

Some properties of this operator can be established (the same applies to ψ_j, as it is equivalent to ψ_{RML} for $M = 2$ in Eq. 13).

Property 2 *ψ_{RML} has the following properties: (i) neither increasing nor decreasing, (ii) idempotent, (iii) not injective, (iv) not invertible, (v) not a morphological filter.*

Proof (i). (Counterexample). Let P_i^t be a partition of $\mathcal{E}$ that consists of three regions, labeled B (background), H (head), and S (shoulders), respectively. Let $P_{i'}^t$ be another partition that consists of two regions, H and $E = B \cup S$ (erroneously merged regions). By construction, $P_i^t \leq P_{i'}^t$. Additionally, assume that there is no motion and that P_{SVO}^{t-1} is correctly composed of the background B and the object $O = H \cup S$. Applying ψ_{RML} to the two partitions, $\psi_{RML}(P_i^t) = P_{SVO}^{t-1}$, and $\psi_{RML}(P_{i'}^t) = P_{i'}^t$, but obviously $P_{SVO}^{t-1} \not\leq P_{i'}^t$, so ψ_{RML} is not increasing. Similarly, it can be proved that ψ_{RML} is not decreasing.

(ii) $\psi_{RML}(\psi_{RML}(P^t)) = \psi_{RML}(P_{SVO}^t)$. But P_{SVO}^t is already the partition of SVOs. A further classification process simply assigns every SVO_j to itself, i.e. $\psi_{RML}(P_{SVO}^t) = P_{SVO}^t$.

(iii) Using the counterexample in (i), $\psi_{RML}(P_i^t) = \psi_{RML}(P_{SVO}^{t-1}) = P_{SVO}^{t-1}$, so P_i^t and P_{SVO}^{t-1} map to the same partition under ψ_{RML}, which shows that the operator is not injective. In general, for a set of partitions $\{P_1^t \leq \cdots \leq P_n^t\}$, the equality $\psi_{RML}(P_1^t) = \cdots = \psi_{RML}(P_n^t)$ will hold.

(iv) Follows from (iii).

(v) Remember that a lattice operator is called a morphological filter iff it is idempotent and increasing. The result immediately follows from (i).

3.2. Statistical Validation of the Partition Operator

To justify the assumptions in the previous subsection, we performed statistical tests on several MPEG-4 video sequences. Indeed, we found that the exponential, symmetrical distribution assumption adequately represents the data. In Table I, we show the ML estimates for the parameters λ_i, for the two-SVO case (foreground object and background). Additionally, the priors $P(svo_j)$ at each time t are estimated from the relative sizes of the SVOs at the previous frame of the video sequence, so that:

$$\widehat{k}^t = P(svo_j^c)/P(svo_j) = card(\mathcal{E} \backslash SVO_j^{t-1})/card(SVO_j^{t-1})$$

TABLE I
Estimated parameters for MPEG-4 sequences

Sequence	$\widehat{\lambda}_1$	$\widehat{\lambda}_2$	$\widehat{k}^0$	$\widehat{T}_{noa}$
Bream	142.21	113.89	2.44	0.45
Foreman	75.17	79.51	1.94	0.51
Hand	97.64	78.83	4.25	0.44

Table I also shows the initial values of $\widehat{k}^t$. It is observed that the assumption that $\lambda_i \gg k^t$ also holds, even for small objects, and that the estimated optimum threshold $\widehat{T}_{noa}$ is actually close to the approximated value. This fact validates the direct use of 1/2 as the value of T_{noa}, which reduces the computational complexity. It is also pointed out that ψ_{RML} can tolerate SVO size changes.

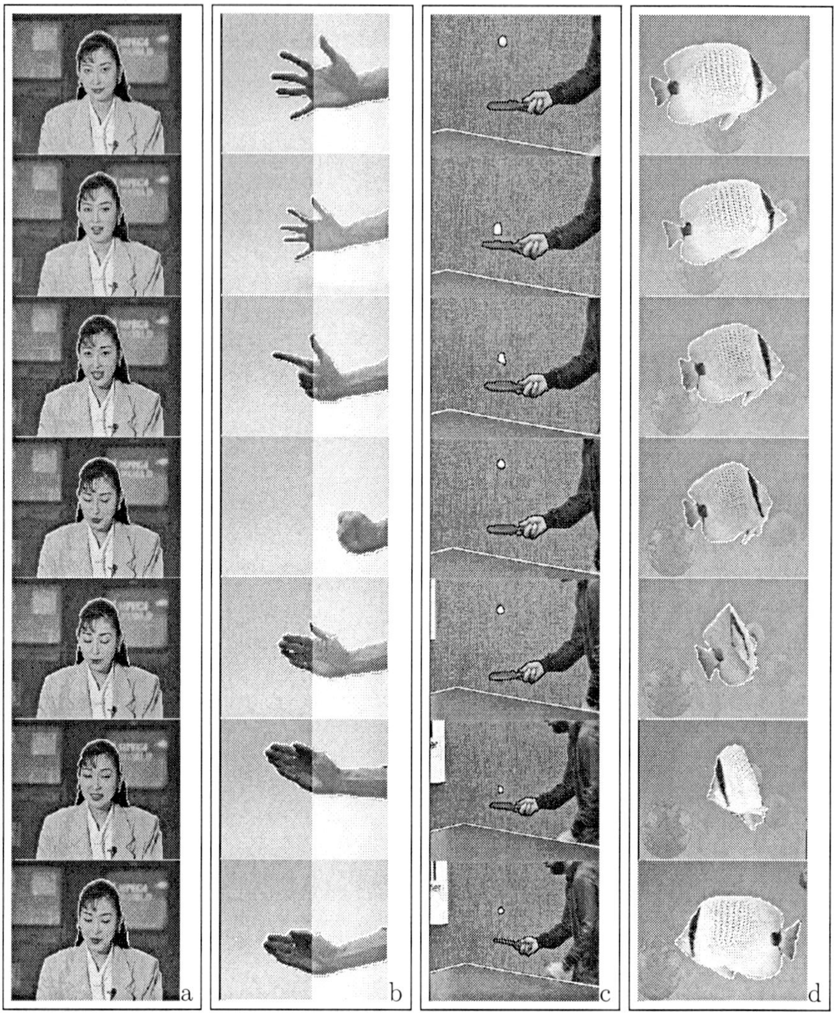

Fig. 2. SVO extraction. (a) *Akiyo.* (b) *Hand.* (c) *Tennis.* (d) *Bream.*

4. Results

Our framework is integrated in an SVO extraction system whose minimum-user interaction model was presented in [5], and consists of four steps:

1. SVO structure definition. A user-defined P^0_{SVO} is generated from $\mathbf{I}^0$.
2. SVO computation by generation of a partition $P^t \; \forall \, t$.
3. SVO tracking by application of our partition operator, $\widehat{P}^t_{SVO} = \psi^t(P^t, P^{t-1}_{SVO})$.
4. SVO postprocessing to refine the object partitions, $\widehat{P}^t_{SVO} \rightarrow P^t_{SVO}$.

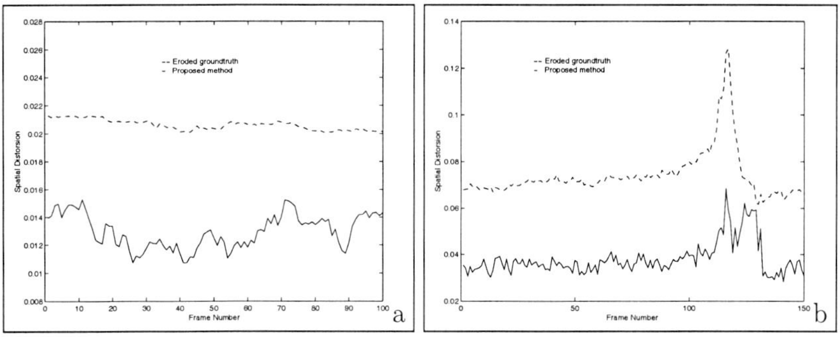

Fig. 3. Generated SVO Spatial Distortion. (a) *Akiyo*, (b) *Bream*. The distortion introduced by the erosion of the ground truth by a 3×3 structuring element is also shown.

Extraction results for various scenarios are shown in Fig. 2[2]. In all cases, we superimpose P^t_{SVO} on $\mathbf{I}^t$. Fig. 2(a) shows the result for the *Akiyo* sequence for two user-defined SVOs: *Akiyo* and *Background*. Even though they have some adjacent regions of similar color, our methodology generated precise SVO contours. Fig. 2(b) illustrates a two-SVO gesture image sequence [1]. The hand presents fast, articulated motion and shades, and the scene has a significant change of illumination. As a third example, the result obtained with the *Tennis* sequence, divided into three SVOs, is shown in Fig. 2(c). The sequence has been correctly partitioned. Finally, the result obtained with the *Bream* sequence, that presents object deformable motion and global camera motion, is shown in Fig. 2(d). In summary, our methodology performs well for different types of object and camera motion.

The computational complexity of our method is low, and adequate for semi-automatic SVO extraction. When fast motion estimation is used, the extraction takes around three seconds/frame in QCIF color images, on an SGI Octane computer; this figure could be significantly reduced by code optimization. Full motion estimation provides the best SVO extraction results at the expense of increasing the processing time, and might be required when tracking tiny objects with large motion.

Objective evaluation of our methodology can be performed for those sequences for which a ground truth is available. The MPEG-4 group has proposed figures for spatial distortion evaluation [16]. In Fig. 3 we present the results obtained for the *Akiyo* and *Bream* test sequences. To provide an idea of the degree of accuracy of the generated SVO partitions, the spatial distortion computed between the ground truth and a 3×3-eroded version of itself, that approximately peels off the ideal SVO partition by one pixel, is also presented, and confirms the obtained quality.

The main limitations of the proposed method arise when extracting SVOs in (i) highly cluttered scenes where the colors of different SVOs are similar, which introduces segmentation errors, and (ii) sequences in which newly uncovered regions have no matches in the previous frame, which produces tracking errors.

[2] Test video sequences are available at `http://hitl.washington.edu/people/danielgp`

We are currently extending our methodology to a multiview SVO representation (i.e., when the partition reference set TR is composed of more than one SVO partition) to address these problems.

5. Conclusions

We described a methodology for multiple SVO extraction based on object contour-preserving spatial partition generation and application of extensive spatio-temporal partitions operators. The use of the partition lattice framework for SVO extraction allows for the modeling of various tracking schemes and leads to the development of optimal algorithms. We have illustrated this with a regional maximum likelihood operator. Experimental results for a variety of real situations in natural video sequences have verified its effectiveness.

Acknowledgements

D.G.P. thanks the support from the National University of Mexico and the Fulbright-CONACyT scholarship program.

References

1. M. J. Black, and A. Jepson. Eigentracking: Robust matching and tracking of articulated objects using a view-based representation. *Int. J. of Comp. Vis.*, 26(1), pp.63-84, 1998
2. P. Correia, and F. Pereira. The role of analysis in content-based video coding and indexing. *Signal Processing*, 66(2):125-142, April 1998.
3. V. Garcia-Garduno. Une approche de compression orientee-objets par suivi de segmentation basee mouvement. Ph.D. dissertation, Universite de Rennes I, France, 1995.
4. L. Garrido, P. Salembier and D. Garcia. Extensive operators in partition lattices for image sequence analysis. *Signal Processing*, 66(2):157-180, April 1998.
5. D. Gatica-Perez, M. T. Sun and C. Gu. Semantic video object extraction based on backward tracking of multivalued watershed. In *Proc. of the IEEE ICIP*, Kobe, Oct. 1999.
6. C. Gu. Multivalued Morphology and segmentation based coding. Ph.D. dissertation, LTS-EPFL, `http://ltswww.epfl.ch/staff/gu.html`, 1995.
7. C. Gu and M.-C. Lee. Semantic Video Object Tracking Using Region-based Classification. In *Proc. of the IEEE ICIP*, pp. 643-647, Chicago, Oct. 1998.
8. H.J.A.M. Heijmans. *Morphological Image Operators.* Academic Press, 1994.
9. H.J.A.M. Heijmans. Connected Morphological Operator for Binary Images. Report PNA-R9708, CWI, Amsterdam, April, 1997.
10. *IEEE Transactions on Circ. and Syst. for Video Tech.*, Special Issue on Segmentation, Description, and Retrieval of Video Content. Vol. 8, No. 5, September 1998.
11. F. Marques and J. Llach. Tracking of generic objects for video object generation. In *Proc. of the IEEE ICIP*, pp. 628-632, Chicago, Oct. 1998.
12. Microsoft Windows Media Player^{TM}, in `http://www.microsoft.com/windows/windowsmedia/ en/default.asp`, 1999.
13. A. Papoulis. *Probability, Random Variables and Stochastic Processes.* McGraw-Hill, 1993.
14. J. Serra. *Image Analysis and Mathematical Morphology, Vol. II: Theoretical Advances.* Academic Press, 1988.
15. J.C. Simon. *Patterns and Operators: the Foundations of Data Representation.* North Oxford Academic, 1986.
16. M. Wollborn and R. Mech. Refined procedure for objective evaluation of video object generation algorithms. Doc. ISO/IEC JTC1/SC29/WG11 M3448, March 1998.

MORPHOLOGICAL GRANULOMETRIC DECONSTRUCTION

PEDRO PINA
CVRM / Centro de Geo-Sistemas, Instituto Superior Técnico
Av. Rovisco Pais, 1049-001 Lisboa, PORTUGAL

Abstract. This paper introduces the notion of morphological granulometric deconstruction. A model based on that notion is proposed and the algorithms developed to simulate images are also described. The model is validated with a set of images of different types of Portuguese granites.

Key words: Morphological Granulometry, Deconstruction, Implantation, Primary Grain, Granite Texture.

1. Introduction

In earth sciences, the information known or available on some structure is normally scarce and only known through a set of sampling points. The way of dealing with this fact and the verification that a variable measured in a point, besides presenting an erratic value, normally exhibits structural features has given origin to the regionalised variables or geostatistics theory [11]. The estimation and the simulation based on this structured information uses a spatial autocorrelation function named variogram [9, 11] which is in the basis of other tools that allow to infer the feature or the structure for the remaining portion of the field where experimental information is not known. However, if the available information on the feature or structure is not so scarce, besides the application of geostatistics, mathematical morphology can also be used to model and to simulate structures. Since the pioneer use of morphological models in geosciences [4, 5, 12] and mineral processing applications [1], a long and consistent way has been built with success in a considerable set of applications [7, 8, 18, 19]. Although these models are relatively simple to apply, they present, however, some difficulty of generalisation, being the choice of the shape of the primary grain a major difficulty. For instance, in a simulation procedure, the random variation of the shape of the primary grain can be helpful [7, 10], but in other cases that situation will not solve the problem of constructing structures, as for instance, the ones presented by natural stones. However, this type of pictures shows a structuring character that, paradoxically, allows thinking about the use of geostatistics. Anyhow, the tools used in this theory do not take explicitly into attention granulometric features of the structures. Thus, in this paper is introduced a model based on granulometric features of the structures and the respective simulation algorithms that are based upon the way the morphological operators deal with their size and shape features.

2. The Model

The underlying idea to this model comes from the computation of the granulometry by openings of increasing size. The application of successive openings of increasing size suppresses the regions or objects of the structures not containing completely the structuring elements. The remaining regions or structures become simpler and simpler and tend to the shape of the structuring element used, as the sequence of opened images of figure 1 illustrates.

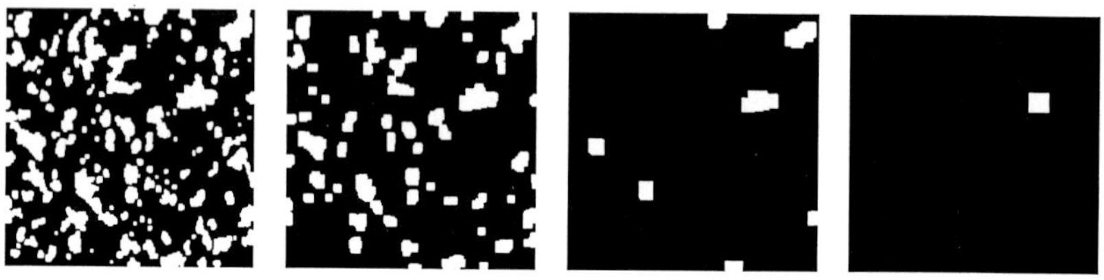

Fig. 1. Openings with a square of increasing size applied to biotite-quartz phase of a Portuguese granite.

This simplification of the image by morphological operators is somehow a decomposition or deconstruction of the initial structure into several opened sets, that through the Lebesgue measure *Meas* of the remaining or of the suppressed successive structures allows to construct the size distribution or granulometric curve [13]. The sequence goes this way from something to nothing, or in this binary case, goes from a non empty set to an empty set. Thus, in order to create something from nothing, why not reverse the order of the sequence, and this way to deconstruct now the granulometric curve in order to create a structure? Having this in mind, the morphological granulometric deconstruction model is proposed. The parameters of the model are extended to other features and are based on the criteria proposed by Serra [18] to characterise a structure: size, dispersion and connectivity.

2.1. Size and shape parameters

Size and shape parameters of a structure X are given by openings γ (or closings φ) with a structuring element B of increasing size λ, being the computation of the respective granulometries in measure $G_M(\lambda)$ given by:

$$G_M(\lambda) = \frac{Meas(X) - Meas[\gamma^{\lambda B}(X)]}{Meas(X)} \tag{1}$$

being $g_M(\lambda) = G_M(\lambda - 1) - G_M(\lambda)$.

2.2. Dispersion and orientation parameters

The dispersion and orientation parameters are given by the covariance $C(h)$, which is the measure of the erosion ε by a pair of points h:

$$C(h) = Meas[\varepsilon^h(X)] \tag{2}$$

or by the variogram $\gamma(h)$ function, that can be deduced from the covariance:

$$\gamma(h) = C(0) - C(h) \tag{3}$$

2.3. Connectivity and neighbourhood parameters

For this set of parameters, the neighbourhood or vicinity functions can be used [6, 3] but, in alternative, one proposes to use the morphological granulometry in number $G_N(\lambda)$:

$$G_N(\lambda) = \frac{N(X) - N[\gamma^{\lambda B}(X)]}{N(X)} \tag{4}$$

where $N(X)$ is the number of particles or objects of X, being the measure of the structure of size λ given by $g_N(\lambda) = G_N(\lambda - 1) - G_N(\lambda)$. But in this case, is crucial to know exactly the number of particles or objects that disappear and are created after the application of each opening γ of size λ. Let $g_{N+}(\lambda)$ be the number of particles or objects that disappear completely in each iteration λ, and $g_{N-}(\lambda)$ the number of particles or objects that are created after each iteration λ, that verify the relation:

$$g_N(\lambda) = g_{N+}(\lambda) - g_{N-}(\lambda) \tag{5}$$

The computation of these quantities is respectively done through the following set of morphological operations:

$$g_{N+}(\lambda) = N[X_{\lambda-1}/R_{X_{\lambda-1}}(X_\lambda)] \tag{6}$$

$$g_{N-}(\lambda) = N[R_{X_{\lambda_i}}(X_\lambda/SKIZ_{X_{\lambda-1}}(X_\lambda)) \bigcap X_\lambda] \tag{7}$$

or simply through eq. 5 when $g_N(\lambda)$ or one of these quantities is known. In eq. 6 and in eq. 7, $R_{X_{\lambda-1}}(X_\lambda)$ is the reconstruction of the marker X_λ in the mask or geodesy $X_{\lambda-1}$ and $SKIZ_{X_{\lambda-1}}(X_\lambda)$ is the geodesic skeleton by influence zones of X_λ in the geodesy $X_{\lambda-1}$.

3. The Algorithms

The algorithms for the simulation of textures using the morphological granulometric deconstruction model consist on the implantation of primary grains in regions previously designed (geodesies) according to the quantities given by the morphological granulometries in measure and in number. Although not inspired on cellular automata methods, there exist some similarities with the developed algorithms, particularly on the definition of the rules for primary grain implantation [17, 20].

The input parameters of the model are the granulometric functions in measure $g_M(\lambda)$ and in number $g_N(\lambda)$, as well as, the associated functions $g_{N+}(\lambda)$ and $g_{N-}(\lambda)$, and also a tolerance t related to the surface to implant in each iteration of size λ. The implantation of a new primary grain at each step can be done according to one of the following conditions:

- the grain to implant cannot be superimposed to already implanted ones (figure 2a);
- the grain to implant can be superimposed to the already implanted ones and
  * must connect them (figure 2b);
  * must not connect them (figure 2c).

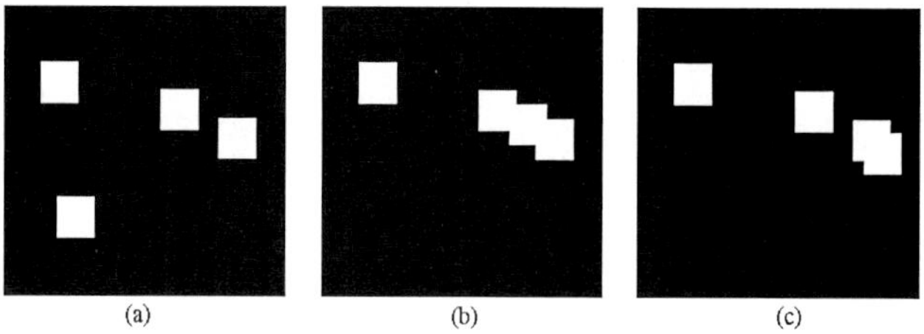

Fig. 2. Possible locations for primary grain implantation: (a) without superimposition (b) with superimposition and connection (c) with superimposition and without connection.

Within each previously defined geodesy, updated after each iteration of the algorithm, the geometrical locus where each primary grain can be implanted is chosen according to a Poisson point process or to other more elaborated point processes.

The sequence of construction of the structures by granulometric deconstruction starts by the implantation of primary grains following the granulometry in number $g_N(\lambda)$ followed by the granulometry in measure $g_M(\lambda)$. The surface of each granulometric class of size λ that is necessary to *spend*, begins by implantation of complete grains (value given by $g_{N+}(\lambda)$) or by superimposed grains in order to connect structures already implanted (value given by $g_{N-}(\lambda)$), and finishes by the implantation of primary grains where is only allowed superimposition with no connections between the new grains and the already implanted ones (value given by $g_M(\lambda)$).

3.1. Implantation of primary grains following $g_N(\lambda)$

At this stage, the procedure consists firstly on the implantation only of complete primary grains (given by the function $g_{N+}(\lambda)$), implanting them over the background without any superimposition, and secondly on the implantation of uncompleted grains (information given by the function $g_{N-}(\lambda)$), by implanting partially the grains in order to connect the ones already implanted.

The procedure begins by the implantation of complete primary grains in the geodesies or masks successively built after each iteration (mask of type A) where is possible to perform that operation. The construction of this type of mask can be simply obtained by the erosion ε of the background X^c with a structuring element of size $(\lambda + 1)B$:

$$M_A = \varepsilon^{(\lambda+1)B}(X^c) \tag{8}$$

An example of the construction of a mask of type A is presented in figure 3. The application of eq. 8 to the union of primary grains already implanted (figure 3a) constructs the mask M_A where is possible to implant a complete primary grain without superimposition (figure 3b).

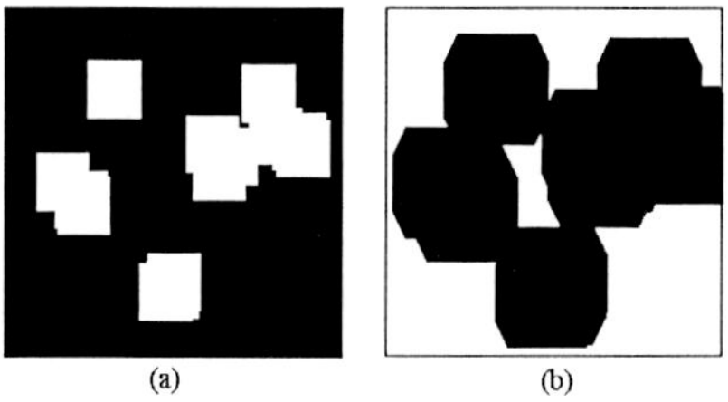

Fig. 3. Construction of mask of type A.

The construction of a mask where is possible to implant grains that will be superimposed to the already implanted ones and that simultaneously establish connections between grains (mask of type B) is more elaborated. It consists on finding the regions where the primary grain fits, *i.e.*, on finding the regions of the complementary set X^c where is possible to implant a primary grain that connects at least two distinct components. The detection of the regions where that topological modification is performed is achieved by a closing φ of size $\lambda/2$, that also detects the concavities of the structures presenting the same size/shape, as can be noticed on figure 4b (regions in grey). To distinguish these two types of regions (concavities and non-concavities) added by the closing, it should be taken into account that the closed concavities are in contact with only one object while the closed non concavities are contacting at least two different objects. Thus, the counting of the number of contacts for each closed region is performed on the image Y resulting from the following set of operations:

$$Y = [\delta^B(\varphi^{\lambda B/2}(X)/X)] \bigcap X \tag{9}$$

where δ is the dilation and φ the closing transforms. The identification of the added regions is obtained by the closing φ with the structuring element $\lambda B/2$ ($\varphi^{\lambda B/2}(X)/X$). This set Y, dilated in the following with the structuring element B of unitary size, followed by its intersection with the initial set X allows to identify the borders or internal contours of X, that are contacting the regions added by the closing. Now, in the new geodesy Y_1:

$$Y_1 = Y \bigcup [\varphi^{\lambda B/2}(X)/X] \tag{10}$$

constituted by the union of the regions added by the closing ($\varphi^{\lambda B/2}(X)/X$) and of the internal contours of X in contact with the regions connected by the closing (set Y), it is necessary to count the connected components of Y within those geodesies. If, within each connected component of Y_1, the number of objects (or contacts) is equal to 1, then the set marked by this object is a concavity, otherwise, for a number superior to 1, this set is not a concavity. If the connected component is not a concavity, it should be investigated if it really connects different objects or if it is only contacting different regions belonging to the same particle. The sequence of operations that allows to discriminate these two types of non concavities demands more algorithmic efforts.

Thus, in order to identify the non concavities that connect different connected components, it is enough to count the number of objects of the initial set X that each connected component Y_{1i} of Y_1 ($Y_1 = \bigcup Y_{1i}$) is marking, and therefore, can reconstruct. By an individual analysis of each Y_{1i} set, the reconstructed Y_{2i} sets are then obtained :

$$Y_{2i} = R_X(Y_{1i}) \tag{11}$$

Analogously to the way of distinguishing concavities from non concavities it suffices to count the number of objects of each set Y_{2i}. If this number is 1, thus this component connects two different regions from the same object, but if the number is superior to 1, then this component connects different objects. After the identification of the connected components that belong to non concavities and where the implantation of a primary grain connects different objects, it is now possible to construct a new type of mask (type B, M_B), by simply reconstructing those components Y_2 in the mask given by the closing $\varphi^{\lambda B/2}(X)$, whose final result is shown in figure 4c:

$$M_B = R_{\varphi^{\lambda B/2}}(Y_2) \tag{12}$$

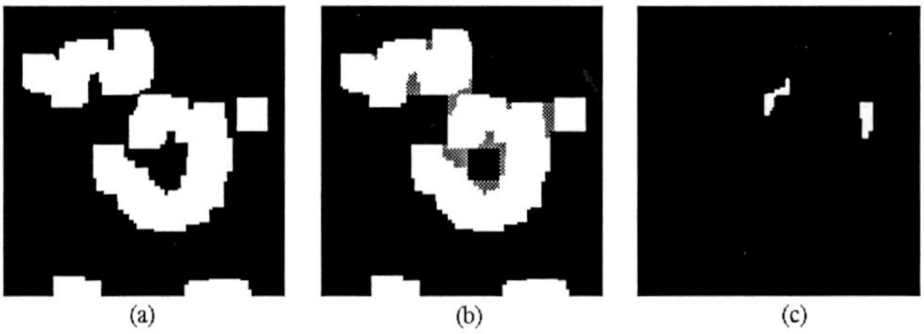

Fig. 4. Construction of mask of type B.

3.2. Implantation of primary grains following $g_M(\lambda)$

When the granulometric distribution function in number $g_N(\lambda)$, and the associated functions $g_{N+}(\lambda)$ and $g_{N-}(\lambda)$) are *spent* for a given class of size λ, *i.e.*, when all the primary grains from those experimental class were implanted (complete or superimposed to connect structures) it may be necessary to implant more primary grains to respect, within the permitted tolerance, the granulometric distribution in measure $g_M(\lambda)$. Obviously, to respect the distribution in number $g_N(\lambda)$, the new primary grain can only touch or be superimposed to the already implanted grains, without connecting them. To keep intact this topological relation, a new type of mask must be built (type C). It takes into account the regions where is possible to implant new superimposed (submask M_{C1}) without connection (submask M_{C2}). Thus, the construction of the submask M_{C1} consists on detecting the interior regions of the implanted grains where is possible to implant a new grain without being completely absorbed (to avoid the unnecessary implantation of a grain within an existent structure) and the exterior regions of the already implanted ones where is possible to implant a grain with superimposition. The first one of these sets is obtained by the logical subtraction between the set of structures and the eroded set. The second set is obtained by the logical subtraction between the dilated structures and the structures. The union of these two sets constitutes the submask M_{C1} (figure 5b):

$$M_{C1} = [X/\varepsilon^{\lambda B}(X)] \bigcup [\delta^{\lambda B}(X)/(X)] \tag{13}$$

Submask M_{C2} is the geodesy where the submask M_{C1} is valid without modifying the homotopy, *i.e.*, where there is no connection. Therefore, submask M_{C2} is constructed through the following set of operations:

$$M_{C2} = \varepsilon^{\lambda B/2}[(SKIZ(X))^c] \tag{14}$$

that corresponds to the influence zones of the already implanted structures obtained by the skeleton by influence zones $SKIZ$, eroded by

$$\lambda B/2(\varepsilon^{\lambda B/2}(SKIZ(X))^c),$$

where it is not possible to establish connections between disjoint structures by primary grains of size λ (figure 5c).

Thus, mask M_C is obtained through the intersection of the two submasks M_{C1} and M_{C2} (figure 5d):

$$M_C = M_{C1} \bigcap M_{C2} \tag{15}$$

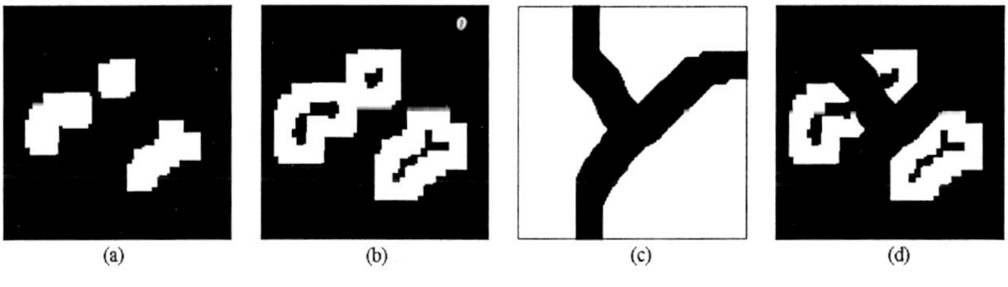

Fig. 5. Construction of mask of type C.

4. Simulation of Binary Textures of Granites

In order to simulate textures using the proposed model, experimental information extracted from a set of 13 types of Portuguese grey granites was used [14, 15]. All these granites are constituted mainly by feldspars, quartz and biotite minerals and present the same textural neighbourhood or vicinity relationships: feldspar (lighter phase) is the matrix or background, where quartz (grey phase) and biotite (darker phase) are included. The biotite occurs mainly within the quartz but always in its boundaries (contacting the feldspar phase) and also, in lower quantities, totally involved in the feldspar phase, being rare the occurrence of biotite completely involved by the quartz phase. Those topological relations are schematically represented in figure 6a. Due to this hierarchy a simplification was introduced by unifying biotite and quartz phases within a single one and keeping feldspar phase unaltered (figure 6b).

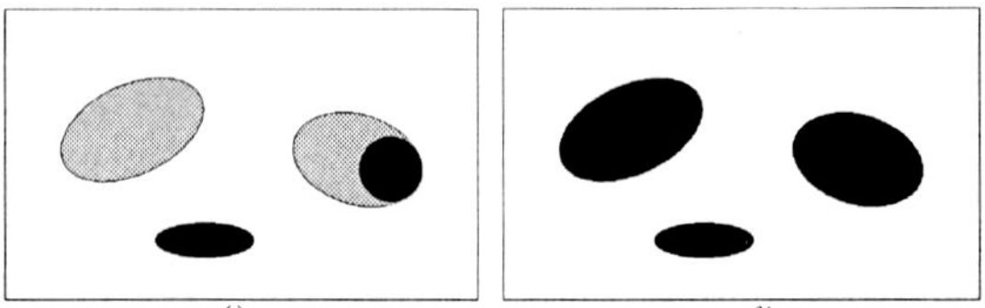

(a) (b)

Fig. 6. Scheme (a) of occurrence of main mineralogical phases in a granite (biotite (black), quartz (grey), feldspar (white)) (b) adopted for the simulation (biotite-quartz (black), feldspar (white)).

In figure 7 an example concerning *Pedras Salgadas* granite is presented (figure 7a shows an experimental or real texture segmented/classified with a morphological approach [16], while figures 7b, 7c and 7d are examples of simulated textures). For each type of granite five realisations of the morphological granulometric deconstruction model, based on size distributions in number and in measure, were performed.

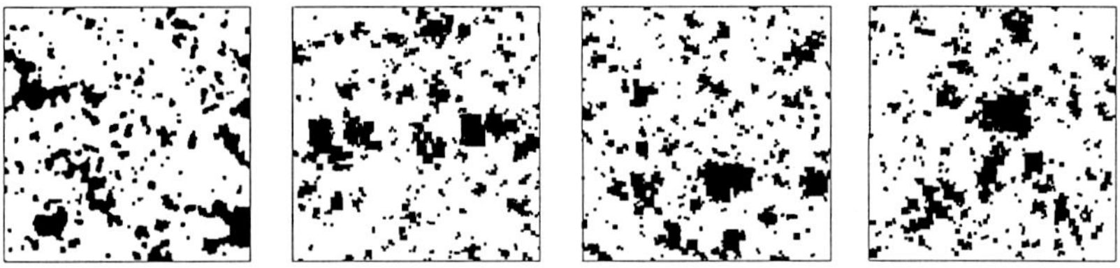

(a) (b) (c) (d)

Fig. 7. Binary textures of granites (biotite-quartz (black) and feldpsar (white)): (a) experimental (b)(c)(d) simulated.

In general, concerning the visual aspect of the simulated images in comparison with the real ones, the similitude is very high, lying the difference on the regularity of the contours of the phases. If the experimental texture presents smoother contours, and does not denunciate the digital character of the image, the simulated textures do not hide that characteristic and present more regular contours. Obviously that a post-processing step can be envisaged, but so far, it is preferable to exhibit this characteristic in order to reconvince us about the veracity of the simulated structure [2].

To validate the proposed model, on each simulated image were performed the same measures that were previously performed on the experimental ones: granulometry in measure and in number for the phase directly simulated (phase biotite-quartz) but also to the complementary phase (phase feldspar), and also a dispersion/anisotropy measure (the variogram function) in two normal directions (0° – 180° and 90° – 270°) for both biotite-quartz and feldspar phases. The granulometric curves (in measure and in number) and the experimental variograms obtained for the *Pedras Salgadas* granite are presented in figures 8 (biotite-quartz phase) and 9 (feldspar phase), corresponding in the simulated case, to average curves of the five realisations of the model performed. In the graphics, each real or experimental value is plotted with a black polygon, while the simulated values are plotted with a circumference.

The granulometries in measure and in number of biotite-quartz phase (phase directly controlled in the simulation algorithms) are highly reproduced. The granulometries in measure and in number for feldspar phase are globally well reproduced, with some few local discrepancies. Anyhow the results can be considered very positive and acceptable. In what concerns the experimental variograms computed in two normal directions, the results obtained present also highly similar behaviours. Globally, the degree of matching for the set of 13 types of granites between the experimental and the simulated curves is very high [15].

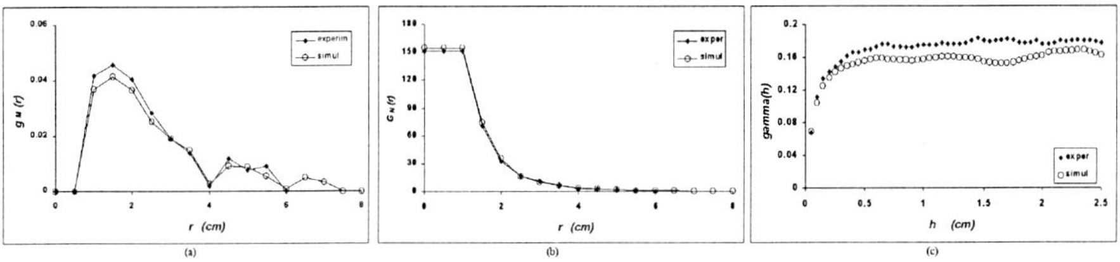

Fig. 8. Validation tests on biotite-quartz phase. (a) granulometry in measure (b) granulometry in number (c) experimental variograms.

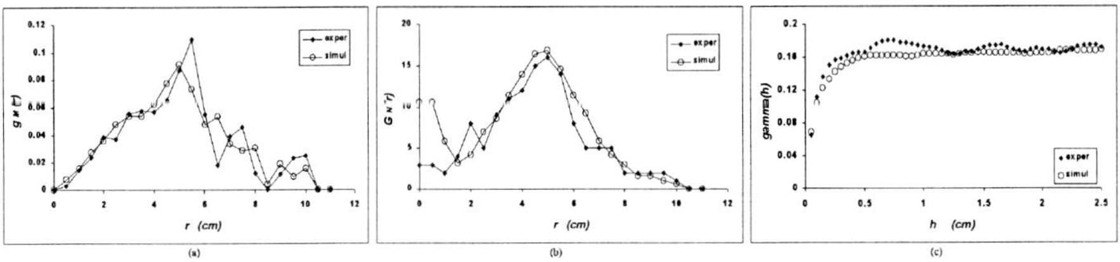

Fig. 9. Validation tests on felspar phase. (a) granulometry in measure (b) granulometry in number (c) experimental variograms.

5. Conclusions and Future Developments

In face of the textures obtained and of the comparison of the experimental and simulated curves, the proposed model can be accepted as a valid one, allowing, so far, to simulate natural structures as the ones presented by the granites (without preferential nor a spatial clustering of its phases) based on the morphological granulometric distributions in number and in measure.

Therefore, the model and the algorithms presented to binary images may be improved taking into account the following points:

– the information about dispersion and orientation should be considered when necessary;
– more phases should be accepted, in a hierarchical or non hierarchical ways;
– a generalisation should be done in order to contemplate grey level and colour images;
– the algorithms should be optimised or fast versions should be created in order to reduce the present high computational costs.

References

1. P. Cauwe. Morphologie mathématique et fragmentation. PhD thesis, Université Catholique de Louvain, Louvain, 1973.
2. J. Derrida and B. Stiegler. *Écographies de la télévision. Entretiens filmés*, Galilée-INA, Paris, 1996.
3. M. Grzebyk. Quantitative study on an evolutionary soil sequence. In A. Soares, editor, *Geostatistics Troia'92*, vol.2, pages 817–828, Dordrecht, 1993. Kluwer Academic Publishers.
4. A. Haas, G. Matheron and J. Serra. Morphologie mathématique et granulometries en place. *Annales des Mines*, XI:734–753, 1967.
5. A. Haas, G. Matheron and J. Serra. Morphologie mathématique et granulometries en place, part II. *Annales des Mines*, XII:767–782, 1967.
6. D. Jeulin. Image analysis and materials image analysis. *Scanning Microscopy*, 2:165–183, 1988.
7. D. Jeulin. Modèles morphologiques de structures alétoires et de changement d'échelle. Thèse de doctorat d'État, Université de Caen, Caen, 1991.
8. D. Jeulin (ed). *Advances in theory and applications of random sets.* World Scientific, Singapore, 1997.
9. A. Journel and Ch. Huijbrechts *Mining geostatistics.* Academic Press, London, 1978.
10. Ch. Lantuéjoul. Iterative algorithms for conditional simulations. In E.Y. Baafi and N.A. Schofield, editors, *Geostatistics Wollongong'96*, vol.1, pages 27–40, Dordrecht, 1997. Kluwer Academic Publishers.
11. G. Matheron *Les variables regionalisées et leur estimation.* Masson, Paris, 1965.
12. G. Matheron. *Éléments pour une théorie des milieux poreux.* Masson, Paris, 1967.
13. G. Matheron. *Random sets and integral geometry.* Johh Wiley and Sons, New York, 1975.
14. P. Pina, V. Ramos and F. Muge. Feature extraction from Portuguese granites for classification purposes. In F. Muge, J. C. Pinto and M. Piedade, editors, *Proc. RecPad'98 - 10th Portuguese Conference on Pattern Recognition*, pages 265–272, Lisboa, Mar. 1998. APRP.
15. P. Pina. Characterisation, modeling and simulation of structures by mathematical morphology (in Portuguese). PhD thesis, Technical University of Lisboa, Lisboa, 1999.
16. P. Pina. Segmentation/classification by mathematical morphology. Application to macroscopic images of Portuguese granites. In D. Jongmans, E. Pirard, Ch. Schroeder, and Ph. Trefois, editors, *Proc. GEOVISION'99 - International Symposium on Imaging Applications in Geology*, pages 193–196, Liège, May 1999. University of Liège.
17. F. Richards, T. Meyer and N. Packard. Extracting cellular automaton rules directly from experimental data. *Physica D*, 45:189–202, 1990.
18. J. Serra. *Image analysis and mathematical morphology.* Academic Press, London, 1982.
19. J. Serra (ed). *Image analysis and mathematical morphology, vol.2: Theoretical advances.* Academic Press, London, 1988.
20. M. A. Smith. Representations of geometrical and topological quantities in cellular automata. *Physica D*, 45:271–277, 1990.

SURFACE TEXTURE CLASSIFICATION FROM MORPHOLOGICAL TRANSFORMATIONS

ANTOINE AUBERT and DOMINIQUE JEULIN
Centre de Morphologie Mathématique Ecole des Mines de Paris
aubert@cmm.ensmp.fr, jeulin@cmm.ensmp.fr

and

RONALDO HASHIMOTO
Departamento de Ciência da Computação
Instituto de Matemática e Estatística. Universidade de São Paulo.
ronaldo@ime.usp.br

Abstract. A morphological tool for accurate automatic texture classification is proposed and applied to a large set of plastering mortars. Starting from measurements of morphological data (covariance, speed of erosion and dilation, size distribution, watershed sizing) on grey level images, correspondence analysis is used in a learning procedure to visualize the surfaces in a reduced space, and to select the most pertinent measurements with respect to the classification of textures.

Key words: Texture Classification, Rough Surfaces, Covariance, Granulometry, Choquet Capacity, Watershed, Multivariate Data Analysis.

1. Introduction

Texture classification by image analysis has many applications in industry. The aim of this study is to provide and demonstrate the usefulness of a texture classification algorithm that uses some morphological measurements and statistical data analysis (Correspondence Analysis) in order to classify random textures into k different classes. This is applied to the development of a fast automatic classification of samples of plastering mortars, in order to replace a standard visual inspection procedure. The presentation which follows is extracted from a more systematic and wider study [1].

The approach is divided in two steps: training and classification. In the training step, morphological measurements on sample images produce a large number of data, which are submitted to a multivariate statistical analysis in order to get the best separation of the representation points into k regions. The classification step is obtained by projecting experimental data in a low dimensional representative subspace of the data.

The main difficulty of this approach is to find the best morphological measurements that should be performed. In the present work, we show that the morphological measurements: (i) covariance curve; (ii) dilation and erosion curve; and (iii) granulometry and antigranulometry distributions, do not give the same accuracy for the classification. We show how to extract the best

(a) (b) (c)

Fig. 1. (a) Q1 quality (too early), (b) Q2 quality (normal), (c) Q3 quality (too late).

morphological functions, from the point of view of a classification, and for this one the more discriminant values. The paper is organized as follows: section 2 describes the samples to be studied. Section 3 presents the morphological measurements used in this work. Section 4 describes the statistical data analysis. Section 5 presents the experimental results and finally, the conclusions are summarized in Section 6.

2. Samples and Image Acquisition

The plastering mortars samples studied in this work are specimens to be given to customers in a sample presenter. The manufactured specimens present three types of texture qualities (noted $Q1, Q2, Q3$), due to some variability in their elaboration. These qualities depend on the drying time of the roughcast before being scrapped. It is possible to define the following categories, from the morphological aspect of samples: "too early", "normal" and "too late" term. Some examples of such samples images are presented in Fig. 1.

The classification problem studied here is, given a collection of sample images of a group G containing different texture types, classify them into classes $Q1$, $Q2$ and $Q3$, according to their texture types.

In the study [1], 3 types of textures were analysed, with 300 samples per type. We restrict the presentation to a single type of texture, which corresponds to 300 samples.

In the process of image acquisition of plastering mortars, a simple device composed of a TV camera and two lamps were used. A normal (and isotropic) lighting flattens the relief and the image has a too low contrast to be tractable. Therefore, an oblique lighting (45 degrees) was used. The projected shadows of the relief give an anisotropic texture, characteristic of the surfaces. In the images, the grey levels do not directly represent the elevation of the considered point, but are however representative of the topography: the obtained images present shadows, due to the presence of a strong relief (Fig. 1).

All acquired images presented a horizontal periodic noise, which was removed by a frequency filter, using a Fast Fourier Transform (FFT). The final images contain 485×285 pixels.

3. Morphological Measurements

When working with textures, it is convenient to consider images as realizations of random structures (random sets for binary images [9, 10], and random functions (RF's) for grey-level images). They can be characterized by means of the Choquet capacity [9, 10, 8], from which are derived most morphological measurements.

3.1. The Choquet capacity

An upper semi-continuous random function $Z(x)$ is characterized by its Choquet capacity, $T(g)$, which can be defined [8] on the lower semi continuous functions (lsc) g with a compact support K

$$T(g) = P\{x \in D_Z(g)\}; D_Z(g)^c = \{x, Z(y) < g(y-x), \forall y \in K\} \tag{1}$$

A particular and usual case is the spatial law defined on a finite number of points $x_1, x_2, ..., x_n$,

$$F(x_1, x_2, ..., x_n, z_1, z_2, ..., z_n) = P\{Z(x_1) < z_1, Z(x_2) < z_2, ..., Z(x_n) < z_n\} \tag{2}$$

Other particular cases give the changes of supports by $\vee$ (supremum), namely a dilation or by $\wedge$ (infimum), namely an erosion, which generate new RF's:

$$Z_\vee(K) = \vee_{x\in K}\{Z(x)\} = Z \oplus K$$

$$Z_\wedge(K) = \wedge_{x\in K}\{Z(x)\} = Z \ominus K$$

As a particular case, when the compact set K is limited to point x and $g(x) = z$, we recover the distribution function $F(z)$:

$$F(z) = P\{Z(x) < z\} = 1 - T(g) \tag{3}$$

3.2. Morphological Measurements

3.2.1. *Covariance Function*

The centered and reduced covariance function of the stationary random function $Z(x)$ is defined by:

$$C(h) = \frac{E\{Z(x)Z(x+h)\} - (E\{Z(x))^2\}}{E\{Z(x)Z(x)\} - (E\{Z(x)\})^2\}}$$

where $E\{\cdot\}$ denotes the mathematical expectation of a random variable. The term $E\{Z(x)Z(x+h)\}$ is estimated by the average of the product $\{Z(x)Z(x+h)\}$ over the pixels of the images. As it can be interpreted in terms of a product of convolution, it is obtained by FFT. We added the value 0.5 to the centered and reduced covariance, to avoid negative values for the Correspondence multivariate data analysis. The measurements were made for separations h ranging from 1 to 100 in the two main directions of the image (horizontal and vertical)

3.2.2. *Closing and Opening Curve (Anti-Granulometry and Granulometry)*
A granulometry is the study of the size distributions of the objects of an image [9, 10]. We compute, at the same time, a granulometry (distributions of sizes of peaks) using morphological openings and an anti-granulometry (distributions of sizes of valleys) using morphological closings. The following estimator for the size density is equivalent to a probability distribution function, while the function $S(k)$ plays the role of a cumulative distribution function [5]:

$$\tilde{S} = S(k) - S(k+1)$$

with

$$S(k) = \frac{1}{N} \cdot \frac{1}{N\mathsf{Vol}\{M \ominus 2kB\}} \cdot \begin{cases} \mathsf{Vol}\{(M \ominus 2kB) \cap ((Z - \mathsf{Av}\{Z\}) \circ kB)\} & \text{if } k \geq 0 \\ \mathsf{Vol}\{(M \ominus 2|k|B) \cap ((Z - \mathsf{Av}\{Z\}) \bullet |k|B)\} & \text{if } k < 0 \end{cases}$$

where:

- $Z \circ kB$ is the opening of Z by kB,
- $Z \bullet |k|B$ is the closing of Z by kB,
- N is the number of levels in the gray-level image (in our case $N = 255$),
- $\mathsf{Vol}\{\cdot\}$ is the volume of image, that is, $\mathsf{Vol}\{X\} = \sum_{w \in X} X(w)$,
- $\mathsf{Av}\{\cdot\}$ is the average constant image, that is, $\mathsf{Av}\{X\}$ is an image of the same size as X with constant height equals to $\frac{1}{S} \cdot \mathsf{Vol}\{X\}$, where S is the number of pixels of X,
- Z is the input Image,
- M is the mask Image,
- B is the structuring element (square; horizontal or vertical segment), and
- $M \cap Z$ means the restriction of image Z to the mask M.

Fig. 2 gives an example of closing opening curves by squares (sizes cover the range 1 to 100 pixels in measurements). The right side (positive values in horizontal axis) is the opening curve while the left side (negative values in horizontal axis) is the closing curve.

3.2.3. *Erosion and Dilation Curve*
We also use a "pseudo-granulometry" and "anti-pseudo-granulometry" obtained by replacing the closing and opening by, respectively, dilation and erosion [5]. More formally,

$$S(k) = \frac{1}{N} \cdot \frac{1}{\mathsf{Vol}\{M \ominus kB\}} \cdot \begin{cases} \mathsf{Vol}\{(M \ominus kB) \cap ((Z - \mathsf{Av}\{Z\}) \ominus kB)\} & \text{if } k \geq 0 \\ \mathsf{Vol}\{(M \ominus |k|B) \cap ((Z - \mathsf{Av}\{Z\}) \oplus kB)\} & \text{if } k < 0 \end{cases}$$

The speed of erosion (namely of dilation) is defined as the difference $|S(k) - S(k+1)|$. In Fig. 3 is given an example of an erosion and dilation curve by vertical segments. As in the case of granulometry, erosions and dilations by squares (with sizes ranging from 1 to 100 pixels), horizontal and vertical segments (with sizes ranging from 1 to 100 pixels) were used for texture classification.

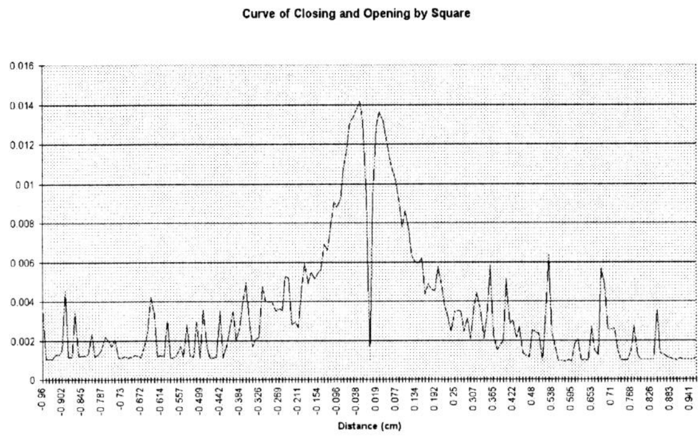

Fig. 2. Example of a closing (left side) and opening (right side) curves by squares.

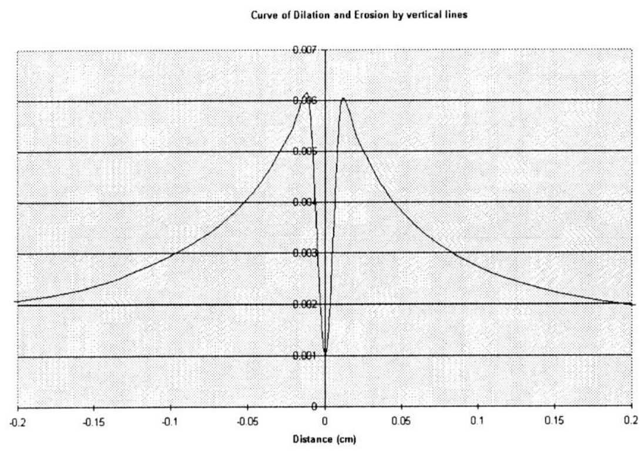

Fig. 3. Example of an erosion (right side) and dilation (left side) curves by vertical segments.

We observe that the computation of the erosion and dilation curve is definitely faster than the computation of the closing and opening one.

3.2.4. *Measurements on peaks and valleys*
In order to get information on the connectivity of peaks and valleys, we extracted watersheds of images and of their negative values (an example is given in Fig. 4). From these segmented images, we measured the number of valleys with a given area S, or with a given volume V (by integration of the grey level values over each watershed), and similarly for the number of peaks. This type of information is illustrated on the curves of Fig. 5, where the left part concerns peaks and the right part concerns valleys. These morphological data were used in a preliminary study of the surfaces, but was not relevant for the

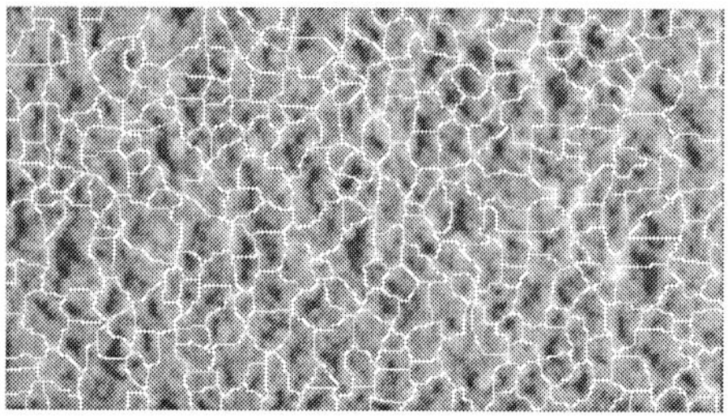

Fig. 4. Example of watersheds.

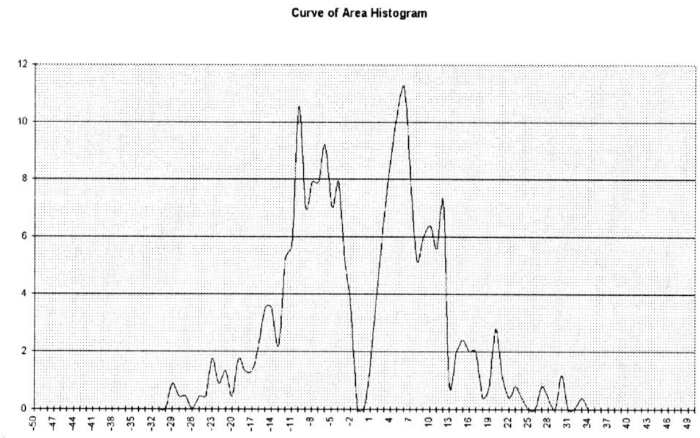

Fig. 5. Example of area distribution of catchment basins: peaks (left part) and valleys (right part).

classification of the present textures.

4. Introduction to Correspondence Analysis

In order to analyze the morphological measurements described in Section 3.2 and to produce a classification of samples, we use Correspondence Analysis [4]. A similar approach was followed in [7] for the automatic classification of non metallic inclusions in steels. It was used more recently for the classification of rough surfaces [2], and of simulations of random sets [3]. The values of measurements are stored in a table, where every line corresponds to a sample (or *observation*) and every column to a measurement (or a *variable*). We define the *line coordinate* of a given surface (observation) as the vector with M dimensions, and the *column coordinate* of a given measurement (variable) as the vector with N dimensions. Alternative factor analysis techniques could be used for this purpose, like Discriminant factor analysis, Canonical Variate Analysis, or even Penalised Discriminant Analysis [6]. These methods share with Correspondence Analysis the fact that an optimal linear combination of variables is looked for the discrimination of data, and in practice they give very similar results. In addition, it can be proved that Correspondence Analysis is a special case of Discriminant or Canonical Analysis, based on the chi square metrics [4]. Correspondence analysis allows to build a map of the line and column coordinates (points with M and N dimensions, respectively) into a reduced dimensional space (for instance a two-dimensional space), generating synthetic criteria (in this sense, the method is a type of factor analysis [4]).

Not all of the variables used in the analysis are important to separate the samples. This analysis allows us to find the most significant variables amongst the initial variables for the separation of samples.

Correspondence analysis allows one representation of the cloud of line points and column points in a reduced space, which separates as well as possible all the points by maximizing the inertia. Moreover this analysis allows the interpretation of the axes (or factors) of the representation plane in terms of the initial variables. That is precisely the main advantage of this method (with respect to the classical factor analysis) for our purpose.

An indication of the quality of the synthetic axes makes it possible to estimate the accuracy of the simplified representation and the amount of lost information, from the corresponding part of inertia that they carry. For example, we will say that the first direction explains 45 % of inertia, or equivalently that it synthesizes 45 % of the total information. Obviously, one representation of points in one plane explains the sum of fractions of inertia for each axis. It is important to notice that we can use the distance between images, as well as the distance between variables.

In the example of the study shown on Fig. 6, we notice that the part of explained inertia is 88.4 % for the first factor and 5.8 % for the second one. That means that we represented a vector with 200 dimensions in a 2D-space, while keeping 95 % of the total information. The visualization of column points allows to identify the most discriminant measurements. To conclude this part, we can remind that the correspondence analysis allows:

- to visualize the surfaces in a reduced space (typically 2 or 3 dimensions) with the help of synthetic factors.
- to give an interpretation of this representation.
- to propose a reduction of the number of variables, useful to decrease the measurement time.

5. Results

The study was made in three steps. At first, the relevance of directional structuring elements is estimated (this choice is justified by a directional lighting). In a second step, the most discriminant variables are determined, in order to reduce the amount of data. These two first steps were done in a reduced database (30 samples). In the final step, we studied the complete data base (300 samples) on a reduced number of measurements.

5.1. First step: surfaces and lighting anisotropy

The first study intends to evaluate if the obtained textures are really characteristic of the different qualities of the product. The opening and closing curves do not give discriminant results in the present case, mainly due to the fact that the experimental curves present many oscillations (see Fig. 3.2.2). Larger images of each sample would be required to proceed to a statistical averaging. In the remaining part of the study, we limit measurements to the erosion and dilation curves (and by the way, abandon opening and closing measurements). Directional structuring elements appear to be a good direction of investigation.

5.2. Second step: reducing the number of variables

The projections of variables in the first factor plane show the influence of each variable on the discrimination. In fact, the most scattered variables in the factor planes are kept (two close variables will have the same influence on the representation and the classification, and therefore a single one can be used). For the remaining part of the study, it is possible to keep the following measurements:

- the covariance (horizontal and vertical), for distances between 1 and 50 pixels;
- the erosion-dilation curve using, vertical or horizontal segments, ranging from 1 to 30 pixels.
- the erosion and dilation curves, using squares with side between 1 and 50 pixels.

These measurements were made for a complete data basis of 100 samples for each quality.

5.3. Final study

At first, a very important result of this study is that the covariance function is not efficient enough to discriminate the three categories, while the morphological operations (erosion and dilation) give a satisfactory discrimination of samples. Using a simple linear regression to define the best limits of the different domains, we obtain for these cases the scores given in Figs 6 and 7. We can notice that in the case of directional operators perpendicular to the incidence of lighting (Fig. 7), the best results give 90 % of well classified surfaces. This is very interesting, because very fast algorithms exist to compute directional erosions and dilations. As reported on Fig. 6 and 7, some misclassified samples are observed. The reasons of these misclassifications are due to the fact that these textures look very similar for specimens corresponding to different families. We can notice than these samples are not easily detectable even with a manual inspection.

6. Conclusion

We have presented in this paper an automatic tool of texture classification. This tool, used here in one practical study, is based on the correspondence analysis of morphological data. It allows a visualization in a synthetic space (here 2D-space), where the meaning of each axis is obtained by projecting the variables in the same space. From examination of the variables in this plane, the most discriminant variables can be extracted. The automatic classification

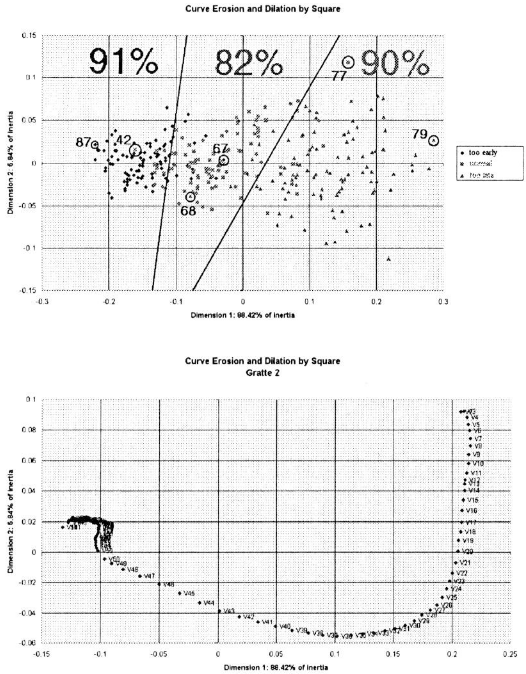

Fig. 6. Classification score for erosion-dilation curves by square and related representation of variables in the first factor plane.

on the basis of the erosion and dilation curves (using a square or directional structuring elements) give very clearly better results than the covariance function in the present case. This is consistent with the theoretical characterization of random structures by means of the Choquet capacity. Working with few optimized data (erosion and dilation curves for structuring elements with sizes between one and 30 or 50) gives a good classification of very similar textures, with a score of 90 %. The control of the misclassified samples shows that they are very difficult to discriminate, even by an experimented observer.

This study is in fact the learning phase of the classification: using a reference database, the best directions (or factors) which discriminate these samples were calculated. Now, to classify a new sample, we only have to project it onto these directions. Using the learning step, it is not necessary to calculate new factors by extracting singular values. As a result of this study, it appears that linear erosions and dilations are very discriminant for very similar textures, where covariances failed to separate samples of rough surfaces.

Acknowledgements

This study was made with the financial support of Weber & Broutin Company, to which we are grateful for supplying the samples. This work has also been

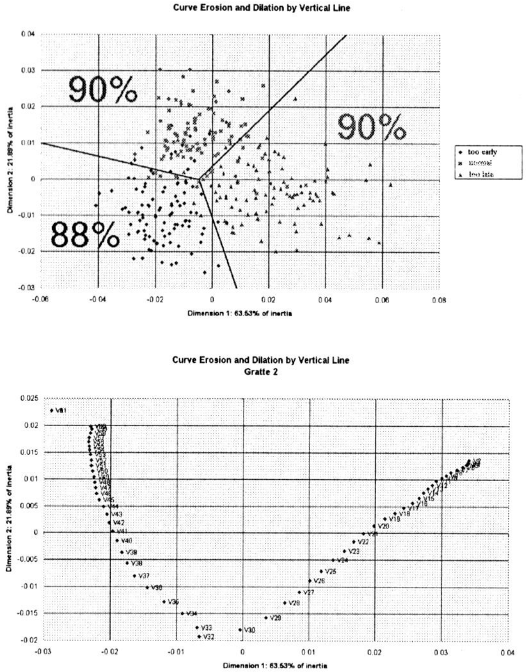

Fig. 7. Classification score for erosion-dilation curves by vertical lines and related representation of variables in the first factor plane.

supported by scholarship from Brazilian Government fellowship CNPq proc. 200973/96-7.

References

1. A. Aubert, R. Hashimoto, and D. Jeulin. Internal Report, Paris School of Mines, July 1998.
2. A. Aubert, and D. Jeulin. Classification morphologique de surfaces rugueuses. to appear in *Revue de Métallurgie*, February 2000.
3. A. Aubert, and D. Jeulin. Estimation of the influence of second and third order moments on random sets reconstructions. *Pattern Recognition*, 33(6): 1083-1103, April 2000.
4. J.P. Benzécri. *L'analyse des données*, volume 2: *L'analyse des correspondances*. Dunod, Paris, 1973.
5. J. Goutsias, K. Sivakumar. Discrete morphological size distributions and densities: estimation techniques and applications. *J. of Electronic Imaging*, 6: 31-53, January 1997.
6. T. Hastie T., A. Buja and R. Tibshirani. Penalized discriminant analysis. *Annals of Statistics*, 23: 73-102, 1995.
7. D. Jeulin and J. Serra. Pour reconnaître les inclusions: chartes ou analyseurs de textures? *Mémoires et Etudes Scientifiques de la Revue de Métallurgie*, 72: 745-751, October 1975.
8. D. Jeulin. *Modèles morphologiques de structures aléatoires et de changement d'échelle*. Thèse de Doctorat d'Etat, University of Caen, April 1991.
9. G. Matheron. *Random sets and integral geometry*. J. Wiley, New York, 1975.
10. J. Serra. *Image analysis and mathematical morphology*. Academic Press, London, 1982.

CONTENT DEPENDENT IMAGE SAMPLING USING MATHEMATICAL MORPHOLOGY: APPLICATION TO MIPMAPPING

ETIENNE DECENCIERE FERRANDIERE,
BEATRIZ MARCOTEGUI and FERNAND MEYER
Centre de Morphologie Mathematique
Ecole Nationale Superieure des Mines de Paris
35, rue Saint Honore, 77305 Fontainebleau, France
Tel: +33 1 64 69 48 09, fax: +33 1 64 69 47 07
Email: decenciere@cmm.ensmp.fr

Abstract. Interactivity has become one of the main objectives of multimedia applications. Texture mapping techniques are essential to allow its development, but they all have one point in common: they are not content dependent. In this paper the development of content dependent texture mapping techniques is studied. A content dependent sampling process, based on a reference image which indicates the importance given to each pixel of the original image, is proposed. Two methods of building the reference image by means of morphological tools are described. This content dependent sampling method is used to build a mipmap, a classical structure used in texture mapping.

Key words: Image Sampling, Content Dependent Sampling, Texture Mapping.

1. Introduction

Multimedia interactive applications are rapidly developing. This enhanced interactivity and new freedom needs to be supported by texture rendering tools. Different texture mapping techniques exist [7], but, as long as we know, they all have one point in common: they are not content dependent. This means that images with different content will be treated the same way, and that inside one image all pixels are also processed without taking into account their "meaning". For example imagine a 3D scene featuring a notice-board hanging from a textured wall. The wall and the text will be rendered exactly in the same way. However, it could be interesting to better preserve the details of the text than the details of the wall texture, for instance.

In this article we focus on the downsampling problem. Note however that content based upsampling algorithms have also been recently developed in the same context by Albiol and Serra [1].

In classical computer graphics, the treatment of a texture pixel depends only on its position. In the present work, we will analyze the content of the texture image and downsample it depending on the importance associated to each pixel. To this end, a content dependent sampling method based on a reference image which indicates the importance given to each pixel of the original image is presented. Two methods for building the reference image that make use of mathematical morphology are proposed. Moreover, we will apply these downsampling techniques to texture mapping using mipmaps.

In order to test the resulting texture rendering algorithms in real time conditions, and to compare them with existing techniques, we have chosen MESA, a freeware implementation of OpenGL (Open Graphics Library), a software interface to graphics hardware.

2. Content Dependent Sampling

We refer the reader interested by a review on texture mapping techniques to an article by Heckbert [7]. The techniques described in this review do not take into account the contents of the image when downsampling.

Mathematical morphology has already been applied to image sampling [5, 6, 8, 9]. In these works the aim was to simplify the image before sampling, in order to keep only those details that could be represented at the lower resolution level. In that sense these non-linear approaches already were content dependent, but in our case we want to develop tools that will enable us to preserve some details that are considered important, even if they are theoretically too small (the theoretical limit being given by the Shannon sampling theorem), while trying to avoid aliasing. Of course, this will only be possible if there is enough place in the image. In that sense, our work is inspired from [3, 4].

2.1. A GENERAL FRAMEWORK FOR IMAGE SAMPLING

The simplest downsampling method is point sampling. It can be analyzed in the following way: the original image I is partitioned into regular 2×2 square blocks, and within each block the first pixel (in the video scanning order) is kept in order to build the sampled image J. We will denote $B(x, y)$ the block composed of pixels $\{(2x, 2y), (2x+1, 2y), (2x, 2y+1), (2x+1, 2y+1)\}$. This procedure is illustrated by figure 1.

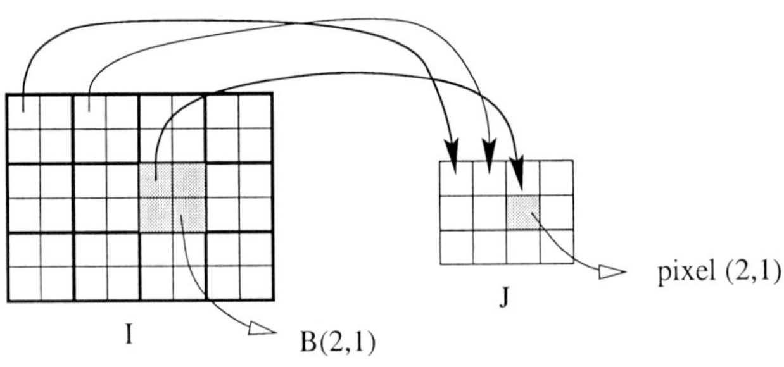

Fig. 1. Point sampling

Let R be a grey level image the same size as I, whose pixel values are defined by:

$$R(x, y) = 1 \quad \text{if } x \text{ and } y \text{ are even,} \tag{1}$$

$$R(x,y) = 0 \quad \text{otherwise.} \tag{2}$$

Then the point sampling operator can be expressed as follows:

$$J(x,y) = \sum_{P \in B(x,y)} R(P)I(P). \tag{3}$$

But why should we choose the first pixel in each block? The only reason is that its coordinates are even. The question then is whether it would not be better to keep the most interesting pixels among the four, or even better to combine the four pixels in some way. This leads us to the definition of a general sampling operator based on a reference image.

Definition 1 *Let Ω be an operator which takes as input one color image I of size $n \times m$ (the original image) and one grey level image R (the* **reference image***) of same size and which gives back an image J of half their size ($\frac{n}{2} \times \frac{m}{2}$):*

$$J = \Omega(I, R). \tag{4}$$

Ω is a reference sampling operator if and only if:

$$J(x,y) = (\Omega(I,R))(x,y) = \begin{cases} \frac{\sum_{P \in B(x,y)} R(P)I(P)}{\sum_{P \in B(x,y)} R(P)} & \text{if } \sum_{P \in B(x,y)} R(P) \neq 0 \\ \frac{1}{4}\sum_{P \in B(x,y)} I(P) & \text{otherwise} \end{cases} \tag{5}$$

This means that in order to compute the value of the sampled image pixel $J(x,y)$ we compute a convolution of the pixels of I belonging to $B(x,y)$ using as weights the respective values of R.

Existing downsampling methods can be described via reference sampling. For example, a classical downsampling method based on convolution can be written as:

$$I_{n+1}(x,y) = \frac{1}{4} \sum_{P \in B(x,y)} I_n(P). \tag{6}$$

Using the above definition, this can be expressed as a reference sampling where the reference image is constant. The fact that this sampling method is not content dependent clearly appears here: the values of the reference image do not depend on the pixel values of the original image.

Precisely, the main interest of reference sampling is that it allows the construction of content dependent downsampling methods. The value in the reference image of a pixel P indicates its importance: the higher its value, the better it will be represented after downsampling.

The next step is the construction of the reference image R. Here is where morphological tools come into play. They will allow us to identify interesting pixels.

2.2. From the morphological gradient to the morphological Laplacian

From now on, for the sake of clarity, we will only consider grey level images. The generalization to color images is easy thanks to the fact that we will only use morphological operators to compute the reference image, and not to process the texture image itself. When applying these operators to a color image, we will in fact apply them to its luminance.

Interesting points in an image (maxima and minima, crest points) belong to high gradient regions. This is why we first tested the morphological gradient as reference image. However the gradient proved to be unsatisfactory for this application for one main reason: not all high gradient points are interesting (for example slanted planes or noise produce high gradients).

Therefore we looked for an operator with a higher discrimination factor, and we found the morphological Laplacian, defined as:

$$L_S(I) = \delta_S(I) + \epsilon_S(I) - 2I, \tag{7}$$

where δ_S and ϵ_S respectively denote the morphological dilation and erosion, and S denotes the used structuring element.

Contrary to the gradient, the Laplacian may have negative values, which are as meaningful as positive ones. In order to build the reference image we take the absolute value of the Laplacian. Hence the sampling operator can be written:

$$J = \Omega(I, |L_S(I)|), \tag{8}$$

where $|.|$ denotes the absolute value operator.

Results are shown in section 4.

2.3. Using the morphological tophat to detect details

Now we are going to detect details in a more explicit way using the tophat transform.

However, we cannot directly use the result of the tophat as a reference image; by doing so we would favor noise. We have to filter the tophat images in order to extract only the meaningful details. To this aim we use a hysteresis threshold followed by an area filter in order to obtain a binary image indicating interesting details. Finally we take the intersection between this binary image and the original tophat image in order to recover the grey levels of the details.

After applying this filtering step to the white tophat and the black tophat of the original image, we obtain two reference images. The first R_w corresponds to the light details in the original image, and the second R_b to the dark details. The final reference image is built by taking the supremum of both images: $R = R_w \bigvee R_b$.

Results are shown in section 4.

2.4. Comments

At this point some comments are necessary. The two downsampling methods that have just been described are unsatisfactory from a strict theoretical point of view because they do not even try to minimize the loss of information. For instance, if we upsampled an image obtained with these methods, the result would be quite different from the original image. But the aim of this study is to produce better looking images, at least for some cases such as images containing text, important details or binary information. From this point of view we believe that the results are interesting.

Before concluding, we explain how to apply these content dependent sampling methods to texture mapping.

3. Application to texture mapping

3.1. Mipmap construction

A classical class of texture mapping methods is based on the pre-calculation of a set of smaller versions of the original texture image I_0. If I_0 is of size $n \times m$, then a pyramid is built with images I_i of size $\frac{n}{2^i} \times \frac{m}{2^i}$. This pyramid is called a MIP map (MIP stands for the Latin phrase *Multum In Parvo* which means many things in a small place). These structures were first applied to texture mapping in [2]. They can be built and used in different ways.

A possible way of building a mipmap is recursively. Let I_i be the i-th level of the mipmap. I_i is thus an image of size $\frac{n}{2^i} \times \frac{m}{2^i}$. With a reference sampling operator Ω we can recursively compute the pyramid using the rule:

$$I_{i+1} = \Omega(I_i, R_i). \tag{9}$$

However if we proceed this way, in some cases image details could survive along several mipmap levels, which could be annoying for some applications. For these applications, we propose here an alternate mipmap construction method which uses content dependent operators but that ensures that small details will be kept only in one mipmap level.

Let I_0 be the original texture image, Ω be the reference sampling operator, and g be the function that computes the reference image from a texture image. We want to build a mipmap starting from I_0: $(I_0, I_1, I_2, \ldots)$. Let K_n be a constant reference image the same size as I_n. Then the recursive procedure to build a mipmap which only preserves small details during one mipmap level is defined by:

$$I_0^{smooth} = I_0, \tag{10}$$

$$I_{n+1}^{smooth} = \Omega(I_n^{smooth}, K_n), \tag{11}$$

$$I_{n+1} = \Omega(I_n^{smooth}, g(I_n^{smooth})). \tag{12}$$

This means that we build a parallel smooth mipmap structure $(I_0^{smooth}, I_1^{smooth}, I_2^{smooth}, \ldots)$ and that we apply a content dependent sampling operator to compute I_{n+1} from I_n^{smooth}.

3.2. Implementation

We have implemented the general reference downsampling method, as well as the two reference image construction methods (Laplacian method, tophat sampling) using XLIM3D, a mathematical morphology library developed at the Center of Mathematical Morphology.

The reference sampling procedure itself is very fast. The computation of the reference image may be slower, depending on the chosen method. For instance the Laplacian method is very fast, requiring only the computation of one dilation and one erosion, but the tophat method can take longer. It is not the computation of the tophat that takes most time, but the filtering step: the reconstruction used in the hysteresis threshold, as well as the area filtering, may require many iterations. However, thanks to the algorithms used in XLIM3D, which are based on hierarchical queues, we can now use these operators in real time. Finally, note that the construction of the mipmap is not as critical as its use during the rendering process.

The resulting mipmaps have been used within MESA, an implementation of OpenGL. The resulting texture mappings preserve the details better than the classical methods.

4. Results

In order to evaluate the new Laplacian and tophat sampling methods, and to compare them with the classical downsampling method, we have computed the mipmaps of two images with the three methods. The first image shows some flat regions, rich textures, and text (figure 2). The second is a binary image (figure 4). Note that, in order to stress the differences between classical and content-dependent downsampling methods, we have not used the alternate mipmap construction method described in the precedent section, which avoids the survival of the same detail through several mipmap levels.

Let us first have a look at figure 3. The three sampling methods perform similarly in smooth regions. Differences appear in non-smooth areas like texture rich regions and near borders. Despite the fact that the representation of these images in this document cannot allow detailed viewing, it can be appreciated that images sampled with the Laplacian method are slightly less blurred than the images obtained with the classical convolution method. As noted previously, this has been achieved without introducing any visible aliasing. The improvement brought by the tophat method is more visible. More details are preserved from one downsampling level to the next. For example text in images downsampled with the tophat method is easier to read.

Differences between these mipmap construction methods appear more clearly when applied to a binary image. For example in figure 5 we can appreciate the improvement brought by the new methods: detail and contrast have been better preserved with these methods than with the classical downsampling method based on convolution.

5. Conclusion

A general sampling method, called reference sampling, has been defined. It generalizes some classical downsampling strategies and moreover it allows the

Fig. 2. Original image (256 × 256).

construction of content dependent downsampling operators.

Using the morphological Laplacian and the tophat operator two content dependent downsampling operators, which aim at keeping meaningful image details, have been proposed.

Finally, it has been explained how to build and use mipmaps based on these content dependent sampling operators. Their implementation under MESA gives interesting results.

A certain number of questions remain. What sampling method should we use? What values should we give to the parameters? The sampling method can be chosen according to the original texture image. If it is very smooth, then the Laplacian method should be used. Otherwise, the tophat method is more interesting. The parameters used by the tophat algorithm can also be adapted to the image: the higher their values, the less the details that will be kept.

These choices can be made on the spot, after analyzing the image, or be somehow included with the image. For example the composer of the 3D scene could choose the sampling technique for the textures he uses.

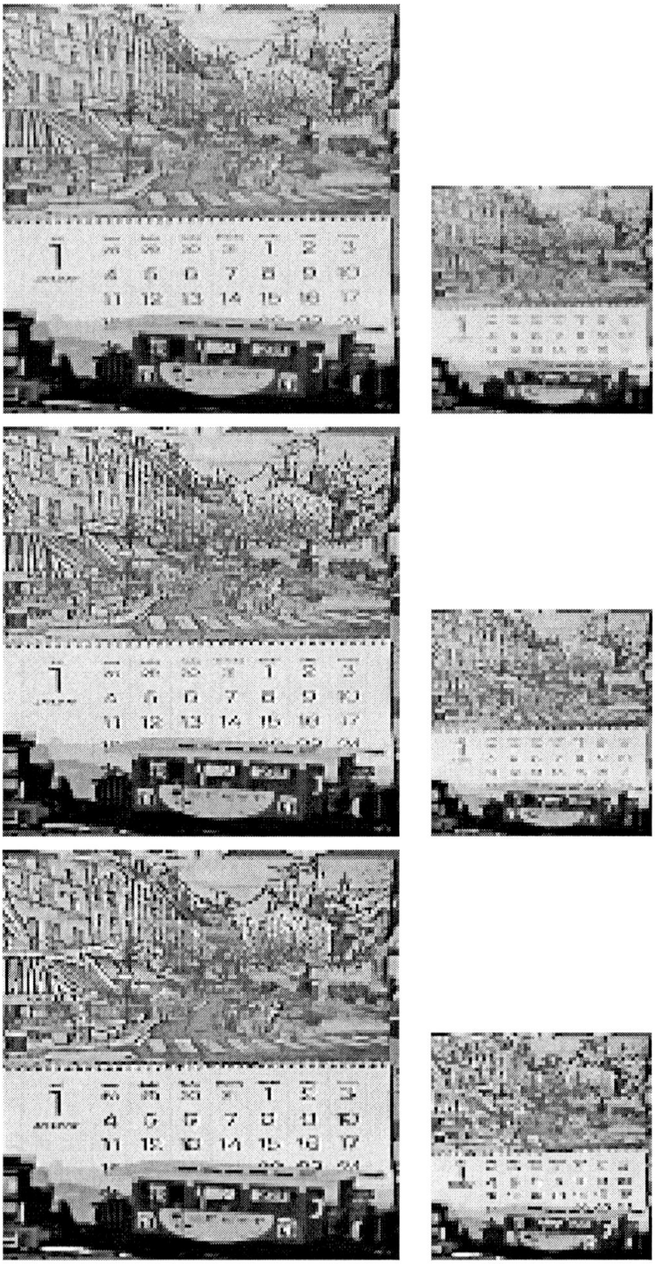

Fig. 3. Comparison of sampling results; the first column corresponds to the first downsampling step (image sizes are 128×128), and the second column to the second downsampling (image sizes are 64×64). First row: constant reference image. Second row: Laplacian sampling. Third row: Tophat sampling.

Fig. 4. Original binary image (256 × 256).

References

1. A. Albiol and J. Serra. Morphological image enlargements. *Journal of Visual Communication and Image Representation*, 8(4):367–383, December 1997.
2. E. Catmull. *A subdivison algorithm for computer display of curved surfaces*. PhD thesis, University of Utah, December 1974.
3. E. Decencière Ferrandière, B. Marcotegui, and F. Meyer. Application de la morphologie mathématique au mipmapping. In *CORESA'98*, June 1998.
4. E. Decencière Ferrandière, B. Marcotegui, and F. Meyer. Content dependent image sampling using mathematical morphology: application to texture mapping. *Accepted for publication by Image Communication*, 2000.
5. D.A.F. Florêncio and R.W. Schafer. Critical morphological sampling and its applications to image coding. In *Mathematical Morphology and its Applications to Image Processing (Proceedings ISMM'94)*, pages 109–116, September 1994.
6. R.M. Haralick, X. Zhuang, C. Lin, and J.S.J. Lee. The digital morphological sampling theorem. *IEEE Transactions on Acoustics, Speech and Signal Processing*, 37:2067–2090, December 1989.
7. P.S. Heckbert. Survey of texture mapping. *IEEE Computer Graphics and Applications*, 6(11):56–67, November 1986.
8. H.J.A.M. Heijmans and A. Toet. Morphological sampling. *Computer Vision, Graphics, and Image Processing: Image Understanding*, 54:384–400, November 1991.
9. J. Serra. A sampling approach based on equicontinuity. In J. Serra and P. Soille, editors, *Mathematical Morphology and its Applications to Image Processing (Proceedings ISMM'94)*, pages 117–124, September 1994.

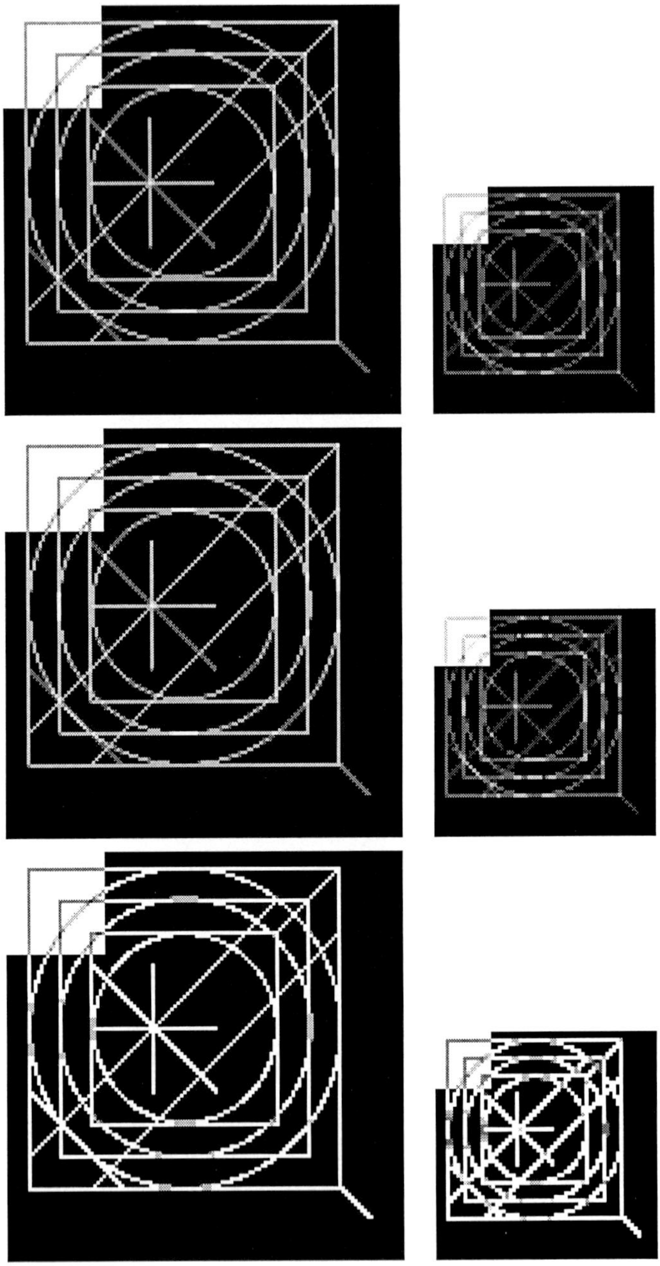

Fig. 5. Comparison of mipmaps computed from a binary image; the first column corresponds to the first downsampling step (image sizes are 128 × 128), and the second column to the second downsampling (image sizes are 64 × 64). First row: constant reference image. Second row: Laplacian sampling. Third row: Tophat sampling.

MORPHOLOGICAL PYRAMIDS AND WAVELETS BASED ON THE QUINCUNX LATTICE*

HENK J.A.M. HEIJMANS
CWI, P.O. Box 94079, 1090 GB Amsterdam, The Netherlands

and

JOHN GOUTSIAS
Center for Imaging Science, Dept. of Electrical and Comp. Eng.
The Johns Hopkins University, Baltimore, MD 21218, USA

Abstract. This paper is concerned with two types of multiresolution image decompositions, pyramids and wavelets. We present an axiomatic approach for both cases, encompassing linear as well as nonlinear decompositions. A wavelet decomposition is more specific in the sense that it always involves a pyramid transform. Both families will be illustrated by means of concrete examples using the quincunx scheme in two dimensions. One nonlinear wavelet transform will be discussed in more detail: it uses the lifting scheme and has the intriguing property that it preserves local maxima over a range of scales.

Key words: Multiresolution Decomposition, Pyramid, Wavelet, Lifting Scheme, Quincunx Grid.

1. Introduction

It is widely accepted that multiresolution approaches are extremely useful in various image processing applications. This is due to the fact that most images contain physically relevant features at different scales. For their proper understanding, multiresolution (or multiscale) techniques are indispensable. Another good reasons to take recourse to multiresolution approaches is that the corresponding algorithms offer various computational advantages.

In this paper we present a brief overview of the axiomatic framework for the pyramid and the wavelet transform, which has been discussed in great detail in [5, 8], and we present worked-out examples for both cases based on the two-dimensional quincunx sampling scheme. An important feature of our framework is that it allows linear as well as nonlinear transforms. This is important to us, since our interest primarily goes to decompositions which are based on morphological operators.

In Section 2 we introduce the pyramid transform and in Section 3 we discuss a particular example based on a morphological adjunction. The general wavelet transform is introduced in Section 4. There we also explain how the lifting scheme can be used to design (linear and nonlinear) wavelet transforms. An

* This work was supported in part by NATO Collaborative Research Grant CRG.971503. John Goutsias was also supported by the Office of Naval Research (U.S.A.), Mathematical, Computer, and Information Sciences Division, under ONR Grant N00014-90-1345.

interesting example, called *max-lifting*, based on morphological operators is discussed in Section 5. As we will show, the major characteristic of this scheme is that it preserves local maxima.

2. The Pyramid Transform

In this section we will formalize the concept of a pyramid, as first introduced by Burt and Adelson [1], using general analysis and synthesis operators. For a comprehensive discussion we refer to [5]. Thus, consider a family of image spaces V_j, $j \geq 0$. Assume that we can go from level j to the next level $j+1$ by means of an *analysis operator* $\psi_j^\uparrow : V_j \to V_{j+1}$. To go back from level $j+1$ to the lower level j we need to dispose of a *synthesis operator* $\psi_j^\downarrow : V_{j+1} \to V_j$. In this scheme, every analysis operator $\psi_j^\uparrow$ is designed to reduce the information contained in images at level j, and as such, they are not invertible in general: the composition $\psi_j^\downarrow \psi_j^\uparrow(x)$ does only yield an *approximation* of the input image $x \in V_j$. On the other hand, we demand that the synthesis operator does not cause a further reduction of the information content of an image. To achieve this, we will make the following assumption to which we refer as the *pyramid condition*: for every $j \geq 0$, the operators $\psi_j^\uparrow$, $\psi_j^\downarrow$ satisfy

$$\psi_j^\uparrow \psi_j^\downarrow(x) = x \text{ for } x \in V_{j+1}. \tag{1}$$

This condition yields that the composition $\psi_j^\downarrow \psi_j^\uparrow$ is idempotent. The pyramid scheme introduced above can be used to obtain an alternative representation of an image x, provided that we have an addition $\dot{+}$ and a subtraction $\dot{-}$ on V_j such that $x_1 \dot{+} (x_2 \dot{-} x_1) = x_2$ for $x_1, x_2 \in V_j$. Given an input signal $x_0 \in V_0$, we consider the following recursive signal analysis scheme:

$$x_0 \to \{y_0, x_1\} \to \{y_0, y_1, x_2\} \to \cdots,$$

where

$$x_{j+1} = \psi_j^\uparrow(x_j), y_j = x_j \dot{-} \psi_j^\downarrow(x_{j+1}) .$$

We refer to this scheme as the *pyramid transform*. The original signal $x_0 \in V_0$ can be *exactly* reconstructed from x_{j+1} and $y_0, y_1, \ldots, y_j$ by means of the backward recursion

$$x_j = \psi_j^\downarrow(x_{j+1}) \dot{+} y_j, \; j \geq 0,$$

the *inverse pyramid transform*. For grey-level images, one can choose the standard addition and subtraction for $\dot{+}$ and $\dot{-}$, respectively. In the case of finitely many grey-levels, say n, one can also use the cyclic addition (i.e. addition modulus n). In the binary case, this corresponds with the 'exclusive or'.

3. Morphological Pyramids

Here we investigate pyramid decompositions based on two basic morphological operators, erosion and dilation. Recall that for two complete lattices $\mathcal{L}$ and

$\mathcal{M}$, an operator pair (ε, δ), with $\varepsilon : \mathcal{L} \to \mathcal{M}$ and $\delta : \mathcal{M} \to \mathcal{L}$, is called an *adjunction* if $\delta(y) \leq x$ if and only if $y \leq \varepsilon(x)$ for $x \in \mathcal{L}$ and $y \in \mathcal{M}$. In that case ε is an erosion and δ a dilation; see [7]. We consider pyramids that satisfy the following constraints:
(*i*) the image domains V_j are complete lattices;
(*ii*) the analysis/synthesis pair $(\psi_j^{\uparrow}, \psi_j^{\downarrow})$ forms an adjunction.
Assume that there exist two sets S, Q and a binary relation on $S \times Q$ denoted by $s \to q$. Given a complete lattice $\mathcal{T}$ (later, $\mathcal{T}$ will have the interpretation of grey-level set), we can define an adjunction $(\psi^{\uparrow}, \psi^{\downarrow})$ between $\mathcal{T}^S$ and $\mathcal{T}^Q$ in the following way:

$$\psi^{\uparrow}(x)(q) = \bigwedge_{s:s\to q} x(s) \quad \text{and} \quad \psi^{\downarrow}(x')(s) = \bigvee_{q:s\to q} x'(q) \tag{2}$$

for $x \in \mathcal{T}^S$ and $x' \in \mathcal{T}^Q$. Here $\bigwedge_{s:s\to q} x(s)$ stands for $\min\{x(s) \mid s \in S \text{ and } s \to q\}$ (or 'inf' if the set is infinite). Indeed, the pair $(\psi^{\uparrow}, \psi^{\downarrow})$ constitutes an adjunction. In general it will not satisfy the pyramid condition, however. Thereto we need an additional assumption [5].

Proposition 1 *The pair $(\psi^{\uparrow}, \psi^{\downarrow})$ given by (2) satisfies the pyramid condition (i.e. $\psi^{\uparrow}\psi^{\downarrow} = \mathsf{id}$ on $\mathcal{T}^Q$) if and only if for all $q \in Q$ there exists an $s \in S$ such that (i) $s \to q$ and (ii) $s \to q'$ where $q' \in Q$, implies $q' = q$.*

In the remainder of this section we are exclusively concerned with morphological adjunction pyramids which correspond with a 2D quincunx sampling approach. Let S denote the square lattice comprising the integer points, i.e., $S = \{(s_1, s_2) \mid s_1, s_2 \in \mathbb{Z}\}$. Furthermore, let Q be the subset of S resulting from a quincunx sampling scheme, i.e., $Q = \{(q_1, q_2) \mid q_1, q_2 \in \mathbb{Z} \text{ and } q_1 + q_2 \text{ even }\}$. Finally, let $S' \subset Q$ be the set resulting after a second quincunx sampling step, i.e., $S' = \{(s'_1, s'_2) \mid s'_1, s'_2 \in 2\mathbb{Z}\}$. We define the following two norms on S:

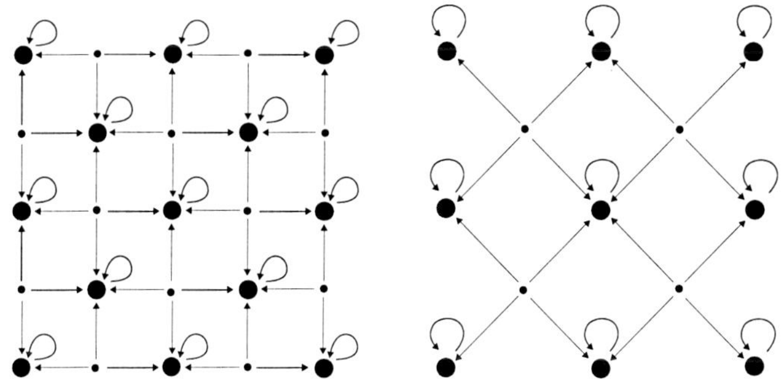

Fig. 1. Arrows express the relations $s \to_0 q$ between $s \in S$ and $q \in Q$ (left), and $q \to_1 s'$ between $q \in Q$ and $s' \in S'$ (right). The larger disks comprise the quincunx grid.

$$\|s\|_1 = |s_1| + |s_2| \text{ and } \|s\|_\infty = \max\{|s_1|, |s_2|\},$$

where $s = (s_1, s_2) \in S$. Consider the binary relations

$$s \to_0 q \text{ iff } \|s - q\|_1 \leq 1 \quad \text{and} \quad q \to_1 s' \text{ iff } \|q - s'\|_\infty \leq 1$$

on $S \times Q$ and $Q \times S'$, respectively. These relations are illustrated in Fig. 1. Observe that both relations satisfy the conditions $(i) - (ii)$ of Proposition 1. Putting $V_0 = \mathcal{T}^S$ and $V_1 = \mathcal{T}^Q$, the analysis and synthesis operator given by (2), i.e.,

$$\psi_0^\uparrow(x)(q) = \bigwedge_{s:s\to_0 q} x(s) \text{ and } \psi_0^\downarrow(x')(s) = \bigvee_{q:s\to_0 q} x'(q),$$

form an adjunction between V_0 and V_1 and satisfy the pyramid condition. Similarly, putting $V_2 = \mathcal{T}^{S'}$ and defining

$$\psi_1^\uparrow(x)(s') = \bigwedge_{q:q\to_1 s'} x(q) \text{ and } \psi_1^\downarrow(x')(q) = \bigvee_{s':q\to_1 s'} x'(s'),$$

we obtain an analysis/synthesis pair $(\psi_1^\uparrow, \psi_1^\downarrow)$ which is an adjunction between V_1 and V_2 and satisfies the pyramid condition. Now, since $S' = 2S$, we can repeat the same procedure by putting $Q' = 2Q$ and $S'' = 2S'$.

In Fig. 2 we compute 2 levels of the corresponding pyramid transform. The odd levels in the pyramid are shown after a 45° clockwise rotation.

We make two important observations regarding this example: (1) since every analysis/synthesis pair constitutes a morphological adjunction, the approximation operator $\psi_j^\downarrow\psi_j^\uparrow$ is a morphological opening [7]. This means in particular that the approximation image $\hat{x}_j = \psi_j^\downarrow\psi_j^\uparrow(x_j)$ is never larger than the original image x_j, and therefore the detail image $y_j = x_j - \hat{x}_j$ is nonnegative. (2) In the expressions for $\psi_j^\uparrow$ and $\psi_j^\downarrow$ we only need to consider those points that lie in the domain of the image, i.e., a square window. This does not affect the validity of the pyramid condition, but it does destroy translation invariance.

4. Wavelets and Lifting

A serious drawback of the pyramid transform is that its output signal comprises more data than the input signal. The wavelet transform (see [9] for a comprehensive account in the linear case), to be discussed below, does not have this drawback. In this case the detail signal is generated by a second analysis operator $\omega_j^\uparrow$, mapping V_j into another space W_{j+1}, in such a way that $x' = \psi_j^\uparrow(x)$ and $y' = \omega_j^\uparrow(x)$ jointly contain the same amount of data as x. Furthermore, there exists a synthesis operator $\Psi_j^\downarrow$ such that the perfect reconstruction condition

$$\Psi_j^\downarrow(\psi_j^\uparrow(x), \omega_j^\uparrow(x)) = x, \quad x \in V_j \tag{3}$$

holds, as well as

$$\psi_j^\uparrow(\Psi_j^\downarrow(x, y)) = x, \quad \text{and} \quad \omega_j^\uparrow(\Psi_j^\downarrow(x, y)) = y, \text{ for } x \in V_{j+1},\ y \in W_{j+1}.$$

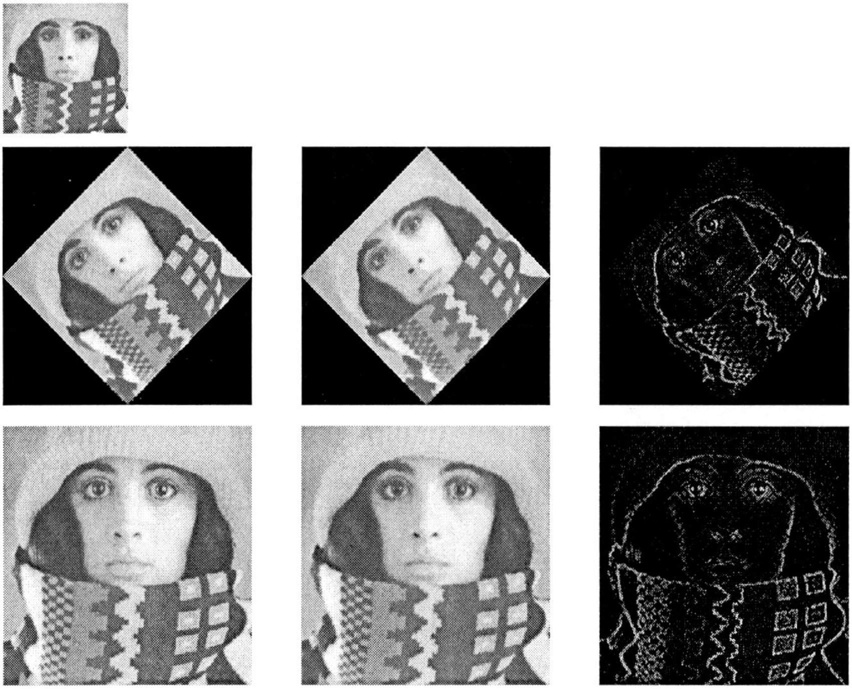

Fig. 2. Morphological adjunction pyramid for the quincunx scheme. From left to right: image x_j (with $j = 0$ at the bottom level), approximation $\hat{x}_j = \psi_j^{\downarrow}\psi_j^{\uparrow}(x_j)$, and detail $y_j = x_j - \hat{x}_j$.

Refer to Fig. 3 for an illustration. Often, e.g. in the linear case, $\Psi_j^{\downarrow}$ is of the form

$$\Psi_j^{\downarrow}(x', y') = \psi_j^{\downarrow}(x') \dot{+} \omega_j^{\downarrow}(y'), \tag{4}$$

and in this case we say that the wavelet transform is *uncoupled* [8].

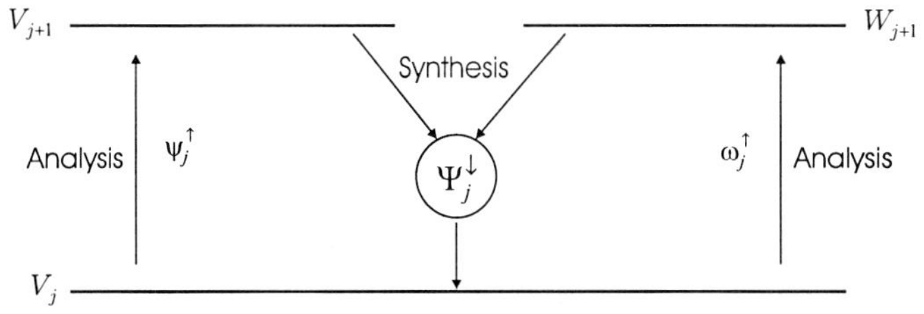

Fig. 3. The general wavelet transform.

The *lifting scheme* introduced by Sweldens [10], provides a simple, flexible, and efficient tool for the construction of linear as well as nonlinear wavelet transforms. A general lifting scheme starts with an invertible transformation Σ of the input data x_0 into two or more channels (or bands). We restrict ourselves to the two-channel case for simplicity. Thus application of Σ to x_0 yields two output signals, a coarse signal x_1 and a detail signal y_1. Application of Σ^{-1} to x_1, y_1 returns the input signal: $x_0 = \Sigma^{-1}(x_1, y_1)$. In practice, Σ will often be a known wavelet transform, for example the *lazy wavelet* which splits the input data into even and odd samples. By a concatenation of so-called *prediction* and *update* lifting steps (see Fig. 4) one arrives at a *lifted wavelet.* In Fig. 4 the prediction operators are denoted by π_i and the update operators by λ_i. If the lifting scheme consists of one prediction step π followed by one

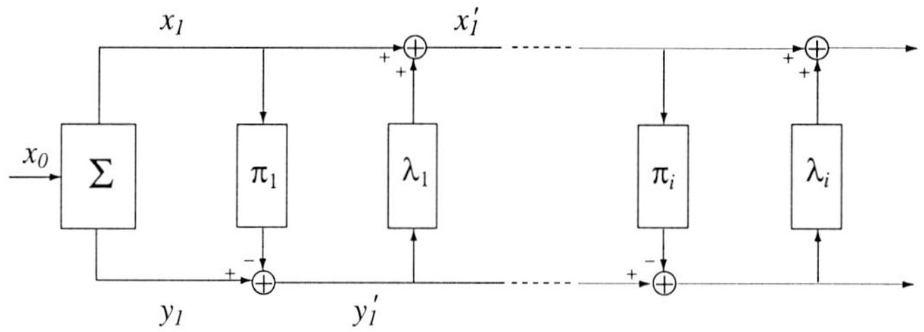

Fig. 4. Lifting scheme.

update step λ, and x_1', y_1' are computed by the following scheme,

$$(x_1, y_1) = \Sigma(x_0), \quad y_1' = y_1 - \pi(x_1), \quad x_1' = x_1 + \lambda(y_1'),$$

then we arrive at a lifted transform $\Sigma'(x_0) = (x_1', y_1')$. This can be inverted by

$$x_1 = x_1' - \lambda(y_1'), \quad y_1 = y_1' + \pi(x_1), \quad x_0 = \Sigma^{-1}(x_1, y_1).$$

Later we use this scheme to obtain a two-dimensional wavelet transform for the case that Σ splits an input image according to the quincunx sampling lattice.

In [8] we have shown that a linear transformation Σ followed by one nonlinear prediction or update step yields an uncoupled wavelet transform. However, two nonlinear lifting steps result in a coupled wavelet transform, in general.

5. Max-lifting for the Quincunx Lattice

In this section we present a particular example of a nonlinear wavelet scheme associated with the two-dimensional quincunx sampling lattice. Consider the partition of the square lattice S into two disjoint subsets, Q and $R = S \setminus Q$. We define an adjacency relation expressing when s is a neighbour of s', i.e., $s \sim s'$ if $\|s - s'\|_1 = 1$. Thus $\sim$ is a symmetric relation on $S \times S$, and $s \sim s'$ can only hold if either s or s' (but not both) is an element of Q. Observe that $r \to_0 q$ if and only if $r \sim q$ for $r \in R$ and $q \in Q$. Now let the operator Σ govern the splitting of a signal x_0 on S into two subsampled signals, x_1 defined on Q

and y_1 defined on R, i.e., $x_1(q) = x_0(q)$ for $q \in Q$ and $y_1(r) = x_0(r)$ for $r \in R$. Consider the coupled wavelet transform obtained by applying first a prediction lifting

$$\pi(x)(r) = \bigvee_{q:q\sim r} x(q) \tag{5}$$

and then an update lifting

$$\lambda(y)(q) = \max\{0, \bigvee_{r:r\sim q} y(r)\} \tag{6}$$

This means that the prediction of the signal at a point r is given by the maximum of its 4 neighbours. The update operator is chosen so that local maxima of the input signal x_0 are mapped to the next level x_1'. The next result, a

Fig. 5. Wavelet decomposition of an image based on the max-lifting scheme. Bottom row: original image x_0. Middle row: wavelet transformed images x_1' and y_1' (after rotation). Top row: images x_1'' and y_1'' resulting from wavelet transform of x_1'. Note that the detail image may contain positive (bright) and negative (dark) grey values.

proof of which can be found in [8], gives a formal statement. But first we need two more definitions. We write $s \sim\sim s'$ if $s = s'$ or $\|s - s'\|_1 = 2$. Given a signal x on S and a point $s \in S$ we denote by $A(x \mid s)$ the set of neighbours s' of s such that $x(s') \geq x(s'')$ for all neighbours s'' of s. Observe that $A(x \mid s)$ contains at least one point.

Proposition 2 *Let x_0 be an input signal on S, let (x_1, y_1) be its splitting into subsignals on Q and R, respectively, and*

$$y_1'(r) = y_1(r) - \pi(x_1)(r), \qquad x_1'(q) = x_1(q) + \lambda(y_1')(q),$$

where π and λ are given by (5)-(6). Then the following holds:
(a) $x_0(q) \leq x_1'(q) \leq \max\{x_0(s) \mid s = q \text{ or } s \sim q\}$, *for $q \in Q$.*
(b) $x_0(r) \leq \max\{x_1'(q) \mid q \sim r\}$, *for $r \in R$.*
(c) *Assume that $r \in R$ is such that $x_0(r) \geq x_0(s)$ for $s \sim r$ and $s \sim\sim r$, then $x_1'(q) = x_0(r)$ for every $q \in A(x_0 \mid r)$.*

Thus the max-lifting scheme preserves local maxima. But we can also show that this scheme will never create new maxima.

Proposition 3 *Suppose that the coarse signal x_1' has a local maximum at $q \in Q$ in the following sense: $x_1'(q') \leq x_1'(q)$ for $q' \in Q$ with $\|q - q'\|_1 \leq 2$ (with every point q there correspond nine such points, including q itself). Then x_0 has a local maximum at some $s \in S$ with $s = q$ or $s \sim q$ and $x_0(s) = x_1'(q)$.*

In Fig. 5 we apply the max-lifting scheme to a particular image.

6. Summary and Conclusions

The pyramid condition given in (1), though seemingly straightforward, imposes relatively strong conditions on the analysis and synthesis operators. In the linear case, this condition is necessary and sufficient in order that the pyramid transform can be extended to a wavelet transform. For the nonlinear case, this problem is open.

Construction of nonlinear wavelets with prescribed properties (in the spirit of vanishing moments conditions in the linear case) is a research area which is almost entirely unexplored; some early work in this direction can be found in [2, 3, 4, 6]. The main tool, and to the best of our knowledge the only tool so far, to address this problem is the lifting scheme. A great deal of future research on nonlinear wavelets will have to face the question how to build schemes that yield decompositions which are useful in applications such as compression, denoising, image fusion, or image retrieval. We believe that the max-lifting scheme will turn out useful for some of these applications. We are currently exploring this issue.

References

1. P. J. Burt and E. H. Adelson, The Laplacian pyramid as a compact image code. *IEEE Transactions on Communications 31* (1983), 532–540.
2. R. Claypoole, R. G. Baraniuk, and R. D. Nowak, Lifting construction of non-linear wavelet transforms. In *IEEE International Symposium on Time-Frequency and Time-Scale Analysis* (Pittsburgh, 1998), pp. 49–52.

3. R. L. de Queiroz, D. A. Florencio and R. W. Schafer, Nonexpansive pyramid for image coding using a nonlinear filterbank. *IEEE Transactions on Image Processing 7*, 2 (1998), 246–252.
4. O. Egger and W. Li, Very low bit rate image coding using morphological operators and adaptive decompositions. In *Proceedings of the IEEE International Conference on Image Processing, volume II* (Austin, Texas, 1994), IEEE, pp. 326–330.
5. J. Goutsias and H. J. A. M. Heijmans, Multiresolution signal decomposition schemes. Part 1: Morphological pyramids. To appear in *IEEE Transactions on Image Processing.*
6. F. J. Hampson and J.-C. Pesquet, M-band nonlinear subband decompositions with perfect reconstruction. *IEEE Transactions on Image Processing 7*, 11 (1998), 1547–1560.
7. H. J. A. M. Heijmans, *Morphological Image Operators.* Academic Press, Boston, 1994.
8. H. J. A. M. Heijmans and J. Goutsias, Multiresolution signal decomposition schemes. Part 2: Morphological wavelets. Research Report PNA-R9905, CWI, 1999.
9. S. Mallat, *A wavelet Tour of Signal Processing.* Academic Press, San Diego, 1998.
10. W. Sweldens, The lifting scheme: A custom-design construction of biorthogonal wavelets. *Appl. Comput. Harmon. Anal. 3*, 2 (1996), 186–200.

MORPHOLOGICAL SCALE-SPACE OPERATORS: AN ALGEBRAIC FRAMEWORK

REIN VAN DEN BOOMGAARD
University of Amsterdam, Kruislaan 403, 1098 SJ Amsterdam, The Netherlands

and

HENK J.A.M. HEIJMANS
CWI, P.O. Box 94079, 1090 GB Amsterdam, The Netherlands

Abstract. This paper proposes a general algebraic definition for image scale-spaces. The basic idea is to first downscale the image by a factor t using an invertible scaling, then apply an image operator at a unit scale, and finally resize the image to its original scale. It is then required that the resulting one-parameter family of image operators satisfies some semigroup property. In this paper only the morphological erosions are considered. In this case, classical tools from convex analysis play an important role.

Key words: Convex Functions, Erosion, Infimal Convolution, Parabolic Structuring Function, Scale-Space, Scaling, Semigroup, Slope Transform, Subpolynomial Functions, Young Fenchel Conjugate.

1. Introduction

Scale-space is an accepted and often used formalism in image processing and computer vision. This formalism makes the choice at what scale visual observations are to be made explicit. In the past the choice for an observation scale was often hidden somewhere in the definition of the operators.

The notion of linear scale-space has a very long history in image processing [1, 2, 3]. Weickert et.al. [4] only recently 'discovered' that the concept of linear Gaussian scale-space dates back to the sixties.

In mathematical morphology the notion of scale (or size) dependent observations was pioneered by Matheron [5] in his study of *granulometries*. Jackway [6] and van den Boomgaard [7] showed that a morphological analogue of the Gaussian linear scale-space does exist: the parabolic erosions and dilations. The morphological parabolic scale-space is the solution of a partial differential equation [8, 9], just like the Gaussian linear scale-space is the solution of the diffusion equation [2].

In this paper we present an algebraic framework for the construction of morphological scale-space operators. Let f be the image at scale zero (i.e. the mathematical concept of an image at infinite resolution) and let $T(s)$ be the operator such that $T(s)f$ is the observation at scale s. The family of operators $\{T(s)\}_{s>0}$ is collectively called a *scale-space*.

We require the scale-space operator to be a macroscopic operator, in the sense that $T(t)$ only needs the zero scale image without the need to calculate all intermediate images $T(s)f$ for $s < t$. We thus take a quite different approach to scale-space compared to Alvarez et. al. [10] who take the evolution of the zero scale image modeled with a partial differential equation as a starting point.

In the proposed framework the notion of scale is explicitly defined. We first downscale the image by a factor t, then apply an image operator at unit scale and finally resize the image to its original scale.

In Section 2 we present our new algebraic definition of a scale-space. In the remainder of the paper we are mainly concerned with the morphological scale-space induced by the erosion (infimal convolution). In that context, tools from convex analysis, such as the Young-Fenchel conjugate (slope transform) play an important role. Such tools are briefly discussed in Section 3. In Section 4 we present a systematic treatment of erosion scale-spaces. Some final conclusions are presented in Section 5.

2. Scalings, Semigroups and Scale-Space

In our definition of scale-space, the scale at which the operators interact with the image is explicitly captured in a scaling operator. The actual image processing operator works at a fixed 'unit' scale. In this paper we consider the scale to be represented with a positive real scalar, i.e. an element of the set $\mathcal{T} = (0, \infty)$.

Definition 1 (Scaling) *A family $S = \{S(t) \mid t > 0\}$ of operators on $\mathcal{L}$ is called a scaling if (i) $S(1) = id$ and (ii) $S(t)S(s) = S(ts)$ for $s, t > 0$.*

A scaling thus forms a commutative group with inverse $S(t)^{-1} = S(1/t)$ and unity element $S(1)$. In a scale-space construction it is the ordering of scales that is much more important than the actual value. The scale is the *free* parameter in the system and we are thus free to reparameterize the scale at will as long as the reparameterization preserves the ordering of the scale values.

Definition 2 (Anamorphic scalings) *Two scalings S and S' are said to be* anamorphic *if there exists an increasing bijection γ on $\mathcal{T}$ such that $S(\gamma(t)) = S'(t)$ for all $t \in \mathcal{T}$.*

If we assume that $S(t) = \text{id}$ if and only if $t = 1$, then it is easy to show that such a bijection γ must satisfy the conditions $\gamma(1) = 1$ and $\gamma(st) = \gamma(s)\gamma(t)$ for $s, t \in \mathcal{T}$. The most important example is given by $\gamma(t) = t^p$, where $p > 0$.

In this paper we only consider images that are mappings from the continuous plane R^d to the real values $\bar{R}$ (R extended with the values $+\infty$ and $-\infty$). This complete lattice of images is denoted as $\mathcal{L} = \text{Fun}(R^d, \bar{R})$. The scalings we will look at are a combination of a spatially isotropic scaling and a grey value scaling:

Definition 3 (Image scalings) *The image scaling $S^\alpha(t) : \mathcal{L} \to \mathcal{L}$ is defined as $S^\alpha(t)f = t^{1-\alpha} f(\frac{\cdot}{t^\alpha})$.*

For $\alpha = 1$ we obtain a pure spatial scaling whereas for $\alpha = 0$ we obtain a pure grey value scaling. For $\alpha = \frac{1}{2}$ there is a perfect balance between spatial scaling and grey value scaling.

The atlas principle introduced by Koenderink [2] states that it should be possible to build a scale-space incrementally. If f denotes the 'zero scale' image, the image observed at scale s is given as $T(s)f$. Observing this image at scale t results in $T(t)T(s)f$ and according to the atlas principle we should have that $T(s)T(t) = T(t)T(s) = T(r)$ for some $r > 0$.

The atlas principle can be associated with a commutative semigroup describing the way in which the scales combine in a composition of scale-space operators. Thus $T(s)T(t) = T(t)T(s) = T(s\dot{+}t)$ where $\dot{+}$ is a binary operation on $\mathcal{T}$ that is associative and commutative, i.e. $(\mathcal{T}, \dot{+})$ is a commutative semigroup. In this paper we only consider a special class of scale semigroups $(\mathcal{T}, \dot{+})$.

Definition 4 ($+_\nu$-scale semigroup) *The binary operator $+_\nu : \mathcal{T} \to \mathcal{T}$ defined by $s +_\nu t = (s^\nu + t^\nu)^{\frac{1}{\nu}}$ forms a commutative semigroup $(\mathcal{T}, +_\nu)$ for $0 < \nu \leq \infty$.*

For $\nu = 1$ we obtain the standard *additive semigroup* $(\mathcal{T}, +)$, whereas $\nu = \infty$ leads to the *supremal semigroup* $(\mathcal{T}, \vee)$. Within the scale-space context this choice for a scale semigroup has particular advantages. It is easy to see that $s \leq t \Rightarrow s +_\nu r \leq t +_\nu r$ for any fixed r, meaning that the ordering of scale values is preserved when we perform a subsequent observation at scale r. Semigroups $(\mathcal{T}, \dot{+})$ that have this property are called *linearly ordered.* For finite ν (i.e. $\nu < \infty$), any observation at scale t following an observation at scale s will increase the scale, i.e. $s, t < s +_\nu t$. For the supremum semigroup we only have $s, t \leq s +_\infty t$ showing that the scale does not decrease by making an observation at any finite scale.

Now we are ready to give a formal definition of a scale-space. In our approach the starting assumption is that the scale-space is a semi-group of operators on the image space $\mathcal{L}$ under composition, compatible with a given linearly ordered semi-group $\dot{+}$ on the range of scale values $\mathcal{T}$ and invariant with respect to a given scaling S.

Definition 5 (Scale-space) *Let $(\mathcal{T}, \dot{+}, \leq)$ be a linearly ordered semigroup and let S be a scaling on the image space $\mathcal{L}$. The family $\{T(t)\}_{t>0}$ of operators on $\mathcal{L}$ is called an $(S, \dot{+})$ scale-space if:*

$$\begin{aligned} T(t)T(s) &= T(t\dot{+}s), \quad s,t > 0, \\ T(t)S(t) &= S(t)T(1), \quad t > 0. \end{aligned}$$

An alternative definition of a scale-space operator follows easily from the above definition:

Definition 6 (Scale-space construction) *Let $(\mathcal{T}, \dot{+}, \leq)$ be a linearly ordered semigroup, let ψ be an image operator on $\mathcal{L}$, and let S be a scaling on $\mathcal{L}$. The family $\{T(t)\}_{t>0}$ defined with $T(t) = S(t)\psi S(t)^{-1}$, defines a $(S, \dot{+})$ scale-space if $T(t)T(s) = T(t\dot{+}s)$, $\quad s, t > 0$.*

3. Convex Analysis and Mathematical Morphology

The morphological operators that we will be looking at, are the classical structural erosions and dilations. The set $\mathcal{L} = \text{Fun}(R^2, \bar{R})$ is a complete lattice under the pointwise ordering. Every translation invariant adjunction on $\mathcal{L}$ is of the form $(\varepsilon_b, \delta_b)$ with:

$$\varepsilon_b(f)(x) = \bigwedge_{h \in R^2} [f(x-h) + b(h)] \tag{1}$$

$$\delta_b(f)(x) = \bigvee_{h \in R^2} [f(x+h) - b(h)]\,. \tag{2}$$

In these expressions, the function b is called the structuring function. We will assume throughout this paper that a structuring function b is lower semi continuous and convex with $b > -\infty$. The expression for the erosion ε_b is a well-known operation in convex function analysis, where it is known under the name *infimal convolution* and denoted by $f \boxminus b$. The dilation δ_b will be denoted as $f \boxplus b$. Our choice for the infimum convolution as the erosion leads to some properties that may look a bit awkward to the seasoned morphologist. For instance it is the erosion that is associative and commutative in this approach, i.e. $(f \boxminus b) \boxminus b' = f \boxminus (b \boxminus b')$ and $f \boxminus b = b \boxminus f$ whereas in the classical approach based on the Minkowski operations it is the dilation that has these properties[1].

This section presents a brief overview of some basic results in convex analysis that play a role in the sequel; refer to [11] for a comprehensive account.

Definition 7 (Convex sets and convex functions) *A set $X \subset R^d$ is convex if $tx + (1-t)y \in X$ when $x, y \in X$ and $0 \leq t \leq 1$. A function $f : R^d \to \bar{R}$ is convex if $f(tx + (1-t)y) \leq tf(x) + (1-t)f(y)$, for $x, y \in R^d$ and $0 \leq t \leq 1$.*

It can be easily shown that the infimal convolution of two convex functions yields a convex function (see [11]). Convexity of sets and of functions are compatible notions; let I_X be the indicator function of a convex set X given by $I_X(x) = 0$ for $x \in X$ and $I_X(x) = \infty$ for $x \notin X$, then I_X is a convex function.

Definition 8 (Subpolynomial functions) *A convex function f is called subpolynomial of degree k where $k \geq 1$ if $f(tx) = t^k f(x), x \in R^d, t > 0$. The family of convex subpolynomial functions of degree k is denoted as $SP(k)$.*

Special cases worth mentioning are the subpolynomial functions of degree 1 that are called sublinear and the subpolynomial functions of degree ∞ being indicator functions of convex sets. An example of a sublinear function is the support function H_B of a set B defined for $x \in R^d$ as $H_B(x) = \sup_{y \in B} \langle x, y \rangle$. It is not difficult to show that $O \in B$ is equivalent with $H_B(x) \geq 0$ for all $x \in R^d$. Not only is the support function sublinear, but it can also be shown that any sublinear function is the support function of some closed convex set B. For a disk $B = \{x \mid \|x\| \leq 1\}$ the support function is the corresponding norm $H_B(x) = \|x\|$.

[1] The classical structural erosion $f \ominus a$ is defined as $(f \ominus a)(x) = \bigwedge_{h \in R^2} [f(x+h) - a(h)]$ showing that $f \boxminus b = f \ominus \check{b}$, where $\check{b}(x) = -b(-x)$.

In convex function analysis the function conjugate being a dual representation of a (convex) function plays an important role. The conjugate function is also known as the Fenchel conjugate, the Young-Fenchel conjugate or the Legendre transform.

Definition 9 (Conjugate function) *The conjugate function f^* of a function $f : R^d \to \bar{R}$ is defined by*

$$f^*(\xi) = \sup_{x \in R^d} \{\langle x, \xi \rangle - f(x)\}$$

Conjugation in the context of mathematical morphology has been emphasized by Dorst and van den Boomgaard [12] and by Maragos [13], who call this operation the *morphological slope transform.* Refer to [14] for a systematic treatment of the slope transform in the complete lattice framework. To a certain extent the slope transform plays a role in mathematical morphology which is somewhat comparable to the role of the Fourier transform in linear signal processing. This is mainly because of the following result.

Proposition 1 *Let f and g be convex functions, then $(f \boxminus g)^* = f^* + g^*$.*

Indicator functions and support functions are linked through conjugation: if B is a nonempty closed convex set, then $I_B^* = H_B$ and $H_B^* = I_B$.

For a number $k \in [1, \infty]$ we define its reciprocal k^* through the relation: $1/k + 1/k^* = 1$.

Proposition 2 *A function f is subpolynomial of degree k if and only if its conjugate f^* is subpolynomial of degree k^*.*

If $k = 1$ then f is sublinear and therefore we know that $f = H_B$ for some closed convex set B. Because the conjugate of a support function is the indicator function of the same set we have $f^* = H_B^* = I_B \in SP(\infty)$, i.e. the conjugate of a sublinear function is an subpolynomial function of degree ∞.

4. Morphological Scale-Space Operators

In this section we are exclusively conserved with the class of scale-space operators on $\mathcal{L} = \text{Fun}(R^d, \bar{R})$ induced by the morphological structural erosion $\varepsilon_b f = f \boxminus b$. Without loss of generality we may concentrate on the erosions only, as it can be shown [15] that in case (ε, δ) is an adjunction and T_ε defines a scale-space then T_δ defines a scale-space as well.

In this paper we will restrict ourselves to scalings of the form $S^\alpha(t)$ as defined in section 2. A straightforward calculation shows that the scaling of the images completely 'carries over' to the scaling of the structuring function that is used:

$$T_\varepsilon(t) f = f \boxminus S^\alpha(t) b.$$

For $T_\varepsilon(t)$ to be a scale-space operator we should have that $T_\varepsilon(t)T_\varepsilon(s) = T_\varepsilon(s \dot{+} t)$. Using associativity of the erosion we obtain a semigroup requirement on the family of scaled structuring functions $S(t)b$ with respect to the erosion operator: $S^\alpha(t)b \boxminus S^\alpha(s)b = S^\alpha(s\dot{+}t)b$. Taking conjugates on both sides we arrive at $(S^\alpha(t)b)^* + (S^\alpha(s)b)^* = (S^\alpha(s\dot{+}t)b)^*$. The conjugate of a scaled function $(S^\alpha(t)b)(\xi)$ equals $t^{1-\alpha}b^*(t^{2\alpha-1}\xi)$ thus:

$$t^{1-\alpha}b^*(t^{2\alpha-1}\xi) + s^{1-\alpha}b^*(s^{2\alpha-1}\xi) = (s\dot{+}t)^{1-\alpha}b^*((s\dot{+}t)^{2\alpha-1}\xi). \tag{3}$$

Erosion scale-space analysis thus boils down to the analysis of the above equation. Without any assumption on b^* (other then b, and thus b^* being convex) we can notice the special case for $\alpha = \frac{1}{2}$, then: $t^{\frac{1}{2}}b^*(\xi) + s^{\frac{1}{2}}b^*(\xi) = (s\dot{+}t)^{\frac{1}{2}}b^*(\xi)$. Regardless of b^* we thus have that:

Proposition 3 *The erosion $\varepsilon(f) = f \boxminus b$ with b convex induces an $(S^{\frac{1}{2}}, +_{\frac{1}{2}})$ scale-space.*

The scaling $S^{\frac{1}{2}}(t)f$ is anamorphic with the umbral scaling $U(t)f$, where $U(t)f = tf(\cdot/t)$ with $t > 0$, more precisely $S^{\frac{1}{2}}(t) = U(\sqrt{t})$. Notice that regardless of the structuring function b we have that $U(t)f \boxminus b = U(t)\left(f \boxminus U(\frac{1}{t})b\right)$ showing the umbral scale invariance of the erosion. It is a well-known result from classical morphology that umbral scales add up in a sequence of erosions (dilations) if and only if the structuring function is convex. From this point of view the above proposition tells us little news.

The algebraic form of the equation governing erosion scale-space (eq.(3)) suggests that choosing a subpolynomial function may lead to other types of erosion scale-spaces. Assuming that $b^* \in SP(k^*)$ we may rewrite eq. (3):

$$t^{1-\alpha+k^*(2\alpha-1)}b^*(\xi) + s^{1-\alpha+k^*(2\alpha-1)}b^*(\xi) = (s\dot{+}t)^{1-\alpha+k^*(2\alpha-1)}b^*(\xi). \tag{4}$$

It should be noted that the case $k^* = \infty$ (i.e. $b \in SP(1)$) has to be treated separately as the above equation only makes sense for $k^* < \infty$, i.e. $k > 1$. First we concentrate on finite k^*. Equation (4) shows that:

Proposition 4 *The erosion $\varepsilon(f) = f \boxminus b$ with $b \in SP(k)$ for $k > 1$ induces an $(S^\alpha, +_\nu)$ scale-space if $\nu = 1 - \alpha + k^*(2\alpha - 1)$.*

A well-known example in this class of erosion induced scale-spaces are associated with the quadratic structuring functions $b_Q(x) = \langle Qx, x\rangle$ where Q is a positive definite diagonal matrix. There $k = k^* = 2$ and thus $\varepsilon(f) = f \boxminus b_Q$ induces an $(S^\alpha, +_{3\alpha-1})$ scale-space in case $\alpha > \frac{1}{3}$. The parabolic morphological scale-space is illustrated in figure 1.

Another example is the case $k^* = 1$, i.e. $k = \infty$ corresponding with $b = I_B$ for some convex set B. Then we have $\nu = 1 - \alpha + 2\alpha - 1 = \alpha$. Thus erosions using flat convex structuring functions induce scale-spaces irrespective of the α parameter defining the balance between the spatial scaling and the grey value scaling. Figure 1 also shows a morphological scale-space induced by a flat erosion using a disk shaped structuring element.

The special case that $k^* = \infty$ is extensively studied in [15], here we only present an important case. The case that $k^* = \infty$ corresponds with a convex

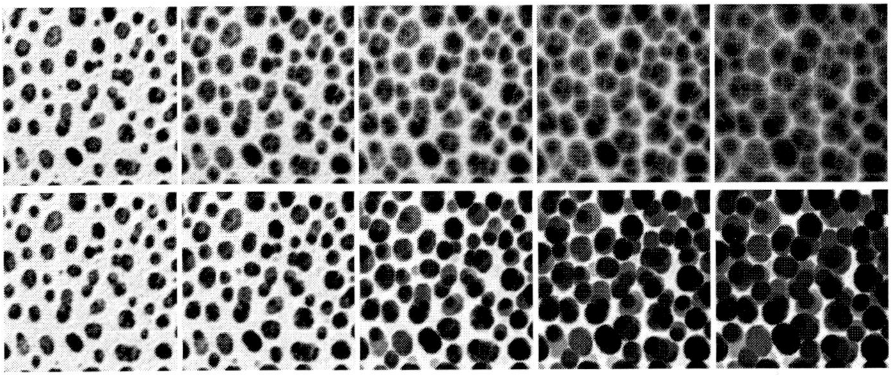

Fig. 1. **Erosion scale-spaces.** The top row shows the parabolic scale space, the second row shows a flat disk scale-space. On the left in both rows the 'zero scale image' f is shown. The four other images on the top row show the erosion $f \boxminus S^1(t)b$ where b is the parabolic function $b(x) = \|x\|^2$ for the first row and $b = I_B$ with B a disk of radius 1 for the second row.

indicator function $b^* = I_B$ where B is a convex set. In case B contains the origin we may rewrite equation (3), leaving out the finite multiplications of the functions $b^* = I_B$ and get $I_B(t^{2\alpha-1}\xi) + I_B(s^{2\alpha-1}\xi) = I_B((s\dot{+}t)^{2\alpha-1}\xi)$. Note that $I_B(\lambda x) = I_{B/\lambda}(x)$, which allows us to rewrite the above into $I_{t^{1-2\alpha}B} + I_{s^{1-2\alpha}B} = I_{(s\dot{+}t)^{1-2\alpha}B}$. Because $I_X + I_Y = I_{X\cap Y}$ we find $I_{t^{1-2\alpha}B} + I_{s^{1-2\alpha}B} = I_{(t^{1-2\alpha}\wedge s^{1-2\alpha})B}$. For $\alpha > 1/2$ we have $t^{1-2\alpha} \wedge s^{1-2\alpha} = (t \vee s)^{1-2\alpha}$ which leads to $s\dot{+}t = s \vee t$. In case $b^* = I_B$, $b = H_B$ i.e. b is the support function of a convex set B containing the origin. Summarizing:

Proposition 5 *The erosion $\varepsilon(f) = f \boxminus b$ with $b = H_B$ where B is convex set containing the origin, induces an $(S^\alpha, \vee)$ scale-space if $\alpha > 1/2$.*

The scale-space operator in this case is $T(t)f = f \boxminus S^\alpha(t)b$. For a sublinear function b we have $(S^\alpha(t))b(x) = t^{1-\alpha}b(x/t^\alpha) = t^{1-2\alpha}b(x)$. Because $\alpha > 1/2$ we have that $t^{1-2\alpha}$ is a decreasing function in t. The erosion $f \boxminus H_B$ resembles the band-pass filter known from linear signal processing. Taking the conjugate (or slope transform) we obtain $(f \boxminus H_B)^* = f^* + H_B^* = f^* + I_B$ which equals $f^*(\xi)$ if $\xi \in B$ and ∞ if $\xi \notin B$. Thus the erosion filters out all slopes that lie outside B.

5. Conclusions

In this paper we proposed an algebraic construction technique for morphological scale-spaces induced by erosions (infimal convolutions) using convex structuring functions. This technique is also valid to analyze the morphological scale-spaces induced by openings and closings and even to analyze linear scale-spaces based on convolutions. A comprehensive account can be found in [15].

Analysis of linear scale-space operators makes heavily use of the Fourier transform. Morphological operators allow for a similar change in representation to facilitate the analysis of scale-space operators: the slope transform or conjugate as it is well-known in convex function theory. Using results from convex function theory we have presented some of the important classes of morphological erosion scale-spaces.

All structuring functions that have been used in the literature to construct morphological scale-spaces, ranging from flat convex structuring functions ('sets') to the parabolic and quadratic structuring functions being the morphological equivalent of the Gaussian function, are member of the class of subpolynomial functions that we have proposed in this paper.

References

1. A. Witkin. Scale space filtering. In *Proc. 8th Int. Joint Conf. on Artificial Intelligence, (IJCAI'83)*, volume 2, pages 1019–1022, Palo Alto, California, 1983. Morgan-Kaufmann.
2. J.J. Koenderink. The structure of images. *Biological Cybernetics*, 50:363–370, 1984.
3. L. M. J. Florack. *Image Structure*, volume 10 of *Computational Imaging and Vision Series*. Kluwer Academic Publishers, Dordrecht, 1997.
4. J. Weickert, S. Ishikawa, and A. Imiya. On the history of Gaussian scale-space axiomatics. In J. Sporring, M. Nielsen, L. Florack, and P. Johansen, editors, *Gaussian scale-space theory*, pages 45–58. Kluwer, Dordrecht, 1997.
5. G. Matheron. *Random Sets and Integral Geometry*. John Wiley & Sons, New York, 1975.
6. P.T. Jackway. Morphological scale-space. In *Proceedings 11th IAPR International Conference on Pattern Recognition*, pages 252–255. IEEE Computer Society Press, 1992.
7. R. van den Boomgaard. *Mathematical Morphology: Extensions towards Computer Vision*. PhD thesis, University of Amsterdam, 1992.
8. R. van den Boomgaard and A. Smeulders. The morphological structure of images: the differential equations of morphological scale space. *IEEE Transactions on Pattern Analysis and Machine Intelligence*, 16:1101–1113, 1994.
9. R.W. Brocket and P. Maragos. Evolution equations for continuous-scale morphological filtering. *IEEE Transactions on Signal Processing*, 42(12):3377–3386, 1994.
10. L. Alvarez, F. Guichard, P-L. Lions, and J-M. Morel. Axioms and fundamental equations of image processing. (multiscale analysis and P.D.E.). *C. R. Acad. Sci., Paris, Ser. I*, 315, No.2., 1992.
11. J.-B. Hiriart-Urruty and C. Lemaréchal. *Convex Analysis and Minimization Algorithms*. Parts I and II. Springer, Berlin, 1993.
12. L. Dorst and R. van den Boomgaard. Morphological signal processing and the slope transform. *Signal Processing*, 38:79–98, 1994.
13. P. Maragos. Morphological systems: slope transforms and max-min difference and differential equations. *Signal Processing*, 38:57–77, 1994.
14. H. J. A. M. Heijmans and P. Maragos. Lattice calculus of the morphological slope transform. *Signal Processing*, 59(1):17–42, 1997.
15. H. J. A. M. Heijmans and R. van den Boomgaard. Algebraic framework for linear and morphological scale-spaces. Research Report PNA-R0003, CWI, 2000.

AN IDEMPOTENT SCALE-SPACE APPROACH FOR MORPHOLOGICAL SEGMENTATION*

NEUCIMAR J. LEITE and MARTA D. TEIXEIRA
Institute of Computing - UNICAMP , C.P. 6176
13083-970, Campinas, SP - Brazil
e-mail: neucimar@dcc.unicamp.br

Abstract. This work constitutes a first approach on image segmentation based on the recently proposed morphological scale-space theory. We introduce an idempotent smoothing operation, in the corresponding scale-space, and analyze some of its main features concerning the monotonicity of the image extrema and the way these extrema merge in a multiscale simplification process. We also define some basic criteria to control the merging of the image extrema across scales to obtain good markers for segmentation. As we will illustrate, these methods take into account only local information of the image and yield sound segmentation results, mainly in those applications where the regions to be segmented can be characterized (marked) by the extrema of the image function.

Key words: Morphological Scale-Space, Dual Reconstruction, Mathematical Morphology.

1. Introduction

This work addresses the oversegmentation problem by means of a multiscale representation of an image. Multiscale approaches have been largely considered in the signal processing theory as an effective way to relate information from different signal representations (scales), and have been extended to many image processing applications such as filtering, segmentation, compression and coding. Examples of this multiscale representations are the wavelets, the pyramid transform, and the granulometric decomposition.

One of the basic problems with any multiscale method concerns the difficulty to relate significant information or features of the signal across the different scales. In [13], Witkin proposed a novel multiscale approach, named the scale-space, in which the representation of a significant feature of a signal describes a continuous path across scales. In this case, we have that if a signal feature is present at a certain coarser scale, then it must be found at all finer scales, up to the original image representation ($\sigma = 0$). This aspect constitutes the monotonic property of the scale-space approach since the number of features across the different representation levels decreases monotonically as a function of scale.

In the Witkin's original work, extrema of a signal and its first derivative constitute the features of interest and the scale-space monotonicity property is given by convolutions of the original signal with Gaussian functions.

* This work was supported by FAPESP - Fundação de Amparo à Pesquisa do Estado de São Paulo and the FINEP/PRONEX/IC project, nº 76.97.1022.00.

The monotonic property of the scale-space approach can be very useful in image segmentation by a hierarchical process [11]. Unfortunately, as discussed in [7], there is no convolution kernel for a two-dimensional signal which can preserve this property when the feature of interest is the signal extrema.

Recently, non-linear filters based on scaled morphological operators have been associated with the scale-space theory [2]. Chen and Yan [3], for example, defined a theorem for zero crossings of binary image by considering openings of its objects with a scaled disk. Park and Lee [8] defined a scale-space for one-dimensional signals based on opening and closing operations. Jang and Chin [6] also considered these operations in the definition of a scale-space in which the interest features are the contour segments of binary images. The extension of these results to gray-scale images is not direct.

In a recent work, Jackway [5] developed a scale-space approach based on a multiscale morphological dilation and erosion (MMDE) smoothing which guarantees the monotonic property for the extrema of an image. He also defines the watershed of a signal (monotonicity for regions) smoothed at a certain scale as the feature of interest. Nevertheless, as stated by the author, the method cannot be directly associated with image segmentation tasks since "the watershed arcs move spatially with varying scale and are not a subset of those at zero scale" [4].

In the following, we analyse some important characteristics of this scale-space smoothing and introduce an idempotent scale-space approach which can yield sound segmentation results when linked to the watershed transform. The first aspect considered here refers to the preservation of the spatial position and gray-scale value of the image extrema guaranteed by the scale-space smoothing. As we will see later, to prevent the watershed lines from shifting spatially, we will use these regional extrema as the set of markers in a homotopic modification of the original image. Finally, we also illustrate how to consider some local aspects of the merging process across scales in order to improve our segmentation results.

This paper is organized as follows. In Section 2 we introduce the morphological scale-space approach and define an algorithm for segmentation based on this approach. We also define an idempotent scale-space that can be used to characterize the set of image extrema considered in the segmentation method. In Section 3 we illustrate the definition of some monotonic-preserving merging criteria used to improve our segmentation results. Finally, some conclusions are drawn in Section 4.

2. The MMDE Scale-Space Approach

The MMDE scale-space is defined for both positive and negative scales. For positive scales, the image is smoothed by dilation, and for negative scales it is processed by erosion.

Let f be an image, $f : R^2 \to R$. A smoothed version of f at scale σ is given by

$$(f \otimes g_\sigma)(x) = \begin{cases} (f \oplus g_\sigma)(x) & \text{if } \sigma > 0 \\ f(x) & \text{if } \sigma = 0 \\ (f \ominus g_\sigma)(x) & \text{if } \sigma < 0, \end{cases} \tag{1}$$

where σ corresponds to the intuitive notion of scale, and $\oplus$ and $\ominus$ stand for gray-scale dilation and erosion [9], respectively. The scaled structuring function $g_\sigma : \mathcal{G}_\sigma \subseteq R^2 \times R \rightarrow R$ is defined as

$$g_\sigma(x) = |\sigma| g(|\sigma|^{-1}x) \ , \quad \forall \sigma \neq 0 \tag{2}$$

One can show that in order to verify the monotonic property, g_σ should be a nonpositive, anticonvex, and even function for all $x \in R$, with $g(0) = 0$ [5]. Thus, by smoothing an image with such a kernel, one can preserve the monotonicity of its regional maxima or minima by performing dilation (for $\sigma > 0$) or erosion (for $\sigma < 0$), respectively. An example of such a kernel used in this work is the circular paraboloid $g(x, y) = -(x^2 + y^2)$ which has been considered as a suitable function for practical reasons such as computational efficiency [5].

In his morphological scale-space, Jackway extends the monotonicity from point sets to regions by establishing a link between the extrema of the image and its gradient, through the watershed transform and the homotopic modification by a reconstruction process [9]. The next section illustrates this aspect and shows how we can define a multiscale segmentation algorithm based on the monotonicity property for regions.

2.1. Morphological Scale-space and Segmentation

Formally, the monotonicity for the image extrema can be stated as follows (here, we consider only the results for negative scales, the extension to the positive ones is obtained from duality) [5].

Theorem 1 *[5] Let the set of points $E_{min}(f) = \{ x \in f : x \text{ is a local minimum}\}$ represent the minima of image f. Then, for any scales $\sigma_2 < \sigma_1 < 0$,*

$$E_{min}(f \otimes g_{\sigma_2}) \subseteq E_{min}(f \otimes g_{\sigma_1}) \subseteq E_{min}(f).$$

A scale-space monotonicity for regions is obtained by associating the watershed transform with the gradient image as follows [5]:

Algorithm 1: FOR each scale σ_k DO:
1. smooth f to obtain $(f \otimes g_\sigma)(x)$ using Eq. 1.
2. find a suitable set of regional minima N_i (for $\sigma_k \leq 0$) or maxima M_i (for $\sigma_k \geq 0$) of $(f \otimes g_\sigma)(x)$ corresponding to a marker function, $g(x)$.
3. compute the magnitude of the gradient image $|\nabla(f \otimes g_{\sigma_k})|$.
4. modify the homotopy of this image by a dual reconstruction process [1, 12] using function $g(x)$ as marker.
5. find the watershed regions of the modified image.
   ENDFOR

Fig. 1(b) illustrates the result of such an algorithm for $\sigma = -5$. Note that due to the smoothing operation in Eq. 1 (erosion, in this case) the gradient watersheds move spatially as a function of scale and do not represent the contours of the image according to a common segmentation model. Thus, for the purpose of segmentation, we need to face the problem of "forcing" these watershed lines to delineate the regions being segmented. The following proposition by Jackway allows us to use the set of extrema, present at a certain scale, as marker in a homotopic modification of the original image. This homotopy modification constitutes the base for the preservation of the image contours across scales.

Proposition 1 *[5] Let the structuring function have a single maximum at the origin, that is, $g(x)$ is a local maximum implies $x = 0$, then:*

If $\sigma < 0$ and $(f \otimes g_\sigma)(x_{min})$ is local minimum, then, $f(x_{min})$ is a local minimum of $f(x)$ and $(f \otimes g_\sigma)(x_{min}) = f(x_{min})$.

Based on this result, which states that the position and amplitude of the extrema in the original and smoothed images do not change, we define a multiscale segmentation algorithm as follows:

Algorithm 2: FOR each scale σ DO

1. smooth f to obtain $(f \otimes g_\sigma)(x)$ using Eq. 1.
2. find a suitable set of regional minima N_i (for $\sigma_k \leq 0$) or maxima M_i (for $\sigma_k \geq 0$) of $(f \otimes g_\sigma)(x)$ corresponding to a marker function, $g(x)$.
3. define a new image f' by modifying the homotopy of the original image f using function $g(x)$ as marker.
4. compute the magnitude of the gradient image $|\nabla(f')|$.
5. modify the homotopy of this image by a dual reconstruction process [1, 12] using function $g(x)$ as marker.
6. compute the watershed of this modified image.

ENDFOR

Fig. 1(c) shows the result of this algorithm representing a better partition of the original image (Fig. 1(a)). Note that, in this case, the set of markers used in the homotopic modification of the original and gradient images (steps 3 and 5 above) is given by the same set of minima as in Fig. 1(b).

The next section discusses some aspects concerning the merging of the image extrema across scales.

2.2. An Idempotent Scale-space

Once we define the smoothed image, it is very difficult to characterize the set of extrema that remains (or should remain) at a certain scale. The following idempotence considerations constitute an important simplification of this set.

Let f be an image function as before and

$$(f \ominus g_\sigma)^n = \underbrace{(((f \ominus g_\sigma) \ominus g_\sigma) \ominus \cdots \ominus g_\sigma)}_{n \ times}$$

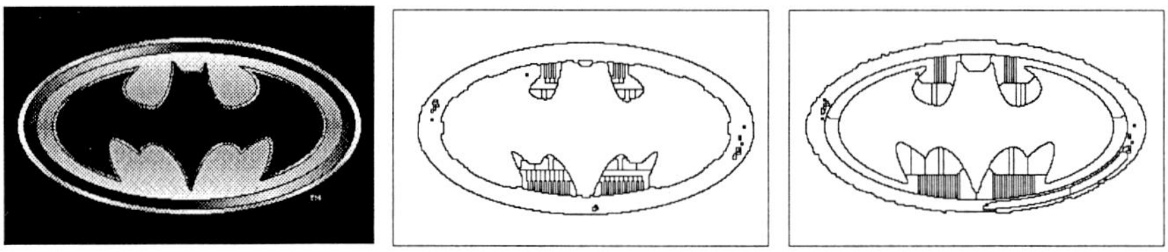

(a) (b) (c)

Fig. 1. (a) Original image. (b) The watershed lines based on the MMDE approach ($\sigma = -5$), and (c) on the homotopy modification by reconstruction of the original image.

(in the sequel, we consider that f and g_σ are defined in the discrete domain). We define an idempotent smoothed version of f, at scale σ, as

$$(f \odot g_\sigma)(x) = \begin{cases} (f \oplus g_\sigma)^n(x) & \text{if } \sigma > 0 \\ f(x) & \text{if } \sigma = 0 \\ (f \ominus g_\sigma)^n(x) & \text{if } \sigma < 0, \end{cases} \tag{3}$$

where n is the number of iterations so that $(f \oplus g_\sigma)^n(x) = (f \oplus g_\sigma)^{n-1}(x)$ for $\sigma > 0$, and $(f \ominus g_\sigma)^n(x) = (f \ominus g_\sigma)^{n-1}(x)$ for $\sigma < 0$. As we have shown in [10], the basic properties of the MMDE approach also hold for this idempotent scale-space. The following proposition concerns the idempotence property of Eq. 1 (The proof of all the propositions discussed in this paper is given in appendix A and in [10]).

Proposition 2 *[10] For any $\sigma < 0$ there exists a value n such that $(f \ominus g_\sigma)^n(x) = (f \ominus g_\sigma)^{n-1}(x)$.*

The set of regional minima obtained after smoothing the image till idempotence constitutes the *minima minimal configuration - MMC* set at scale σ. Shortly, this set represents a simplification of the image defined at scale σ by the original morphological scale space method (Eq. 1). It contains less non-significant minima that can be used as markers in our multiscale segmentation algorithm. The next proposition considers other aspects of the merging process in the definition of the MMC set.

Proposition 3 *[10] Let x_i and $x_j \in E_{min}(f)$ denote two points of the image f with $f(x_i) < f(x_j)$. For a 4-connectivity and $\sigma < 0$, we can show that pixel x_j will belong to the influence zone [9] of x_i, $\mathcal{Z}(x_i)$, if $\exists x_k \in \mathcal{Z}(x_i)$ so that*

$$f(x_j) - f(x_k) \geq D_\sigma \times (d(x_j, x_k) - 1) \tag{4}$$

where d denotes the city-block distance and $D_\sigma = |sup_{t \in \mathcal{G}_\sigma}(g_\sigma(t))|$, $t \neq 0$.

Equation 4 relates the value of a minimum, x_j, and the influence zone of another minimum, x_i, lower than x_j for merging. According to the above proposition, two minima with small height difference can merge only at coarser scales (e.g., minima with the same gray-scale will merge only when $\sigma \to -\infty$, that is, $D_\sigma \to 0$). To illustrate such a point, let us assume the segmentation of the electrophoresis image shown in Fig. 2. Besides the dark blobs to be segmented, this image also shows the aspect related to the distance between the image minima concerning Eq. 4. As we can see in Fig. 2-(b), for $\sigma = -1/2$, some significant minima have been merged with their nearest "deeper" (darker) neighbors. Such a merging constitutes a filtering of the image components used as markers in our segmentation algorithm. For a finer scale, $\sigma = -1/4$ (Fig. 2-(c)), some non-significant minima do not merge either because of their high distance to a significant minima, or due to the small gray-scale difference between them. These non-significant minima that remain at a certain scale define an oversegmentation of the original image (Fig. 2-(a)). This oversegmentation can be reduced by a post-processing operation taking into account the gradient value of the non-significant regions of the image. This operation, which is followed by the watershed computation, is given by

$$g'(x) = \begin{cases} 0 & \text{if } g(x) \leq K \\ g(x) - K & \text{if } g(x) > K, \end{cases} \tag{5}$$

where $g(x)$ is the gradient of the reconstructed image, $f'(x)$, and K is a threshold value. Note that besides eliminating the non-significant minima which do not merge with the significant ones, this post-processing yields a merging of the minima with similar gray-scale and small gradient between them.

Fig. 2-(d) illustrates such an operation applied to the image in Fig. 2-(a).

The next section improves this segmentation results based on the definition of some monotonic-preserving merging criteria used in the smoothing process.

3. Extrema Merging Criteria

One way to improve the segmentation results obtained here is to define some criteria to locally control the merging of the image extrema during the smoothing process. These criteria can be based on information such as the altitude difference between minima (for negative scales), and the gray-scale values of the pixels separating these minima (as we can see from proposition 3, the value of the pixels between the image minima is irrelevant from the point of view of the MMC set definition, which takes into account only the gray-scale value of the minima and their distances).

Through these merging criteria, we can detect pixels that should remain unchanged during the smoothing process, thereby preventing meaningful minima from merging. To do it, we need to guarantee that the non-smoothing of a point at a certain scale will not create a new minimum in the smoothed image. The next proposition concerns this aspect:

Proposition 4 *For a discrete structuring function g_σ of size $N \times N$, N odd, with radius $R = \lfloor \frac{N}{2} \rfloor$, the non-smoothing of a point x_k of the image by Eq. 3 can introduce a new minimum in the neighborhood of x_k with radius R-1.*

Finally, we can also prove the following statement regarding computation time.

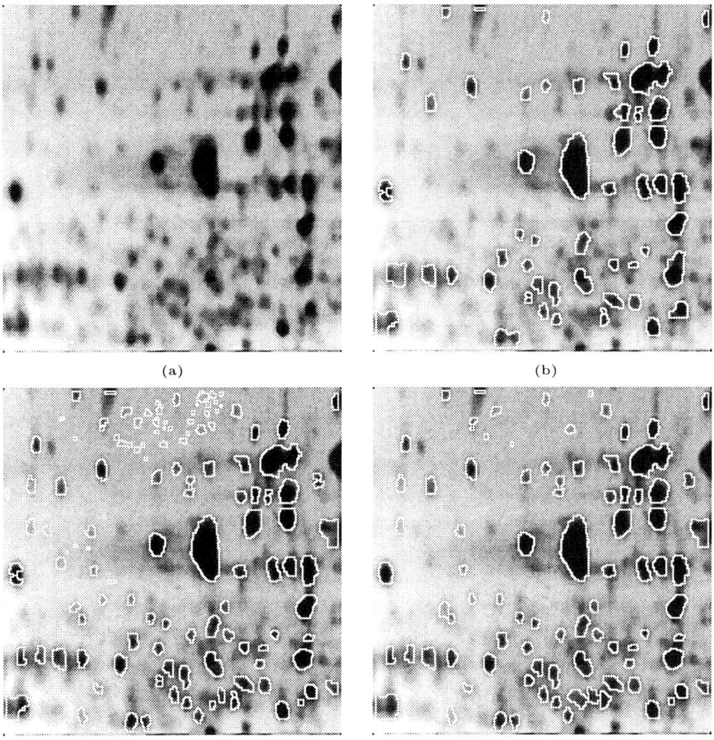

Fig. 2. Segmentation based on the idempotent scale-space. (a) Original image; watershed lines for scales (b) $\sigma = -1/2$ and (c) $\sigma = -1/4$. (d) Image (c) after the post-processing operation (K=10).

Proposition 5 *[10] For discrete images, the MMC set can be obtained from Eq. 1 by considering a small* 3×3 *structuring function* g_σ.

According to propositions 4 and 5 above, we can define local merging criteria based on the unchanged state of a pixel, while preserving the monotonic property of the transformation. Some examples of these criteria are given below.

3.1. The Altitude Difference between Minima

Suppose that in some applications the meaningful regions to be segmented can be associated with a set of minima with high altitude difference between them. One way to avoid these significant minima (markers) from merging is to consider that some points nearby these minima will not be eroded across scales. Topographically, the aim here is to define a kind of "barrier" along a regional minimum which will prevent it from merging with other minima of very different depths.

Let H be a constant representing a height parameter. The smoothed value, $f_\sigma(x_k)$, of a point $x_k \notin E_{min}(f)$, at scale σ, is given by

$$f_\sigma(x_k) = \begin{cases} x_k, \text{ if } x_k \in N_{3x3}(x_{min}) \text{ and } (f(x_{min}) - (f \odot g_\sigma)(x_k) \geq H) \\ (f \odot g_\sigma)(x_k), \text{otherwise} \end{cases} \quad (6)$$

where $N_{3x3}(x_{min})$ denotes the 3×3 neighborhood of a point $x_{min} \in E_{min}(f)$. In this case, the eroded value $f_\sigma(x_k)$ can be related to the value of another regional minimum, x'_{min}, that would merge with x_{min} if $f_\sigma(x_k) \leq f(x_{min})$. The difference $(f(x_{min}) - (f \odot g_\sigma)(x_k))$ then, gives a local information about the altitude difference of these minima.

Figure 3.(a)-(c) shows some segmentation examples by considering Eq. 6.

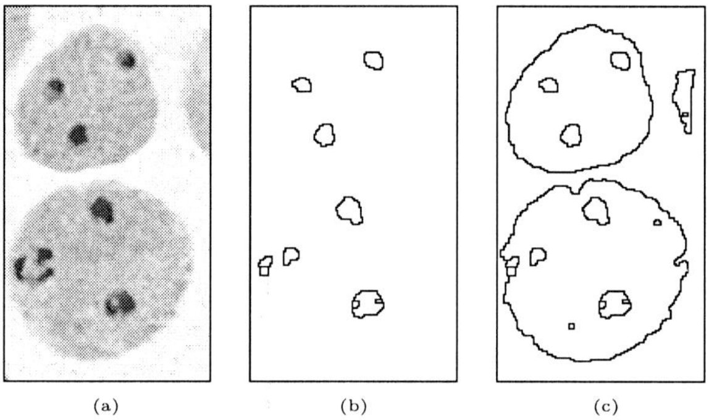

Fig. 3. (a) Original image. The result of the scale-space segmentation for (b) $\sigma = -1$, and (c) $\sigma = -1$ and $H = 20$.

3.2. The Altitude of the Points Separating Regional Minima

Another information that can be taken into account to locally control the merging of the image extrema is the altitude of the points separating these extrema. For the regional minima, for example, we can define the following conditional smoothing, based on the original and the transformed values of the pixels at scale σ.

$$f_\sigma(x_k) = \begin{cases} (f \odot g_\sigma)(x_k) + H & \text{if } (f(x_k) - (f \odot g_\sigma)(x_k)) \geq H \\ (f \odot g_\sigma)(x_k) & \text{otherwise,} \end{cases} \quad (7)$$

where H is a constant representing a height parameter. This equation defines a value $f_\sigma(x_k)$ such that $(f \odot g_\sigma)(x_k) < f_\sigma(x_k) \leq f(x_k)$, when the difference between $f(x_k)$ and its smoothed value, at scale σ, is greater or equal to H. As we have seen before, the transformed value of the pixel x_k can be related to the value of a minimum x_{min} in its neighborhood. In such a case, if $(f(x_k) - (f \odot g_\sigma)(x_k)) \geq H$ and x_k belongs to a path between two regional minima, x_{min} and x'_{min}, then smoothing x_k to a value greater than $(f \odot g_\sigma)(x_k)$ can prevent these minima from merging through this path. Figure 4.(a)-(b) shows some segmentation examples by using Eq. 7.

Both of the above methods consider basic configurations of the significant extrema we seek as markers in our segmentation process. Based on these criteria, we can avoid merging (even at a coarse scale) two minima with high altitude difference, as in Figure 3, where regions which are not characterized by significant extrema are segmented (the nuclei of the cells), or separated by pixels with high gray-scale values, as in Figure 4, where minima of different depths do not merge.

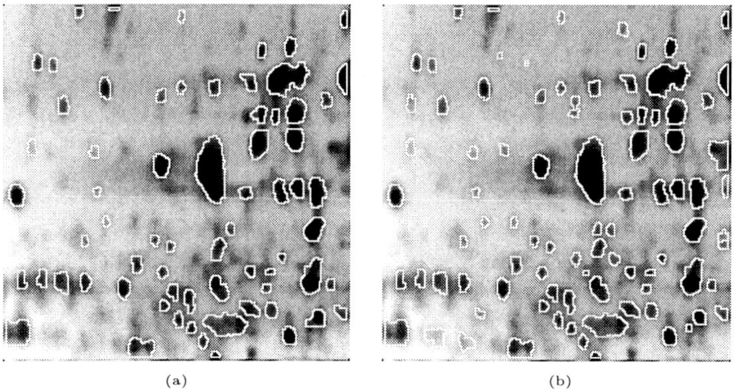

Fig. 4. Watershed lines based on Eq. 7 for $\sigma = -1/2$. (a) $H = 15$ and $K = 10$; (b)$H = 10$ and $K = 10$.

4. Conclusions

In this work we consider the problem of image segmentation by means of the morphological scale-space theory. Our approach is based on the simplification of the extrema of an image which smoothed versions are characterized by a monotonic filtering of these extrema. Among other results, we define an idempotent morphological scale-space smoothing, which associated to the morphological reconstruction algorithm, can be used in a multiscale segmentation model. Also, we proved that by keeping unchanged a point of the image smoothed by a 3×3 structuring function, we do not add any new minima in the final transformation. As we have seen before, this result is very important if we want to define segmentation criteria taking into account the value of the image extrema and the monotonic property, inherent to the scale-space approach.

Appendix

A. Proof of Propositions

Proof of Proposition 4: For two points x_i and x_p of an image f, if $(f \ominus g_\sigma) = f(x_i) - g_\sigma(x_i - x_p)$, then, from the proposition 6 below, we have that $\exists p_*(x_l, x_p)$ such that $(f \ominus g_\sigma)(x_l) < \cdots < (f \ominus g_\sigma)(x_m) < \cdots < (f \ominus g_\sigma)(x_{p-1}) < (f \ominus g_\sigma)(x_p)$, for some $x_l \in p_*(x_i, x_p)$.

Therefore, if a point $x_m \in p_*(x_{p-1}, x_l)$ is not eroded during the smoothing process, then a minimum can be created in x_{m+1} and the set of points $\{x_p, \cdots, x_{m+1}\}$ will belong to an influence zone different from the one of x_p.

Since the longest path occurs when $x_l = x_i$ and we have that $\#p_*(x_i, x_p)$ [1] $= R + 1$, for a structuring element with radius R, the new defined influence zone will have, at most, $R - 1$ points of this path.

Since for $R = 1$, $\#p_*(x_p, x_i) = 2$, then no new minimum is created, and the monotonic property holds for such a case. □

Proposition 6 *Let us consider two points x_i and x_j of an image f, such that $(f \ominus g_\sigma)(x_j) = f(x_i) - g_\sigma(x_i - x_j)$. Then $\exists p_*(x_i, x_j)$ so that for any $x_k \in p_*(x_i, x_{j-1})$, $(f \ominus g_\sigma)(x_k) < (f \ominus g_\sigma)(x_j)$.*

Proof: Since $x_i \in N_*(x_k, \mathcal{G}_\sigma)$, $(f \ominus g_\sigma)(x_k) = inf_t\{f(x_k - t) - g_\sigma(t)\} \leq f(x_i) - g_\sigma(x_k - x_i)$. Thus, since $d(x_k, x_i) < d(x_j, x_i)$, we have that $g_\sigma(x_k - x_i) > g_\sigma(x_j - x_i)$. Therefore, $(f \ominus g_\sigma)(x_k) \leq f(x_i) - g_\sigma(x_k - x_i) < f(x_i) - g_\sigma(x_j - x_i) = (f \ominus g_\sigma)(x_j)$, and so, $(f \ominus g_\sigma)(x_k) < (f \ominus g_\sigma)(x_j)$. □

References

1. S. Beucher and F. Meyer. The morphological approach to segmentation: The watershed transform. In E. R. Dougherty, editor, *Mathematical Morphology in Image Processing*, chapter 12, pages 433–481. Marcel Dekker, Inc., 1993.
2. R. W. Brockett and P. Maragos. Evolution equations for continuous-scale morphological filtering. *IEEE Transactions on Signal Processing*, 42(12):3377–3386, December 1994.
3. M.-H. Chen and P.-F. Yan. A multiscale approach based on morphological filtering. *IEEE Transactions on Pattern Analysis and Machine Intelligence*, 11(7):694–700, July 1989.
4. P. T. Jackway. Gradient watersheds in morphological scale-space. *IEEE Transactions on Image Processing*, 5(6):913–921, June 1996.
5. P. T. Jackway and M. Deriche. Scale-space properties of multiscale morphological dilatation-erosion. *IEEE Transactions on Pattern Analysis and Machine Intelligence*, 18(1):38–51, January 1996.
6. B. K. Jang and R. T. Chin. Morphological scale space for 2d shape smoothing. *Computer Vision and Image Understanding*, 70(2):121–141, May 1998.
7. L. M. Lifshitz and S. M. Pizer. A multiresolution hierarchical approach to image segmentation based on intensity extrema. *IEEE Transactions on Pattern Analysis as Machine Intelligence*, 12(4):529–540, 1990.
8. K. Park and C. Lee. Scale-space using mathematical morphology. *IEEE Transactions on Pattern Analysis as Machine Intelligence*, 18(11):1121–1126, November 1996.
9. J. Serra. *Image Analysis and Mathematical Morphology*. Academic Press, 1982.
10. M. D. Teixeira and N. J. Leite. Morphological scale-space theory for segmentation problems. In *Proc. of IEEE-EURASIP Workshop on Nonlinear Signal and Image Processing*, pages 364–368, Antalya, Turkey, 1999.
11. R. van den Boomgaard and A. Smeulders. The morphological structure of images: The differential equations of morphological scale-space. *IEEE Transactions on Pattern Analysis and Machine Intelligence*, 16(11):1101–1113, November 1994.
12. L. Vincent. Morphological grayscale reconstruction in image analysis: Applications and efficient algorithms. *IEEE Transactions on Image Processing*, 2(2):176–201, April 1993.
13. A. P. Witkin. Scale-space filtering. *Proc,. Int'l Joint Conf. Artificial Intelligence, Pablo Alto, Calif.: Kaufmann*, pages 1019–1022, August 1983.

[1] $\#p_*(x_i, x_p)$ = number of points of the path $p_*(x_i, x_p)$.

EFFICIENT DILATION, EROSION, OPENING AND CLOSING ALGORITHMS

JOSEPH (YOSSI) GIL and RON KIMMEL
Department of Computer Science
Technion–Israel Institute of Technology
Technion City, Haifa 32000, Israel

Abstract. We propose an efficient algorithm for computing the dilation and erosion filters. For a p-element sliding window, our algorithm computes the 1D filter using $1.5 + o(1)$ comparisons per sample point. Our algorithm constitutes improvements over the best previously known such algorithm by Gil and Werman [5]. The previous improvement on [5] offered by Gevorkian, Astola and Atourian [2] was in better expected performance for random signals. Our result improves on [5] result without assuming any distribution of the input. Further, a randomized version of our algorithm gives an expected number of $1.25 + o(1)$ comparisons per sample point, for *any* input distribution. We deal with the problem of computing the dilation and the erosion filters simultaneously, and again improve the Gil-Werman algorithm in this case for independently distributed inputs. We then turn to the *opening* filter, defined as the application of the min filter to the max filter, and give an efficient algorithm for its computation. Specifically, this algorithm is only slightly slower than the computation of just the max filter. The improved algorithms are readily generalized to two dimensions for rectangular structuring element, as well as to any higher finite dimension for a hyper-box structuring element, with the number of comparisons per window remaining constant.

Key words: Max-Filter, Min-Filter, Running Window.

1. Introduction

In signal and image analysis one often encounters the problem of min (or max) computation in a window with p elements in the one-dimensional (1D) case, or $p \times p$ elements in the 2D case. In mathematical morphology [7], the result of such an operator is referred to as the erosion (or dilation) of the signal with a *structuring element* given by a flat ramp of width p.

An important application is the morphological edge detector, obtained by applying both the min filter and the max filter, and then subtracting the results. Denoising is yet another example of using the min and max filters. The *opening* (respectively *closing*) filter is obtained by feeding the results of the max (resp. min) filter to the min (resp. max) filter. In image processing the opening filter eliminates small white regions, while closing eliminates small dark regions. In both filters, the size of the window determines the size of the regions that can be removed. Some other applications of the min and max filters in pattern analysis, adaptive signal processing and morphological analysis are mentioned in [2]. These applications call our interest to the problem of efficiently computing the min and max filters for a wide range of p.

The one-dimensional version of the problem can be formulated as follows:

1D MAX-FILTER: *Given a sequence* $x_0, \ldots, x_{n-1}$, *and an integer* $p > 1$, *compute* $y_i = \max_{0 \le j < p} x_{i+j}$, *for* $i = 0, \ldots, n-p$.

The 1D MIN-FILTER problem is similarly defined.

As usual in filtering, we assume that $p \ll n$. As an efficiency measure of algorithms for this problem we use C_1, defined as the number of comparison operations per sample (or output) point as n goes to infinity.

Since any max filter computation must examine every input element at least once, we have that $C_1 \ge 1$. A trivial algorithm for the 1D MAX-FILTER problem gives $C_1 = p-1$. On the other hand, since it is impossible to compute the filter without examining each input point at least once, there is a trivial information theoretical lower bound for the problem of $C_1 \ge 1$.

Two non-trivial algorithms for the problem were published in [6]: The first achieves $C_1 = O(\lg p)$[1] and the second $C_1 = 3 + o(1)$ for uniformly distributed independent input signals. The worst case performance for both of these algorithms depends on the window size.

Gil and Werman, in their work on computing the median filter [5], gave the first algorithm for computing the max filter whose performance does not depend on p. Their algorithm is more general since it can compute any semi-ring operation, $\diamond$, filter of size p while using $3-4/p$ applications of $\diamond$ per sample point. Since max is a semi-ring operation, their result gives $C_1 = 3 - 4/p$.

Gevorkian, Astola and Atourian [2] observed that in the special case when the semi-ring operation is max, the Gil-Werman algorithm can be improved, assuming locally uniform distributed signals, to achieve $E(C_1) = 2.5 - 3.5/p$. The expectation here is respectively to input distribution. In the worst input case, the performance of the algorithm of [2] is the same as the Gil-Werman algorithm. Here we describe an algorithm achieving further reduction, $C_1 = 1.5 + \frac{\lg p}{p} - O(1/p)$. This improvement is *deterministic* and does not make any assumptions on the input distribution.

Further, we also describe a randomized algorithm which comes even closer to the lower bound in achieving $E(C_1) = 1.25 + \frac{\lg p}{p} - O(1/p)$, where the expectation is w.r.t. random selections made by the algorithm; i.e., this expected performance is obtained for any input.

The optimal L-filter, the morphological edge detector, and other applications call for the simultaneous computation of the min and max in each window, as summarized in the following problem definition.

1D MAX-MIN-FILTER: *Given a sequence* $x_0, \ldots, x_{n-1}$, *and an integer* $p > 1$, *compute* $y_i = \max_{0 \le j < p} x_{i+j}$ *and* $z_i = \min_{0 \le j < p} x_{i+j}$ *for* $i = 0, \ldots, n-p$.

We give an algorithm that solves the 1D MAX-MIN-FILTER problem faster than solving the 1D MAX-FILTER and the 1D MIN-FILTER. Let C_1^m be the number of comparisons per input sample for solving 1D MAX-MIN-FILTER. Then the algorithm achieves $E(C_1^m) \approx 2 + 2.3466\frac{\lg p}{p}$, for the special case of independent input distribution, i.e., the expectation is with regard to input distribution. In the worst case this algorithm does not improve on the independent computation of the Min- and Max filters. However, for natural images,

[1] We use $\lg(\cdot)$ to denote $\log_2(\cdot)$

the algorithm makes such an improvement.

The problem posed by the opening filter is similar to 1D MAX-MIN-FILTER, since in both it is required to compute both a Min-Filter and a Max-Filter. However, the fact that in the opening filter these filters are computed sequentially, where the results of one filter are fed to the other, makes it much easier. Let C_1^o be the number of comparisons per input sample for computing the opening filter. Then, we show that $C_1^o \leq C_1 + O(\frac{\lg^2 p}{p})$. Clearly, the same result holds for the closing filter.

As described in [5], a 1D max filter can be extended to square (or rectangular) window 2D max filter. This is done by first applying the 1D filter along the rows, and then feeding the result to a 1D filter running along the columns. Let C_2 be the number of comparison operations required per input point for computing the 2D max filter. We have that $C_2 = 2C_1$, and more generally, $C_d = dC_1$, where C_d is defined accordingly for the d-dimensional filter. We similarly have that $C_d^m = dC_1^m$ and $C_d^o = dC_1^o$.

Outline. The remainder of this paper is organized as follows. Section 2 reviews the Gil-Werman algorithm. The deterministic and randomized algorithms improving it are described in Section 3. In Section 4 we give our algorithm for the 1D MAX-MIN FILTER PROBLEM. The efficient algorithm for computing the opening (and closing) filter is described in Section 5, and conclude with Section 6.

2. The Gil-Werman Algorithm

The Gil-Werman algorithm is based on a partitioning of the input signal to overlapping segments of size $2p - 1$, centered at $x_{p-1}, x_{2p-1}, x_{3p-1}, \ldots$. Let j be the index of an element at the center of a certain segment. The maxima of the p windows which include x_j are computed in one *batch* of the Gil-Werman algorithm as follows: First, define R_k and S_k for $k = 0, \ldots, p-1$:

$$R_k = \max(x_j, x_{j-1}, \ldots, x_{j-k}), \quad \text{and} \quad S_k = \max(x_j, x_{j+1}, \ldots, x_{j+k}). \quad (1)$$

Now, the R_k's and the S_k's can be merged together to compute the max filter:

$$\max(x_{j-k}, \ldots, x_0, \ldots, x_{j+p-k-1}) = \max(R_k, S_{p-k-1}), \quad (2)$$

for $k = 1, \ldots, p-2$. In addition, we have $\max(x_{j-p-1}, \ldots, x_j) = R_{p-1}$ and $\max(x_0, \ldots, x_{j+p-1}) = S_{p-1}$.

There are two steps to the Gil-Werman algorithm:

Preprocessing Computing all R_k and S_k from their definition (1) which is done in $2(p-1)$ comparisons.

Merge Merging the R_k and S_k together using (2), for which another $p-2$ comparisons are required.

Since this procedure computes the maximum of p windows in total, we have that the amortized number of comparisons per window is $3 - 4/p$. For large p, we have that the preprocessing step requires two comparison operations per element, while the merge step requires one more comparison.

3. The Improved Algorithms For the Max-Filter

Let us show how the two steps of the Gil-Werman algorithm can be carried out more efficiently.

3.1. An efficient merge procedure

We first show how to improve the merge step, by reducing the number of comparisons from 1 to $\lg p/p + o(1)$. In this step, we compute

$$\max(R_k, S_{p-k-1}), \tag{3}$$

for $k = 1, \ldots, p-2$. Observing that $R_{p-2} \geq R_{p-1} \geq \ldots \geq R_1$, and $S_{p-2} \geq S_{p-1} \geq \ldots \geq S_1$, we can eliminate most of these comparisons. Suppose that for some specific i it was found that $R_i \geq S_{p-i-1}$, then for all $k > i$, we have that $R_k \geq R_i \geq S_{p-i-1} \geq S_{p-k-1}$, and therefore there is no need to do the comparisons of (3) for all $k > i$. Similarly, if it is determined that $R_i \leq S_{p-i-1}$, then we do not need to do the comparisons of (3) for all $k < i$.

The optimized procedure for the merge step is therefore a binary search. We start by setting $i = \lceil (p-2)/2 \rceil$, and then continue with the remaining half of the problem size. The number of comparisons is thus reduced from $p-2$ to $O(\lg p)$. In fact, it can be easily checked that the number of comparisons in the binary search of the merge step is exactly $\lceil \lg p - 1 \rceil$. The amortized contribution of the improved merge step to the complexity is $\frac{\lceil \lg p - 1 \rceil}{p}$.

3.2. An efficient preprocessing computation

Let us now deal with the preprocessing step of the Gil-Werman algorithm. Gevorkian, Astola and Atourian [2] observed that preprocessing computation can be made more efficient for randomized input, using the fact that in the Gil-Werman algorithm, the suffixes S_k of one segment overlap with the prefixes R_k of the following segment. Specifically, the problem that needs to be solved is

Prefix-Suffix Max: *Given a sequence $x_0, \ldots, x_p$, compute all of its* prefix maxima*: $s_k = \max(x_0, \ldots, x_k)$, for $k = 0, \ldots, p-1$, and all its* suffix maxima*: $r_k = \max(x_k, \ldots, x_p)$, for $k = 1, \ldots, p$.*

Note that this problem does not call for computing the overall maximum of the input $s_p = r_0 = \max(x_0, \ldots, x_p)$.

The original Gil-Werman algorithm makes $2(p-2)$ comparisons in solving the Prefix-Suffix Max problem. We propose the following efficient solution for this problem. Let $q = \lfloor (p+1)/2 \rfloor = p/2 + (p \bmod 2)/2$. In the *first part* of the modified implementation, compute all s_k, for $k = 0, \ldots, q-1$ and r_k for $k = q, \ldots, p$. This is carried out using $p-1$ comparisons.

The *second part* of the modified implementation of the preprocessing stage begins in comparing s_{q-1} and r_q. If $r_q \geq s_{q-1}$, then we know that the overall maximum falls is one of $x_q, \ldots, x_p$. Therefore, it is unnecessary to further compute the value of $r_{q-1}, r_{q-2}, \ldots, r_1$. Instead, the algorithm outputs

$r_{q-1} = r_{q-2} = \ldots = r_1 = r_q$, and continues to compute $s_q, \ldots, s_{p-1}$. A similar situation occurs if $r_q \le s_{q-1}$, in which case it is unnecessary to compute $s_q, \ldots, s_{p-1}$. In both cases, the number of comparisons that remain to be done is $\lfloor (p+1)/2 \rfloor$. The total number of comparisons in the more efficient algorithm for PREFIX-SUFFIX MAX is $(p-1)+1+\lfloor \frac{p+1}{2} \rfloor = 1.5p + \frac{p \bmod 2}{2}$. Noting that each batch requires (on amortization) solving one instance of PREFIX-SUFFIX MAX, we can combine our results so far to obtain:

Theorem 1 *There exists a deterministic algorithm for the* 1D MAX-FILTER *problem, achieving* $C_1 = 1.5 + \frac{\lceil \lg p - 1 \rceil}{p} + \frac{p \bmod 2}{2p} \le 1.5 + \frac{\lg p}{p}$.

Can we improve on this result? An information theoretical lower bound for the number of comparisons required to solve PREFIX-SUFFIX MAX, is $p + \lg p - O(1)$. This bound is derived as follows. A compact output of an algorithm for the problem uses $p + \lg p - O(1)$ bits comprised as follows:

1. $\lg p$ bits to designate the location of the overall maximum (for simplicity, we assume that p is a power of 2),
2. one bit for each location prior to the maximum, designating whether the corresponding element changes the prefix maxima, and
3. one bit for each location following to the maximum, designating whether the corresponding element changes the suffix maxima.

Moreover, there are distinct inputs which produce all the bit combinations of this compact representation. Thus, in order to make the distinction between these inputs, the algorithm is forced to make at least $p + \lg p - O(1)$ comparisons.

Although we are unable to meet this lower bound, we can come close to it in an important special cases. Suppose that in an input to the PREFIX-SUFFIX MAX problem, the overall maximum is located at a random location ℓ in the input sequence. (This does not necessarily mean that the input is uniformly and independently distributed). Then, once the comparison between s_{q-1} and r_q is made, all that remains is to proceed to compute outputs $s_q, s_{q+1}, \ldots, s_{\ell-1}$ in the case that $s_{q-1} < r_q$, or $r_{q-1}, r_{q-2}, \ldots, r_{\ell+1}$ otherwise. The expected number of comparisons in this completion stage is

$$\frac{1}{p+1}\left(\sum_{i=0}^{q-1} i + \sum_{i=0}^{p+1-q} i\right) = \frac{p^2-(p \bmod 2)}{4(p+1)} \le \frac{p}{4} - \frac{1}{4} + \frac{1}{4(p+1)} \le \frac{p}{4}. \quad (4)$$

In general, it cannot be assumed that an arbitrary input to the PREFIX-SUFFIX MAX problem will have its maximum at a random location. However, in using this procedure as part of an algorithm for solving the 1D MAX-FILTER problem, we can achieve this effect by choosing at random the starting point for segmentation. Thus, the segments will be centered at positions indexed $\tau, \tau + p, \tau + 2p, \ldots$, where τ is an integer selected at random in the range $[0, \ldots, p-1]$. Such a random selection does not degrade the efficiency due to the assumption that $p \ll n$. We have thus obtained:

Theorem 2 *There exists a randomized algorithm for the* 1D MAX-FILTER *problem, achieving* $E(C_1) \le 1.25 + \frac{\lceil \lg p - 1 \rceil}{p} + \le 1.25 + \frac{\lg p}{p}$.

The interested reader is referred to e.g., [3] and the references thereof for examples of applying the template mechanism for non-trivial compile-time computation and code generation that are useful for implementation of logical cases involved in the proposed algorithms.

4. Efficient Algorithm for Computing the Max and Min Together

Let us deal with the 1D MAX-MIN FILTER problem, and show how the min and max filters together can be computed more efficiently than an independent computation of both. We start again from the Gil-Werman algorithm. The gain comes from partitioning the input signal into pairs of consecutive elements, and comparing the values in each pair. The greater value in each pair carries on the maximum computation while the lesser one carries one to the minimum computation.

4.1. THE PREFIX MAX-MIN PROBLEM

Let us first consider the following problem,

PREFIX MAX-MIN: *Given a sequence* $x_0, \ldots, x_{q-1}$, *compute* $M_k = \max(x_0, \ldots, x_k)$, *and* $m_k = \min(x_0, \ldots, x_k)$, *for* $k = 0, \ldots, q-1$.

The straightforward solution for PREFIX MAX-MIN uses a total of $2(q-2)+1$ comparisons. Analyzing this problem from an information theoretic point of view we find that for all $i > 2$, there are three cases for element x_i. It either increases the running prefix maximum, or it decreases the running prefix minimum, or makes no changes to those. There are only two possible cases for x_1, while there is exactly one case for x_0. Thus, we obtain $1+(q-2)\lg 3 \approx 1.58496q$, as an information theoretic lower bound for the number of comparisons for this problem.

We do not know of a general way of bringing the amortized number of comparisons from $2 - o(1)$ closer to the $\lg 3$ lower bound, or proving a stronger lower bound. However, if it is known that the distribution of input elements is independent, we can even do better than the lower bound! This improvement is carried out as follows. Suppose that M_i and m_i were already computed. Then, to compute M_{i+1}, M_{i+2}, m_{i+1} and m_{i+2}, we apply the following *incorporate-next-input-pair* algorithm.

Algorithm incorporate-next-input-pair: Extend the result of a solution to PREFIX MAX-MIN to include input elements x_{i+1} and x_{i+2}, using the four following comparisons:

1. *Compare* x_{i+1} *and* x_{i+2}. *Assume, without loss of generality, that* $x_{i+1} \geq x_{i+2}$.
2. *Compare* M_i *with* $x_{i+1} = \max(x_{i+1}, x_{i+2})$.
3. *Compare* m_i *with* $x_{i+2} = \min(x_{i+1}, x_{i+2})$.
4. *At this stage, the algorithm has determined both* M_{i+2} *and* m_{i+2}. *Specifically,* $M_{i+2} = \max(x_{i+1}, M_i)$ *and* $m_{i+2} = \min(x_{i+2}, m_i)$. *There are four cases to consider in computing* m_{i+1} *and* M_{i+1}.
   a) No changes: $x_{i+1} \leq M_i$ *and* $x_{i+2} \geq m_i$. *No more comparisons need to be done in this case, and the algorithm outputs* $M_{i+2} = M_{i+1} = M_i$ *and* $m_{i+2} = m_{i+1} = m_i$.

b) Changes to both the maximum and the minimum: $x_{i+1} \geq M_i$ *and* $x_{i+2} \leq m_i$. *Again, no more comparisons need to be done in this case, and the algorithm outputs* $M_{i+2} = M_{i+1} = x_{i+1}$, $m_{i+1} = m_i$, *and* $m_{i+2} = x_{i+2}$.

c) Change to the maximum: $x_{i+1} \geq M_i$ *and* $x_{i+2} \geq m_i$. *The algorithm outputs* $M_{i+2} = M_{i+1} = x_{i+1}$ *and* $m_{i+2} = m_{i+1} = m_i$. *without additional comparisons.*

d) Possible change to the minimum: $x_{i+1} \leq M_i$ *and* $x_{i+2} \leq m_i$. *This is the only case in which an additional comparison is required: The algorithm first outputs* $M_{i+2} = M_{i+1} = M_i$, $m_{i+2} = x_{i+2}$ *and then determines* m_{i+1} *by comparing* x_{i+1} *with* M_i. *If* $x_{i+1} < m_i$ *then* $m_{i+1} = x_{i+1}$, *otherwise,* $m_{i+1} = m_i$.

Thus, in the worst case, the algorithm makes four comparisons for each pair x_{i+1} and x_{i+2}, where $i > 0$ is odd, which does not improve on the two comparisons for element by the trivial algorithm. The fourth comparison however is needed only in case

$$x_{i+2} < m_i = \min_{0 \leq j \leq i} (x_i), \tag{5}$$

or in the dual case, namely when the first comparison yields $x_{i+1} \leq x_{i+2}$, and

$$x_{i+2} < M_i = \max_{0 \leq j \leq i} (x_i). \tag{6}$$

With i.i.d. the probability of (5) or (6) holding is $1/(i+3)$, for all $i > 0$. Let $u = \lfloor q/2 \rfloor - 1 = (q - (q \bmod 2)) - 1$. Then in the last application of the above algorithm we deal with the pair x_{2u} and x_{2u+1}. In total, F_q, the *expected* (with regard to input distribution) number of times the fourth comparison is made is given by

$$F_q = \frac{1}{4} + \frac{1}{6} + \frac{1}{8} + \cdots + \frac{1}{2u+2} = (H_{u+1} - 1)/2, \tag{7}$$

where H_u is the uth harmonic number. It is well known that

$$\lim_{u \to \infty} H_u = \ln u + \gamma \quad \text{and} \quad \ln u + \gamma \leq H_u \leq \ln u + 1, \tag{8}$$

where $\gamma \approx 0.577216$ is Euler's constant (also called Mascheroni's constant). Combining (7) and (8) we have

$$\begin{aligned} F_q &= \tfrac{\ln(u+1)}{2} + \tfrac{\gamma}{2} - 0.5 + o(1) \\ &\approx \tfrac{\ln(u+1)}{2} - 0.211392 + o(1) \leq \tfrac{\ln(u+1)}{2} \leq \tfrac{\ln q - 1}{2}. \end{aligned} \tag{9}$$

Other than these, in solving PREFIX MAX-MIN, there are u applications of *incorporate-next-input-pair*, in which $3u$ comparisons are made, one comparison in which x_0 is compared with x_1 to determine M_0, M_1, m_0 and m_1, and finally, and only if q is odd, two comparisons to determine M_{q-1} and m_{q-1}. The number of these comparisons is

$$1 + 3u + 2(q \bmod 2) = \frac{3q}{2} - 2 + \frac{q \bmod 2}{2}. \tag{10}$$

Adding (9) and (10) we have that the expected total number of comparisons in our solution to PREFIX MAX-MIN is at most

$$\frac{3q}{2}+\frac{\ln q}{2}-2, \tag{11}$$

and the expected amortized number of comparisons per element is $1.5+\frac{\ln q}{2q}-2/q$. It should be noted that one cannot hope to improve much on this result. The reason is that solving PREFIX MAX-MIN also yields the maximum and the minimum of the whole input. However, computing both these values cannot be done in less than $\lceil 3p/2 \rceil$ comparisons [1, page 187] even for randomized inputs.

4.2. COMPUTING THE MIN-MAX FILTER

We now employ algorithm *incorporate-next-input-pair* in the pre-processing stage of the modified Gil-Werman algorithm adapted for finding both the minimum and the maximum filters. Specifically, we are concerned in this stage in finding an efficient algorithm to the PREFIX-SUFFIX MAX-MIN problem, defined as computing the maximum and the minimum of all prefixes and all suffixes of an array of size $p+1$. Such an efficient algorithm is obtained by partitioning the input array into to two halves. In the lower half which comprises $q=\lfloor (p+1)/2 \rfloor = p/2+(p \bmod 2)/2$ elements we repetitively apply *incorporate-next-input-pair* to compute the prefix maxima and the prefix minima in this half. A similar computation is carried out in the upper half with $p-q+1=\lceil (p+1)/2 \rceil$ elements of the input array, except that algorithm *incorporate-next-input-pair* is mirrored to compute the *suffix* minima and the suffix maxima in this half. The total expected number of comparisons so far can be computed from (11):

$$\tfrac{3q}{2}+\tfrac{3(p+1-q)}{2}+\tfrac{\ln q}{2}+\tfrac{\ln(p+1-q)}{2}-4 \le \tfrac{3p}{2}+\ln p-2.5. \tag{12}$$

Once this computation is done, we carry on as before to produce the rest of the required output. In two more comparisons we find out where the maximum and the minimum of the whole array occur. If the maximum occurs in the lower (resp. upper) half then it remains to compute the suffix (resp. prefix) maxima from the mid-point down-to (resp. up-to) the location of the maximum. From (4) we have that this computation costs another 0.25 comparison per input element. A similar completion stage must be carried out for the minimum prefixes or suffixes, using another 0.25 amortized comparisons. All that remains is the merge step, which has to be carried out twice, once for the minimum and once for the maximum. The number of comparisons for the merge is at most $2\lg p$. Combining this with (12) we obtain:

Theorem 3 *There exists an algorithm for the* 1D MIN-MAX FILTER *problem, that at the worst case makes twice the number of comparisons as that of Theorem 2. For independently distributed inputs, the amortized number of comparisons that the algorithm makes is*

$$C_1^m < 2+2\frac{\ln p+\lg p}{p} = 2+(2+\frac{\ln 2}{2})\frac{\lg p}{p} \approx 2+2.3466\frac{\lg p}{p}.$$

Stated differently, we have that asymptotically for large p, and for i.i.d. one comparison per element is required to compute each of the minimum and the maximum filters, provided they are computed together. We refer to [4] for further details like performance on natural images.

5. An Efficient Algorithm for the Opening and Closing Filters

In this section we describe how the opening (and closing) filter can be computed more efficiently than a mere sequential application of the Max-Filter and then the Min-Filter.

To understand the improvement, consider the problem of computing the prefix-minimum, in the case that the input of length p is given as a sequence of L monotonically increasing or decreasing segments. Suppose that the prefix-minimum has been computed up to a point i, i.e., that the value of $m_i = \min(x_0, \ldots, x_i)$ is known, and that $x_{i+1}, \ldots, x_{i+k}$ is a monotonically decreasing segment of the input of length k. Then, in order to compute $m_{i+1}, \ldots, m_{i+k}$, all that is required is to find the smallest ℓ such that $m_\ell < m_i$. This ℓ can be easily found using a binary search in $\lceil \lg k \rceil$ comparisons. We then have

$$m_{i+j} = \begin{cases} m_i & \text{if } j < \ell \\ x_{i+j} & \text{if } \ell \leq j \leq k. \end{cases}$$

If on the other hand $x_{i+1}, \ldots, x_{i+k}$ is a monotonically *increasing* sequence, all that is required in order to compute $m_{i+1}, \ldots, m_{i+k}$ is to compare x_{i+1} and m_i. In this case we have that $m_{i+1} = m_{i+2} = \cdots = m_{j+k} = \min(x_{i+1}, m_i)$. Using Lagrange multipliers we obtain that the number of comparisons is bounded above by

$$L \left\lceil \lg \frac{p}{L} \right\rceil. \tag{13}$$

Recall now the improved merge step described in Section 3.1. Each iteration of the binary search algorithm generates about half of the outputs of the max-filter that remained to be computed. Note that all values generated in one such iteration are consecutive in the output. Further, since these values are obtained from computing either R_i or S_i, they are either monotonically increasing or monotonically decreasing. Thus an application of the modified max filter algorithm also partitions each stretch of p outputs into at most $\lceil \lg p \rceil$ monotonic segments.

The improved opening filter algorithm is thus obtained by first applying the modified Gil-Werman max-filter algorithm, while preserving this partitioning of the output. Then, the results are fed into the modified Gil-Werman min-filter algorithm. The partitioning information is then used for an efficient implementation of the preprocessing stage in which prefix- and suffix-minima are computed. It follows from (13) that the preprocessing stage can be done in at most $O(\lg^2 p)$ comparisons. Since the merge step can be done in $O(\lg p)$ comparisons, we obtain:

Theorem 4 *There exists an algorithm which computes the opening filter, achieving* $C_1^o = C_1 + O(\frac{\lg^2 p}{p})$.

In other words, asymptotically computing the opening filter is not more expensive than computing just the max-filter.

6. Conclusions

We presented improvements of the Gil-Werman algorithm for running min and max filters. The average computational complexity was shown to be $1.25+o(1)$ per element for a randomized algorithm, without any assumption on the distribution of the data, and $1.5 + o(1)$ for a deterministic algorithm. These improvements, which come close to the best known lower bound for the problem, were enabled by careful examination of the redundancies in the preprocessing and the merge steps of the Gil-Werman algorithm.

We continued to study a related problem, namely the computation of the min and the max filter together. We found that for independently distributed input elements, it is possible to compute the minimum and the maximum filters together in $2 + o(1)$ comparisons per data point. This is less than $2.5 + o(1)$ comparisons required by applying twice the best max filter algorithm.

The opening and closing filters which are similar to the problem of computing the min- and max-filters together, can be computed much more efficiently. We found algorithms for these filters using $1.5 + o(1)$ comparisons deterministically, or $1.25 + o(1)$ comparisons randomly, for worst case inputs.

All algorithms are readily extendible to higher dimensions.

Acknowledgements

Stimulating discussions of both authors with Reuven Bar-Yehuda of the Technion during the writeup of this paper are gratefully acknowledged. The second author is grateful to Renato Keshet from HP Labs. Israel, for intriguing discussions on efficient morphological operators.

References

1. T. H. Cormen, C. E. Leiserson, and R. L. Rivest. *Introduction to Algorithms.* MIT Press, Cambridge, Massachusetts, 1990.
2. D. Z. Gevorkian, J. T. Astola, and S. M. Atourian. Improving Gil-Werman algorithm for running min and max filters. *IEEE Transactions on Pattern Analysis and Machine Intelligence*, 19(5):526–529, May 1997.
3. J. Y. Gil and Z. Gutterman. Compile time symbolic derivation with C++ templates. In *Proceedings of the fourth Conference on Object-Oriented Technologies and Systems (COOTS'98)*, Santa Fe, New Mexico, May 1998. USENIX.
4. J. Y. Gil and R. Kimmel. Further improvements to the Gil-Werman's min and max filters. Submitted to IEEE Transactions on Pattern Analysis and Machine Intelligence, 2000.
5. J. Y. Gil and M. Werman. Computing 2-D min, median, and max filters. *IEEE Transactions on Pattern Analysis and Machine Intelligence*, 15(5):504–507, May 1993.
6. I. Pitas:1989:FAR. Fast algorithms for running ordering and max/min recalculations. *IEEE Transactions on Circuits and Systems*, CAS-36(6):795–804, June 1989.
7. J. Serra. *Image analysis and mathematical morphology.* Academic Press, New York, 1982.

FAST MORPHOLOGICAL ATTRIBUTE OPERATIONS USING TARJAN'S UNION-FIND ALGORITHM

MICHAEL H. F. WILKINSON and JOS B. T. M. ROERDINK
Institute for Mathematics and Computing Science,
University of Groningen, P.O. Box 800, 9700 AV, Groningen,
The Netherlands

Abstract. Morphological attribute openings and closings and related operators are generalizations of the area opening and closing, and allow filtering of images based on a wide variety of shape or size based criteria. A fast union-find algorithm for the computation of these operators is presented in this paper. The new algorithm has a worst case time complexity of $O(N \log N)$ where N is the image size, as opposed to $O(N^2 \log N)$ for the existing algorithm. Memory requirements are $O(N)$ for both algorithms.

Key words: Area Operators, Attribute Operators, Granulometries, Union-find Algorithm.

1. Introduction

Morphological attribute openings, thinnings and granulometries were introduced by Breen and Jones [1] as a generalization of morphological area operators proposed by Vincent [8, 9]. Attribute openings are most easily understood in the binary case. Unlike structural openings, attribute openings are shape preserving, because they simply test whether a connected component satisfies some increasing criterion T. If it does, it is retained, if not, it is removed. In the case of the area opening, the area of each component is compared to some threshold value λ, and if the area of the component is larger, it is retained. The flexibility of this methodology is shown in Figure 1. In this figure a binary image of bacteria is filtered using three attribute openings, each of which would remove all squares smaller than 11×11 pixels. The first is the area opening, with $\lambda = 121$. All small bacteria have been removed in the resulting image. By contrast, the attribute opening using the criterion that the moment of inertia I must be larger than $\lambda = 11^4/6$, removes most of the smaller components, but not the elongated ones. Attribute opening using the length of the diagonal of the minimum enclosing rectangle as criterion, with $\lambda = \sqrt{242}$, has similar results.

The algorithm Breen and Jones derive for their wider class of operators is based on Vincent's pixel queue algorithm for area operators. Recently, a new algorithm for area openings and closings has been developed [5], which is based on Tarjan's union-find algorithm [7]. It was found that the union-find based algorithm was between 2 and 10 times faster than the original algorithm on the images tested. Furthermore, the computational burden of the new algorithm was practically independent of the size criterion λ used, or the image content. By contrast, Vincent's algorithm is particularly sensitive to the presence of

(a) (b) (c) (d)

Fig. 1. Attribute openings of an image of bacteria: (a) a binary image of 256 × 256 pixels; attribute using (b) area $A \geq 121$; (c) Moment of inertia $I \geq 11^4/6$, corresponding to that of an 11 × 11 square, and (d) length of diagonal of minimum enclosing rectangle $D \geq \sqrt{242}$. Structural opening of (a) by an 11 × 11 square structuring element removes all objects.

linear structures in the image, in which case the computing time rises almost linearly with λ.

In this paper we extend the union-find algorithm to the wider class of attribute openings and closings. Later work will focus on extension to thinnings and thickenings, and granulometries or size distributions.

2. Attribute Morphology: Theory

The theory of attribute operators is given only briefly here. For a more thorough discussion the reader is referred to [1]. Here we will first discuss binary attribute openings and closings, and then the extension to the grey scale case. Binary attribute openings are based on binary connected openings. Let the set $X \subseteq \mathbf{M}$ denote a binary image with domain $\mathbf{M}$. The binary connected opening $\Gamma_x(X)$ of X at point $x \in \mathbf{M}$ yields the connected component of X containing x if $x \in X$, and $\emptyset$ otherwise. Thus Γ_x extracts the connected component to which x belongs, discarding all others. Breen and Jones then use the concept of trivial openings Γ_T, which use an increasing criterion T to accept or reject connected sets. A criterion T is increasing if the fact that C satisfies T implies that D satisfies T for all $D \supseteq C$. The trivial opening Γ_T of a connected set C with increasing criterion T is just the set C if C satisfies T, and is empty otherwise. Furthermore, $\Gamma_T(\emptyset) = \emptyset$. The binary attribute opening is defined as follows.

Definition 1 *The binary attribute opening Γ^T of set X with increasing criterion T is given by*

$$\Gamma^T(X) = \bigcup_{x \in X} \Gamma_T(\Gamma_x(X)) \tag{1}$$

It can be shown that this is an opening because it is increasing, idempotent, and anti-extensive [1]. The attribute opening is equivalent to performing a trivial opening on all connected components in the image.

A generalization to grey scale can be made by first defining thresholded images $X_h(f)$,

$$X_h(f) = \{x \in \mathbf{M} | f(x) \geq h\} \tag{2}$$

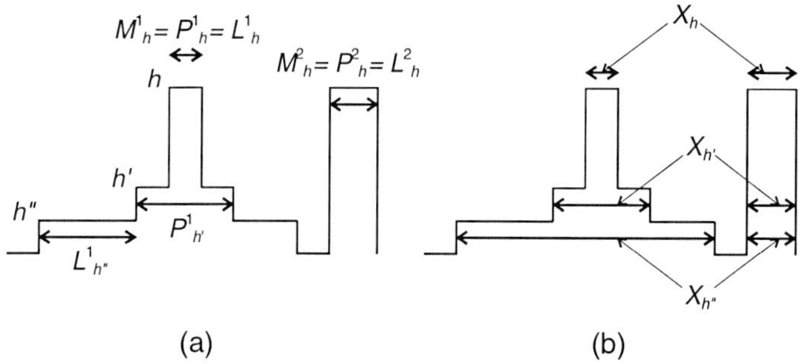

Fig. 2. One dimensional discrete image with grey levels $h > h' > h''$ to illustrate the definitions of level components, regional maxima, peak components, and the threshold images: (a) double arrows indicate three level components L^1_h, L^2_h and $L^1_{h''}$; the former two are also both peak components P^1_h and P^2_h and regional maxima at level h; a further peak component $P^1_{h'}$ at level h' is also shown; (b) shows the threshold sets $X_h, X_{h'}$, and $X_{h''}$ in relationship to the grey scale image.

where the grey scale image f is a mapping from the image domain $\mathbf{M}$ to $\mathbb{Z} \cup \{-\infty, \infty\}$.

Definition 2 *The grey scale attribute opening γ^T of image f with increasing criterion T is given by*

$$(\gamma^T(f))(x) = \max\{h | x \in \Gamma^T(X_h(f))\}. \tag{3}$$

Grey scale attribute closings can easily be defined by a duality relationship with the grey scale attribute openings [5].

3. Algorithms

Before going into the details of the algorithms, we first define a *level component* L_h at level h of a grey scale image f as a connected component of the set of pixels $\{p \in \mathbf{M} | f(p) = h\}$. A *regional maximum* M_h at level h is a level component no members of which have neighbors larger than h. A *peak component* P_h at level h is a connected component of $X_h(f)$. At each level h there may be several such components, which will be indexed as L^i_h, P^j_h and M^k_h, respectively, with i, j, and k from some index set. It can be seen that any regional maximum M^k_h is also a peak component, but the reverse is not true. Examples of these three types of components, and of the threshold sets $X_h(f)$ are given in Figure 2.

All level components L^i_h at level h are of course subsets of some peak component $P^j_h \subset X_h(f)$ at the same level h. However, for a given criterion T, not necessarily all $L^i_h \subset \Gamma_T(P^j_h)$, because not all P^j_h need meet the criterion T. It can be seen from (3), that not all level components are necessarily affected by a grey scale attribute opening. Only those L^i_h which are *not* subsets of a peak component P^j_h which meets the criterion T must be changed in grey level by γ^T. In other words, all $L^i_h \subset \Gamma_T(P^j_h) \subset \Gamma^T(X_h(f))$ must be left unaltered. If

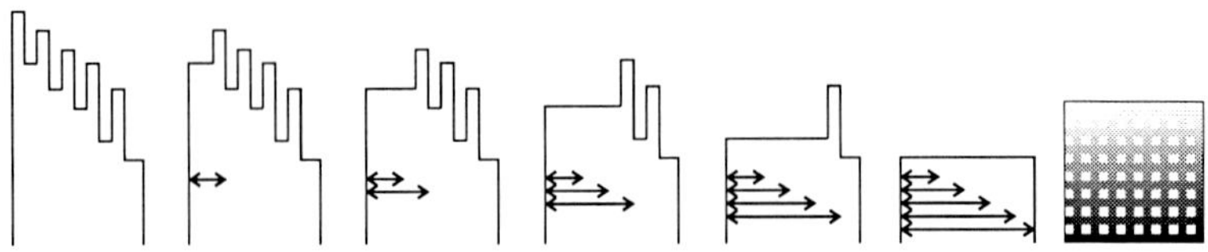

(a) (b) (c) (d) (e) (f) (g)

Fig. 3. Processing nested maxima by the pixel queue algorithm, assuming only the peak at full width meets the criterion: (a) original image; (b-f) situation after processing the first, second, third, fourth, and fifth maximum from the left. At each stage the pixels indicated by the double arrows have been inspected. After visiting the current maximum, the algorithm first inspects the pixels to the left of the maximum, because the valley to the right is lower. This results in a frequent rescanning of pixels of the left-most regional maxima. (g) 16×16 pixel image showing nested maximum structure on which the computational burden is expected to be $O(N^2 \log N)$.

we assume that the peak component $P^1_{h'}$ in Figure 2 meets the criterion, the level component labeled $L^1_{h''}$ in the same figure will remain unaltered. This is because $h'' < h'$, so that $P^1_{h'} \subset P^1_{h''}$ (the latter is not shown in the figure), and since T must be increasing in the case of a grey scale attribute opening, $P^1_{h''}$ must meet T. By contrast, assume that $P^1_h = M^1_h = L^1_h$ does not meet the criterion, and therefore $L^1_h \not\subset \Gamma_T(P^1_h)$ since $\Gamma_T(P^1_h) = \emptyset$. Then the grey level of L^1_h must be altered to h', because $P^1_{h'} \supset L^1_h$ is the smallest peak component containing L^1_h which meets T.

3.1. The Pixel Queue Algorithm

The pixel queue based algorithms for morphological area and attribute operators are given in some detail elsewhere [1, 5, 8, 9], so we will describe them only briefly. The source code of our implementations is available on request.

Briefly, the image is first scanned using a pixel queue to create a list of all regional maxima M^k_h. After this, all M^k_h are processed sequentially. This is done by growing a peak component $P^j_{h_1}, h_1 \leq h$ around a seed pixel within the maximum M^k_h using a priority queue. As each pixel is added to the growing region, its neighbors which do not (yet) belong to the region are put in the priority queue, from which they are retrieved in reverse grey level order. The process of adding pixels pauses whenever the next pixel taken from the priority queue has a grey level h'' different from the current level h'. If $h'' > h'$, the region grown so far is not a peak component $P^j_{h'}$ at level h'. All the grey levels of pixels found so far are set to h', and the maximum M^k_h from which the region was grown is removed from the list. If $h'' < h'$, the region grown so far is a peak component at h', which is subsequently checked against the criterion. If the criterion is met, the grey level of all pixels $p \in P^j_{h'}$ are set to h', and M^k_h is removed from the list. Otherwise, the routine continues adding new pixels at level h''. The algorithm terminates when all maxima have been processed.

One problem which occurs is that pixels may be visited more than once if nested maxima exist, especially if the attribute threshold λ is large. This effect can be seen in Figure 3. In this one-dimensional example, the algorithm processes the maxima from left to right, and each time only detects that the growing region is not a peak component *after* having re-visited all pixels visited by the previous region-growing loop. If λ is chosen so that the entire image (of N pixels) is the smallest set satisfying the criterion, it is possible to construct an image in which each pixel is processed $O(N)$ times. A two-dimensional example can be seen in Figure 3g. At each visit a pixel has to be inserted into and retrieved from a priority queue of length of order $\sqrt{N}$. In that case we arrive at a worst case running time of $O(N^2 \log N)$. Strictly speaking, this effect is due to awkward arrangement of the depths of the valleys between the maxima, not to the heights of the maxima. Had all the maxima in figure 3(a) been given the same height (as is the case in figure 3(g)), the problem still remains. Processing the maxima in order of grey level does not solve the problem.

The algorithm requires a label image of N pixels, and a (priority) queue also of N pixels in the worst case. Therefore its memory requirements are $O(N)$.

3.2. The Union-Find Method

Tarjan [7] presents the union-find algorithm which provides a general method for keeping track of disjoint sets. It allows performing set-union operations on sets which are in some way equivalent, while ensuring that the end product of such a union is disjoint from any other set. Since connected components and level components in an image are by definition disjoint sets, the union-find algorithm lends itself to any image processing method which is defined by such image components. Dillencourt et al. [2] have shown that the union-find algorithm can be used for efficient connected component labeling of arbitrary image representations. Fiorio and Gustedt propose a similar algorithm [3], and Meijster and Roerdink [4] adapted the algorithm to level-component labeling. Since attribute openings and closings are connected filters, their operation can be defined directly in terms of connected components in the binary case and level components in the grey scale case. This means Tarjan's algorithm can be adapted to attribute openings. This is born out by the application of the algorithm to area openings [5].

Tarjan uses tree structures to represent sets. Each non-root node in a tree points to its parent, while the root is flagged in some way. Two objects x and y are members of the same set if and only if x and y are nodes of the same tree, which is equivalent to saying that they share the same root. There are four important operations.

- `Makeset(x)`: Create a new singleton set $\{x\}$.
- `FindRoot(x)`: Return the root element of the set containing x.
- `Union(x,y)`: Compute the union of the two sets containing x and y.
- `Equiv(x,y)`: determine whether x and y satisfy some equivalence criterion.

For level component labeling the algorithm becomes:

```
for pixels p do
  { MakeSet(p);
    for all neighbors n<p do
      if ( Equiv( n, p ) )
        Union( n, p );
  }
```

Note that in this context the condition `n<p` means that `n` is a pixel which has been processed before `p`. In this case `Equiv(n,p)` is true if the image value `I[n]` equals `I[p]`. `Union` calls `FindRoot` internally to determine the root nodes of the trees containing `n` and `p`. After this scan, a second "resolving" scan assigns each root pixel a unique label, and to each non-root pixel the label of its root.

Before going into the details of the attribute opening algorithm itself, we will discuss the general framework for storing the disjoint sets, and the auxiliary functions needed for attribute openings and closings.

The disjoint sets we have to find are all level components $L_h^i \subset \Gamma^T(X_h(f))$, which are not altered by the attribute opening γ^T, and, for all other L_h^i, the smallest peak component $P_{h'}^j \supset L_h^i$ which meets T. Each of these sets is represented as a tree, with each pixel containing a pointer to its parent pixel. To store the trees for the entire image, we use an integer array `parent` of the same size as the image (i.e., N), in which `parent[p]` is the parent of pixel `p`. Pixels are stored as `width*y+x`,with `x` and `y` the pixel's x and y coordinates, and `width` the image width. If a pixel is a root of a tree, i.e. it has no parent, we flag this by setting `parent[p] < 0`. If `p` is the root of a *peak* component which does not meet the criterion, we call the pixel an *active root*, which is flagged by `parent[p] = ACTIVE`< 0. All other roots are labeled `INACTIVE` (< 0). An array `auxdata` of N void pointers is used to store pointers to any auxiliary data about the peak component (e.g., area size, centroid location, etc.) needed for computation of the attribute. Only if `p` is an active root does `auxdata[p]` point to valid data. This allows us to process different extrema simultaneously, rather than sequentially.

To perform any kind of attribute opening using a single routine, pointers to four functions must be passed to the attribute opening procedure:

- `NewAuxData`, which initializes the auxiliary data,
- `DisposeAuxData`, which discards them,
- `MergeAuxData`, which merges two sets of auxiliary data,
- `Attribute`, which computes the attribute based on the auxiliary data.

The pseudo-code for the `MakeSet`, `FindRoot`, `Equiv`, and `Union` routines is shown in Figure 4. In this case, as in the case of the area opening [5], the `Equiv` and `Union` routines are asymmetrical. This is done to ensure that if the set we are dealing with is a peak component P_h at level h, the root element `r` has a grey level `I[r]`$= h$. Therefore, we process the pixels in decreasing grey level order, and always make the last pixel processed the root of the new tree. We do this by radix-sorting the pixels, and storing the coordinates in an array `SortPixels` of length N. Pixels of the same grey level are processed in scan line order. Scanning of peak components from high to low grey levels is guaranteed, without finding regional maxima explicitly.

As each pixel `p` is processed, we first check whether its grey level `I[p]` is different from its predecessor's greylevel `I[p-]`. If so, all active roots with grey level `I[p-]` `q` are inspected, checking whether `Attribute(auxdata[q])` $\geq \lambda$. If so, they are labeled as `INACTIVE` and their auxiliary data are discarded. After this clean up, the `MakeSet` routine labels `p` as a singleton set, setting `parent[p]` to `ACTIVE`, and calling `NewAuxData`, passing the pixel `p` to it (see

```
void MakeSet ( int x )
{ parent[x] = ACTIVE;
  NewAuxData(x);
}

void Link ( int x, int y )
{ if ( (parent[y] == ACTIVE) and (parent[x] == ACTIVE) )
    { auxdata[y] = MergeAuxData(auxdata[x],auxdata[y]);
      DisposeAuxData(auxdata[x]);
    }
  else if (parent[x] = ACTIVE)
         DisposeAuxData(auxdata[x]);
       else
         { DisposeAuxData(auxdata[y]);
           parent[y] = INACTIVE; }
  parent[x] = y;
}

int FindRoot ( int x )
{ if ( parent[x] >=0 )
    { parent[x] = FindRoot( parent[x] );
      return parent[x];
    }
  else return x;
}

boolean Equiv ( int x, int y )
{ return ( (I[x] == I[y]) or (parent[x] == ACTIVE) );
}

void Union ( int n, int p )
{ int r = FindRoot(n);
  if ( r != p )
    if ( Equiv(r, p) )
      Link( r, p );
    else if (parent[p] == ACTIVE)
           { parent[p] = INACTIVE;
             DisposeAuxData(auxdata[p]); }
}
```

Fig. 4. The basic operations for attribute openings and closings using the union-find method. The negative constants ACTIVE or INACTIVE flag active and inactive roots in the parent array, and `auxdata[p]` contains pointers to auxiliary data. The variable `lambda` is equal to the parameter λ. The parameters of *Equiv* must be root nodes. Linking is done if either the image values `I[x]` and `I[y]` are identical, or if `x` is a root of an active peak component.

figure 4). The Union procedure is now called for each neighbor n which has already been processed. We briefly describe this procedure here. Since the p is *always* a root, FindRoot is only called to find the root pixel r of n. Next, Equiv is called with r and p as parameters. If the grey level I[r] of r is equal to that of p *or* if r is active, Equiv returns "true" and the two trees are merged using the Link routine (see figure 4). If Equiv returns "false", a neighbor has a root grey level higher than I[p] *and* is inactive, so p $\in L_h^i \subset \Gamma^T(X_h(f))$. Therefore, p is set to inactive, and its auxiliary data are discarded. The Link routine always assigns p to parent[r]. Before that, Link inspects both roots. If both r and p are active, MergeAuxData is called on the auxiliary data of r and p, storing the result in the auxiliary data of p, and discarding the auxiliary data of r. If either r or p are inactive, the active root is set to inactive, and its auxiliary data are discarded.

In pseudo code this part of the algorithm now becomes:

```
for pixels p do
  { if ( I[p] != I[p-] )
      for all pixels q with I[q]==I[p-] do
         if ( ( parent[q] == ACTIVE ) and
              ( Attribute(auxdata[q]) >= lambda ) )
           { parent[q]=INACTIVE;
             DisposeAuxData(auxdata[q]);
           }
    MakeSet(p);
    for all neighbors n<p do
        Union( n, p );
  }
```

Here p- denotes the pixel processed immediately before p. This part of the algorithm requires $O(N \log N)$ operations in the worst case [5, 7].

At the end of this part of the algorithm, we have found two kinds of disjoint sets: (i) those with constant grey level, which are level components $L_h^i \subset \Gamma^T(X_h(f))$, and (ii) those with varying grey level, which are peak components P_h^j, with h the maximum grey value for which the criterion is satisfied. Because the root r of these peak components is always the last pixel processed, its grey level in the input image satisfies $f(r) = h$. Therefore, if we set the grey level of each pixel in the output image to that of its root in the input image, all $L_h^i \subset \Gamma^T(X_h(f))$ remain unchanged, whereas all P_h^j are filled uniformly with a grey level of h, as in the previous algorithm. Assigning all pixels the grey level of the root of their component can be done in linear time [5]. The simplest approach is to store the output image in the parent array:

```
For each pixel p in reverse sort order do
  if (parent[p] < 0) then
    parent[p] = I[p];
  else
    parent[p] = parent[parent[p]];
```

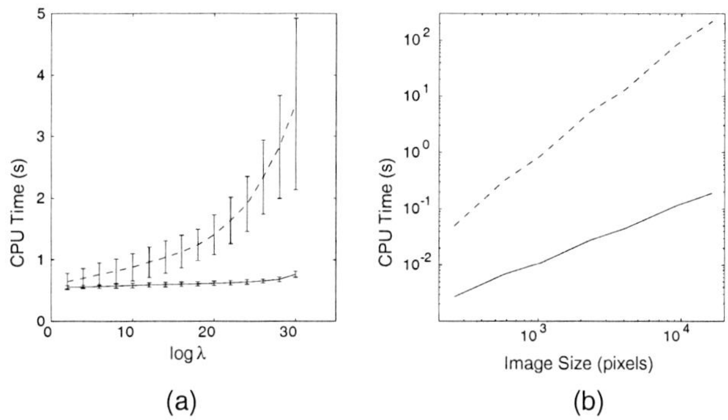

Fig. 5. Timing of algorithms for mean and pathological cases: (a) Mean CPU times and standard deviations of moment of inertia closings as a function of λ for the union-find (solid line) and pixel queue (dashed line) algorithms, for 20 natural images of 256×256 pixels; (b) log-log plot of timing results as a function of image size N for both algorithms, for the case shown in Figure 3a, showing at least $O(N^2)$ complexity for the pixel queue method (dashed), versus $O(N)$ for union-find (solid).

Thus, each pixel which has a negative `parent[p]` is a root, and is assigned its image value. Every image value which is not a root has a `parent[p]` which always points to a pixel which was processed *later* in the first phase of the algorithm, and which will therefore always have been assigned the correct image value *before* the current pixel in the reverse order scan. At the end of this phase, the array `parent` contains the output image.

It might be thought that finding all components $L_h^i \subset \Gamma^T(X_h(f))$ as well as those L_h^i which need to be changed is wasteful, compared to the region-growing phase of the pixel queue approach, which only grows the peak components, without visiting those L_h^i which need not be altered. However, during the phase in which the maxima are sought, the pixel queue algorithm also visits all L_h^i, regardless of whether they should be altered or not.

4. Timing Results

To compare the computational complexities of both algorithms on real images as a function of the attribute threshold λ, we computed moment of inertia openings with increasing λ for 20 natural images of 256×256 pixels, including microscopic images, buildings, portraits, aerial photographs and astronomical images. The results are shown in figure 5a. Similar results were obtained using the diagonal of the minimum enclosing rectangle as attribute (data not shown). As in the case of area openings and closings [5], CPU times for the pixel queue algorithm depend strongly on λ, and on image content, hence the large standard deviations of the timings. The union-find approach is faster in all cases, except for small λ in a few images. The coefficient of variance of the timings is also much smaller, indicating a far smaller dependence on image contents.

A far more dramatic difference in CPU times was observed in a set of artificial images, designed to demonstrate the $O(N^2 \log N)$ worst case behavior of the pixel queue algorithm. The images consist of maxima nested in such a way that each time a new maximum is flooded, *all* previously processed pixels must be visited again, in a two-dimensional variant of the case shown in Figure 3. A 16×16 pixel image from this set is shown in figure 3g. A log-log plot of CPU times versus image size for both algorithms for images ranging from 16×16 up to 128×128 pixels is shown in figure 5b. The pixel queue algorithm shows a quadratic dependency of CPU time on image size, and at a size of 256×256 the CPU time was 4340 s. By contrast, the union-find algorithm shows a linear dependency on image size, and needs only 0.7 s for an image of 256×256 pixels. It appears that the pixel queue approach does not handle nested extrema efficiently, contrary to what has been claimed [1, 8, 9].

5. Conclusions

It has been shown that the union-find algorithm is a fast method for computing attribute openings, especially at high values of λ. Its theoretical worst case is an order of magnitude smaller than that of the pixel queue algorithm, though it may require more memory, depending on the attribute used. However, both algorithms require $O(N)$ memory in the worst case (apart from the image itself).

Further work is in progress to extend the algorithm to computation of attribute granulometries. It is expected that the speed gains will be considerable, since the pixel queue methods are particularly slow when large connected components must be scanned, i.e. at large λ. The union-find algorithm does not suffer from this drawback. A comparison to the MAX-tree approach for computation of anti-extensive connected filters [6] must also be made.

References

1. E. J. Breen and R. Jones. Attribute openings, thinnings and granulometries. *Computer Vision and Image Understanding*, 64(3):377–389, 1996.
2. M. Dillencourt, H. Samet, and M. Tamminen. A general approach to connected-component labeling for arbitrary image representations. *Journal of the ACM*, 39:253–280, 1992.
3. C. Fiorio and J. Gustedt. Two linear time union-find strategies for image processing. *Theoretical Computer Science (A)*, 154:165–181, 1996.
4. A. Meijster and J. B. T. M. Roerdink. A disjoint set algorithm for the watershed transform. In *Proc. EUSIPCO'98, IX European Signal Processing Conference*, pages 1665–1668, Rhodes, Greece, 1998.
5. A. Meijster and M. H. F. Wilkinson. An efficient algorithm for morphological area operators. Technical Report 99-9-07, Institute for Mathematics and Computing Science, University of Groningen, submitted.
6. P. Salembier, A. Oliveras, and L. Garrido. Anti-extensive connected operators for image and sequence processing. *IEEE Transactions on Image Processing*, 7:555–570, 1998.
7. R. E. Tarjan. Efficiency of a good but not linear set union algorithm. *Journal of the ACM*, 22:215–225, 1975.
8. L. Vincent. Grayscale area openings and closings, their efficient implementation and applications. In *Proc. EURASIP Workshop on Mathematical Morphology and its Application to Signal Processing*, pages 22–27, Barcelona, Spain, 1993.
9. L. Vincent. Morphological area openings and closings for grey-scale images. In Y.-L. O, A. Toet, D. Foster, H. J. A. M. Heijmans, and P. Meer, editors, *Shape in Picture: Mathematical Description of Shape in Grey-level Images*, pages 197–208. NATO, 1993.

A CHANGE DETECTOR BASED ON LEVEL SETS*

F. GUICHARD, S. BOUCHAFA and D. AUBERT
INRETS, F-94114 ARCUEIL, FRANCE.
author@inrets.fr
Fax (33) 1 40 43 29 31

Abstract. This paper presents a local measurement based on the level lines within an image. Its most important feature is that it separates local geometry (the shape of the level lines) from local contrast (the grey-levels). Using only the first of these we have derived two types of motion detection one of which relates to the disappearance of local level lines and the other to a change in their local geometry. The nature of the measurement allows us to use both a short term and long term time reference and therefore detect objects that are moving or that were not present a few minutes (for example) before. We have used this technique in a number of applications. Appraisals by transportation operators have provided encouraging results.

Key words: Change Detection, Motion Detection, Level Sets, Video Surveillance.

1. Introduction

The aim of detecting motion or change is to locate parts of the image that move (or that appear or disappear). In this paper we shall assume that the video camera is fixed with respect to the background.

When lighting conditions are constant or known, the problem involves detecting changes in intensity. The first algorithms (see [5, 10], and for a good survey of this, [18]) generally started by analyzing change over time in grey-levels (or any measurements based on them, gradient, wavelet coefficients,...), as for the background these changes will be zero, if we ignore noise.

However, intensity-based measurements are contrast dependent, so the method is sensitive to lighting changes. Of course, with a high enough frame rate the change in contrast between successive images is small, and up to a certain limit comparisons between such measurements can still be effective.

The use of edge maps, such as zero crossings of the Laplacian or a Canny-Deriche edge detector, is an important step towards achieving robustness against contrast change [5, 7]. As edges (generally) correspond to large intensity variations, an edge at a pixel is more stable than the grey-level value. However, such edge selection is performed on the basis of intensity (and intensity-derived) criteria. As it is contrast-dependent it will either be sensitive to contrast change or discard low contrast zones (or both). Image level lines can be considered as being roughly equivalent to an edge map where no selection has been made.

Other methods are based on comparisons between estimated motion and

* A part of the work presented is undertaken in the CROMATICA project. It is granted by the EC in the 4th PCRD framework.

an estimated dominant velocity model (in our case 0 as the camera is fixed) [13, 4, 9]. A good survey of this can be found in [9]. Such methods are effective on condition that the motion estimate is reliable. (Note that motion estimates are also generally based on contrast-sensitive image measurements). However, when the camera is fixed, as in our cases, it is not necessary to compute motion in order to detect changes.

The nature of the detection problem is also slightly different if the aim is to construct a change detector rather than an instantaneous motion detector. For example, counting the cars waiting at a pedestrian crossing involves detecting objects that are not necessarily moving but which were not present a few minutes before. This problem is generally addressed by constructing reference image which gives the average grey-level measurement for each pixel over previous minutes (see, for example [7, 17] and the references therein). This clearly assumes that under changing lighting conditions measurements will remain stable, or be stabilized (for example by histogram modification or other more precise methods).

This paper will deal essentially on the determination of a measurement that separates geometry from contrast with a view to designing a change detector. We shall begin by giving a brief description of a "contrast invariant" representation of the image as proposed in [6, 11, 12] which will provide us with our framework. We shall then investigate possible ways of making local measurements where contrast information is separated from geometrical information. The nature of local detection will be specified. Thirdly, we shall propose a simple "change" detection algorithm that uses only geometrical data. This will aim to identify objects that have appeared in the image within a given historical time threshold. With a short historical time threshold the device becomes an instantaneous motion detector. Finally, we will show some of the real time experiments done on real situations.

2. Sensor Modelization and a "Contrast Invariant" Measurement

2.1. Representation of the sensor

We can represent image acquisition in a very approximate manner by the following sequence of operations: scene contrast, smoothing, sensor contrast adjustment, quantization, sampling. This representation is very minimal, and is certainly open to criticism, but it will provide us with a framework and help to show the limits of the proposed method. We shall use the term "global contrast change" to describe a global change in the intensity values of the entire image in which the relative levels of illumination are retained. That is to say that an image J can be deduced from an image I by a contrast change if there is non-decreasing function g from $\mathbb{R}$ to $\mathbb{R}$ such that $J = g(I)$. In our representation, we shall consider that the "scene contrast" and "sensor contrast adjustment" represent a global contrast change. We shall also assume that the role of smoothing, which is mainly due to the lens, is negligible, in that it occurs at a level smaller than one pixel. It should now be noted that quantization can be viewed as a contrast change. Indeed, quantization between a value of 0 and

255 can be written as a contrast change with $g_q(x) = E(Max(0, Min(x, 255)))$ where E is the truncation function and g_q is a non-decreasing function. We shall ignore noise for the time being and consider its impact in section 4.

These assumptions imply that the image I can be deduced from the scene luminance by a change in contrast. This stage is followed by sampling. If our hypothesis that smoothing is negligible is correct I can be compared to a sampled version of the scene which we shall refer to as S_d . For the function g that represents the combined effect of all the contrast changes we have

$$I = g(S_d) \tag{1}$$

It should be noted that g may change over time, so that if there are a large number of observations of I_i and S each of them will satisfy (1) ($I_i = g_i(S_d)$). Let us now consider what common features are shared by all the images I_i.

2.2. Representation of the images by means of level-sets

Let $I(i, j)$ denote the intensity of the image I at the pixel location (i, j). The level set λ of I is the set of pixels with an intensity equal or greater than λ, that is to say:

$$\mathcal{X}_\lambda I = \{\mathbf{x} = (i, j), \text{such that } I(\mathbf{x}) \geq \lambda\}. \tag{2}$$

We shall refer to the boundary of this set as the λ level line. In the case of a digital image, a level line is formed by a finite number of Jordan curves. As a consequence of (2), the level sets of an image are included in others. (If $\lambda \geq \mu$ then $\mathcal{X}_\lambda I \subset \mathcal{X}_\mu I$). Therefore the level lines do not cross each other. As $I(\mathbf{x}) = \sup\ \{\lambda\ /\ \mathbf{x} \in \mathcal{X}_\lambda\}$, The data of the family of the $\mathcal{X}_\lambda$ is sufficient to reconstruct the image [12, 6]. Therefore, the family of the level sets provides a complete representation of the image.

2.3. The impact of contrast change

We shall now examine the effects of global contrast changes in order to distinguish between them and changes due to motion and new objects, etc. Let us return to our representation of the sensor and consider the relationship between the scene S_d and the observation of this image I. With g a contrast change,

$$\text{If } I = g(S_d) \text{ then } \{\mathcal{X}_\lambda I\}_{\lambda \in \{0,\ldots,255\}} \subset \{\mathcal{X}_\lambda S_d\}_{\lambda \in I\!R} \tag{3}$$

This means that all the level sets in the observed image are present in the scene. The operation by which the image is produced from the scene can be considered as the straightforward removal of some level sets with a possible change in their levels. However, the geometry of the remaining level lines is not affected. Let us now consider what occurs between two observations I_1 and I_2. From (3) we can obtain:

$$\{\mathcal{X}_\lambda I_1\}_{\lambda \in \{0,\ldots,255\}} \subset \{\mathcal{X}_\lambda S_d\}_{\lambda \in I\!R} \text{ and } \{\mathcal{X}_\lambda I_2\}_{\lambda \in \{0,\ldots,255\}} \subset \{\mathcal{X}_\lambda S_d\}_{\lambda \in I\!R} \tag{4}$$

A level set in I_1 is not necessarily a level set in I_2 or vice-versa. All we can state for certain is that both families of level sets are subsets of the family of level sets in S_d. Therefore, the level lines in I_1 and I_2 do not cross each other, on condition that I_1 and I_2 are observations of the same scene.

This already provides us with a useful basis for motion detection. The detection of level lines that cross in successive images is an indicator that something other than a change in contrast has occurred between the two images. However, a change detection algorithm based only on this criterion will only provide information about those parts of the images which have level lines, being effectively blind in homogeneous areas.

2.4. Outline of an algorithm based on reference data

Comparing two observations of a scene is far more limiting than comparing an observation directly with the scene S_d. For this reason, in order to improve motion detection we have to construct a reference data which represent the background image of the scene. This reference data will be built up by combining all the level lines in the observed images. If there is no motion and no noise, the reference data will provide a good representation of all the level lines in the background of the scene. From (3), the following possibilities exist for a section of level lines in a new image. Firstly, this section exists (to within a certain approximation) in the reference data, in which case no detection occurs. Secondly, some of the level lines can cross a level line which is contained in the reference data, in which case detection occurs (in what follows we shall refer to this type of detection as "strong detection"). Finally, it is possible that the area of level lines does not exist in the reference image. This absence may either occur when the reference image is sufficiently complete (in which case detection of a type we shall refer to as "weak detection" occurs) or when the reference image is not sufficiently complete (in which case we do not know if a detection or merely a change in contrast has occurred).

3. Local Measurement Based on Level Lines

Measurements based on level lines have already been explicitly or implicitly proposed. For example, in [3] or, a slight variant of it, [14] the authors defined characteristics based on the shape of the connected sections of the level set. However, for our application the non-local nature of these makes them difficult to use. In [2] a motion detector using the variation over time of a gradient-based detector measurement was evaluated, and the authors noted that their results were considerably improved if gradient amplitude was discarded. They therefore proposed to use gradient orientation alone (in a non-digitized image, the orientation of the gradient is contrast independent). Let us now consider what possible contrast invariant local measurements exist. It is apparent from the previous section that image representation using level lines separates image information into contrast (its levels) and geometry (its shapes). The reason for making local measurements is to avoid mixing measurements which relate to moving objects with measurements which relate to the background. To achieve this we have defined "local" by two conditions. Firstly, we have measured only the geometry of all level lines that pass through the current point. Secondly,

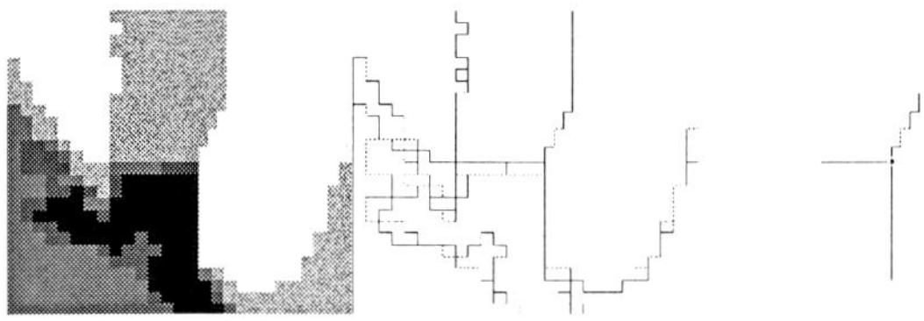

Fig. 1. The observation window around a point is here defined as the straight part of the set of level lines passing through it. The directions of these lines are used in change detection.

starting from the current point, we have taken measurements along each of these lines until the orientation of the tangent changes. At each point we have thus defined an adaptive neighbourhood which depends solely on the geometry of the level lines and is therefore contrast invariant. A simple measurement that we have used is thus to list the direction of the half-tangent of each of the level lines passing through each point.

4. A Change Detection Algorithm

4.1. Local direction extraction

Extraction of the local direction of the level lines can be performed recursively. Starting at each point, all level lines are followed until a direction change, up to the discrete grid, is encountered. We have considered that the direction changes when the path covered is no more the sampling on the grid of a line segment. Once all the recursions have been completed we have, for each point, a description of the straight sections (half-tangents) of the level lines passing through it (See Figure 1). The precision to within which the orientation of each half-tangent is known depends on the path length (the longer this is the higher the precision). Path lengths of less than three steps are discarded at this stage, which means that a precision of $\pm\pi/8$ is achieved. The only parameter we have used at this stage of processing is the **minimum of 3 steps**. The measurement performed at each pixel therefore consists of a list of angles $\theta_0, ..., \theta_i$ and the number of level lines for each of these $\delta_0, ..., \delta_i$. (i depends on the pixel). By this means we have separated geometry (the angles) from the contrast (number of level lines).

4.2. Constructing the reference data

The aim of the reference data is to show the normal configuration of the directions at each point. The directions are quantized to give a discrete set of η directions. If the duration for which a direction occurs exceeds a certain threshold it will be included in the reference data. The occurrence of the direction θ_i at pixel $\mathbf{x}$ is represented by the number $O(\mathbf{x}, \theta_i)$. As each image is received this number is increased if the direction is present. The reference data for this point and this direction are obtained by thresholding this number.

It should be noted that different scene configurations may be present simultaneously in the reference data. Some objects, for example traffic signals, change over time but are present for a sufficiently large proportion of the time to be included in the reference data and therefore not detected as changes.

The parameters at this stage are as follows:

Direction quantization. We have chosen to quantify the direction at intervals of $\pi/8$, which gives $\eta = 32$. This is the consequence of an experimental trade-off between memory use (one counter per direction per pixel $O(\mathbf{x}, \theta_i)$) and the number of different geometrical configurations considered (more than 30,000 per pixel).

Historical time threshold T. The number of occurrences of each quantized direction at each point during a time period T are recorded. T is the "historical time threshold" for the reference data.

Occurrence threshold T_0. A direction with more than T_0 occurrences will be considered to belong to the background. T_0 must be fixed on the basis of experience, depending on the characteristics of the object (average speed, maximum waiting time at a pedestrian crossing, etc.). However, T_0 has a minimal bound in order for directions generated by noise not to be considered as belonging to the background. Such a bound can be computed using probability.

4.3. DETECTION.

We have seen above that there are two types of detection, strong detection (the detection of directions which are incompatible with those stored in the reference image because level lines cross) and weak detection (the detection of a direction that is not present in the reference image. We shall begin by considering the latter.

Weak Detection. When a direction is detected at a point we can check whether this direction occurs in the reference data (within the quantization precision of the directions stored in the reference data). If it is not, we state that a weak detection has occurred at this point, which means either that the point is not part of the background or that the point is part of the background but the reference image is incomplete. It should be noted that several simultaneous weak detections, corresponding to several directions that are not in the reference image, can occur for a single point.

Strong Detection. This occurs when the level line directions in the current image and the reference image are incompatible (i.e. cross each other). In practice, strong detection is often ineffective, as it cannot occur whenever the intensity of the background is uniform (see figure 3) In this situation, a possible way of making it operative would be to place a textured pattern on the ground.

No parameters are involved in the detection stage.

4.4. FILTERING OF DETECTION.

As detection is based on a comparison between the current image and reference data it is sensitive to noise. In the same way that a criterion is required in order to decide if a direction has been present long enough to become part of the reference image, we need a criterion in order to decide if a detection is valid. To perform this we have defined our final parameter, which is a **minimum size threshold**. This involves removing connected detections that are too

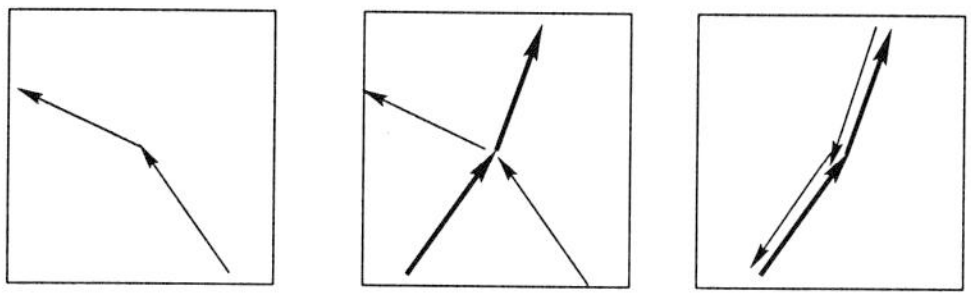

Fig. 2. Typical examples of weak and strong detections. Left: some directions (thin arrows) in the current image are not in the reference, we get a weak detection. Middle: directions in the current image cross directions in the reference (thick arrows), we get a weak and a strong detection. Right: directions in the current image are opposite to the directions in the reference (which means a local contrast inversion), we get again both detections.

Fig. 3. Middle: weak detections (only detected areas of more than 10 pixels are shown). Right: poor strong detections on a non textured background.

small (e.g. [20]). The reasoning behind this is that a moving object has a minimum spatial and temporal presence which is greater than that of noise. As we were unable to estimate formally the maximum spacial and temporal presence that could be generated by noise, we empirically set a value of ten pixels in our trials.

4.5. Quality of motion detection.

Is the algorithm robust to contrast changes?

As we have seen the algorithm mainly relies on weak detections. However, spurious weak detections can occur when there is a contrast change and the reference image is incomplete. Typically, such a situation arises when there is a sudden increase in contrast which has not occurred previously during the historical time threshold T. It should be noted that this type of event will not affect strong detections.

It would have been possible to approach this problem in the following ways:
1. using a long historical time threshold T and a low occurrence threshold T_0.
2. discarding frames in which there is a large increase in weak detections and none in strong detections.
3. Considering only those detections that create trajectories (see [18]).

In practice, even in outdoor applications, we found that not using directions associated with only a single level line was sufficient. The procedure is that all directions are used to construct the reference image but only those with at least two levels are considered for detection (the θ_i for a $\delta_i \geq 2$). This approach is both simpler and more flexible than the other possible solutions we have listed. However, it adds another parameter "$\geq$ **2**") and means that

Fig. 4. Left: daytime image of a crossing, next: number of level line directions observed for each pixel (the greater the number of directions the darker the image, white indicates no level lines), next: black pixels show the presence of at least one direction in the reference data, right: detected moving parts (weak detection with an area of more than 10 pixels).

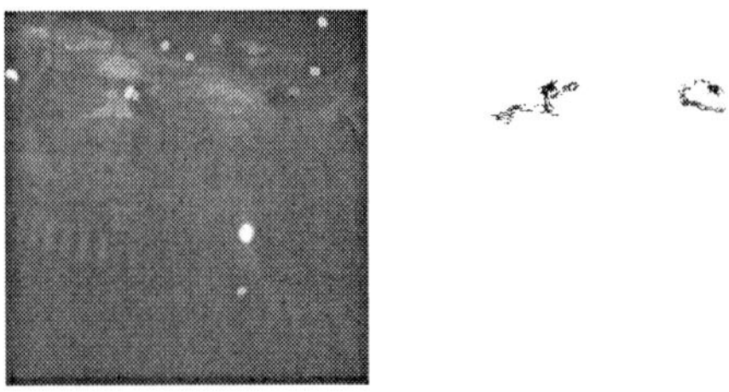

Fig. 5. left: night-time image of the crossing shown figure 4, right: detected moving parts. Note that the night presents particular problems (Canny edge detector often missed the boundaries of cars due to very low contrast. The signal/noise ratio is also quite low).

detection ignores very low contrast objects (i.e. those whose contrast is less than twice the quantization threshold). It can been seen from Figure 5 that the technique is still susceptible to low contrast.

Does the algorithm detect shadows? In its present form the algorithm makes no distinction between shadows and objects. It will therefore detect shadows that are not present in the scene for longer than the occurrence threshold. If their nature requires it these might be removed as described, e.g, in [15].

5. Trials

We are currently using this algorithm in a number of transport applications, for example vehicle location at an intersection and detection of abnormal human behaviour in a subway environment. The aim of the first of these is to count moving and stationary vehicles at an intersection in order to improve traffic signal control. To be of benefit, a system of this type must be effective 24 hours a day under all atmospheric and lighting conditions (see Figure 5). The system runs on a 166MHz Pentium-based PC at a rate of 7 frames per second. The historical time threshold has been set at 10 minutes and the occurrence threshold at 3 minutes. The results obtained are quite satisfactory (at the 7 frame per second frame rate 98.5% of queues are detected and queue length measurements are accurate to within 90%) and this system has now been deployed. It should be noted that the measurement errors are largely due to some shadows the system has been unable to remove.

Fig. 6. Left: image recorded a few months after the images and reference data shown in Figures 4 and 5. There have been some changes in vegetation and road markings and the lighting conditions are different. The black areas in the middle image show those pixels in this image which, after the best histogram modification we could achieve, have a grey-level difference of more than 32 with those in the top left image in Figure 5. Right: unfiltered detection achieved by comparing the image with the reference data for the image on the right in Figure 4. This example shows that, for this application, measurements based on level lines are sufficiently robust to contrast changes.

The second type of application aimed to detect abnormalities generated, for instance, by a person or object that has remained stationary for too long in a subway corridor. In this environment systems have to cope with the problems resulting from obstructions, poor contrast, and the fact that people change shape and move erratically. The detection of stationarities is performed by discarding from a novelty map (obtained through the change detector with a large historical time threshold) parts in motion (obtained through the change detector with a short historical time threshold). The system has been tested for over 240 hours covering 400 different stationarity situations under real conditions: with a frame rate of 7 images per second 98% of abnormal stationarities are detected, with no false detection. For our applications the number of directions memorized in the reference image was set at 32. One level line for a direction was sufficient for it to be used to construct the reference image, but at least two were required for detection. The minimum area in the noise filtering stage was 10 pixels. The historical time threshold and occurrence threshold were chosen with reference to the application and the characteristics of the scene (for example pedestrian crossing signal timings).

Fig. 7. Detection of stationarities in subway corridors.

6. Summary and Conclusions

In this paper we have described a new measurement technique that can be used to detect moving parts in an image. Its main original feature is that local geometry is separated from local contrast information. Discarding contrast information, we have developed two types of motion detection, one of which relates to the disappearance of local level lines and the other to a change in their local geometry. The temporal reference employed in the technique can be either long or short and objects that were not present a few minutes (for example) before can be detected. The technique has been used in a number of applications. Appraisals by transportation operators have provided encouraging results.

References

1. M. Black, P. Anandan, *A Model for the Detection of Motion over Time*, 3rd Int. Conf. on computer vision, 1990.
2. J.M. Blosseville, S. Bouzar, F. Lenoir and R. Glachet, *Traitement d'image pour la mesure et la surveillance du trafic: une avancée significative*, Transport Environment et Circulation. 1994
3. V. Caselles, B. Coll, J.M. Morel, *Topographic Maps and Local Contrast Changes in Natural Images.* Int. Journ. of Comp. vision, September 1999.
4. A. Elnagar, A. Basu, *Motion detection using background constraints*, Vol. 28, No. 10, pp. 1537-1554, Pattern Recognition, 1995.
5. M. Fathy, M.Y. Siyal. *An image detection technique based on morphological edge detection and background differencing for real-time traffic analysis.* Vol. 16, Pattern Recognition Letters, pp 1321-1330. 1995
6. F. Guichard, J.M. Morel, *Partial Differential Equations and image iterative filtering.* Tutorial ICIP 95, Washington DC, 1995.
7. N. Hoose, Computer Image Processing in Traffic Engineering. Research Studies Press, Taunton, 1991.
8. Y.Z. Hsu, H.H. Nagel, G. Rekers, *New likelihood test methods for change detection in image sequences.* Comp. Vision Graphic and Image Processing, Vol. CVGIP-26, 1984.
9. M. Irani, P. Anandan, *A unified approach to moving object detection is 2D and 3D scenes* IEEE transactions on pattern analysis and machine intelligence. vol.20, No. 6, June 1998.
10. R. Jain, *Dynamic scene analysis.* Pattern recognition 2, vol. 1, North-Holland, pp. 125-167, 1985.
11. P. Maragos, *A representation theory for morphological image and signal processing*, vol. 11, No. 6, June, 1989.
12. G. Matheron, *Random Sets and Integral Geometry*, John Wiley, N.Y, 1975.
13. F. Meyer, P. Bouthemy, *Region-based tracking in image sequences*, ECCV, p 476-484, Santa Margarita Ligure, 1992.
14. P. Monasse, F. Guichard, *Fast computation of a contrast-invariant image representation.* To be publish in IEEE Trans. on image processing, 1998.
15. K. Onoguchi, *Shadow Elimination Method for Moving Object Detection*, Int. Conf. Pattern Recognition. 1998.
16. J. Serra, *Image analysis and mathematical morphology*, Academic Press, 1982.
17. M.Y. Siyal, M. Fathy, C.G. Darkin, *Image processing algorithms for detecting moving objects*, ICARCV'94, Vol3. pp1719-1723, 1994.
18. A. M. Tekalp, Digital video processing, by Prentice-Hall Inc, 1995.
19. W.B. Thompson, T.C. Pong, *Detecting moving objects*, IJCV, vol 4. p29-57, 1990.
20. L. Vincent, *Morphological Area Openings and Closings for Grey-scale Images.* Math. Description of Shapes and Grey-level Images, Springer-Verlag, p197-208, 1994.

A GENERAL ALGORITHM FOR COMPUTING DISTANCE TRANSFORMS IN LINEAR TIME

A. MEIJSTER*, J.B.T.M. ROERDINK† and W.H. HESSELINK†
University of Groningen
P.O. Box 800, 9700 AV Groningen, The Netherlands
email: a.meijster@rc.rug.nl, {roe,wim}@cs.rug.nl

Abstract. A new general algorithm for computing distance transforms of digital images is presented. The algorithm consists of two phases. Both phases consist of two scans, a forward and a backward scan. The first phase scans the image column-wise, while the second phase scans the image row-wise. Since the computation per row (column) is independent of the computation of other rows (columns), the algorithm can be easily parallelized on shared memory computers. The algorithm can be used for the computation of the exact Euclidean, Manhattan (L_1 norm), and chessboard distance (L_∞ norm) transforms.

Key words: Distance Transforms, Row-Column Factorization, Parallelization.

1. Introduction

Distance transforms play an important role in many morphological image processing applications. They have been extensively studied and used in computational geometry, image processing, computer graphics and pattern recognition, e.g., [1, 2, 3, 7]. The two-dimensional distance transform can be described as follows. Let B be a set of grid points taken from a rectangular grid of size $m \times n$. The problem is to assign to every grid point (x, y) the distance to the nearest point in B. If we use the Euclidean metric for computing distances, and represent B by a boolean array $b[\cdot,\cdot]$, we thus want to compute the two dimensional array $dt[x, y] = \sqrt{EDT(x, y)}$, where

$$\mathrm{EDT}(x, y) = \mathtt{MIN}(i, j : 0 \leq i < m \wedge 0 \leq j < n \wedge b[i, j] : (x - i)^2 + (y - j)^2).$$

Here we use the notation $\mathtt{MIN}(k : P(k) : f(k))$ for the minimal value of $f(k)$ when k ranges over all values that satisfy $P(k)$.

Since the exact Euclidean distance transform is often regarded as too computationally intensive, several algorithms have been proposed that use some mask which is swept over the image in two scans, to compute approximations like the Manhattan (city-block) distance, the chessboard distance, or chamfer distances (see [1, 2, 3, 7]). The time complexity is linear in the number of pixels of the image (i.e. $O(m \times n)$), but it does not yield the exact Euclidean distance, which is required for some applications. Another drawback of these

* A. Meijster works at the Computing Centre of the University of Groningen.

† J.B.T.M. Roerdink and W.H. Hesselink work at the Institute for Mathematics and Computing Science.

algorithms is that they are hard to parallelize for parallel computers since previously computed results are propagated during the computation, making the process highly sequential. A recursive algorithm of order $mn \log m$ for the exact EDT is given in [5]. In [6] a recursive algorithm of order mn for the exact EDT is given by reducing the problem to a matrix search algorithm.

In this paper, which is based upon [4], we present a new algorithm that also computes distance transforms in linear time, is simpler and more efficient than [6], and is easy to parallelize. It can compute the Euclidean (EDT), the Manhattan (MDT), and the chessboard distance (CDT) transform, defined by

$$\begin{aligned} \mathrm{EDT}(x,y) &= \mathtt{MIN}(i,j : 0 \leq i < m \wedge 0 \leq j < n \wedge b[i,j] : (x-i)^2 + (y-j)^2), \\ \mathrm{MDT}(x,y) &= \mathtt{MIN}(i,j : 0 \leq i < m \wedge 0 \leq j < n \wedge b[i,j] : |x-i| + |y-j|), \\ \mathrm{CDT}(x,y) &= \mathtt{MIN}(i,j : 0 \leq i < m \wedge 0 \leq j < n \wedge b[i,j] : |x-i| \ \mathbf{max}\ |y-j|). \end{aligned}$$

If we define the minimum of the empty set to be ∞, and use the rule $z+\infty = \infty$ for all z, we find with some calculation

$$\begin{aligned} \mathrm{EDT}(x,y) &= \mathtt{MIN}(i : 0 \leq i < m : (x-i)^2 + G(i,y)^2), \\ \mathrm{MDT}(x,y) &= \mathtt{MIN}(i : 0 \leq i < m : |x-i| + G(i,y)), \\ \mathrm{CDT}(x,y) &= \mathtt{MIN}(i : 0 \leq i < m : |x-i| \ \mathbf{max}\ G(i,y)), \end{aligned}$$

where $G(i,y) = \mathtt{MIN}(j : 0 \leq j < n \wedge b[i,j] : |y-j|)$.

The algorithm can be summarized as follows. In a first phase each column C_x (defined by points (x,y) with x fixed) is separately scanned. For each point (x,y) on C_x, the distance $G(x,y)$ of (x,y) to the nearest points of $C_x \cap B$ is determined. In a second phase each row R_y (defined by points (x,y) with y fixed) is separately scanned, and for each point (x,y) on R_y the minimum of $(x-x')^2 + G(x',y)^2$ for EDT, $|x-x'| + G(x',y)$ for MDT, and $|x-x'|$ **max** $G(x',y)$ for CDT is determined, where (x',y) ranges over row R_y.

2. The First Phase

The object of the first phase is to determine the function G. We first observe that we can split G into two functions GT (top) and GB (bottom), such that $G(i,y) = \mathrm{GT}(i,y)$ **min** $\mathrm{GB}(i,y)$, where

$$\begin{aligned} \mathrm{GT}(i,y) &= \mathtt{MIN}(j : 0 \leq j \leq y \wedge b[i,j] : y-j) \\ \mathrm{GB}(i,y) &= \mathtt{MIN}(j : y \leq j < n \wedge b[i,j] : j-y) \end{aligned}$$

We start with the computation of GT by introducing an array g to store its values. It is easy to see that $\mathrm{GT}(i,y) = 0$ if $b[i,y]$ holds, and that, otherwise, $\mathrm{GT}(i,y) = \mathrm{GT}(i,y-1) + 1$ (or ∞ if $y = 0$). We can therefore compute $g[x,y] := \mathrm{GT}(x,y)$ using only $g[x,y-1]$ in a simple column scan from top to bottom. Similarly, we find $\mathrm{GB}(i,y) = \mathrm{GB}(i,y+1) + 1$. The second scan runs from bottom to top, and computes $G(x,y)$ directly, using GT from the previous scan, and GB from the current one. After some simplification, this results in the code fragment given in Fig. 1. Clearly, the time complexity is linear in the

```
forall x ∈ [0..m − 1] do
  (* scan 1 *)
  if b[x, 0] then
    g[x, 0] := 0
  else
    g[x, 0] := ∞;
  endif
  for y := 1 to n − 1 do
    if b[x, y] then
      g[x, y] := 0
    else
      g[x, y] := 1 + g[x, y − 1];
    endif
  (* scan 2 *)
  for y := n − 2 downto 0 do
    if g[x, y + 1] < g[x, y] then
      g[x, y] := (1 + g[x, y + 1])
    endif
end forall
```

Fig. 1. Program fragments for the first phase.

number of pixels (i.e. $O(m \times n)$). In actual implementations it is convenient to replace ∞ by $m+n$, since all distances in the images are less than $m+n$ if the set B is non-empty.

3. The Second Phase

In the second phase we want to compute EDT, MDT, or CDT row by row, i.e. for all x with fixed y. Therefore, in this section we regard y as a constant and omit it as a parameter in auxiliary functions, and introduce $g(i) = G(i, y)$. Instead of developing an algorithm for each metric separately, we aim at a more general algorithm for

$$\mathrm{DT}(x, y) = \mathtt{MIN}(i : 0 \leq i < m : f(x, i)). \tag{1}$$

The choice of the function f depends on the metric we wish to use, i.e.

$$f(x, i) = \begin{cases} (x-i)^2 + g(i)^2 & \text{for EDT,} \\ |x-i| + g(i) & \text{for MDT,} \\ |x-i| \ \mathbf{max} \ g(i) & \text{for CDT.} \end{cases}$$

It is helpful to introduce a geometrical interpretation of the minimization problem of Eq. (1). For any i with $0 \leq i < m$, denote by F_i the function $x \mapsto f(x, i)$ on the real interval $[0, m-1]$. We call i the *index* of F_i. In the case of EDT, the graph of F_i is a parabola with vertex at $(i, g(i))$. In the case of

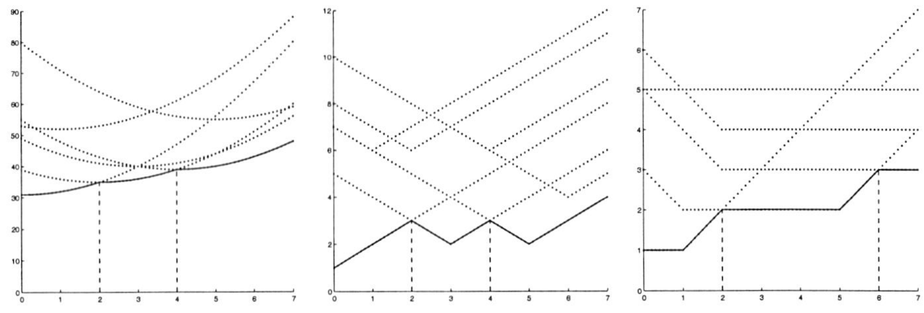

Fig. 2. DT as the lower envelope (solid line) of curves F_i, $0 \leq i < m$ (dotted lines). The dashed vertical lines indicate the transitions between regions.

MDT the parabolas are replaced by V-shaped approximations, while in the case of CDT we deal with 'topped off' V-shaped approximations (see Fig. 2). We can interpret DT geometrically as the *lower envelope* of the collection $\{F_i | 0 \leq i < m\}$ evaluated at integer coordinates, cf. Fig. 2. The lower envelopes consist of a number of consecutive curve segments, whose index we denote by $s[0], s[1], \ldots, s[q]$ counting from left to right. The projections of the segments on the x-axis are called *regions*, and form a partition of the interval $[0, m)$ by consecutive segments. The computation of *DT* now consists of two scans. In a forward (left-to-right) scan the set of regions is determined using an incremental algorithm. In a backward (right-to-left) scan the values $\mathrm{DT}(x, y)$ are trivially computed for all x.

We start by replacing the upper bound m in (1) by a variable u and define

$$\mathrm{FL}(x, u) = \mathtt{MIN}(i : 0 \leq i < u : f(x, i)).$$

The geometric interpretation is that we restrict the set B to the half plane to the left of u. Clearly, $\mathrm{DT}(x, y) = \mathrm{FL}(x, m)$.

For given upper bound $u > 0$, we define an index h to be a *minimizer* at x if, in the expression for $\mathrm{FL}(x, u)$, the minimal value of $f(x, i)$ occurs at h. In general, x may have more than one minimizer. defined as the least index h with $0 \leq h < u$ such that $f(x, h) \leq f(x, i)$ for all i in the same range, i.e.

$$H(x, u) = \mathtt{MIN}(h : 0 \leq h < u \wedge \forall(i : 0 \leq i < u : f(x, h) \leq f(x, i)) : h). \quad (2)$$

We clearly have $\mathrm{FL}(x, u) = f(x, H(x, u))$, hence $\mathrm{DT}(x, y) = f(x, H(x, m))$. Therefore, the problem reduces to the computation of $H(x, m)$.

We consider the sets $S(u)$ of the least minimizers that occur during the scan from left to right, and the sets $T(h, u)$ of points with the same least minimizer h. We thus define

$$\begin{aligned} S(u) &= \{H(x, u) \,|\, 0 \leq x < m\}, \qquad (3) \\ T(h, u) &= \{x \,|\, 0 \leq x < m \ \wedge \ H(x, u) = h\} \text{ if } 0 \leq h < u. \end{aligned}$$

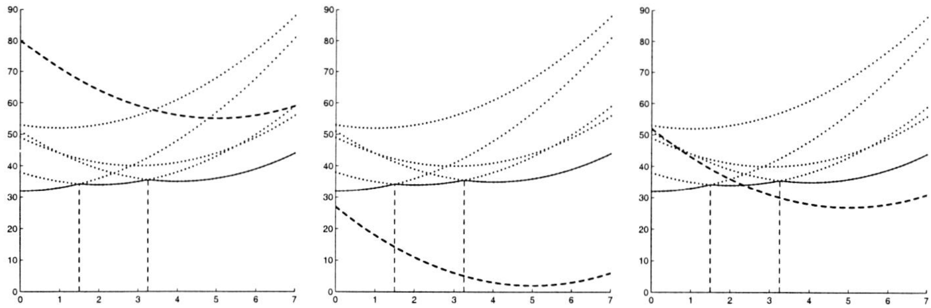

(a) above (b) below (c) intersection

Fig. 3. Location of F_u (dashed curve) w.r.t. the lower envelope (solid line).

Clearly, $S(u)$ is a nonempty subset of $[0, u)$, and $S(u) = \{h \mid T(h, u) \neq \emptyset\}$. We define the *regions* for u to be the sets $T(h, u)$ that are nonempty. It is easy to see that the regions for u form a partition of $[0, m)$.

The aim is the case where $u = m$. Indeed, for $x \in T(h, m)$, we have $H(x, m) = h$ and hence $\mathrm{DT}(x, y) = f(x, h)$. The second phase of the algorithm therefore consists of two scans: scan 3 computes the partition of $[0, m)$ that consists of the regions for m and scan 4 uses these regions to compute DT. For given u, only the curves with indices from 0 to $u-1$ are taken into account. The minimizer of x corresponds to the index of the curve segment whose projection on the horizontal axis contains x. Let the current lower envelope consist of $q+1$ segments, i.e. $S(u) = \{s[0], s[1], \ldots, s[q]\}$, with $s[\ell]$ the index of the ℓ-th segment. Consider what happens when F_u is added. Three situations may occur:

(a) F_u is *above* the current lower envelope on $[0, m-1]$, cf. Fig. 3(a). Then $S(u+1) = S(u)$, since the set $T(u, u+1)$ is empty.

(b) F_u is *below* the current lower envelope on $[0, m-1]$, cf. Fig. 3(b). Then $S(u+1) = \{u\}$, i.e., all old regions have disappeared, and there is one new region $T(u, u+1) = [0, m)$.

(c) F_u *intersects* the current lower envelope on $[0, m-1]$, cf. Fig. 3(c). The current regions will either shrink or disappear, and there is one new region $T(u, u+1)$.

We start searching from right to left for the current region which is intersected by F_u. This can be determined by comparing the values of F_u and F_ℓ at the begin point $t[\ell]$ of each current region $\ell = q, q-1, \ldots$, until we find the first $\ell = \ell^*$ such that $F_u(t[\ell^*]) \geq F_{s[\ell^*]}(t[\ell^*])$. Then F_u is not the least minimizer at $t[\ell^*]$, and there must be an intersection of F_u with F_{ℓ^*} in region ℓ^*. Let x^* be the horizontal coordinate of the intersection. If $\ell^* = q$ and $x^* \geq m$ we have case (a); if $\ell^* < 0$ we have case (b); otherwise case (c) pertains.

```
forall y ∈ [0..n − 1] do
  q := 0; s[0] := 0; t[0] := 0;
  for u := 1 to m − 1 do (* scan 3 *)
    while q ≥ 0 ∧ f(t[q], s[q]) > f(t[q], u) do
      q := q − 1;
    if q < 0 then
      q := 0; s[0] := u
    else
      w := 1 + Sep(s[q], u);
      if w < m then
        q := q + 1; s[q] := u; t[q] := w
      end if
    end if
  end for
  for u := m − 1 downto 0 do  (* scan 4 *)
    dt[u, y] := f(u, s[q]);
    if u = t[q] then q := q − 1
  end for
end forall
```

Fig. 4. Program fragments for the second phase.

To find x^*, we introduce a function Sep, where $\mathrm{Sep}(i, u)$ is the first integer larger or equal than the horizontal coordinate of the intersection point of F_u and F_i with $i < u$, i.e.

$$F_i(x) \leq F_u(x) \quad \Leftrightarrow \quad x \leq \mathrm{Sep}(i, u). \tag{4}$$

We thus have $x^* = \mathrm{Sep}(s[l^*], u)$. Clearly, the function Sep is dependent on which distance transform we want to compute. In the next section we will derive the expressions for the function Sep, but in the remainder of this section we simply assume that Sep is available.

We introduce an integer program variable u. It is convenient to represent $S(u)$ by an increasing sequence of elements. Since the regions form a partition of $[0, m)$ by consecutive segments, we can represent them by the sequence of their least elements. According to the case analysis above, the regions are to be adapted at their end. We can therefore implement these sequences in two integer arrays, s and t, with an integer variable q as index of the end point.

We start with the forward scan, see scan 3 in Fig. 4. We have $S(1) = \{0\}$, and $T(0, 1) = [0, m)$, and thus start with $q = 0$, $s[0] = 0$, and $t[0] = 0$. In a loop, variable u is incremented, and thus the representations of S and T must be updated by means of the case analysis above. For details, we refer to our report [4].

To investigate the complexity of the forward scan, we consider the expression $q + 2(m - u)$, which is initially $2m$. In every execution of the body of the outer loop (scan 3 in Fig. 4), and also in every execution of the body of its inner loop,

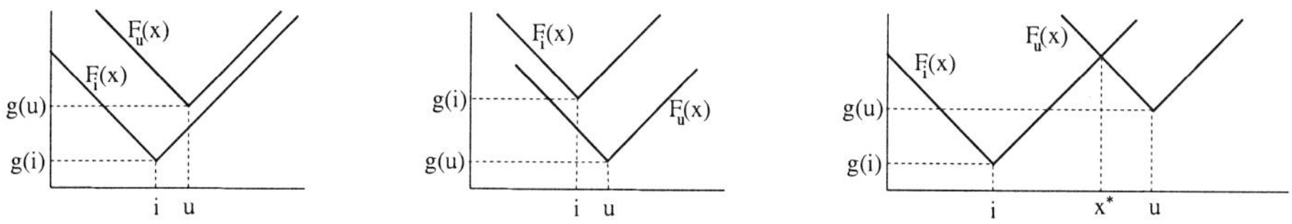

(a) $g(u) \geq g(i) + u - i$ (b) $g(i) > g(u) + u - i$ (c) otherwise

Fig. 5. Cases for finding Sep for MDT.

the value of the expression decreases. This implies that the time complexity of the scan is linear in m. Note that, the *average* number of iterations of the inner loop is at most two. The algorithm uses less than $2m$ comparisons of f values, and function Sep is evaluated less than m times.

When the forward scan is finished, we have completely determined the partition of $[0, m)$ in regions. Given these regions, we can trivially compute dt-values in a simple backward scan (see scan 4 in Fig. 4).

4. Derivation of the Function Sep

The derivation in the previous section was independent of the actual metric used. The functions dependent on the metric are f and Sep. In this section we compute expressions for Sep for EDT, MDT, and CDT. The easiest is EDT. We find for $i < u$

$$
\begin{aligned}
& F_i(x) \leq F_u(x) \\
\Leftrightarrow & \{\text{definition of } F_i,\ F_u\} \\
& (x-i)^2 + g(i)^2 \leq (x-u)^2 + g(h)^2 \\
\Leftrightarrow & \ \{\text{calculus, } i < u;\ x \text{ is an integer}\} \\
& x \leq (u^2 - i^2 + g(u)^2 - g(i)^2) \ \mathbf{div} \ (2(u-i)).
\end{aligned}
$$

Here, we denote integer division with rounding off towards zero by **div**. Thus, we find for EDT that

$$
\mathrm{Sep}(i,u) = (u^2 - i^2 + g(u)^2 - g(i)^2) \ \mathbf{div} \ (2(u-i)).
$$

If we use the Manhattan metric, the analysis is slightly more complicated. Since we have to deal with absolute values in the expressions, awkward case analysis is necessary if we want to compute Sep analytically. Therefore we prefer a geometric argument. We have to consider three cases (see Fig. 5).

If $g(u) \geq g(i) + u - i$, the graph of F_u lies entirely above the graph of F_i for all x, thus we choose $\mathrm{Sep}(i,u) = \infty$. If $g(i) > g(u) + u - i$, the graph of F_i lies entirely above the graph of F_u, so $F_i(x) \leq F_u(x)$ for no x at all. Thus, we must choose $\mathrm{Sep}(i,u) = -\infty$ to satisfy (4). In all other cases, F_u intersects F_i

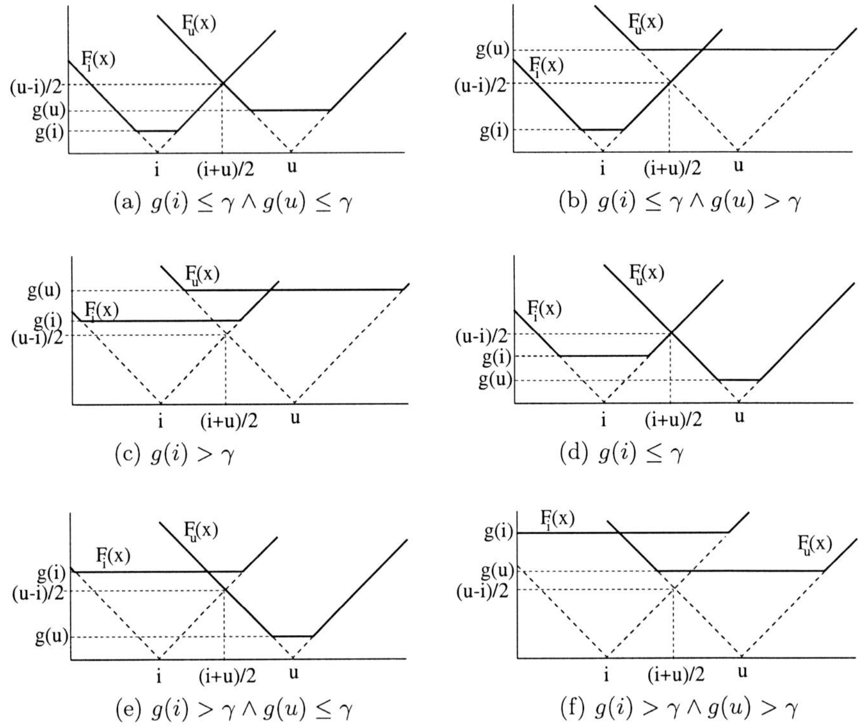

Fig. 6. Cases for finding Sep for CDT, where $\gamma = (u-i)/2$. Cases (a)-(c): $g(i) \leq g(u)$. Cases (d)-(f): $g(i) > g(u)$.

at $x^* = (g(u) - g(i) + h + i)/2$. So, if we want to compute MDT we use

$$\text{Sep}(i,u) = \begin{cases} \infty & \text{if } g(u) \geq g(i) + u - i, \\ -\infty & \text{if } g(i) > g(u) + u - i, \\ (g(u) - g(i) + h + i) \textbf{ div } 2 & \text{otherwise.} \end{cases}$$

For the case of CDT we have $|x-i| \textbf{ max } g(i) \leq |x-u| \textbf{ max } g(u)$. We consider two main cases, which each can be split up in three sub-cases. First we consider the case $g(i) \leq g(u)$. From Fig. 6(a)-(c), we see that the increasing segment of F_i $(y = x - i)$ intersects the decreasing part of F_u $(y = u - x)$, or the constant part $(y = g(u))$. Let γ be the vertical coordinate corresponding with the middle of i and u $(x = (i+u)/2)$, i.e. $\gamma = (u-i)/2$. From Fig. 6(a), we see that if $g(i) \leq \gamma \wedge g(u) \leq \gamma$, we have $F_i(x) \leq F_u(x)$ if $x \leq (i+u)/2$. From Fig. 6(b)-(c), we see that the increasing part of F_i intersects the constant segment of F_u at $i + g(u)$, and thus we have $F_i(x) \leq F_u(x)$ if $x \leq i + g(u)$. Putting the three cases together, we can conclude

$$g(i) \leq g(u) \Rightarrow \left(F_i(x) \leq F_u(x) \Leftrightarrow x \leq \frac{i+u}{2} \textbf{ max } (i + g(u))\right).$$

TABLE I
Timing results in ms. From left to right: EDT, MDT, and CDT.

size	p=1	p=2	p=3	p=4	p=1	p=2	p=3	p=4	p=1	p=2	p=3	p=4
256	12	7	5	4	11	6	4	3	12	6	4	3
512	69	35	25	19	63	34	24	17	67	35	25	18
1024	307	156	104	79	281	147	97	74	298	152	101	77
2048	1542	780	517	389	1407	709	476	357	1501	753	506	381
4096	6251	3137	2098	1577	5753	2886	1929	1451	6073	3053	2041	1530

The other main case is $g(i) > g(u)$. Again, in Fig. 6(d), we see that if $g(i) \leq \gamma$, the intersection at $(i+u)/2$ is the separator. If $g(i) > \gamma$ (see Fig. 6(e)-(f)), the horizontal segment of F_i intersects the decreasing part of F_u at $x = u - g(i)$. Just like in the previous case, we can put these cases together. This results in the following expression for Sep:

$$\text{Sep}(i,u) = \begin{cases} (i + g(u)) \ \textbf{max} \ ((i+u) \ \textbf{div} \ 2) & \text{if } g(i) \leq g(u), \\ (u - g(i)) \ \textbf{min} \ ((i+u) \ \textbf{div} \ 2) & \text{otherwise.} \end{cases}$$

5. Parallelization, Timing Results, and Conclusions

Since the computation per row (column) is independent of the computation of other rows (columns), the algorithm is well suited for parallelization on a shared memory machine. In the first (second) phase, the columns (rows) are distributed over the processors. The two phases must be separated by a *barrier*, which assures that all processors have completed the first phase before any of them starts with the second phase. The theoretical time complexity of the parallel algorithm for p processors (where $p \leq m$ **min** n) is $\mathcal{O}(mn/p)$.

We ran experiments on an Intel Pentium III based shared memory parallel computer with 4 cpu's, running at a 550MHz clock frequency. We performed time measurements using several binary images, and found that the execution time is almost independent of image content, and scales well w.r.t. the number of processors. This is as expected, since the amount of work per row and column is almost the same. In table I the timings for square images are given for $p = 1$ to $p = 4$ processors. Note that the computation of MDT and CDT is only slightly faster than the exact EDT. We also implemented the sequential algorithm of [7] for CDT, and found that our algorithm is less than a factor of 2 slower, which can easily be overcome by parallel processing.

The algorithm can be easily extended to d-dimensional distance transforms by separating the problem into d phases, each solving a one-dimensional problem, as carried out above for the case $d = 2$.

References

1. G. Borgefors. Distance transformations in arbitrary dimensions. *Computer Vision, Graphics, and Image Processing*, 27:321–345, 1984.
2. G. Borgefors. Distance transformations in digital images. *Computer Vision, Graphics, and Image Processing*, 34:344–371, 1986.
3. P. Danielsson. Euclidean distance mapping. *Comput. Graphics Image Process.*, 14:227–248, 1980.
4. W. H. Hesselink, A. Meijster, and J. B. T. M. Roerdink. An exact Euclidean distance transform in linear time. Technical Report IWI 99-9-04, Institute for Mathematics and Computing Science, University of Groningen, the Netherlands, Apr. 1999.
5. M. Kolountzakis and K. Kutulakos. Fast computation of the Euclidean distance maps for binary images. *Information Processing Letters*, 43:181–184, 1992.
6. S. Pavel and A. Akl. Efficient algorithms for the Euclidean distance transform. *Parallel Processing Letters*, 5:205–212, 1995.
7. A. Rosenfeld and J. Pfaltz. Distance functions on digital pictures. *Pattern Recognition*, 1:33–61, 1968.

THE ORDERED QUEUE AND THE OPTIMALITY OF THE WATERSHED APPROACHES

ROBERTO LOTUFO and ALEXANDRE FALCAO
Universidade Estadual de Campinas, Campinas, SP, Brazil
Email: lotufo@dca.fee.unicamp.br afalcao@dcc.unicamp.br

Abstract. This work reviews the watershed in the graph framework of a shortest-path forest problem using a lexicographic path cost formulation. This formulation reflects the behavior of the ordered queue-based watershed algorithm. This algorithm is compared with our proposed shortest-path forest (IFT–**I**mage **F**oresting **T**ransform), concluding that the watershed is a special case of that. Recently many different watershed approaches are being used. We point out that in some cases the watershed algorithm does not keep the optimality of the shortest-path forest solution unless the IFT algorithm is used. The main difference between the algorithms is related to permanently labeling a pixel when inserting or removing it from the queue. The watershed based on the pixel dissimilarity using IFT can segment one-pixel width regions while keeping the optimality of the shortest-path forest solution.

Key words: Watershed, Hierarchical Queue, Shortest-Path Forest, Image Foresting Transform, Graph-Based Image Processing, Color Image Segmentation.

1. Introduction

The watershed is one of the most powerful tools for image segmentation in Mathematical Morphology. Although it was first reported in 1979 [1], its development is still active in the community.

In this paper we propose a watershed formulation, based on a particular case of the Image Foresting Transform framework [7], that reflects the behavior of watershed algorithm using the ordered queue. We also point out a very special formulation of arc weights in the watershed algorithm which we believe was the cause of many difficulties concerning the analysis of the queue-based watershed algorithm, mainly related to the way a pixel is labeled before or after queueing. Finally we show that the resolution problem of the watershed on gradient, described in [4] can be solved when using different weight assignments as in [10] but applying the IFT algorithm.

The outline of this paper is as follows. Section 2 reviews the graph definitions and presents the watershed as a shortest-path forest problem. Section 3 presents the IFT shortest-path forest algorithm using the ordered queue. Section 4 compares this algorithm with the watershed algorithm using ordered queue. Section 5 presents the extension of the watershed and the importance of using the IFT algorithm. Section 6 discusses the watershed where there are catchment basins without markers.

2. Definitions and Notation

A *graph* $G = (V, A)$ is composed of two sets V and A. V is the set of *nodes*, and A is the set of *arcs* $(p, q), p, q \in V$ associated to a pair of adjacent nodes. The graph is *weighted* if a *weight* $w(p, q)$ is associated to each arc, and it is called *digraph* if the arcs are directed, i.e., (p, q) and (q, p).

A *path* from v_1 to v_n is a list of unique adjacent nodes $(v_1, v_2, \ldots, v_n)$, $(v_i, v_{i+1}) \in A$. The *path Cost* $C(v_1, v_2, \ldots, v_n)$ in a weighted graph can be given by a non-decreasing function of the arcs weights in the path. Two particular formulations to define the path costs are of interest: C_s, the sum of each arc weight in the path, and C_m, the maximum of the arc weight in the path.

$$C_s(v_1, v_2, \ldots, v_n) = \sum_{i=1}^{n-1} w(v_i, v_{i+1}) \quad C_m(v_1, v_2, \ldots, v_n) = \bigvee_{i=1}^{n-1} w(v_i, v_{i+1})$$

The first formulation is traditional in the majority of the shortest-path problems and is applied in image processing mainly for distance transform computation [9] and edge tracking [8]. The second formulation of the path cost will be mainly used in this paper as it is related to the flooding simulation of the watershed transform.

The *watershed from markers* can be described by flooding a topographical relief model of the gray-scale image. The markers are holes in the image relief where colored water can enter as the relief is flooded. There is one color associated to each set of markers. As the relief is uniformly flooded, different colored water may meet but cannot be mixed. When all the relief is flooded, each colored water region defines the *catchment basin* CB associated to the marker. The classical watershed transform is when the markers are the regional minima of the image.

To mimic this flooding process, we need a path cost formulation with two components $C = [Md]$ with a lexicographic order. The first and most priority is M as the maximum arc weight in the path, representing the flooding process. The second component d, of less priority, is the distance from the end of the path to the nearest node with a lower path cost. This component reflects the constant evolution of the flooding process when water reaches a plateau in the relief.

$$\begin{aligned} C_m(v_1, v_2, \ldots, v_n) &= [M(n), d] \\ M(n) &= \bigvee_{i=1}^{n-1} w(v_i, v_{i+1}) \\ d &= min\{j : M(i) = M(i-j), j = 0, 1, 2, \ldots, i-1\} \end{aligned}$$

The shortest cost between two nodes $C_m^*(p, q)$ is given by the smallest lexicographic cost of all the paths between p and q.

$$C_m^*(p, q) = \bigwedge C_m(v_1, v_2, \ldots, v_n), p = v_1, q = v_n \tag{1}$$

For the watershed, the arc weight is the height of the wall between nodes and the shortest cost is the minimal height where the water coming from two points merge. When water comes from different CBs in a plateau, the second component of the lexicographic cost allows the partition to be at the medial line of the plateau.

The *shortest-path forest* problem is a problem in graph theory that finds, for each node, the shortest path connecting it from its nearest root node. The result is a disjoint set of shortest-path trees. It is important to note that in the forest partitioning, an arc cannot be used simultaneously to compute the path cost related to two different catchment basins. This characteristic makes this formulation very different from other watershed definitions based on the SKIZ (SKeleton of Influence Zone) formulation [11] [13].

Next we present the formulation of the watershed transform based on the shortest-path forest framework that is very efficiently implemented by the ordered queue watershed algorithm.

2.1. WATERSHED FORMULATION IN THE SHORTEST-PATH FOREST FRAMEWORK

The watershed from markers L_k is computed on a gray-scale image $g(p)$. The image is modeled as a connected digraph with each pixel as a node and the arcs defined by the neighborhood connectivity. All the incident arcs to p have the same weight, given by the pixel value at p, $w(i,p) = g(p)$. The lexicographic path cost between two points is given by the equation 1 (C_m^*). The path cost from a region L to a point p, $C_m^*(L,p)$, is given by the smallest path cost between any pixel of the region L to the pixel p:

$$C_m^*(L,p) = Min\{C_m^*(l,p), l \in L\}$$

The catchment basin CB_k associated to the marker L_k is given by the nodes with smaller or equal path cost from this marker than from any other marker.

$$CB_k = \{p | C_m^*(L_k,p) \leq C_m^*(L_j,p)\}, k \neq j$$

Fig 1 shows with a numerical example the watershed as a forest shortest path problem in an image with 5 rows and 6 columns. This image has many plateaus to illustrate the role of the second component (d) in the lexicographic path cost.

3. The IFT Algorithm to Compute the Shortest-path Forest

The first shortest-path forest algorithm was due to Moore [12] in 1957. This algorithm is very similar to the well known Dijkstra's shortest-path algorithm [6] and is valid for any path cost using a non-decreasing function of the arc weights (See this proof in [7]). Dial proposed in 1969 the first implementation of the Moore's shortest-path forest using an ordered queue [5]. We present next the IFT shortest-path forest algorithm which uses an ordered queue to find the catchment basins of the watershed based on the definition just presented.

An ordered, hierarchical, or priority queue, with a FIFO restriction is a data structure very popular in some morphological image processing algorithms such as gray-scale reconstruction and watershed. A node p, associated with a priority value c, can be inserted in the ordered FIFO queue (`Enqueue(p,c)`). When a node is de-queued (`DeQueueMin`), it is selected the oldest one from

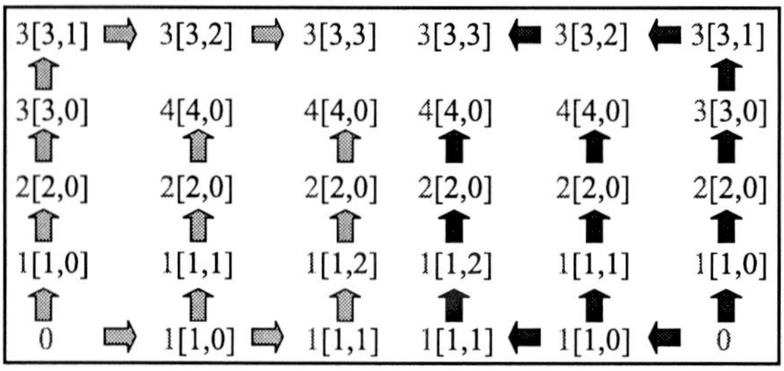

PixelValue[Lexicographic cost (M,d)]

Fig. 1. Watershed as the shortest-path forest. The markers (roots) are the two corner pixels at the lowest row with value 0. The arrows indicate the shortest lexicographic paths. Their gray intensity (light or dark) indicate the forest partitioning.

the lowest priority queue. The following algorithm also needs an operation to remove randomly any node p from the queue (`DeQueue(p)`). An important property of this data structure when used in the IFT algorithm below, is to keep the data implicitly sorted following the lexicographic path cost defined by equation 1. The First-In-First-Out behavior associated with the nature of the IFT algorithm to propagate the lower cost paths first (ordered queue) are responsible for the intrinsic lexicographic sorting.

The watershed can be implemented using the IFT algorithm below if w(p,q) is substituted by g(q). Later, in section 5, we will generalize w(p,q) to other watershed approaches. In the algorithm, C(p) is the cost path from p to its nearest marker; L(p) is the input marker image and also the result of the watershed partitioning with the catchment basins.

IFT ALGORITHM

1. Initialization
   a) flag(p)=TEMP; p in all nodes
   b) C(p)=∞; L(p)=0; p: non-marker nodes
   c) C(p)=0; Enqueue(p,0); L(p)= label of marker; p: marker nodes
2. Propagation
   while Queue is not empty
   a) v = DeQueueMin
   b) flag(v)=DONE
   c) **for** each p neighbor of v and flag(p)==TEMP
      if Max{C(v), w(v,p)} < C(p)
         C(p) = Max{C(v),w(v,p)}; L(p)=L(v);
         if p is in queue **then** Dequeue(p);
         Enqueue(p,C(p));

The algorithm works with two set of nodes: temporary (TEMP) and permanent (DONE). Initially all nodes are set as temporary (line 1a) and as the algorithm evolves, the nodes are transformed in permanent (line 2b). An important property of this algorithm is that once a node is permanent, its path cost is the final optimal shortest-path. As the FIFO priority queue keeps intrinsically sorted the second component d of the lexicographic cost, we can work only with the first component M. For sake of simplicity, we will call simply by path cost this first lexicographic cost component in the description of the algorithm.

In the initialization phase, all nodes are set as temporary, the markers have their path cost assigned to 0 and all other nodes have costs assigned to infinity. The marker nodes are labeled and non-marker nodes have label 0. The propagation step works until there is a temporary node. The node with the minimum temporary cost is selected by removing it from the ordered queue and it is transformed in a permanent node. The temporary nodes p which are neighbors of the new permanent node v are processed. If the path cost computed through the permanent node v is smaller than the temporary cost associated with node p, its cost and label are updated. If the node was already in the queue, it is removed. Finally the node is enqueued with the priority of the new path cost.

4. The IFT Algorithm and the Watershed Using Ordered Queue

The best performance sequential watershed algorithm is achieved using an ordered FIFO queue [2].

WATERSHED ALGORITHM

1. Initialization

    L(p)=0 for non-marker nodes

    for p:marker nodes, Enqueue(p,g(p))

2. Propagation

    while Queue is not empty

        v = DeQueueMin

        for each non-labeled p neighbor of v

            L(p)=L(v);

            Enqueue(p,g(p))

There are two important differences between the watershed and the IFT. 1) In the watershed, a node is labeled when entering into the queue whereas in the IFT the node is permanently labeled only when it leaves the queue and while in the queue, its label can change; 2) The priority assigned to a node in the queue is the cumulative path cost in the IFT algorithm as opposed to the value of the pixel associated with the node, in the watershed algorithm.

For simplicity, in the following explanation, it is assumed that the image has no plateaus and therefore the lexicographic cost path can be reduced to

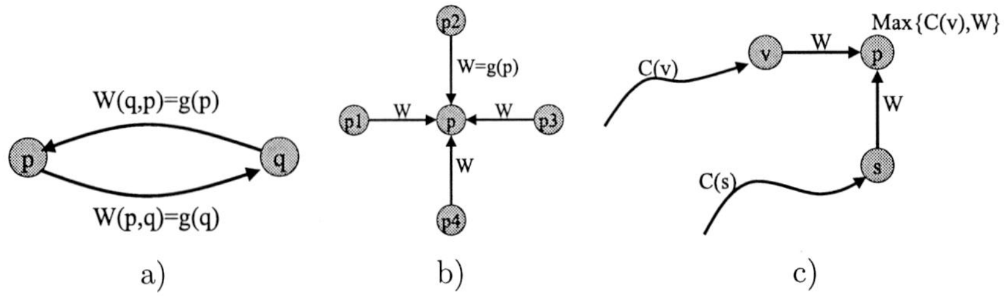

Fig. 2. a) arc weights between two pixels in the watershed; b) Incident arc weights in the watershed; c) Cumulative cost when all incident arc weights are the same

the first component $M(n)$ which is the maximum arc weight in the path. Nevertheless the conclusions are also valid for situations with plateaus using both components of the lexicographic cost formulation.

First we explain why in the last algorithm a node can be labeled permanently when entering into the queue without loosing its shortest-path optimality. According to our definition of the arc weight in the watershed formulation, the graph is directed with $w(p,q) = g(q)$ and $w(q,p) = g(p)$. The weight from node p to q is different from the weight of going from q to p (See Fig 2a). With this formulation, all the arcs arriving at node p have the same weight g(p). Fig 2b illustrates this.

It is not difficult to see that when a digraph has this property, the IFT algorithm does not require pixel queue re-evaluation. We can see this with the help of Fig 2c. Suppose that pixel v has just become permanent and pixel $\mathtt{p}$ is its non-labeled neighbor. Pixel $\mathtt{v}$ is the only permanently labeled in the neighborhood of $\mathtt{p}$. This means that the cumulative path cost until pixel v is less or equal to any other neighbor of $\mathtt{p}$, so $C_m(v) \leq C_m(s)$, v and s neighbors of p. As all the arc weights going to $\mathtt{p}$ have the same W, so $\bigvee\{C_m(v), W\} \leq \bigvee\{C_m(s), W\}$, meaning that $C_m(p)$ can be assigned a permanent label as it will not change any further. As far as we know no one has call the attention to this point, and in our view, this very particular aspect of the watershed algorithm was a cause of great difficulty in its analysis and extension to new classes of watershed algorithms. To guarantee the optimality of the shortest-path forest solution, it is important to understand that making the label permanent when inserting the pixel in the queue is only possible when the incident arc weights are the same. When the graph is not directed, this is not the case as it will be explained in the next section.

Next, we discuss why the watershed uses as priority the pixel value (arc weight) instead of the cumulative path cost $C_m(p)$. When we consider the watershed from regional minima, the arc weights in the catchment basins leaving the minima are monotonically non decreasing. In this sense, the path cost computed using the maximum arc is equivalent to the path cost computed from the last arc in the path as this last arc will always be equal or larger than all the previous arcs: $C_m(v_1, v_n) = \bigvee_{i=1}^{n-1} w(v_i, v_{i+1}) = w(v_{n-1}, v_n) = g(v_n)$ as $w(v_i, v_{i+1}) \leq w(v_{i+1}, v_{i+2})$ in the solution path.

These considerations make us conclude that the IFT algorithm gives the same result as the watershed using the ordered queue if the markers are the regional minima of g(p). The case where the markers are not in the regional minima is commented in section 6.

5. Extension of the Watershed Based on Dissimilarity

The watershed defined in subsection 2.1 can be generalized using other arc weights. The first modification of the watershed algorithm using arc weights in an undirected graph was reported by Meyer [10]. The arc weights from p to q and from q to p are the same: $w(p,q) = w(q,p)$, and in one of the simplest form could be computed from the absolute difference of the pixel gray levels: $w(p,q) = |f(p) - f(q)|$. The advantages of this approach are mainly two: first, it achieves a higher resolution than the use of morphological gradient [14]; second, it can be used for color images.

In this case, the watershed algorithm, labeling a pixel when queueing, fails and the IFT algorithm must be used to guarantee the optimality solution of the shortest-path forest solution. There is another variation of the IFT algorithm in updating the queued node that was reported in [3] which is the following. When a shorter cumulative cost is found to a pixel already in the queue, instead of removing it and inserting it again with a lower priority value, which requires a more complex queue data structure to be able to remove any pixel randomly, one can insert the pixel again with the lower priority. In this situation, a pixel can be in the queue more than once and when the pixel is de-queued a test is included to certify if it is already labeled or not. In this situation the first instance of the same pixel removed from the queue will be the one with the lowest priority and will label permanently the pixel. The second and further instances of the pixel in the queue will have no effect when de-queueing as the pixel is already labeled.

To illustrate the advantage of higher resolution of the watershed based on dissimilarity and the requirement of queue re-evaluation of the shortest-path algorithm, we show an experiment where a numeric image of two blobs connected by an one-pixel thick region is segmented using three approaches: 1) watershed on gradient; 2) watershed on dissimilarity, labeling the pixel when entering the queue; and 3) the watershed on dissimilarity using the IFT algorithm. In all three cases we use one inside and one outside marker. Fig 3 shows this illustrative example. The correct segmentation is achieved only with the IFT algorithm. The use of higher resolution segmentation is of particular importance in digital video segmentation [14].

6. Case Where the Markers Are Not in the Regional Minima

To overcome the oversegmentation problem when using the watershed from its natural regional minima, a minima imposition operator can be used to change the homotopy of the image in such a way that the markers are the only regional minima in the image [2]. Later it was realized that the watershed from

```
0 0 0 0 0 0 0 0 0     0 0 0 0 0 0 0 0 0     0 0 0 0 0 0 0 0 0
0 1 1 0 0 0 1 1 0     0 0 0 0 0 0 0 0 0     0 1 1 0 0 0 1 1 0
0 1 1 0 0 0 1 1 0     0 0 0 0 0 0 0 0 0     0 1 1 0 0 0 1 1 0
0 1 1 1 1 1 1 1 0     0 1 0 0 0 0 0 0 0     0 1 0 1 1 1 0 1 0
0 1 1 0 0 0 1 1 0     0 0 0 0 0 0 0 0 0     0 1 1 0 0 0 1 1 0
0 1 1 0 0 0 1 1 0     0 0 0 0 2 0 0 0 0     0 1 1 0 0 0 1 1 0
0 0 0 0 0 0 0 0 0     0 0 0 0 0 0 0 0 0     0 0 0 0 0 0 0 0 0
a) Original image        b) Marker         c) Internal gradient

1 1 1 1 1 1 1 1 2     1 1 1 2 2 2 2 2 2     2 2 2 2 2 2 2 2 2
1 1 1 1 1 1 1 2 2     1 1 1 1 2 2 2 2 2     2 1 1 2 2 2 1 1 2
1 1 1 1 1 1 1 2 2     1 1 1 1 2 2 2 2 2     2 1 1 2 2 2 1 1 2
1 1 1 1 2 2 2 2 2     1 1 1 1 2 2 2 2 2     2 1 1 1 1 1 1 1 2
1 1 1 2 2 2 2 2 2     1 1 1 2 2 2 2 2 2     2 1 1 2 2 2 1 1 2
1 1 2 2 2 2 2 2 2     1 1 2 2 2 2 2 2 2     2 1 1 2 2 2 1 1 2
1 2 2 2 2 2 2 2 2     1 1 2 2 2 2 2 2 2     2 2 2 2 2 2 2 2 2
 d)  Watershed on     e)  Watershed on      f) Watershed on
     Gradient       Dissimilarity without      Dissimilarity
                      Q re-evaluation          based on IFT
```

Fig. 3. Result of several watershed approaches

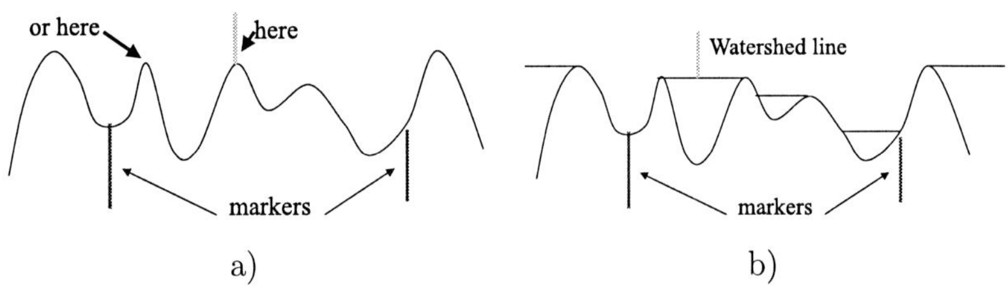

Fig. 4. Watershed boundary with a catchment basin with no markers and same peak height. a) watershed with ordered queue; b) watershed with IFT

markers using the ordered queue did not require this change of the homotopy to impose the minima on the markers. The behavior of this algorithm is best illustrated using Fig 4. The figure shows a situation where a catchment basin with no marker is surrounded by peaks of same height. As the algorithm evolves during its propagation step, the markers fill their own catchment basins and both arrive simultaneously at the peaks of the same height. The first that enqueues a neighbor in this markless catchment basin with a smaller priority level will flood the entire basin with its label. The final watershed boundary will be at the crest line of one of the peaks (randomly chosen as shown in Fig 4a). If the minima imposition operator were used, the image with new homotopy will present a plateau closing the catchment basin without a maker and the watershed boundary will be at the medial line of this plateau. The use of the cumulative path cost (C_m) in the watershed definition of subsection 2.1 is equivalent of making intrinsically this change of homotopy. Thus the IFT algorithm implements this definition as illustrated in Fig 4b.

7. Conclusions and Discussions

The watershed can be seen as the solution of the shortest-path forest problem (IFT) in the graph theory framework. The lexicographic path cost is based on two components, the first with the highest priority, is related to the normal flooding (maximum arc weight in the path) and the second, is related to the watershed behavior at the image plateaus. This formulation has the advantage of being consistent with the most efficient watershed algorithm based on the ordered queue. We have called the attention that this algorithm is a particular case of the IFT shortest-path forest algorithm. We have explained why the FIFO characteristic of the ordered queue is important. We have also explained the main difference of both algorithms which is related to the way a pixel is labeled before or after its inclusion in the ordered queue. Finally, we have shown that the IFT algorithm guarantees the optimality of the shortest-path forest solution when other watershed approaches are used. We have illustrated this with a simple numeric example of segmenting a one-pixel width region.

The watershed as a shortest-path forest can be used both at pixelwise or region level. This paper described the approach at pixelwise level, where each pixel is a node in the graph and the arcs are pixel neighborhood relations. The same model can be applied at the region level, where each node is a region of a segmented image, the arcs are obtained from the region adjacency information, and the arc weights are computed as a measurement related to these regions. As long as the cost path is a non-decreasing function of the arc weights, the shortest-path forest partition can be found with the IFT algorithm. We note that a commonly path cost used in many region-growing algorithms, based on the absolute difference between the mean gray-scale value of the last node region and the mean gray-scale of all the previous nodes in the path is not a non-decreasing function and does not lead to shortest-path forest partition.

Acknowledgements

The project is partially funded by FAPESP, process n. 97/13306-6. The second author is also supported by CNPq, process n. 300156/97-9. Dr. Alexander Tuzikov from the Institute of Engineering Cybernetics, Belarus, is gratefully acknowledged for valuable comments, specially in respect to the lexicographic path cost formulation.

References

1. S. Beucher and C. Lantuéjoul. Use of watersheds in contour detection. In *International Workshop on Image Processing, Real-Time Edge and Motion Detection/Estimation*, Rennes, France, 1979.
2. S. Beucher and F. Meyer. The morphological approach to segmentation: the watershed transformation. In E. R. Dougherty, editor, *Mathematical Morphology in Image Processing*, chapter 12, pages 433–481. Marcel Dekker, New York, 1993.
3. J. Crespo. The flat zone approach: A general low-level region merging segmentation method. *Signal Processing*, 62(1):37–60, 1998.
4. J. Crespo and R. Schafer. The flat zone approach and color images. In J. Serra and

P. Soille, editors, *Mathematical morphology and its applications to image processing*, pages 85–92. Kluwer Academic Publishers, 1994.
5. R. Dial. Algorithm 360: Shortest-path forest with topological ordering. *Comm. Assoc. Comput. Mach.*, 1969.
6. E. W. Dijkstra. A note on two problems in connexion with graphs. *Numerische Mathematik*, 1959.
7. A. Falcão, R. Lotufo, and G. Araujo. Image foresting transform. *IEEE Trans. on Pattern Analysis and Machine Intelligence, submitted*, 1999.
8. A. Falcão, J. Udupa, S. Samarasekera, S. Sharma, B. Hirsch, and R. Lotufo. User-steering image segmentation paradigms: Live wire and live lane. *Graphical Models and Image Processing*, 60(4):233–260, 1998.
9. R. Lotufo, A. Falcão, and F. Zampirolli. Fast Euclidean distance transform using a graph-search algorithm. *submitted*, 1999.
10. F. Meyer. Color image segmentation. In *4th Intern. Conf. on "Image Processing and its Applications"*, page 4, Maastricht, the Netherlands, April 1992.
11. F. Meyer. Topographic distance and watershed lines. *Signal Processing*, 38:113–125, 1994.
12. E. Moore. The shortest path through a maze. *Proc. Internat. Symposium on the Theory of Switching, Part II, Harvard university press, Cambridge, MA*, 1957.
13. F. Preteux. On a distance function approach for gray-level mathematical morphology. In E. R. Dougherty, editor, *Mathematical Morphology in Image Processing*, chapter 10, pages 323–349. Marcel Dekker, New York, 1993.
14. P. Salembier, P. Brigger, J. R. Casas, and M. Pardas. Morphological operators for image and video compression. *IEEE Transactions on Image Processing*, 5(6):881–898, 1996.
15. L. Vincent. Morphological algorithms. In E. R. Dougherty, editor, *Mathematical Morphology in Image Processing*, chapter 8, pages 255–288. Marcel Dekker, New York, 1993.

DISCRETE 3D WAVE PROPAGATION FOR COMPUTING MORPHOLOGICAL OPERATIONS FROM SURFACE PATCHES AND UNORGANIZED POINTS

FREDERIC F. LEYMARIE and BENJAMIN B. KIMIA
LEMS, Div. of Engineering, Brown University, Box D
Providence, RI 02912, USA
Correspondence: [leymarie,kimia]@lems.brown.edu
Phone (Fax): +1.401.863.2760 (9039)

Abstract. We present a discrete framework for 3D wave propagation to support morphological computations with an emphasis on the recovery of the medial axis of a 3D solid, a collection of surface patches, or a data set of unorganized points. The wave propagation is implemented on a discrete lattice, where initial surfaces are considered as sources of propagation. Three classes of discrete rays are designed to cover the propagation space with a minimal number of computations. These pencils of rays represent a "compromise" view between Huygens and Fermat principles. The 3D medial axis points are then found at the collision of wavefronts. This method has linear time complexity in the number of nodes of the lattice used to discretize the propagation medium, i.e., it is independent of the topological complexity of the initial data. As such, it is highly efficient for the extraction of symmetries, as well as for implementing 3D morphological filters based on erosions and dilations, from large 3D data sets. The wave propagation scheme permits to implement the effect of various metrics including the Euclidean one.[1]

Key words: 3D Morphology, Skeletons, Euclidean Distance Transform, Wave Propagation, Eikonal, Unorganized Datasets, Cellular Automata.

1. Introduction

The significance of morphological operations and symmetry-based representations of shape is well-established in computer vision [6, 18], computer graphics [2], and computational and solid geometry [11, 20]. Mathematical morphology is concerned in particular with "probing" operations on discrete sets, through the use of structural elements aimed at emphasizing or filtering some structure of the probed sets. This probing is performed for a range of scales dictated by the structural element's size and shape. A particular result of such probing operations, is the "medial axis" or "skeleton" which consists of points equidistant to the boundary, or alternatively is the locus of maximal bitangent circles in 2D or spheres in 3D. Dilation/erosion, closing/opening, hat-filters and other morphological probing operations are useful in many practical applications, including image and video signal filtering, topography feature extraction (*e.g.*, watershed segmentation) and Computer Aided Design (*e.g.*, offsets for numerical machine tooling). In this paper we concentrate on the probing of 3D discrete sets, which has proved difficult in the past to perform efficiently and accurately.

[1] We gratefully acknowledge the support of this research by the NSF grant IRI-9700497.

We cast our 3D mathematical morphology work within the framework of geometrical optics, where the equivalence between the continuous wave propagation representation of optics and the discrete set probing of morphology can be exploited to design useful algorithms. This equivalence has started being exploited in the processing of 2D sets and grey-level images in the recent years (e.g., see [3, 21, 15, 22]). It directly generalizes to 3D problems and can be stated as follows [3]:

Theorem 1 (Wave propagation as a morphological (set) operation) *Let $\mathcal{S}$ be a 3D object or set of points in the metric space $I\!R^3$. Let $\mathcal{B}(t)$ be a family of convex structural elements in $I\!R^3$ which can be continuously scaled in size as a function of a "time" t. Then, we can describe morphological operations on $\mathcal{S}$ via the following evolution equation of its boundary $\partial\mathcal{S}$:*

$$\frac{\partial \mathcal{S}}{\partial t} = \beta(\theta) \cdot \mathcal{N} \ , \tag{1}$$

where β is the amount of deformation (e.g. due to a dilation or erosion) by a structural element $\mathcal{B}$, θ encodes the relative orientation of $\mathcal{B}$ with respect to the boundary $\partial\mathcal{S}$ of our object, and $\mathcal{N}$ is the normal to $\partial\mathcal{S}$.

At any time t, one can show that this is a "normal mapping", weighted as a function of the structural element shape and orientation with respect to a given "wavefront" $\partial\mathcal{S}(t)$. Equation 1 gives us a continuous interpretation of morphological operations via a differential deformation, β [21]. We also recognize Equation 1 as a Hamilton-Jacobi equation describing a general wave propagation. This equivalence with geometrical optics identifies β as the *refractive index* of our space, into which the wavefront $\partial\mathcal{S}$ evolves. Consider now the gradient field to our object's boundary, $\nabla\mathcal{S}$, at any time t, and let us denote the time evolving boundary $\partial\mathcal{S}(t)$ as a wavefront ϕ, which in $I\!R^3$ is a surface: $\phi(x, y, z) = 0$ (in implicit form), for a given time t.

Corollary 1 (Morphological operations via the Eikonal equation) *Dilations and erosions can be modeled in 3D via the Eikonal equation of geometrical optics [19]:*

$$d\phi \cdot \nabla\mathcal{S} = \beta \ , \tag{2}$$

where the (refractive) index, β, provides the metric structure of the medium of propagation.

β may vary in shape locally, defining a Riemannian geometry. In this paper we will restrict ourselves to the special, but important case, where β is taken to be homogeneous, that is, where our structural elements do not vary in "shape", but only in size.

2. Continuous Propagation

Consider a point $\mathbf{q}_0$ in Euclidean space E^3 at time t_0, taken to be the source of a disturbance transmitting itself "locally" [4, p. 250]. At time $t_0 + \Delta t$ the disturbance is propagated through space in the form of a surface or wavefront, $\phi(x, y, z, t) = 0$. At any later time, the new wavefront can be obtained as a collection of disturbances from a previous wavefront; this is the famous geometrical construction of Huygens [4, p. 250]:

Theorem 2 (Huygens' principle) *Let $\phi_{\mathbf{q}_0}(t)$ be the wavefront of the point after time t. For every point $\mathbf{q}$ of this front, consider the wavefront after time t_1: $\phi_{\mathbf{q}}(t_1)$. Then, the wavefront of the point $\mathbf{q}_0$ after time $t + t_1$: $\phi_{\mathbf{q}_0}(t + t_1)$, will be the envelope of the fronts $\phi_{\mathbf{q}}(t_1)$ such that $\mathbf{q} \in \phi_{\mathbf{q}_0}(t)$.*

The wavefront is thus the boundary of the set of all points $\mathbf{q}$ to which "information" from a given source can travel in time less than or equal to t (Fig. 1.(b)). Huygens' metaphor can be transfered to the discrete domain without difficulty, *e.g.*, by considering the **Minkowski sum** of small discrete spherical sets with a discrete front at time t. This is precisely the realm of 3D Mathematical Morphology, where the spherical sets are called structural element [18]. The problem is one of inefficiency: Minkowski sums with isotropic structural elements - required to ensure Euclidean results - have too much (maximal) overlap.

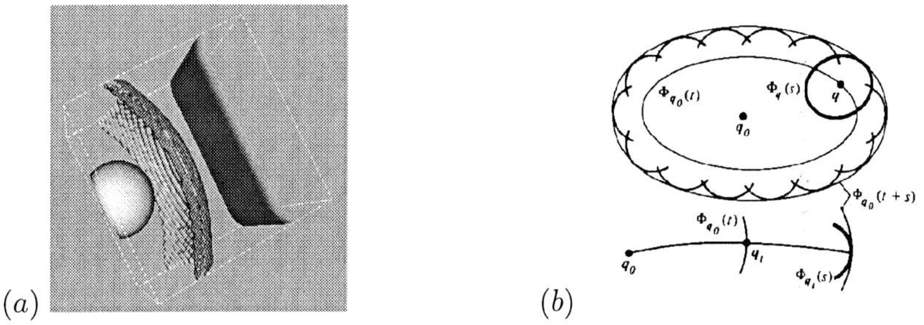

Fig. 1. (a) Example of 3D medial axis computed through wave propagation from a spherical cap and a rectangular plane filing a cubical portion of space. The symmetry sheet, in the middle, is a paraboloid of revolution. (b) A wavefront as the envelope of fronts of wavelets and an equivalent minimal path along a ray (after [4, p. 249]).

Theorem 3 (Fermat's principle) *Information travels along* rays *from a point $\mathbf{q}_0$ to a point $\mathbf{q}$ in the shortest possible time [19].*

Fermat's principle represent the alternative viewpoint to Huygens', where **rays** are such that their gradient vector coincides with the normal to the wavefront, *i.e.*, the direction of a ray is given as: $p = \nabla S$. Rays and wavefronts are related precisely through the refractive index β, via Equation (2). Integration of this index from a source $\mathbf{q}_0$ to $\mathbf{q}$ result in a minimal time path. One interesting special case is where the index is taken to be both isotropic and homogeneous. In this case, the direction of motion of the rays and the wavefront coincide [4]; thus,

$$d\phi \cdot \nabla S = \|d\phi\| \; \|\nabla S\| = 1 \ .$$

Note that, considering a (light) particle traveling at constant speed c, since the index is fixed, we also have that $\|p\| = \|\nabla S\| = c$, leading to a simplified form of the Eikonal equation:

$$\phi_x^2 + \phi_y^2 + \phi_z^2 = 1/c^2 \ .$$

This evolution equation is equivalent to taking for our family of structural elements, $\mathcal{B}(t)$, *spheres* of continuously varying radius. In other words, the underlying metric of the medium of propagation is Euclidean.

3. Discrete Propagation: Three Beam Classes

Each data point belonging to the initial surface (closed or not, unorganized and possibly isolated) is considered a *source*, as in the "grassfire" model of Blum [6], where skeletons are computed as the "quench points" of the fire or wavefronts. The dual views of (continuous) wave propagation imply two distinct approaches to their simulation in the discrete domain. On the one hand, *Huygens' principle* suggests propagating waves from the wavefront to all its neighbors. This method is clearly redundant and inefficient, but it covers the entire discrete space. On the other hand, *Fermat's principle* suggests propagating waves from the wavefront along discrete set of directions, *e.g.*, along grid axes and diagonals. This method is efficient, but does not cover the whole space and leaves gaps behind. We adopt an intermediate view where, in addition to regular (1D) discrete rays, normally achieved by propagating along grid axes and diagonals, we also fill the remaining gaps of propagation with 2D beams (planar sectors) along each grid plane of propagation, namely the XY, YZ, and XZ planes. There are still some gaps left behind by this approach which are completely filled by 3D beams (volumetric sectors). Thus, we have reached a compromise: the three sets of beams represent neither rays nor spherical wavelets, but discrete conical beams which more efficiently represent the continuous wave propagation it simulates.

In order to formally specify the above model, namely, the three classes of beams and their propagation on a discrete lattice, we adopt the notion of **cellular automata**. A **cell** is defined as a voxel with a set of directions of propagation, and **rules** for propagating these directions to neighbor cells are established. Specifically, four concepts are required [1]: *(i)* a discrete lattice or array of cells: the cellular space; *(ii)* a set of neighborhoods (or templates or masks): cellular neighborhoods; *(iii)* a definition of possible cell states; *(iv)* a set of rules for local transitions (or cell-state transitions function).[2]

First, the discrete lattice is provided by the 3D grid and a set of voxels. Second, we take for the CA neighborhood, the 3x3x3 set of closest voxels centered around each cell. Third, the cell state indicates which of the discrete directions of propagation are to be followed. These directions of propagation maintain the contiguity of the wave front and come in three classes: 1D, 2D and 3D. Finally, the cell-state transitions function is uniquely determined by

[2] Together *(ii)* and *(iii)* correspond to the notion of structural element, and thus represent the embodiment of the index (or deformation), β.

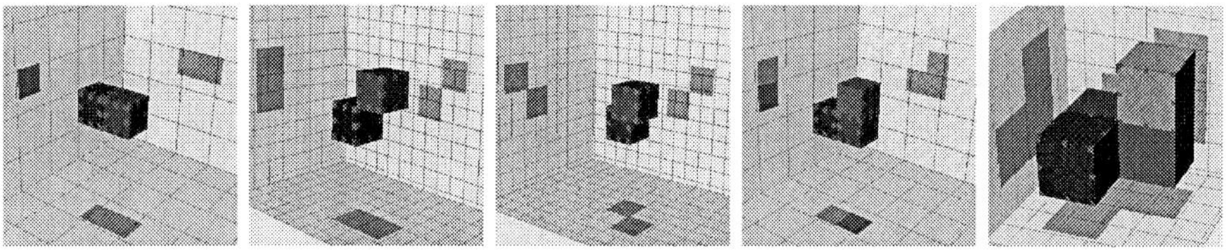

Fig. 2. Examples of neighborhoods, from left to right, for the 1DPR: 1D axial, 2D diagonal, and 3D diagonal, for the 2DPR, and, finally, for the 3DPR.

its initial state and the application of the local transition. Note that by "local transition" we mean that the state of a cell at time $t+1$ is only a function of its own state and the states of its immediate "neighbors" at time t.

Formally, let c_{ijk} denote the cell at location (i, j, k). Consider a cubic (3x3x3) neighborhood consisting of the 26 nearest cells. Let $c_{ijk} \oplus \{\delta_{lmn}\}$ denote a propagation (or dilation) from cell c_{ijk} to a cell with relative position (l, m, n) in the 3x3x3 neighborhood, *i.e.*, $l, m, n \in \{-1, 0, 1\}$. Furthermore, let

$$c_{ijk} \oplus \left\{\delta_{l_\eta m_\eta n_\eta}\,,\, \eta = 1, \ldots, N\right\} \ , \tag{3}$$

denote simultaneous propagation from cell c_{ijk} to all cells $(i+l_\eta\,,\, j+m_\eta\,,\, k+n_\eta)$ for $\eta = 1, \ldots, N$ $(N = 26)$, and where l_η, m_η, n_η can take values $-1, 0, +1$.

The propagation of a spherical wave in the discrete domain takes place along 6 axial directions, where l_η, m_η and n_η take on values from Table I (LHS), for a total of 26 discrete, but 1D, directions of propagation. We call this first class of propagation rays the *1D pencil of rays* (*1DPR*), Figure 2. While this set of rays does cover the immediate neighborhood correctly (*i.e.*, as a correct approximation of continuous wavefront propagation), it does not cover all neighbors of neighbors correctly (and therefore beyond), and gaps result. This is well known from the "ordered propagation" ideas of Ragnemalm in 2D [17]. An efficient solution is to augment the 1DPR by **pairs** of neighboring propagation directions taken along each plane of the grid. This leads to the following transition rule:

$$c_{ijk} \oplus \left\{(\delta_{l_\eta m_\eta n_\eta}\,,\, \delta_{o_\eta p_\eta r_\eta})\,,\, \eta = 1\,, \ldots\,, N\right\} \ , \tag{4}$$

where the indices $(\delta_{l_\eta m_\eta n_\eta}\,,\, \delta_{o_\eta p_\eta r_\eta})$ can only take values as described in Table II, for a total of 24 2D Pencils of Rays (*2DPR*), Figure 2. Similarly, a 3D pencil of rays (*3DPR*) is needed to cover the solid angle limited by a (3D) diagonal and grid axes, Figure 2. These are represented as

$$c_{ijk} \oplus \left\{(\delta_{l_\eta m_\eta n_\eta}\,,\, \delta_{o_\eta p_\eta q_\eta}\,,\, \delta_{r_\eta s_\eta t_\eta})\,,\, \eta = 1, \ldots, N\right\} \ , \tag{5}$$

where the indices take on values described in Table I (RHS), for a total of 48 combination of 3D Pencils of Rays (*3DPR*).

The above discussion implicitly addresses the state transition function. Each state inherits information regarding the source of wave propagation and can thus determine its own set of pencils of rays to propagate. Ideally, beams

1DPR					3DPR		
Axial	Plane diagonal		3D diagonal		Axial	2D diagonal	3D diagonal
1,0,0	1,1,0	-1,1,0	1,1,1		1,0,0	1,1,0	1,1,1
-1,0,0	1,-1,0	-1,-1,0	-1,1,1		1,0,0	1,1,0	1,1,-1
0,1,0	1,0,1	-1,0,1	1,-1,1		1,0,0	1,-1,0	1,-1,1
0,-1,0	1,0,-1	-1,0,-1	-1,-1,1		1,0,0	1,-1,0	1,-1,-1
0,0,1	0,1,1	0,-1,1	1,1,-1		1,0,0	1,0,1	1,1,1
0,0,-1	0,1,-1	0,-1,-1	-1,1,-1		1,0,0	1,0,1	1,-1,1
			1,-1,-1		1,0,0	1,0,-1	1,1,-1
			-1,-1,-1		1,0,0	1,0,-1	1,-1,-1

TABLE I
(LHS) Indices of the 26 1DPR: l_η, m_η, n_η. (RHS) Indices of the 3DPR, by triplets (l_η, m_η, n_η), (o_η, p_η, q_η), (r_η, s_η, t_η), associated to the positive X direction. Taken together with indices for the other 5 directions, gives a total of 48 possible 3DPR.

X-axis	Y-axis	Z-axis
(1,0,0), (1,1,0)	(0,1,0), (1,1,0)	(0,0,1), (1,0,1)
(1,0,0), (1,-1,0)	(0,1,0), (-1,1,0)	(0,0,1), (-1,0,1)
(1,0,0), (1,0,1)	(0,1,0), (0,1,1)	(0,0,1), (0,1,1)
(1,0,0), (1,0,-1)	(0,1,0), (0,1,-1)	(0,0,1), (0,-1,1)
(-1,0,0), (-1,1,0)	(0,-1,0), (1,-1,0)	(0,0,-1), (1,0,-1)
(-1,0,0), (-1,-1,0)	(0,-1,0), (-1,-1,0)	(0,0,-1), (-1,0,-1)
(-1,0,0), (-1,0,1)	(0,-1,0), (0,-1,1)	(0,0,-1), (0,1,-1)
(-1,0,0), (-1,0,-1)	(0,-1,0), (0,-1,-1)	(0,0,-1), (0,-1,-1)

TABLE II
Indices of the 24 2DPR, for pairs (l_η, m_η, n_η), (o_η, p_η, q_η).

proceed in parallel. Simulation of this process on a sequential machine, however, requires a notion of time, or distance from source with constant speed propagation.

The state of a cell c_{ijk} is described either as active ("on" a wavefront) or as quiescent ("off"). A label is used to identify membership to a particular "beam" type, namely a selection among $1DPR$, $2DPR$ and $3DPR$ directions. Each cell maintains a distance map (from sources) in the form of (L_X, L_Y, L_Z) which reflects the (signed) coordinates of the vector from the current cell to its source(s). If we compute a single distance map, only one (minimal) distance set is kept.[3] A symmetry label is also associated to each cell, depending on the type and number of its sources.

The 3D local state *transition* function is used to update an active cell's state at any given time-step. These transitions are implemented using a discrete approximation to the Euclidean metric [12, 17]. Specifically, an active cell transmits its *signed* Euclidean distance from a source vector, (L_X, L_Y, L_Z),

[3] Multiple distance maps and interpenetrating waves in 2D have been proposed in [26, 23].

and thus updates all the neighbors contained in its beam.[4] The remaining quiescent cells, which typically form a much larger number, do not require computations.

Note that, at initialization, since we do not assume any a-priori knowledge of the object surface geometry other than sampled point positions, the full 26 neighbors are used for the propagation from the sources. To further optimize the propagation and reduce the possibility of multiple updates of each cell, ordered propagation through distance is then simulated [17], *i.e.*, on the list of active cells, we always first process those with lowest (squared) Euclidean distance, $L^2 = L_x^2 + L_y^2 + L_z^2$. Quiescent cells when first visited automatically accept the distance value L^2 written by an active neighboring cell. Later updates may occur if a new distance is of lower value. When a quiescent cell's distance vector is updated, it is added to the active list, and inherits some "label information" from the cell updating it, *e.g.*, initial source and pencil of rays. Active cells use their associated pencil of rays to change the state of neighboring cells.

Numerical complexity and exactness: Our implementation of wave propagation performs the Euclidean Distance Transform (EDT) with respect to the initial source points. We integrate ideas from ordered propagation in 2D [16, 24, 17] to obtain a nearly optimal algorithm. The EDT's numerical complexity is linear in the number of cells visited, that is $O(M)$, where M is the number of cells used to sample the medium of propagation, and, thus, is independent of the object's outline shape. The constant of linearity is a function of the neighborhoods used and the way overlaps are handled. Since the larger beams, 2DPR and 3DPR, are all made by the concatenation of simpler 1DPR, one may eliminate overlap between nearby beams by constant bookkeeping, a refinement recently proposed by Eggers [8].

EDTs are nearly exact methods [7], with problems occurring only for those Voronoi regions which have sharp ends going between lattice nodes (in 3D, these are cones of sub-voxel width). Some authors proposed partial solutions to this problem, *e.q.*, using delayed removal of cells from the active list [25, 17]. A complete solution to this problem, however, requires the propagation of shock waves [22], which we will implement in the future. Despite these minute errors, the results are much more accurate than all other integer approximation (methods based on so-called Chamfer DTs as well as thinning methodologies), together with similar or lower numerical complexity [12].

4. Symmetry Detection

Symmetry loci can be directly detected during our process of wave propagation, when wavefronts *meet.* In the discrete domain, two cases are possible: either different automata reach the same cell at the same time-step/distance-value, or an automaton reaches a cell with a lower minimal distance value, which signifies that a *crossing* of waves has occurred *between* voxels. In the latter case, we update the symmetry status of all intervening cells. In general this

[4] A signed EDT is useful when computing angles between intersecting waves, § 4.

involves either two or three voxels, associated to either a skeletal point on a sheet or on an axial curve. Hence the computed symmetry sheets appear thick when we locally register many cells in the vicinity of a sub-voxel symmetry site. This is only an apparent "defect" which should be tackled by using an adapted sub-voxel tracer. In the examples provided in this paper, we have used the now classical marching-cube algorithm [13], which generates bounding iso-surface on each sides of the skeletal sheets.

Pruning of the 3D skeleton graph: Skeletal symmetries are known for their sensitivity to any perturbation of the original surface shape, *e.g.*, each little bump is responsible for a skeletal feature. In 3D, this usually generates a new surface sheet. This is a drawback of the skeletal representation, which, however, can be tackled using a degree of significance first hinted at by Blum [6].

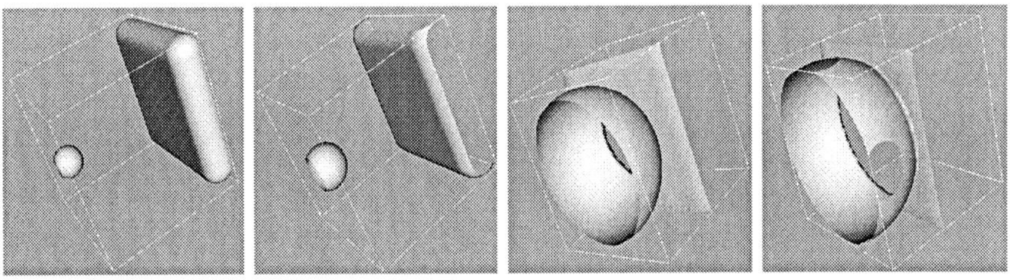

Fig. 3. Accurate 3D wave propagation from a dot and a square plane.

Two simple criteria may be combined to remove most effects due to noise (small perturbations of the boundary) or sampling-effects of the grid (a rotated plane becomes a rippled surface). First, one may consider a minimum distance of propagation before retaining any wave crossing as indicating a skeletal symmetry. We call it "minimal thickness", and denote it by ρ. The second criterion is the angle made by pairs of crossing waves, which we denote by α. The closer to π is α the higher the "local" significance of the symmetry; for $\alpha = \pi$ waves are exactly facing each other as their originating sources. As $\alpha \to 0$ the corresponding surface elements become parallel. Small perturbations or ripples on the surface may generate large wave crossing angles α initially, near the originating surface, but those then rapidly decrease as they travel away from the boundary. This is simply due to the fact that noise or ripples have a relatively small support on the surface shape. Recently, some authors have shown how to select *a-posteriori* empirical values for ρ and α, for any given sampled object [5, 9]. In our wave propagation scheme, ρ and α are easily computed using distance values, L and the signed distance vectors of crossing waves (L_x, L_y, L_z), respectively.

The above criteria, however, prove insufficient if one must guarantee the homotopy of the final result, such that no holes are created in the skeleton due to a too high threshold on α. Therefore, one must combine these with a classical homotopy preserving criteria, such as used by thinning algorithms, *e.g.*, see [14]. Again, such a criterion is best applied in the course of propagation, in order to remove any insignificant skeletal point (according to α) without creating holes.

Indeed, if one waits to perform such deletions as a post-processing step, one must start from low ρ values and in fact re-do much of the propagation already performed once.

Finally, a post-processing step may be performed in order to retrieve parts of the skeletal sheets and rims which are either near or linked to the object surface, but which were removed because of the minimum ρ criterion set to remove the ripple effects. This is relevant in particular for objects where *ridges* or *valleys* are required to be precisely located as surface shape features, *e.g.*, for the nose line of a face [10].

Results: We illustrate the accuracy of the propagation and the topological correctness of this scheme in a few examples. Figure 1a represents propagation from a single point (left) and a rectangular plane (right) for a cubic domain of size 400x400x400. The intermediate symmetry sheets, in the expected form of a paraboloid of revolution, are also shown. In Figure 3 the intermediate steps of that propagation are shown. The computation is performed in a sub-minute time scale, on a typical UNIX workstation, which indicates the algorithm's efficiency. We expect parallel hardware to bring such computations to real-time applications.

In Figure 4a,b we show the symmetries detected for a parabolic gutter, which has the special feature of containing a generic axis of symmetry at the meeting of three planar sheets. Note that we use a symmetry significance criterion to remove insignificant symmetries *as we propagate waves*, hence insuring the correct topology for the final skeleton. Finally, in Figure 4c,d are shown the computed medial axes for an open and closed cylinder.[5]

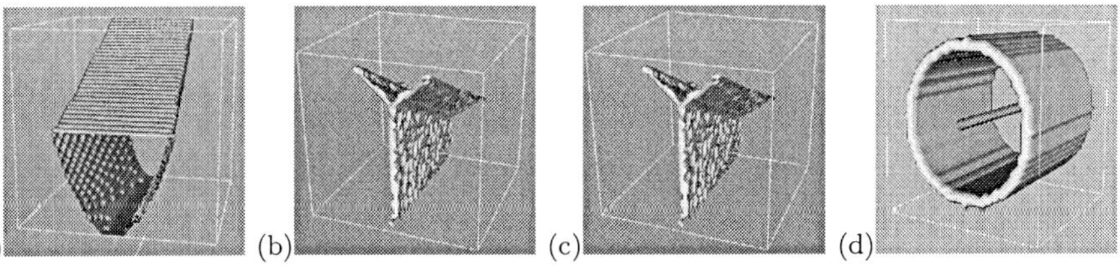

(a) (b) (c) (d)

Fig. 4. (a) Parabolic gutter and (b) its symmetry sheets. (c) *Open* cylinder and computed medial axis. (d) Computed medial axis for the associated, *closed* cylinder.

5. Conclusion

We have introduced an efficient algorithm for 3D morphological probing operations under the Euclidean metric, including dilations/erosions, and for symmetry detection of shapes. This is accomplished based on a discrete wave propagation scheme, which has *linear time complexity* in the number of cells used to sample space. Our method is nearly optimal as we restrict potential

[5] Visit the web site: http://www.lems.brown.edu/~leymarie, under the *Mathematical Morphology* page/link, for more results, in particular for filtering and surface interpolation applications.

multiple updates of cells' state to only those beams involved in wave crossings. A fully optimal algorithm, permitting to retrieve an exact distance map and a continuous sub-voxel trace of the shape offsets and skeleton, will require the use of an analytic distance propagation, a topic we are exploring; for 2D results, see [23, 21, 22].

We emphasize that the extracted skeletons are *invariant* under rigid transformations, due to the use of Euclidean metric to update cell states, and can operate on unconnected surface patches or even unorganized points.

References

1. A. Adamatzky. *Identification of cellular automata.* Taylor and Francis, London, 1994.
2. N. Amenta et al. A new Voronoi-based surface reconstruction algorithm. *Computer Graphics (SIGGRAPH '98)*, pages 415–421, July 1998.
3. A. Arehart, L. Vincent, and B. Kimia. Mathematical Morphology: The Hamilton-Jacobi connection. In *Proc. of ICCV*, pages 215–219, 1993.
4. V. I. Arnold. *Mathematical Methods of Classical Mechanics.* Springer V., 1989.
5. D. Attali and A. Montanvert. Computing and simplifying 2D and 3D continuous skeletons. *CVIU*, 67(3):261–273, 1997.
6. H. Blum. Biological shape and visual science. *J. of Theor. Bio.*, 38:205–287, 1973.
7. P.-E. Danielsson. Euclidean distance mapping. *CGIP*, 14:227–248, 1980.
8. H. Eggers. Two fast Euclidean distance transformations in Z^2 based on sufficient propagation. *CVIU*, 69(1):106–116, 1998.
9. G. Malandain and S. Fernandez-Vidal. Euclidean skeletons. *Image and Vision Computing*, 16(5):317–327, 1998.
10. P. Halliman, G. Gordon, A. Yuille, P. Giblin, and D. Mumford. *Two- and Three-Dimensional Patterns of the Face.* A. K. Peters, 1999.
11. C. Hoffmann and J. Rossignac. A road map to solid modeling. *IEEE Trans. on Visualization & Computer Graphics*, 2(1):3–10, March 1996.
12. F. Leymarie and M. Levine. Fast raster scan distance propagation on the discrete rectangular lattice. *CVGIP-IU*, 55(1):84–94, 1992.
13. W. Lorensen and H. Cline. Marching cubes. *Computer Graphics*, 21(4):163–169, 1987.
14. C. Ma and M. Sonka. A fully parallel 3D thinning algorithm and its applications. *CVIU*, 64(3):420–433, 1996.
15. P. Maragos and M. Butt. Advances in differential morphology. In *ISMM'98*, pp. 167–174. Kluwer Academic, 1998.
16. U. Montanari. Continuous skeletons from digitized images. *JACM*, 16(4):534–549, 1969.
17. I. Ragnemalm. Neighborhoods for distance transformations using ordered propagation. *CVGIP: Image Understanding*, 56(3):399–409, 1992.
18. J. Serra. *Image Analysis & Mathematical Morphology.* Acad. Pr., 1982.
19. O. Stavroudis. *The Optics of Rays, Wavefronts, and Caustics*, vol. 38 of *Pure and Applied Physics.* Academic Press, 1972.
20. A. Sudhalkar, L. Gürsöz, and F. Prinz. Box-skeletons of discrete solids. *Computer-Aided Design*, 28:507–517, 1996.
21. H. Tek and B. Kimia. Curve evolution, wave propagation, & mathematical morphology. In *ISMM'98*, pp. 115–126. Kluwer Academic, 1998.
22. H. Tek and B. Kimia. Symmetry maps of free-form curve segments via wave propagation. In *Proc. of the 7th ICCV*, pp. 362–369, Kerkrya, Greece, Sept. 1999.
23. H. Tek, F. Leymarie, and B. Kimia. Interpenetrating waves & multiple generation shocks via the CEDT. In *Adv. in Visual Form Analysis*, pp. 582–593. World Sci., 1997.
24. B. Verwer, P. Verbeek, and S. Dekker. An uniform cost algorithm applied to distance transforms. *IEEE Trans. on PAMI*, 11(4):425–429, 1989.
25. L. Vincent. Exact Euclidean distance function by chain propagations. In *Proc of CVPR'91*, pages 520–525, 1991.
26. M. Wright et al. Skeletonization using an extended Euclidean distance transform. *Image & Vision Comp.*, 13(5):367–375, 1995.

TWO-STAGE LOSSY/LOSSLESS COMPRESSION OF GRAYSCALE DOCUMENT IMAGES

KRIS POPAT and DAN S. BLOOMBERG
Xerox Palo Alto Research Center
3333 Coyote Hill Road
Palo Alto, CA 94304
{popat,bloomberg}@parc.xerox.com

Abstract. This paper describes a two-stage method of document image compression wherein a grayscale document image is first processed to improve its compressibility, then losslessly compressed. The initial processing involves hierarchical, coarse-to-fine morphological operations designed to combat the noiselike variability of the low-order bits while attempting to preserve or even improve intelligibility. The result of this stage is losslessly compressed by an arithmetic coder that uses a mixture model to derive context-conditional graylevel probabilities. The lossless stage is compared experimentally with several reference methods, and is found to be competitive at all rates. The overall system is found to be comparable with JPEG in terms of mean-square error performance, but appears to outperform JPEG in terms of subjectively judged document image intelligibility.

Key words: Document Image Compression, Image Morphology, Arithmetic Coding, Multiresolution, Gaussian Mixtures.

1. Introduction

Grayscale scanning offers several advantages over binary scanning in terms of image quality and downstream flexibility, even when the documents being scanned are binary. The downside is the need to capture, process, and store much more information. Compression of the scanned grayscale document images is therefore of great importance.

JPEG [11] is the most widely used method for compressing natural scenes, but introduces undesirable artifacts around the sharp edges found in document images, particularly within text and line-art regions. Recently, methods have been proposed in which different segments of a document image are encoded using different techniques; a good example is the DjVu format developed by AT&T [6].

While it is advantageous to use segmentation information to adapt compression locally, it is not necessary to encode the segments separately. In particular, the approach proposed here is to use the segmentation information to switch among probability models used with arithmetic coding, thus allowing the encoding to be carried out on a single raster layer. The segmentation information must be transmitted, but this requires few bits relative to those required for the image itself. We propose a two-stage approach: (a) morphologically-based segmentation and a one-time lossy transformation to improve compressibility and maintain intelligibility, and (b) lossless compression by arithmetic coding. Section 2 describes the lossy transformation and segmentation method, Section 3 describes the lossless compression, and Section 4 presents some experimental results.

2. Lossy Stage

Unlike conventional lossy compression techniques, such as JPEG, which performs lossy quantization and lossless encoding of the coefficients atomically, here we separate the lossy step from the lossless one. Because the lossy step is performed in the image domain, rather than the transform domain, the change in visual appearance can be controlled by minimizing the maximum pixel value change.

Most of the entropy in a scanned grayscale image is in the low-order bits (LSBs) of each pixel. Because these bits are the least visible, they can be set to zero by rounding. For example, for 8 bit pixels, three LSBs can be set to zero by adding 4 (binary 100) and truncating to the 5 MSBs, taking care to avoid overflow. To take account of neighboring correlations, we use a multiscale approach were four pixels at one scale are compared and are either left unchanged or averaged with post-rounding. The thresholds used for the comparison at each scale can be chosen either to preserve low-contrast features (such as bleed-through) or to remove them.

2.1. 'Pyramid Scheme'

For document images, it is important to represent the pixels in transition regions between light and dark (i.e., at edges) with fidelity. This is accomplished in the following manner. The lossy stage first rounds the n_r LSBs of all pixels at full resolution to 0. Then it generates a pyramid of n_d reduced images of dimension 2^{-n_d} relative to the original. At each $2 \times 2 \rightarrow 1 \times 1$ stage, the image is tiled into 2×2 pixels and the maximum deviation from the average within the tile is compared to a level-dependent threshold. If the deviation is smaller than the threshold, a single pixel is saved with the average value (again with n_r LSBs rounded to 0); otherwise, the pixel is marked with the value 1, which is distinguished from all possible rounded average values. Beyond the first stage in the pyramid, pixels with value 1 can be encountered and are ignored in the averaging process.

After the reduced images are generated, the average values are propagated back up the chain. Consider the propagation from level m to $m-1$. If a 1 is encountered at level m, then the four corresponding pixels at $m-1$ are left unchanged. Otherwise, any of the four pixels at $m-1$ that are not 1 are set to the pixel value from level m.

The parameters in the encoder are thus: n_r, the number of LSBs rounded at each stage to 0; n_d, the number of reduced images generated in the pyramid; and $\{t_m, m = 1, \ldots, n_d\}$, the thresholds set for each level. The thresholds and n_r cannot be chosen independently, and a workable choice is $t_m = 2^{n_r}$ for $m \leq 2$ and $t_m = 2^{n_r - 1}$ for $2 < m \leq n_d$.

It may be desirable to choose different encoder parameters for text and halftone regions. For example, with large values for n_r, n_d and the thresholds, the text regions are highly compressible, with smoothed background and removal of bleed-through. However, such parameters can cause visible contouring and smearing in halftone regions. Instead, we may want a larger number of gray levels, albeit at low resolution, and might choose smaller values of n_r, n_d and thresholds. Unlike the text regions, where significant added compressibility is achieved by the multiresolution operations, in halftone regions most of the compressibility is due to the initial rounding.

2.2. Segmentation

To apply different parameters to halftone and text regions, and to allow the lossless stage to switch to a probability model appropriate for the category and choice of parameters, it is necessary for the encoder to generate a segmentation mask. There are many methods for generating a mask covering the halftone regions, and we describe a particularly efficient morphologically-based one that uses a binarized version of the image, generated from a global threshold.

The threshold can be chosen from a (subsampled) histogram of image pixels. If there is a significant amount of text or line-art, the set of background pixels will be evident in the histogram. A global threshold for projecting the foreground pixels can then be chosen at the dark edge of this set, by placing the pixels in overlapping histogram bins, finding the darkest bin containing a sufficient fraction of all pixels, and choosing a value near the minimum (dark) boundary of this bin.

If no threshold value is found, no segmentation is performed, and the image is compressed as halftone. Otherwise, the image is pixelwise lowpass filtered using the threshold, giving a *foreground mask* binary image. From this binary image, a *halftone seed* is derived morphologically, and a *halftone mask* is generated by binary reconstruction into the foreground mask from the seed. The seed is generated by a series of closings and openings, which is efficiently carried out on an image pyramid using a sequence of threshold reductions [3], along with closings and openings. Threshold reductions with threshold values of 1 and 4 are equivalent to dilations and erosions with a 2×2 structuring element, respectively, followed by subsampling.

3. Lossless Compression

The processing described in the previous section modifies a document image to make it more compressible, without changing it so much as to detract from its aesthetic appeal or its intelligibility. The remaining problem is to represent the resulting array of cleaned-up pixels in a compact manner. Although in some circumstances we may wish to consider encoding methods which incur further loss, for simplicity we assume that the lossy stage has resulted in precisely the image we wish to represent, and accordingly restrict consideration to lossless compression methods. Throughout this section, the unqualified terms "image" and "pixel" will refer to the result of the lossy stage.

3.1. Why Not Generic Lossless Compression?

One possibility would be to use a generic compression routine such as the *gzip* program [5]. This approach has the appeal of simplicity and robustness, but does not take full advantage of what is known in advance about the statistics of the specific class of images being encoded. Still, this approach can be effective, provided that the two-dimensional pixel array is represented as a one-dimensional sequence in an appropriate manner. For instance, by encoding differences between pixels values in an appropriate way, the Portable Networks Graphics (PNG) format can achieve good compression on graphics images; see the discussion about "filters" in [1].

Generic compressors like *gzip* function by discovering patterns in the data presented to them, but they do not interpret the data in order to draw reasonable generalizations about similar patterns. Specifically, they do not exploit the following *smoothness* property of the joint probability law that can be thought to govern pixel neighborhoods: *small changes in pixel amplitudes in a pattern correspond to small changes in the probability assigned to that pattern.* This property suggests that an advantage is to be had by sharing statistics among patterns that are deemed similar.

3.2. Conditioning Contexts and Mixture Models

We thus consider techniques wherein approximate pattern matches are used when learning the statistical regularity upon which the compression will be based. Arithmetic coding [14, 13] offers a convenient means of separating the statistical modeling task from the actual compression task without giving up performance, thereby allowing the use of specialized statistical models capable of exploiting the above-mentioned smoothness property. For practical reasons, the pixels are processed sequentially rather than in the aggregate, but within that constraint, the use of arithmetic coding allows us to freely specify any statistical model. The remaining constraint is that the statistical model may be conditioned only on preceding pixels in the chosen ordering.

The conditioning structure of the statistical model we consider is patterned after the grayscale extension [10] of the causal-neighborhood context model originally proposed in [7] for binary images. Specifically, for every pixel location in the sequence, a set of nearby but strictly preceding pixel locations is specified as a conditioning context. We consider the simplest case, wherein the pixels are encoded in raster order and the set of conditioning pixels is specified relative to each encoded pixel by a constant causal context neighborhood template (see Figure 1).[1] An estimate of the conditional density is obtained by appropriately normalizing a Gaussian mixture estimate of the joint density of the conditioning pixels and the pixel being encoded. Although normalizing a joint mixture estimated in this way generally does not result in the best conditional density estimate of comparable complexity [9], this approach is conceptually simple,

[1] Hierarchical, coarse-to-fine sequencing is also possible, but multiple statistical models must then be employed, and the conditioning neighborhoods required become more complex. Furthermore, preliminary results have not demonstrated a clear performance advantage in the present application that might offset this added complexity.

and in practice has often been found to perform adequately relative to more involved estimation methods.

Fig. 1. Two examples of causal context neighborhoods. Solid dots indicate conditioning pixels, while the unfilled dots indicate the pixel currently being encoded or decoded.

After the lossy stage, the image pixels assume values in a relatively small set. For example, for eight-bit original images and when $n_r = 3$, there are only thirty-one possible values for each pixel: 0, 8, 16, …, 248. Such coarse discretization is slightly at odds with the smoothness property mentioned earlier, but we wish to exploit smoothness for the generalization benefit it offers, and resort to the following artifice. We imagine that the value of each pixel represents an independent quantization of a hypothetical continuous valued pixel. The mixture is used to model the conditional density of this continuous valued pixel, conditioned on specific previous quantized pixel values. The probability mass function provided to the arithmetic coder is obtained by integrating this conditional density over each quantization region. Since the hypothetical continuous-valued pixel is unavailable for training the mixture, the quantized values are used instead, after adding to each a small amount of uniformly distributed noise. Independent quantization of individual pixels is an imperfect model of the lossy stage, as it does not account for the important spatial interaction that occurs there. Nevertheless, we have found that it is a useful model for deriving a probability mass function for arithmetic coding, as is borne out in the results presented in Section 4.

3.3. Details of the Lossless Compression Method

The pixels are always processed in raster order. Let x denote the current pixel being encoded, and let $(y_1, \ldots, y_N)$ denote a vector of preceding conditioning pixels specified by a fixed context neighborhood of the type shown in Figure 1. A set of training images, each deemed similar in nature to the image segment to be encoded and each processed by the lossy stage using the same parameter values, is determined. For instance, if the cleaned-up image segment to be encoded is a line drawing, the training images selected should also be line drawings, processed using the same parameters in the lossy stage.

These training images are scanned by sliding the context neighborhood along the image and, at every pixel location for which the entire neighborhood lies within the image boundaries, assembling the values indicated by the neighborhood into a vector. In this way, a collection of training vectors $\{(x, y_1, \ldots, y_N)_i, \, i = 1, \ldots, T\}$ is obtained. The number of training vectors T is targeted to be $T \approx 100K$, where K is the number of components to be used in the mixture model. To control T, the training images are chosen to be of sufficient size and number to yield a number of vectors somewhat larger than the desired T, then subsampling is used to reduce the number to the desired value. Pseudorandom noise distributed uniformly on $(-\alpha\delta/2, \alpha\delta/2)$ is added to each coordinate of every vector thus obtained, where δ is the minimum spacing between any two graylevels in the training images, and $\alpha \in [0,1]$ is a parameter that controls the amount of noise added. Note that this noise is added only to the training vectors for the purpose of more robustly fitting the mixture model; it is not used after the model has been fit, i.e., when images are actually encoded and decoded.

The mixture model consists of K separable Gaussian components, where K is a parameter that controls model complexity. Because the training and test image sets are disjoint, choosing too large a K (i.e., one that causes overfitting) would be signaled by poor performance on the test images. The estimation of the mixing proportions and component density parameters is accomplished via the expectation-maximization algorithm [12]. After training, the mixture model $\hat{p}(x_c, y_1, \ldots, y_N)$ is used to provide a conditional density estimate

$$\hat{p}(x_c|y_1, \ldots, y_N) \propto \hat{p}(x_c, y_1, \ldots, y_N)$$

for each (hypothetical) continuous-valued pixel in a new image, where the $y_1, \ldots, y_N$ are now regarded as fixed. To obtain a probability mass estimate for each actual (i.e., quantized) pixel value x', $\hat{p}(x_c|y_1, \ldots, y_N)$ is integrated over the quantization region that supports that value. Specifically, we use the estimate

$$\hat{\Pr}(x = x'|y_1, \ldots, y_N) = \int_{x'-\delta/2}^{x'+\delta/2} \hat{p}(x_c|y_1, \ldots, y_N) dx_c$$

for each quantized pixel value x', with the exceptions that the lower integration limit for the smallest x' is set to $-\infty$ and the upper integration limit for the largest x' is set to ∞.

For a given conditioning vector $(y_1, \ldots, y_N)$, let $p(x)$ denote the probability mass function $\hat{\Pr}(x = x'|y_1, \ldots, y_N)$. For use in arithmetic coding, $p(x)$ is approximated by a fixed-precision probability mass function $q(x)$ such that $q(x) > 0\,\forall x$. For a given $p(x)$, we choose $q(x)$ from the feasible set to minimize the expected ideal bit rate $-\sum_{x'} p(x') \log_2 q(x')$, which is equivalent to minimizing the relative entropy [4] between $p(x)$ and $q(x)$.

Near the borders of the image, some of the conditioning pixels will lie outside the image. In order to still have a conditional density estimate in such cases, the estimate $\hat{p}(x_c, y_1, \ldots, y_N)$ is integrated over the coordinates corresponding to the unavailable conditioning pixels, and the result is used to obtain the desired conditional density estimate as described above.

Although we explicitly account for the quantized nature of x when obtaining $\hat{\Pr}(x = x'|y_1, \ldots, y_N)$, we do not make any adjustment for the fact that the conditioning pixels are also quantized. However, little is at stake here: the consequence of quantizing the conditioning pixels is to perturb the location of the line parallel to the x-axis in $(x, y_1, \ldots, y_N)$-space along which the conditional density is evaluated. If the perturbation is small, then the smoothness property mentioned in Section 3.1 suggests that the resulting change in the estimated pixel probabilities will likewise be small. In contrast, accounting for the quantized nature of x is essential, as the probability mass assigned to the various values that x assumes directly determines the number of bits required for encoding those values.

Arithmetic coding is carried out as described in [8], using a code-register precision of 16 bits and a 15-bit-wide carry-control register.[2] This allows the two registers to coexist as fields in a single 32-bit hardware register without involving the sign bit. When an image is encoded, a header is first created indicating: the dimensions of the image, the minimum spacing δ between graylevels, and the minimum and maximum graylevels that occur in the image. Since each of these quantities is a relatively small integer, the number of bits needed for this header is negligible compared with the number needed for the rest of the image. After the last pixel has been encoded, the code- and carry-control registers are flushed into the code bit stream and the procedure terminates. The bits thus produced are packed into bytes and written into a binary file; this file constitutes the compressed representation of the image. It has been confirmed experimentally that decoding this compressed representation results in an exact bit-for-bit replica of the image.

4. Results and Conclusions

Experiments were performed on a set of nine 512×512 subimages of 300 dpi, 8 bpp grayscale scans of selected pages of the March 1998 issue of the IEEE Transactions on Image Processing. For clarity, we compress segments of various types separately, rather than switching models on the basis of the segmentation mask as would be done in a practical system. Accordingly, three subimages were selected in each of the following categories: text, line drawings, and halftone. For each image in each category, the compression technique described in this paper was applied, using the other two images in the category to train the mixture model.

To find suitable combinations of parameter values for the lossy stage, all combinations of the parameter values $n_r \in \{2, \ldots, 6\}$, $n_d \in \{3, 4, 5\}$, and threshold sequences among those listed in Table I were applied to each of the nine images. The compressibility of each result was estimated by the minimum of the bit rates obtained over all techniques described in Appendix A. The minimum rate and mean-square error (MSE) together define a point in the rate-MSE plane for each of the sixty candidate parameter combinations for each image. The convex hull of these points was determined using the *qhull* program [2] for each original image, and the parameter value combinations that appeared most frequently among vertices of the facets facing the origin were adopted for use in testing the lossless stage. The $(n_r, n_d, \{t_m\})$ triples found in this way were: (5,3,3), (4,3,1), (3,3,2), (3,3,3), and (2,3,3). Several context

[2] Other versions of arithmetic coding could be used as well, but care must be taken to properly interface the statistical model described above to the coder, particularly if conversion to an intermediate binary source representation is required for the coder to operate.

TABLE I
Threshold sequences used in lossy stage.

Label	t_m, $m = 1, \ldots, 6$
1	$\{16, 8, 8, 8, 8, 8\}$
2	$\{8, 8, 4, 4, 4, 4\}$
3	$\{4, 4, 4, 4, 4, 4\}$

neighborhoods and values of K and α were tried for the lossless stage; no one set worked best across all images. Figure 2 shows the MSE performance of the overall approach, averaged across images in each category using leave-one-out crossvalidation, for $K = 256$, $\alpha = 0.5$, and the context neighborhood shown in Figure 1a. Significantly larger values of K were found to result in overfitting. Also presented in Figure 2 are results for the lossless methods described in Appendix A, and for JPEG applied to the original images using several different quality factors. It can be seen that the proposed lossless technique is competitive at all rates, and outperforms most of the reference techniques at all but the highest bit rates.

JPEG can be seen to generally outperform the proposed technique in Figure 2. However, MSE is not the whole story, as larger MSE can correspond to improved intelligibility. An example of this is in Figure 3, where JPEG is more faithful to the original scan, but the proposed technique results in greater intelligibility by removing bleed-through and preserving edges.

Based on these results, we tentatively conclude that the proposed technique can be an attractive alternative to established methods when compressing grayscale document images, as it offers good MSE performance while maintaining or even improving intelligibility. The lossless stage is interesting in its own right, as it appears to outperform several standard approaches, particularly when the image supplied to it has low complexity.

Appendix

A. Lossless Image Compression Techniques used for Reference

The following lossless compression techniques for grayscale images were used for reference purposes in this study.

- BTPC: Binary Tree Predictive Coding (Version 4.1) by John A. Robinson. The program used (in lossless mode) was `cbtpc`, available from http://www.engr.mun.ca/~john/btpc.html.
- CALIC: Context-based, Adaptive, Lossless Image Coder (arithmetic coding version) by Xiaolin Wu and Nasir Memon. The program used was `enCALICa.sun`, available from ftp://ftp.csd.uwo.ca/pub/from_wu/v.arith.
- HIST: A rough measure of compressibility obtained by computing the entropy of the normalized image histogram, treating each pixel indepen-

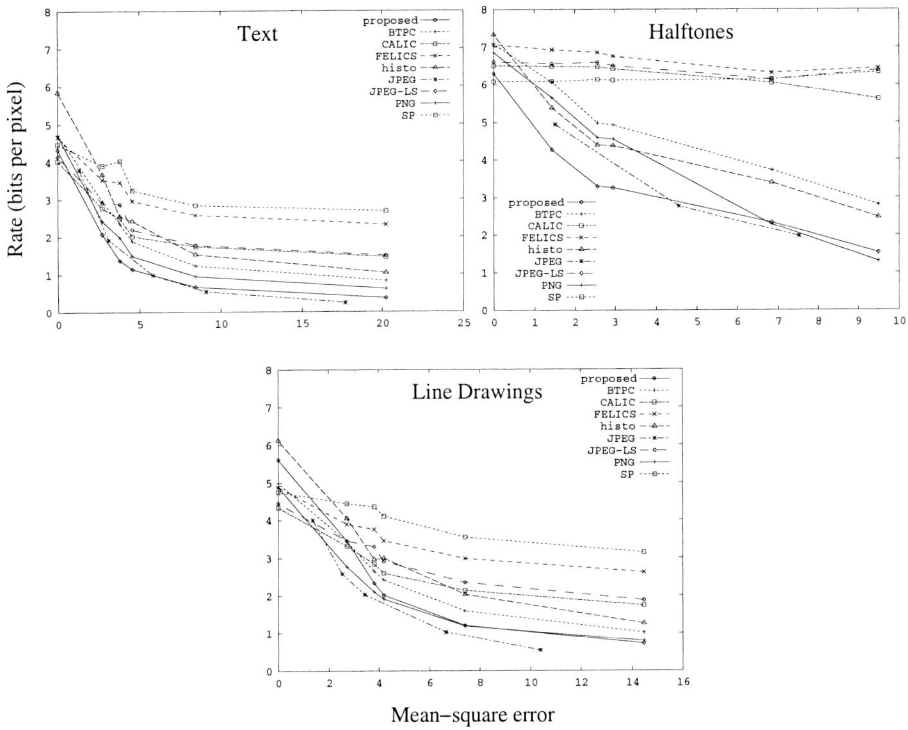

Fig. 2. Average crossvalidated MSE-Rate performance for each of the three image categories.

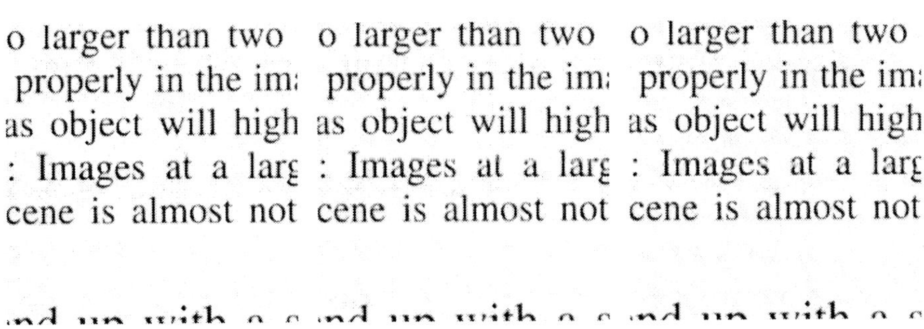

Fig. 3. Left: A 256 × 256 patch from one of the original text images. Middle: result of JPEG compression at 0.97 bpp (MSE = 5.84). Right: result of proposed scheme at 0.78 bpp (MSE = 7.58).

dently.

- FELICS: Fast, Efficient, Lossless Image Compression System by Paul G. Howard and Jeffrey Scott Vitter. The program used was `mgfelics`, available from http://www.cs.mu.oz.au/mg/mg-1.2.1.tar.gz.
- JPEG-LS: JPEG-LS Reference Encoder, Hewlett-Packard LOCO-I imple-

mentation. The program used was `locoe`, available from http://www.hpl.hp.com/loco/software.htm

- PNG: The `pnmtopng` open-source program (version 2.37.1) by Alexander Lehmann, Willem van Schaik, and Greg Roelofs. Available in ftp://swrinde.nde.swri.edu/pub/png/applications. This program is based on the Portable Networks Graphics (PNG) library [1].
- SP: Arithmetic coding version of the lossless image compression program by Amir Said and William A. Pearlman. The program used was `sp_compress`, available from http://www.cipr.rpi.edu/research/SPIHT/EW_Code/lossless.tar.gz.

References

1. M. Adler, T. Boutell, C. Brunschen, A. M. Costello, L. D. Crocker, A. Dilger, O. Fromme, J. Gailly, C. Herborth, A. Jakulin, N. Kettler, T. Lane, A. Lehmann, C. Lilley, D. Martindale, O. Mortensen, K. S. Pickens, R. P. Poole, G. Randers-Pehrson, G. Roelofs, W. van Schaik, G. Schalnat, P. Schmidt, T. Wegner, and J. Wohl. Png (portable network graphics) specification, version 1.0. Technical report, World Wide Web Consortium (W3C), October 1996. http://www.w3.org/TR/REC-png.
2. C. Barber, D. Dobkin, and H. Huhdanpaa. The quickhull algorithm for convex hulls. *ACM Trans. on Mathematical Software*, December 1996. http://www.geom.umn.edu/software/qhull.
3. D. S. Bloomberg. Multiresolution morphological analysis of document images. In *SPIE Conf. 1818, Visual Communications and Image Processing*, pages 648–662, Boston, 1992.
4. T. M. Cover and J. A. Thomas. *Elements of Information Theory.* John Wiley and Sons, 1991.
5. J. Gailly and M. Adler. The gzip homepage, December 1999. http://www.gzip.org/algorthm.txt.
6. P. Haffner, L. Bottou, P. Howard, P. Simard, Y. Bengio, and Y. LeCun. Browsing through high quality document images with djvu. In *Proc. of Advances in Digital Libraries 98*. IEEE, 1998.
7. G. G. Langdon and J. J. Rissanen. Compression of black-white images with arithmetic coding. *IEEE Trans. Comm.*, COM-29:858–867, June 1981.
8. A. C. Popat. Scalar quantization with arithmetic coding. Master's thesis, Dept. of Elec. Eng. and Comp. Science, M.I.T., Cambridge, Mass., 1990. ftp://ftp.media.mit.edu/pub/k-arith-code.
9. A. C. Popat. *Conjoint Probabilistic Subband Modeling.* PhD thesis, Massachusetts Institute of Technology, 1997.
10. K. Popat and R. W. Picard. Cluster-based probability model applied to image restoration and compression. In *ICASSP-94: 1994 International Conference on Acoustics, Speech, and Signal Processing*, pages 381–384, Adelaide, Australia, April 1994. IEEE.
11. M. Rabbani and P. W. Jones. *Digital Image Compression Techniques.* SPIE Optical Engineering Press, Bellingham, Washington, 1991.
12. R. A. Redner and H. F. Walker. Mixture densities, maximum likelihood, and the EM algorithm. *SIAM Review*, 26(2):195–239, April 1984.
13. J. J. Rissanen and G. G. Langdon. Arithmetic coding. *IBM J. Res. Develop.*, 23(2):149–162, March 1979.
14. I. Witten, R. Neal, and J. Cleary. Arithmetic coding for data compression. *Communications of the ACM*, 30(6):520–540, June 1987.

BOOSTING OCR CLASSIFIER BY OPTIMAL EDGE NOISE FILTERING*

JUNIOR BARRERA, MARCEL BRUN and ROUTO TERADA
Universidade de São Paulo, Departamento de Ciência da Computação
CP 66281 - CEP 05389-970 - São Paulo, SP, Brazil
e-mail: jb@ime.usp.br

and

EDWARD R. DOUGHERTY
Texas A&M University, Department of Electrical Engineering,
College Station, TX 77843-3128,
E-mail: edward@ee.tamu.edu

Abstract. The problem of Optical Character Recognition (OCR) can be solved by set operators implemented as programs for a Morphological Machine (MMach). In this paper, we present two techniques to boost such programs: (1) Anchoring and (2) Edge Noise Filtering by Stamp. The power of these techniques is demonstrated by some impressive experimental results.

Key words: Morphological Machines, OCR, Edge Noise Filtering, Set Operators.

1. Introduction

Optical Character Recognition (OCR) refers to a process in which printed documents are transformed into ASCII files mainly for the purpose of editing. A key problem in OCR is the recognition of characters by their shapes.

A model of a procedure for shape recognition is a set operator applied on a Discrete Random Set [8]. Mathematical Morphology (MM) is a general framework to study set operators [1].

An important aspect of MM is the description of set operators by a formal language that is complete and expressive [3]. Since the sixties special machines, the Morphological Machines (MMachs), have been built to implement this language.

The automatic programming of MMachs has been modeled as an statistical optimization problem. In this approach, the goals of the user are represented as a collection of input-output of image pairs and the target operator is estimated from these data.

In this paper, we present two normalization techniques to boost MMach programs for character recognition: (1) Anchoring and (2) Edge Noise Filtering by Stamp.

* The authors have received partial support from FAPESP and CNPq. The authors also thanks the students Teofilo E. Campos, Rogerio Feris, Archias A. de A. Filho, Franklin C. Flores and Fabiano C. Sousa, that have developed part of the software used in this paper.

An image sampling is necessary to reduce the design complexity of the MMach program that recognizes characters. However, a problem with regular sampling is that the shape of the low resolution character realization generated depends on the position on the grid of the corresponding high resolution realization. The anchoring solves that.

Usually, text images have edge noise, due to physical phenomena of the scanning process, and to degradations of paper and ink of the text. This creates badly distributed random shapes, that affects seriously the quality of the recognition program. We have developed a new technique for designing noise edge filtering, that performs a nice normalization. The principle of this technique is to design a MMach program that transforms the characters realization in a kind of "average" character, the *stamp*.

We have gotten impressive experimental results applying anchoring and edge noise filtering before recognition. These experiments were realized on images with ten classes of characters, the ten digits, and a severe edge pepper noise.

Following this introduction, section 2 recalls how to design discrete set operators. Section 3 recalls the principles of the OCR recognition program used. Section 4 introduces the Anchoring technique. Section 5 describes the stamp technique for designing Edge Noise Filters. Section 6 presents some experimental results. Finally, in Section 7, we discuss the technique proposed and present some future steps of this research.

2. Statistical Optimal Operator Design of W-Operators

The formulation of the set operator design process as a statistical optimization problem depends on an algebraic representation of a family of operators [7] [2] and on a statistical optimization procedure [5] on the space of parameters of this representation.

Let E denote the integer plane Z^2. Let $\mathcal{P}(E)$ be the powerset of E. A *binary image* X is an element of $\mathcal{P}(E)$.

The set E is an Abelian group with respect to the vector addition, denoted by $+$. The zero element of $(E, +)$ is the *origin* of E and is denoted by o.

Let $X \in \mathcal{P}(E)$ and $u \in E$. The *translation* of X by u is the element of $\mathcal{P}(E)$ defined by $X_u = \{x \in E : x - u \in X\}$.

A mapping ψ from $\mathcal{P}(E)$ to $\mathcal{P}(E)$ is called a *set operator*. A set operator ψ is called *translation invariant* if and only if (iff), for any $u \in E$ and $X \in \mathcal{P}(E)$, $\psi(X_u) = \psi(X)_u$.

Let W be a finite subset of E. A set operator ψ is called *locally defined* in the window W iff, for any $u \in E$ and $X \in \mathcal{P}(E)$, $u \in \psi(X) \Leftrightarrow u \in \psi(X \cap W_u)$.

A set operator that is both translation invariant and locally defined in W is called a *W-operator*. The family of W-operators is denoted Ψ_W.

Let $\mathcal{P}(W)$ be the power set of W and h_ψ be the Boolean function from $\mathcal{P}(W)$ to $\{0,1\}$ defined by, for any $\mathbf{x} \in \mathcal{P}(W)$, $h_\psi(\mathbf{x}) = 1 \Longleftrightarrow 0 \in \psi(\mathbf{x})$

A W-operator ψ can be represented in terms of h_ψ by, for any $X \in \mathcal{P}(E)$, $\psi(X) = \{x \in E : h_\psi(X_{-x} \cap W) = 1\}$ The Boolean function h_ψ is called the *characteristic function* of ψ.

Set operators will be optimally designed to estimate an image when it is observed after going through a system. Images are modeled as discrete random sets. If ψ is a set operator, then for each random set $\mathbf{S}$ there is an output random set $\psi(\mathbf{S})$. If S is an observed realization of $\mathbf{S}$, then $\psi(S)$ is a realization of $\psi(\mathbf{S})$. The task is to design an operator ψ such that, given the input process $\mathbf{S}$, $\psi(\mathbf{S})$ is statistically close to a desired process $\mathbf{I}$. We call $\mathbf{S}$, $\mathbf{I}$ and $\psi(\mathbf{S})$ the *observation*, *ideal* and *estimator processes*, respectively. The distance between $\mathbf{I}$ and $\psi(\mathbf{S})$ is measured by a probabilistic error measure $Er[\mathbf{I}, \psi(\mathbf{S})]$. Assuming that operators belong to an operator family $\mathcal{F}$, an *optimal operator* relative to $\mathcal{F}$ is an operator $\psi_{opt} \in \mathcal{F}$ such that $Er[\mathbf{I}, \psi_{opt}(\mathbf{S})] \leq Er[\mathbf{I}, \psi(\mathbf{S})]$, for any $\psi \in \mathcal{F}$.

Assuming that $\mathbf{I}$ and $\psi(\mathbf{S})$ are jointly stationary, the optimal operator is a W-operator. Therefore, it is enough to search in the equivalent space of Boolean functions

In practice, the joint probability is unknown and should be estimated from a collection of input-output image pairs. Thus, the error is computed from the estimated probabilities.

3. Character Recognition by Shape

The purpose of *character recognition* is to assign a unique classification code to each distinct character so each lower or uppercase letter, digit and punctuation symbol will have a code that is unique.

Characters can be recognized by a family of set operators [5]. However, this is not computationally efficient and there is not an easy way to generalize all classes together, when we deal with many classes of characters. In this case, a more practical approach is multi-classification [6].

Multi-classifiers are operators ψ characterized by functions h_ψ from $\mathcal{P}(W)$ to $\{0, 1, \ldots, n\}$, that classify the shapes $\mathbf{x} \in \mathcal{P}(W)$ in n classes. 0 is reserved for the background.

In this case, the training images present all the correctly classified characters (i.e., labeled by a unique code). Therefore, when the multi-classifier is applied to an image, it generates an image in which the pixels are coded with values between 0 and n, according to the characters they form.

The multi-classifiers can be implemented by a slight modification of the ISI algorithm [4], in the following way :

1. Group the examples by their codes in decreasing order of frequency. This results in n groups $\mathcal{X}_1, \mathcal{X}_2, \ldots, \mathcal{X}_n$, where $\mathcal{X}_i$ contains all examples with the assigned code i.
2. Initialize the process with the unitary interval set $B_n = \{[\{0\}, W]\}$ (i.e., the basis of the identity operator). Recall that the desired operator ψ_i, with basis B_i, is anti-extensive, i.e., $\psi_i(X) \subseteq X$.
3. Extract from B_n all examples in $\mathcal{X}_n$. Let B_{n-1} be the set of the resulting intervals. Note that each example in $\mathcal{X}_1, \ldots, \mathcal{X}_{n-1}$ is in some interval in B_{n-1}. Then, extract from the intervals in B_{n-1}, all examples in $\mathcal{X}_{n-1}$, and put the resulting intervals in B_{n-2}. Again, note that the examples $\mathcal{X}_n$ and

$\mathcal{X}_{n-1}$ are not in B_{n-2} while all the other examples are in it. Repeat this process down to $n = 2$ (i.e., until the extraction of all examples in $\mathcal{X}_2$ to generate B_1 is done). Note that $B_1 \subseteq B_2 \subseteq \ldots \subseteq B_n$.

4. Given a configuration $\mathbf{x}$, to apply the operator one can use the following algorithm:
   if $\mathbf{x} \in [A, B] \in B_1$ then $h_\psi(\mathbf{x}) = 1$
   else if $\mathbf{x} \in [A, B] \in B_2$ then $h_\psi(\mathbf{x}) = 2$
   else if...
   else if $\mathbf{x} \in [A, B] \in B_{k-1}$ then $h_\psi(\mathbf{x}) = k - 1$
   else if $\mathbf{x} \in [A, B] \in B_k$ then $h_\psi(\mathbf{x}) = k$
   else $h_\psi(\mathbf{x}) = 0$

4. Normalization of Characters

The complexity of designing an operator from examples can be measured in terms of the amount of training data that the learning algorithm needs to learn an operator with a given precision. This measure is called *sample complexity*.

The sample complexity depends on the cardinality of the window W and on the stochastic process representing the family of images considered. For a given stochastic process the complexity grows with the cardinality of W. Symmetrically, for a given window W, the complexity grows with the variety of shapes occurring in the family of images considered.

Therefore, to design easily good character classifiers, we need tools to decrease the images resolution and to normalize the characters shape. The resolution reduction will reduce the size of the window and the shape normalization will simplify the stochastic process (i.e., family of images) considered.

In the following, we introduce two simple operators for resolution reduction and shape normalization: anchoring and edge noise filtering by stamp. Figure 1 presents the complete process proposed.

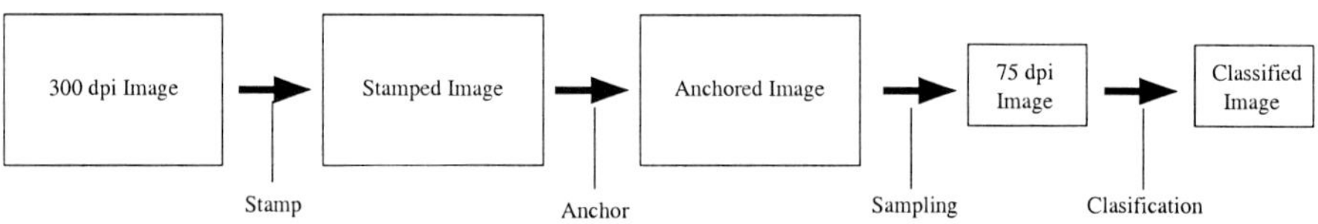

Fig. 1. Block diagram of the system.

4.1. Anchoring

Resolution reduction is performed by a process of sampling the high resolution image. A regular sampling depends on two positive integers m and n, respectively, the horizontal and vertical sampling steps. Only the pixels that are in the columns $0, m, 2m, \ldots$ and in the lines $0, n, 2n, \ldots$ will occur in the low resolution image.

A problem with regular sampling is that, if we have two objects with the same shape in different positions in the high resolution image, they may turn into different shapes after the sampling. Figure 2 shows an example that illustrates this fact. Note that all the characters 1 and all the characters 2 have the same shape (Figure 2a) in the high resolution image, but not in the low resolution image (Figure 2b). Even after applying a morphological filter as closing (Figure 2c) or close-open (Figure 2d) the problem still persists.

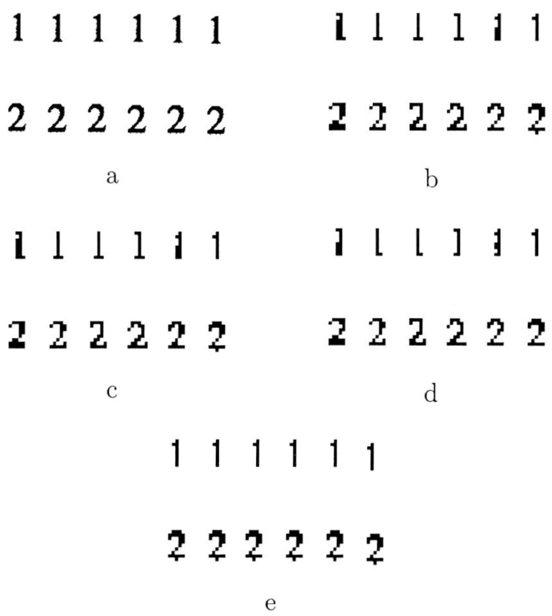

Fig. 2. Sampling of identical objects: (a) original image; (b) Direct sampling ; (c) Sampling after 7×7 box closing; (d) Sampling after 7×7 box closing-opening; (e) Sampling after anchoring.

To avoid that, we have introduced the anchoring process. It consists of constructing a new high resolution image by fixing a reference point for each object and translating each object of the image in such way that their reference point be on the nearest point of the low resolution grid. A reference point easy to compute is the center of the minimum rectangle that contains the object.

If we apply an (m, n) anchoring, followed by the corresponding (m, n) sampling, objects of the same shape in the high resolution image keep the same shape in the low resolution image. For example, note that all characters 1 and 2 have the same shape in the high and low resolution images (respectively, Figures 2a and 2e).

Figure 3 shows the sampling, with $m = n = 2$, of the character "o" in different positions of the greed. Figure 3a, 3b and 3c show, respectively, the origin image, the sampling and the representation of the sampling in a low resolution image. Figure 3d, 3e and 3f show, respectively, the anchoring, the sampling of the anchored image, and the low resolution representation of this sampling. Note that when the anchoring is applied, all characters remain identical after sampling.

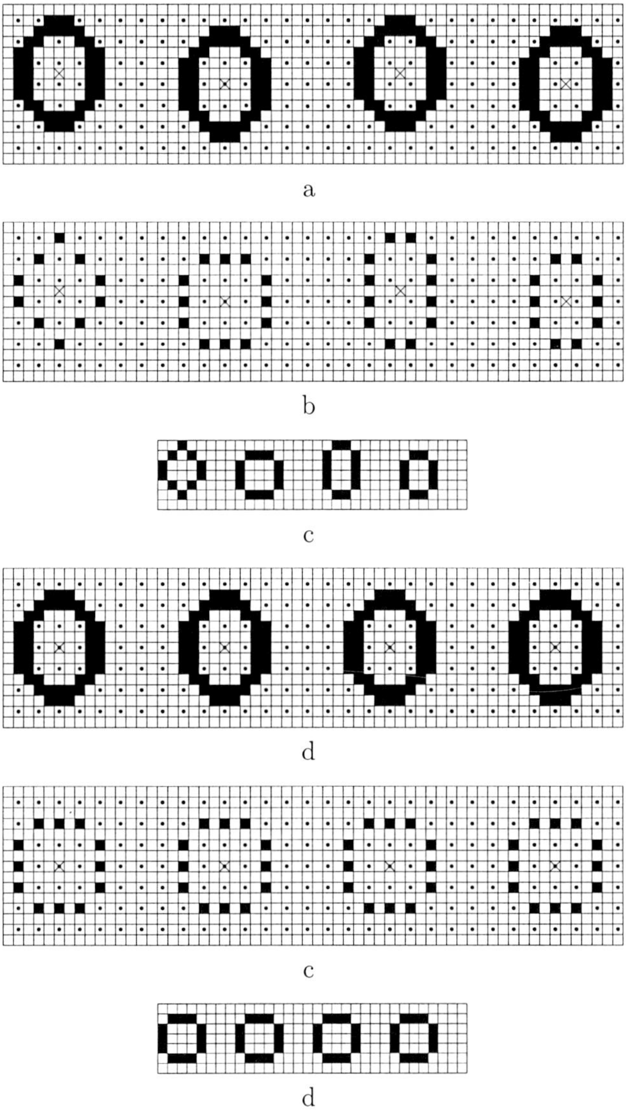

Fig. 3. Anchor: (a) Original image; (b) Direct sampling ; (c) Sampled image (d) Anchoring of the original image; (e) Sampling after anchoring; (f) Sampled image with anchoring.

4.2. Edge noise filtering by stamp

Usually, text images have edge noise, due to random physical phenomena of the scanning process and to degradations of the paper and ink of the text. This noise creates shape differences between realizations of the character and these differences affect badly the characters classifier.

A solution to reduce these differences is to design an edge noise filter. However, this is not an easy task, since usually there are not ideal images available for training the filter.

To overcome this difficulty, we have developed the notion of stamp. The stamp for a given character is an "average" character obtained by adding and thresholding registered realizations of the character. The registration is done by fixing a reference at the same coordinate. As in the anchoring, an easy to compute reference is the center of the minimum rectangle that contains the character. A more robust alternative is the character center of mass.

Using the notion of stamp, we can create an ideal image to train the filter. This image is built by substituting each realization of a character by its corresponding stamp, registered through the reference point adopted.

Figure 4 shows an example that illustrates the application of the stamp. Observe the edge noise image (Figure 4a) with a cross denoting the centers of the digits, the corresponding ideal image via the stamp (Figure 4b) and the new image obtained by replacing the digits by the stamp (Figure 4c). The images of Figure 4a and 4c will be used to train the stamp operator [4]. Figure 4b shows the count of black pixels for each digit. The threshold is chosen as half of the number of digits. that is, for each pixel is attributed its most frequent value (0 or 1) in the observed character realizations. In the next example, the threshold calculated is three.

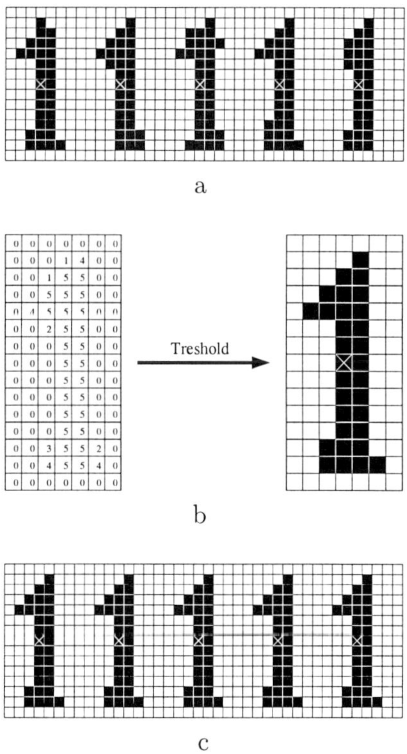

Fig. 4. Stamp: (a) Original image; (b) Stamp generated from image; (c) Image replaced by stamp.

Figure 5 shows an example of edge noise filtering. Figure 5a is the noise image. Figure 5b is the symmetrical difference of the image of Figure 5a with the corresponding stamp image. Figure 5c is the result of applying the filter designed on the image of Figure 5a. Figure 5d is the symmetrical difference of the ideal image (stamp image) and Figure 5c (recovered image). Note that the edges in Figure 5d are less dense than the ones in Figure 5b, because of the correction done by the filter.

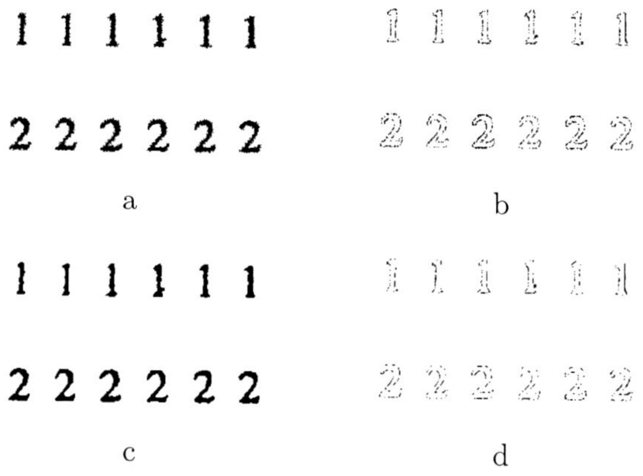

Fig. 5. Stamp trained operator: (a) original image; (b) Symmetrical difference between original and ideal image (c) Stamp with the designed operator (d) Symmetrical difference between the designed operator and the ideal image .

5. Experimental Results

We have realized some experiments to test the proposed normalization technique.

The source images are composed just of digits $(0, 1, 2, ..., 9)$. These images were obtained by scanning text printed in a laser printer and adding pepper noise on the external edges of the character realizations. The density of the noise imposed was 50%. Each image is composed of 400 characters, 40 of each digit. The images were scanned in 300 dpi.

In all the experiments, the classifier was trained using a window 3×5 and only one training image. When a noise edge filtering was designed, it was designed based on only one training image. The classification error was evaluated in 5 different images, containing together 2000 character realizations. All the resolution reduction applied were from 300dpi to 75dpi.

In the first experiment, we have created a stamp image, reduced its resolution and designed a classifier from it. The empirical error measured for the classifier applied on the stamp image itself was 0%.

In the second experiment, we have reduced the resolution of a noise image by pure sampling and then the classifier was trained. The empirical error measured by the application of the classifier designed in the test images was

13.5%.

In the third experiment, we have applied the anchoring on a noise image before reducing its resolution. The empirical error measured for the classifier was 9.2%.

In the fourth experiment, we have designed an edge noise filter, applied the anchoring, reduced the resolution and designed a classifier. The empirical error measured for the classifier was 2.35%. The filter designed was a two stage filter [9], where each stage depends on a 3×3 window.

We have also tried to do the image recovery by some filters designed heuristically. Applying the median, with a 3×3 window and the close-open alternate sequential filter by the 3×3 square, the empirical error for the classifier was, respectively, 4.95% and 4.35%

Observe that our experimental results confirm our theoretical expectation. The empirical errors measured were bounded by the non normalized and completely normalized cases. Each normalization applied decreased the measured error in a significant amount. Improvements in edge noise filter design will lead the error asymptotically to zero.

6. Conclusion

Classically OCR is considered as a problem of Pattern Recognition, where the main task is to design sophisticated shape classifiers. We have presented another view of this problem: to design simple shape classifiers for filtered images. Under this view, the main task becomes the design of the image filter.

The potential of the new approach was demonstrated by some impressive experimental results. In our experiments, the correctness of pure classification was about 90%, while the one after filtering was about 97%.

The technique used for designing filters was inspired in an old technique of information theory: to send a signal several times by a noise channel and recover the most frequent value for each bit. The filter designed emulates such recognition process, with the scanner acting as the noise channel.

Note that the goal of the filter is character normalization and this is exactly modeled by an idempotent operator. Therefore, the filter designed could have even better performance, if the estimation of the optimal operator were constrained to the family of idempotent operators.

The next priority of our research in this subject is the development of a technique for the design of (non-increasing) idempotent W-operators.

References

1. J. Serra. *Image Analysis and Mathematical Morphology.* Academic Press, New York, 1982.
2. G. J. F. Banon and J. Barrera. Minimal representation for translation invariant set mappings by Mathematical Morphology. *SIAM J. Appl. Math.*, V. 51, pp. 1782-1798, 1991.
3. J. Barrera and G. J. F. Banon. Expressiveness of the Morphological Language. In *Image Algebra and Morphological Image Processing III*, V. 1769 of SPIE Proceedings, pp. 264-275, San Diego, CA, 1992.
4. J. Barrera, E. R. Dougherty, and N. S. Tomita. Automatic programming of binary

morphological machines by design of statistically optimal operators in the context of computational learning theory. *Electronic Imaging*, 6(1):54–67, January 1997.

5. J. Barrera and R. Terada and F. S. C. da Silva and N. S. Tomita Automatic programming of morphological machines for OCR. In *Mathematical Morphology and its Applications to Image and Signal Processing*, pages 385,392, Atlanta, GA, May 1996, ISMM, Kluwer Academic Publishers
6. J. Barrera and R. Terada and R. A. Lotufo and N. S. T. Hirata and R. Hirata Jr. and F. A. Zampirolli. An OCR based on Mathematical Morphology. In *Nonlinear Image Processing IX*, V. 3304 of SPIE, pages 197-208, San Jose, CA, January 1998.
7. J. Barrera and G. P. Salas. Set operations on collections of closed intervals and their applications to the automatic programming of morphological machines In *Electronic Imaging*, 5(3):335–352, July 1996.
8. J. Goutsias. Morphological analysis of discrete random shapes. *Journal of Mathematical Image and Vision*, V. 2, pp. 193-215, 1992.
9. E. R. Dougherty, Y. Zhang and Y. Chen. Optimal Iterative Increasing Binary Morphological Filters. *Optical Engineering*, V. 35, N. 12, pp. 3495-3507, December 1996.

MORPHOLOGICAL BANK CHECK LOGO SEGMENTATION WITH FEW A PRIORI KNOWLEDGE

ANDREIA DE JESUS and JACQUES FACON
PPGIA-PUCPR - Pontificia Universidade Catolica do Parana
Rua Imaculada Conceicao 1155, Prado Velho, 80215-901 Curitiba-Pr, Brazil
{jesus,facon}@ppgia.pucpr.br

Abstract. This paper presents a method to locate and extract Brazilian bank check logos automatically by employing mathematical morphology. The objective is to minimize the number of heuristic parameters to obtain the most precise possible segmentation in every situation, thereby allowing this approach to be reused in other applications. Trials and results are presented.

Key words: Mathematical Morphology, Bank Check Logos, Fillhole.

1. Introduction

One of the major challenges in image processing might be image segmentation of significant data. The major problem lies in the absence of a priori knowledge of the number and type of structures the image will present. Such structures are identified on the basis of their geometry, shape, topology, texture, color or brilliance, those characteristics allowing better identification being chosen. Image segmentation is a process that typically partitions the spatial domain of an image into mutually excluding subsets, called regions. Each region is uniform and homogeneous with regards to some properties such as hue or texture, and the values of those properties vary in some aspects and meanings from the properties of the each neighboring region.

What is a logo? It is typically a perceptually salient marking, larger than individual text components [2]. In the literature very little is said about automatic logo segmentation. [7], [5] and [1] focused their efforts on the recognition phase, but the location and segmentation of logos was performed manually. [2] employed the page segmentation algorithm in the logo segmentation phase of his work, but the complete and precise segmentation was not possible in all situations because in some cases the logos generated showed parts missing, corrupted regions, or the presence of components that were extraneous to them.

In segmentation many heuristics are generally used for the size and position of the components to be extracted. Logos (typically confined to a compact and isolated part of the document, and not linked to the text structure as other graphs) are frequently separated from the other graphic components by their position in the page and by using the previous knowledge of the type of document (memorandum, letter, etc.).

The use of many heuristics ends up by limiting segmentation for a given type of application. The approach presented here employs mathematical morphology to automatically locate and extract logos from the image of Brazilian bank checks. Such approach is supported by erosion, dilation, binary granulometry, reconstruction and Fillhole operators, described in [6].

The objective is to minimize the number of heuristic parameters in order to obtain the most precise possible segmentation in every situation. We think that, by reducing the number of heuristic parameters, the segmentation is more general and, thereby this approach can be reused in other applications. This paper is organized as follows: Section 2 will describe the database used in the experiments. Section 3 will describe the assumptions underlying the approach presented here. Section 4 will describe the initial phases of Brazilian bank checks segmentation. Section 5 will describe the Brazilian bank check logo location and segmentation method. Section 6 will present trials and results.

2. Logo Acquisition

To carry out our approach on bank check logos, initially a bank of 478 colored Brazilian bank checks of 18 institutions was set up (150 empty and 328 printed digitized at 300 DPIs with 256 indexed colors). The horizontal and vertical sizes of each bank check image are 2053×898 pixels and its average size is 1.100 kb.

3. Logo Segmentation Assumption

Segmenting without previous knowledge and employing just a few heuristics is a challenge. Since the purpose of a logo is to provide a symbolic link to an organization, it is typically a perceptually salient marking, larger than individual text components [2]. In the case of Brazilian bank checks, despite the complexity of images, just the single assumption that the logo is bigger (in term of area) than any other information present in the check will be employed.

4. Initial Segmentation Phases

Because of the blending of relevant logo aspects with the artistic background, the extraction of information from colored images of Brazilian bank checks is a complex task. Acknowledging that fact, the transformation of originally colored images into grayscale images was chosen.

The initial steps of automatic logo segmentation consist in:

- Removing the artistic background from the grayscale image of Brazilian bank checks using the Fillhole process;
- Thresholding the grayscale image;
- Removing the noise from the binary image;

The initial phases of the Brazilian bank check logo location and extraction methodology are described and justified below.

4.1. Removal of the artistic background by the Filllhole

Since the logo blends with the artistic background, up-front removal of the check background becomes necessary. The direct background removal was rejected because it affects the logo (some components have disappeared, some parts became too thin, the process has modified the logo structure). The inverse approach, maintaining the background, was chosen. The Fillhole morphological operator was deemed the best tool because it is automatic and not heuristic, and allows the maintenance of most background components. The removal is performed by subtraction between the bank check grayscale image and the image resulting from the Fillhole morphological operator. The Fillhole operator consists in removing holes of an image using a minima technique imposition [6]. The Fillhole operator is based on the reconstruction by erosion (or dual reconstruction) where the mask image is the original image and the marker image is the maximum value of the original image except along the borders where the values of the original image are kept. In the case of bank checks, the connectivity used was the 8-neighborhood.

4.2. Thresholding phase

The thresholding process consisting in transforming grayscale images into binary images not always provides good quality black-and-white images. The assessment of that quality is always a challenge. [4] studied the efficacy of several thresholding algorithms on Brazilian bank check images. Many threshold algorithms were tested and the Otsus algorithm was chosen because it is automatic and it presents the best results in a short time.

4.3. Noise Removal Phase

The means chosen to remove noise originating from background residues and the Otsu algorithm is the morphological erosion operator. The binary granulometric approach was employed to carry out that removal objectively, and not heuristically. Granulometry was applied to check images by varying the type of structuring elements (horizontal, vertical, cross, square, rhombus). It was concluded that the most adequate solution is 2 erosion iterations with the cross structuring element, because it eliminates undesirable information without affecting relevant information.

5. Logo Location and Extraction Methodology

Because of the wide variety of logos in bank checks, their location and extraction by eliminating irrelevant information and reconstructing the logos is extremely heuristic.

The automatic logo extraction is supported by a single assumption: the logo is the component which has the largest area found in a bank check image. On the basis of this assumption, one can think of sieving the image until its largest component is found. By using the size criteria in the absence of heuristic parameters, binary granulometry was once more employed (Fig 1). Theoretically, the last class generated in the granulometric process must contain the whole logo. In practice, however, because of wear sustained during the thresholding process and because of the fact that the size of the logo component is irregular, this not always happens, and just a portion of the logo is obtained as a result (Fig 2-(e)). To reduce the influence of these problems and consequently the impairment of the efficacy of the method, the assumption was adapted to sieve the logo in the last classes and not only in the last one. The crucial aspect of the sieving process is to only extract parts of the logo or the entire logo. The logo classification is performed through the sieving in the last classes and not only in the last one. It was verified (Table II) that the two last classes are the only ones where it is possible to only sieve a portion of the logo.

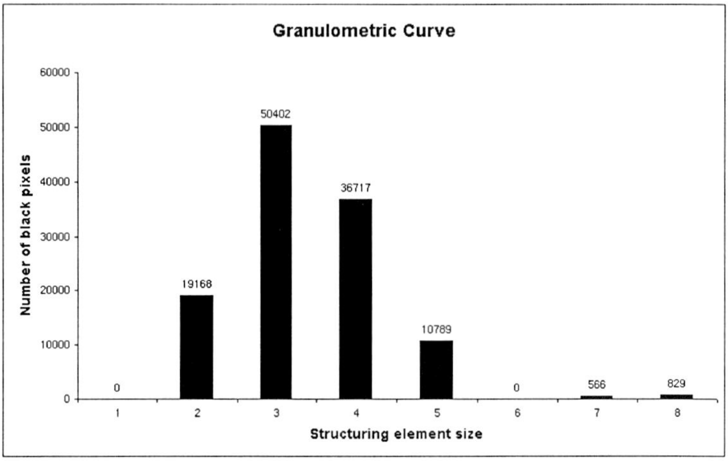

Fig. 1. Granulometric curve relative to the Fig 2-(d)

The logo location and extraction process is carried out as follows (Fig 2 and 3):

- Extracting the largest part of the logo contained in the last or in the two last classes by granulometric sieving;
- Performing the heuristic horizontal dilation of the filtered binary image of the check with 20 iterations so as to connect all the symbols and letters that make up the logo;
- Extracting a dilated version of the logo by means of binary reconstruction, having the part of the logo contained in the last or in the two last classes as marker, and the dilated image as mask;
- Extracting the logo in its original shape by intersecting the filtered binary image and the image containing the dilated logo version.

6. Trials

Logos are complex patterns, consisting in several text and image patterns that can be divided into four types [3]:

- word-in-mark, containing only the characters or words in the mark;
- device-mark, containing only graphic or figurative elements;
- composite-mark, consisting in characters or words and graphic elements;
- complex-mark, containing a complex image.

As a rule, works found in the literature process just one type of logo, such as the one by [3]. In this study a database of Brazilian bank check images mixing word-in-mark and composite-mark logos was processed. The database contains 478 grayscale bank checks filled in by hand, typewritten and blank, from eighteen banking institutions.

The main problems found were (Table I):

(i) Unsatisfactory thresholding of bank check images generating degradation and reducing the size of some logos, that consequently are not isolately sieved in the last class;

(ii) Extraction of other check components besides the logo because of overlapping handwritten strokes. This overlap leads to the linkage of those elements, blending them into a sole, larger component that is therefore placed with a part of the logo in the last class by the granulometry process sieving;

(iii) Some other components than the logo were placed in the last classes by the granulometry process sieving.

Table I shows that, by sieving a portion of the logo into the two last classes, in 8.99% of the cases the logo was not segmented and in 11.93% of the cases, other elements besides the logo were segmented.

TABLE I
Problems found in the proposed approach of automatic Brazilian bank check logo segmentation technique

Problem	**Number of checks**	**% of checks**
(i)	43	8.99
(ii)	29	6.07
(iii)	28	5.86

TABLE II
Results obtained by the proposed approach of automatic Brazilian bank check logo segmentation technique

Assumption	**Segmentation**	**Number of checks**	**% of checks**
The last class	Partial	82	17.15
	Full	285	59.62
The two last classes	Partial	83	17.36
	Full	295	61.72

Table II groups the numerical results obtained by the proposed approach of automatic Brazilian bank check logo segmentation. From the assumption that the logo is placed in the last class, the logo was partially segmented in 17.15% of the cases. And the rate of full logo segmentation was 59.62%. From the assumption that the logo is placed in the two last classes, in 17.36% of the cases the logo was partially segmented. And the rate of full logo segmentation was 61.72%. These results mean that it is possible in 79.08% of the cases to only segment Brazilian bank check logos with few a priori knowledge.

7. Conclusion

This paper presents a method to locate and extract Brazilian bank check logos automatically by employing mathematical morphology in order to minimize heuristic parameters. The feasibility of the granulometric approach to reduce the number of heuristic parameters in automatic logo location and extraction was evidenced in a database of 478 Brazilian bank checks. The results were shown that it is possible in 79.08% of the cases to only segment Brazilian bank check logos with no a priori knowledge. This approach was also shown to allow the extraction of different types of logos, in our case word-in-mark and composite-mark logos.

These results have shown that by improving some basic steps as the thresholding process for instance, it is possible to increase the logo segmentation rates. To reduce the influence of overlapping handwritten strokes and consequently the impairment of the efficacy of the method, future work may involve the extraction of handwritten or pre-printed strokes (by applying the Hough transform, for instance) and/or a process of handwritten or pre-printed patterns recognition.

References

1. F. Cesarini and al. A neural-based architecture for spot-noisy logo recognition. *4th International Conference on Document Analysis and Recognition, Ulm, Germany*, vol 1, 18–20, August 1997.
2. D.S.Doermann and E. Rivlin and I.Weiss. Logo recognition using geometric invariants. *Technical Report CS-TR-3145, University of Maryland College Park, MD*, 894–897, 1993.
3. Y.S. Kim and W.Y. Kim. Content-based trademark retrieval system using a visually salient feature. *Image and Vision Computing, Korea*, 931–939, November 1998.
4. M. F. Mattana and A.S. Britto Jr and J. Facon. Evaluation by recognition of thresholding-based segmentation techniques on Brazilian bank checks. *OPTILAS'98: 6th Latin American Meeting on Optics, Laser and their Applications, Colombia*, 1–4, September 1998.
5. A. Soffer and H. Samet. Using negative shape features for logo similarity matching. *14th International Conference on Pattern Recognition (ICDAR'98)*, vol 1, 571–573, August 1998.
6. P. Soille. Geodesic transformations and metrics: an overview. *Brazilian Workshop'97 on Mathematical Morphology, Mini-Course II, INPE, So Jos dos Campos, SP, Brazil*, 1997.
7. P. Suda and C. Bridoux and G. Manderlechner. Logo and word matching using a general approach to signal registration. *4th International Conference Documents Analysis and Recognition (ICDAR'97)*, 61–65, 1997.

(a)

(b)

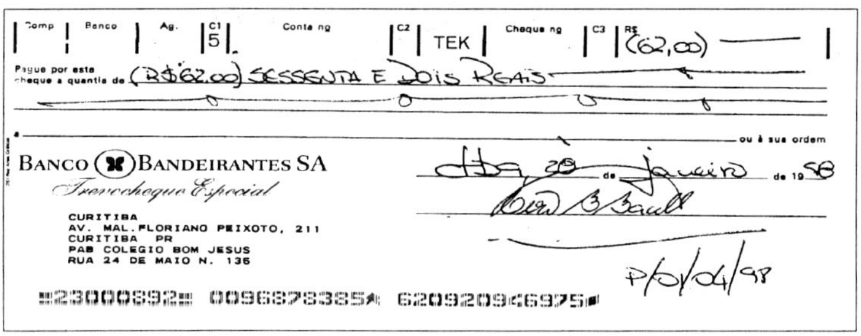

(c)

(d)

Fig. 2. (a) original Grayscale image (b) Elimination of background bank check image (c) Binary image (d) Noise elimination

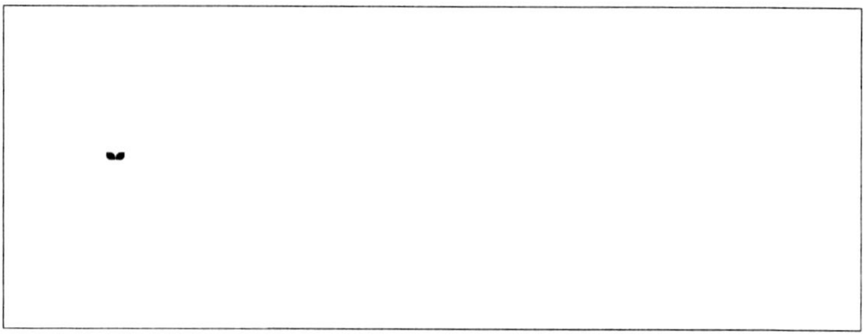

(e)

(f)

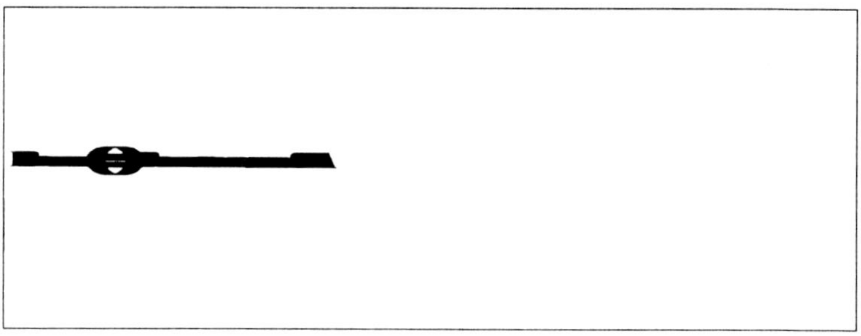

(g)

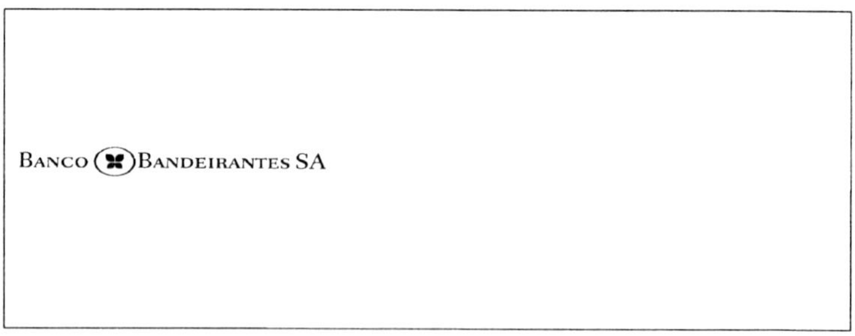

(h)

Fig. 3. e) Portion of a logo extracted by granulometry(f) Connected set image (g) Dilated logo image (h) Segmented logo

MORPHOLOGICAL PSEUDO CONVEX HULL TO CORRECT HANDWRITING BASELINE SKEW

ANDERSON ROCHA, JACQUES FACON and
FLAVIO BORTOLOZZI
PPGIA-PUCPR - Pontificia Universidade Catolica do Parana
Rua Imaculada Conceicao 1155, Prado Velho, 80215-901 Curitiba-Pr, Brazil
{anderson,facon, fborto}@ppgia.pucpr.br

and

ROBERT SABOURIN
LIVIA-Ets Ecole de technologie superieure
1100, rue Notre Dame Quest, Montreal, Quebec, H3C IK3, Canada
sabourin@gpa.etsmtl.ca

Abstract. An approach to correct the baseline handwritten word skew in the image of Brazilian bank check dates is presented in this article. The main goal of such approach is to reduce the use of empirical parameters. The weighted least squares approach is used on the morphological pseudo-convex hull.

Key words: Mathematical Morphology, Pseudo-Convex Hull, Handwritten Word, Skew.

1. Introduction

There are several research works in the area of handwritten word recognition in search of still undiscovered solutions. Such research is not a trivial pursuit, in view of several factors, among which we may mention the complexity inherent to human handwriting. Variations introduced by the human writer, for instance the slant of characters, the different sizes of height and width of characters, can decrease the recognizer performances. The paper deals here with the problem of the skew of the handwriting line, named the handwriting baseline skew. To reduce the variability, many handwritten word recognition systems correct the handwriting baseline skew, so that the extraction of that line may be performed correctly, minimizing error in the recognition phase, as proposed in El Yacoubi's work in [12].

The correction of the handwriting baseline skew consists in detecting the lower pixels of each word character (named minima of a word baseline) and finding the best straight joining them. Because of the complexity of handwriting, several works demand the use of empirical parameters, rendering analysis and implementation more complex. It will be shown that it is possible to reduce the use of heuristics in the correction of the handwriting baseline skew by employing a pseudo-convex hull from the mathematical morphology.

The state of the art in terms of handwritten word skew correction will be presented in section 2. The morphological pseudo-convex hull technique to reduce the use of heuristics will be presented in section 3. The approach developed to detect and eliminate undesirable minima in the handwriting baseline on the basis of the weighted least squares approach will be presented in section 4. The approach to correct the handwriting baseline skew in bank check images on the basis of pseudo-convex hulls [10] will be presented in section 5. Some results achieved will be presented in section 6.

2. State of the Art

Only a few authors in the literature described in detail the approach used to correct the handwriting baseline skew. On the basis of the papers studied, it was perceived that the use of heuristics is frequent. Many works [1] [2] [6] [8] [12] are found in the literature that deal with detection and correction of the handwriting baseline skew employing several empirical parameters (for instance 8 empirical parameters in [12]). This reality might adversely impact the development of automatic processes regardless of the type of application (dates, addresses, amounts in writing, etc..). The main goal of this work is to reduce the use of parameters when filtering undesirable minima to correct the handwriting baseline skew.

3. Morphological Pseudo-Convex Hull

By definition, the convex hull of a set X corresponds to the smallest convex set containing X. There are several ways to obtain the convex hull [4] [5]. [11] has shown that an equivalent definition of the convex hull consists in carrying out the intersection of all half-planes containing X.

The effect of building the convex hull of X is wrapping X. In the case of handwritten word processing, since its application considerably reduces the quantity of minima in a word baseline, the convex hull cannot be used, just its concept will be applied. Therefore, we define what we call the "pseudo-convex hull" as follows: the pseudo-convex hull represents a not convex set which wraps the word without loosing the main baseline minima. In a practical way, the pseudo-convex hull of X may be obtained simply through the basic binary dilation and reconstruction. Only horizontal and vertical directions are considered. This is corresponding to the intersection of 4 half-planes [11]. The pseudo-convex hull operation that is being used may be given by:

$$\text{pseudo-convex-hull}(X) = \rho^{B_2}_{(\delta^{B_1}(X))}(X)$$

where B_1 and B_2 represent, respectively, the structuring elements $B_{hor} = \{\bullet\bullet\bullet\}$ and $B_{ver} = \left\{\begin{matrix}\bullet\\ \bullet\\ \bullet\end{matrix}\right\}$ or vice-versa. And where $\rho^{B}_{S}(Z)$ is the reconstruction of the binary set S from Z with the structuring element B.

By using the complete reconstruction on dilated set, the pseudo-convex hull becomes convex. As may be observed in Fig. 1, since the number of iterations is reduced, the number of minima increases accordingly. To keep the main

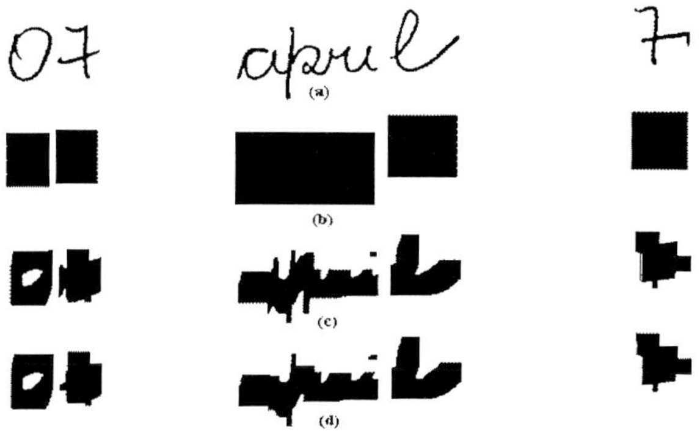

Fig. 1. (a) Original image, (b) Convex hull from complete reconstruction, (c) and (d) Pseudo-convex hulls.

minima of handwritten words, an incomplete reconstruction is carried out by using the conditional dilation.

In the approach developed, the pseudo-convex hull employed is being extracted in two steps. The process utilized to obtain the first pseudo-convex hull in Fig. 2(b) consists in the horizontal dilation of the word in the original image (set X), followed by the vertical conditional dilation. The second pseudo-convex hull is obtained by inverting only the structuring elements in the operations described above, as shown in Fig. 2(c). In Fig. 2 , 10 iterations of dilation and conditional dilation were applied. Both pseudo-convex hull images extracted provided similar but not identical aspects as may be observed in Fig. 2(b) and (c). Therefore, the intersection of those two images (Fig. 2(d)) provides a pseudo-convex hull image with more details than the horizontal and/or vertical axis.

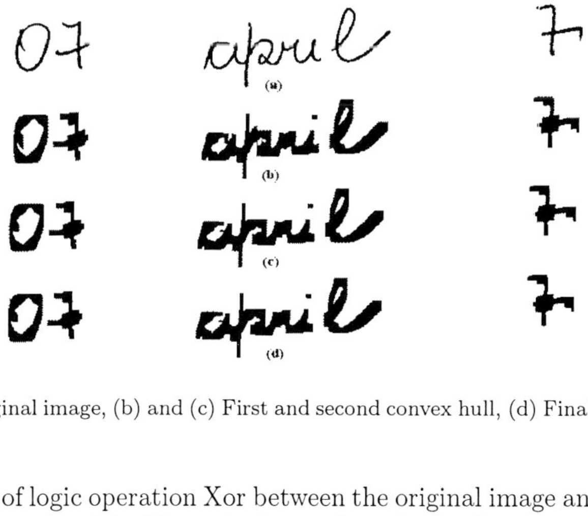

Fig. 2. (a) Original image, (b) and (c) First and second convex hull, (d) Final Pseudo-convex hull.

The result of logic operation Xor between the original image and the pseudo-convex hull, presented in Fig. 3, shows that the application of the pseudo-convex hull wraps the words, without altering the vertical positioning of the

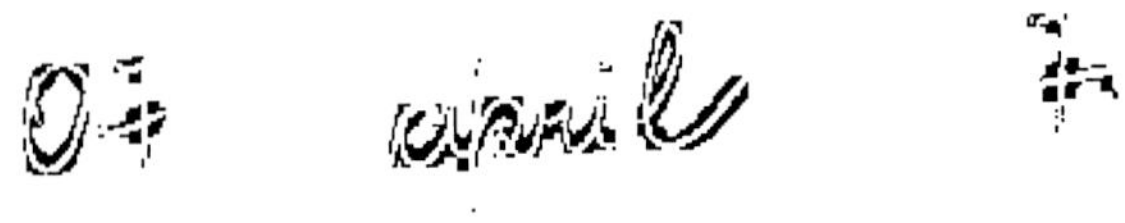

Fig. 3. Logic operation Xor between the handwritten words and their pseudo-convex hulls.

most relevant word minima and maxima originating from the lower and upper baselines. This is very important because if the minima alter their vertical positions, the correction of the word undulation may be adversely affected.

4. Detection and Elimination of Minima

The main objective of employing the pseudo-convex hull is to decrease the use of empirical thresholds in developing this approach. This technique is being used in a way that reduces the minima in a word so that, when filtering undesirable minima, few empirical thresholds will have to be defined. Of course, the number of minima depends on the format of pseudo-convex hull which depends on applied dilation and conditional dilation number. The influence of dilation iteration numbers is studied in section 6.

Because of the likely fragmentation of the words, from now on the pseudo-convex hull of parts of a word or of a whole word will be called connected component. Fig. 4(a) shows an example where the connected components minima were detected in the inferior contour of the pseudo-convex hull. In the case of Fig. 4(b), the minima were detected in the original inferior contour of words. The utilization of the pseudo-convex hull leads to the detection and storage of just the most relevant minima, so that the insignificant (desirable or undesirable) minima are eliminated, thereby avoiding the use of heuristics to eliminate the undesirable ones. Furthermore, the regular aspect of the pseudo-convex hull helps detecting minima.

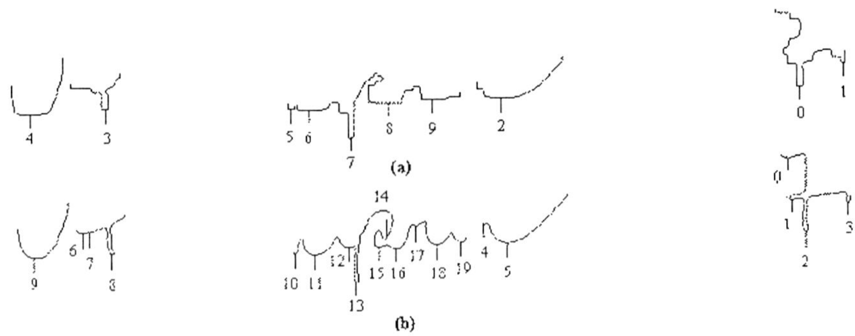

Fig. 4. (a) Detection of minima in connected components employing the inferior pseudo-convex hull contour, (b) Example of detection of minima in words, using their original contour.

Minima are detected in fall/rise transitions from the inferior contour of connected components. Once the minimum points are determined in each connected component using the pseudo-convex hull, such minima must then be adjusted to a straight line. The usual would be to apply the least squares method using the straight line mathematical model $y = ax + b$, where a and b are parameters, a is the straight line skew and b the intercept, x and y are the coordinates of the word minimum points.

The criterion of minimizing the sum of the squares of the remainder, v_k, must be applied when there are more than two minimum points in the handwriting, i.e., $n > 2$. In mathematical notation, the least squares method is defined by:

$$\text{minimize}(\sum_{k=1}^{n} v_k^2) = \text{minimize}_{(a,b\in\Re)}(\sum_{k=1}^{n} [ax_k + b - y_k]^2)$$

When different weights are applied to the remainder, the weighted least squares method is defined by minimize $\sum_{k=1}^{n} v_k^2 p_k$, where p_k is the weight of the corresponding remainder v_k. The parameter estimates are obtained, both for the weighted case and for the case in which the weights are considered units, by deriving the above expressions in relation to the parameters and equaling to zero. That procedure will not be demonstrated here because it is considered too trivial.

The undesirable minima detection methodology proposed in this paper is known in Geodesy as the "Danish Method" [9] or "Changing Weights" [7]. It consists in decreasing, at each iteration, the weight of the points the remainder of which surpasses a given pre-fixed iteration value. One of the weighting function methods [7] is:

$$p_{k+1} = \begin{cases} p_k e^{-(\frac{|v_k|}{2\sigma})} & \text{if } |v_k| \geq 2\sigma \\ p_k & \text{if } |v_k| < 2\sigma \end{cases} \quad \text{with } \sigma = \sqrt{\frac{\sum_{k=1}^{n} p_k v_k^2}{n - 2}}$$

where σ is the standard deviation.

As a rule, the first iteration is performed without weights. In our event, in which handwritten words are processed, a different approach was adopted to increase the efficiency of the method. Such approach consists initially in defining and employing weights since the first iteration. The criterion adopted to define the weights p_k, with $k = 1, ..., n$, is:

$$p_k = \frac{1}{d_k^2} \quad \text{where } d_k = \delta_k / min(\delta_k), \text{with } k = 1, ..., n$$

$$\text{and } \delta_k = (y_k - y_{k-1})^2 + (y_k - y_{k+1})^2$$

Then the weight of the remainders surpassing the standard deviation value is decreased for the first two iterations. For the ensuing iterations, the criterion adopted is to decrease the weight of the remainders surpassing 2σ.

5. Correction of the Handwriting Baseline Skew

Once the straight line is adjusted by the weighted least squares methods, the global word skew is evaluated and corrected, by rotating the original image utilizing the one-pass implementation [3]. That aims to decrease the losses originating mainly from the pixel fractional addresses, avoiding gaps in the target image.

If necessary, the minima detected in the month of the date are filtered to eliminate the undesirable minima, again utilizing the weighted least squares approach. In the case of figures (day/year), the lowest minimum is preserved.

The use of Hook's transformation [12] to correct the skew of each connected set was eliminated, as opposed to the previous studies [10]. The correction of the undulation of each connected component is now carried out, if needed, by positioning all the minima belonging to the connected component in the same vertical position of its first minimum, employing El Yacoubi's algorithm [12]. Once the undulation is corrected, the connected components are vertically repositioned so as to be aligned with the first connected component. To ensure a good correction of the undulation using this algorithm, the elimination of the undesirable minima is fundamental.

6. Results Achieved

The tests were applied to 1600 images of Brazilian bankcheck dates. The influence of dilation and conditional dilation numbers into the baseline skew correction was studied and may be observed in Fig 5. The item "Few minima" represents the problem of obtaining less minima than the expected number. It is increasing when the dilation and conditional dilation numbers are also increasing. The item "Unwanted minima" represents the number of wrong minima. This factor is constant. The item "Threshold" represents the influence of the bad binarization of bank check dates. Its influence increases when the dilation and conditional dilation numbers are also increasing. The item "Segmentation" represents the errors occurred in the database. This factor is constant. The item "Number error" represents the errors occurred with the numbers. It is also constant. The item "Other errors" represents the unknown factors which have introduced some errors.

By combining all these factors with the percents of correctly deskewed images, the best results are obtained with 6 iterations of dilation and conditional dilation, with 78% of Brazilian bank check dates correctly processed. In Fig. 6, interesting results in a complex image may be observed.

Unsolved problems remain when it occurs of fragmentation of word (when a connected word is broken in two or more connected sets). The elimination of only the really undesirable minima is fundamental, so as not to impair the handwriting baseline skew correction.

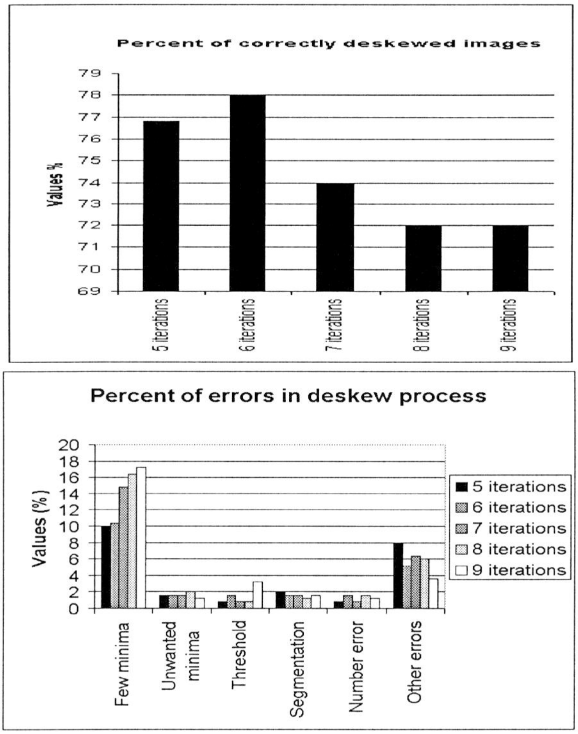

Fig. 5. Numerical results of correctly and incorrectly deskewed dates.

7. Conclusion

The main goal of this approach was to demonstrate that it is possible to reduce the utilization of empirical thresholds in handwriting baseline skew correction. In this article it was shown that, by associating binary mathematical morphology techniques for the determination of the pseudo-convex hull to the weighted least squares method, the reduction of the empirical thresholds was made possible. In the case of handwritten words, though, the complete elimination of heuristics is still very difficult because of the complexity of handwriting. The utilization of the weighted least squares method allowed the reduction of such problems, leading to better results. Nevertheless, because of word fragmentation and the sometimes small number of minima extracted by the pseudo-convex hull, some problems in the elimination of undesirable minima occur, impairing the handwriting baseline skew correction. The association of this approach to classical techniques such as projection profiles, etc... is envisaged for future works.

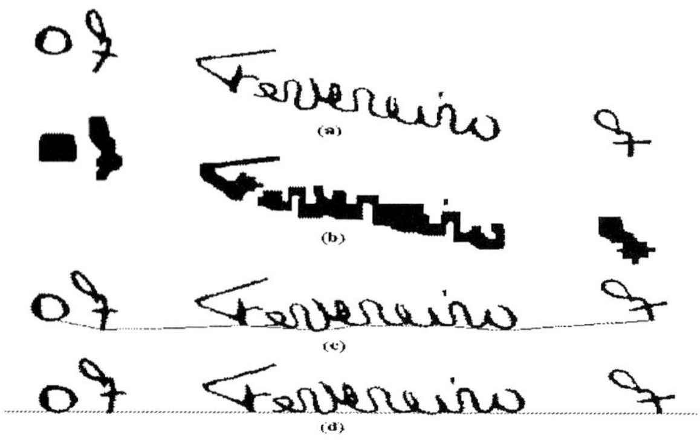

Fig. 6. (a) Original image, (b) Pseudo-convex hull image, (c) Connection of the pre-filtered minima segments, (d) Image with the corrected skew.

References

1. R.M Bozinovic and S. N. Srihari Off-line cursive script word recognition, *IEEE Trans PAMI*, vol. 11, 68–83, 1989.
2. M. Côté Utilisation d'un modèle d'accés lexical et de concepts perceptifs pour la reconnaissance d'images de mots cursifs. *PhD Thesis, École Nationale Supérieure des Télécommunications*, June 1997.
3. P.E. Danielsson and M. Hammerin. High-Accuracy Rotation of Images. *Graphical Models and Image Processing*, 54(4), 340–344, July 1992.
4. R.C Gonzalez and P. Wintz, Digital Image Processing, 2nd Edition. *Addison-Wesley Publishing Company*, 1987.
5. R.L Graham and F.F Yao, Finding the Convex Hull of a simple polygon. *J. Algorithms*, vol. 4,324–331, 1983.
6. W. Guerfali and R. Plamondon. Normalization and Restoring On-Line Handwriting. *Pattern Recognition*, 26(3),419–431, 1993.
7. A. Leick. GPS satellite surveying. 2 edition. *John Wiley and Sons, New York*, 1995.
8. S. Madhvanath and S. N. Srihari. A technique for local baseline determination, *Proceedings of the fifth International Workshop on frontiers in Handwriting Recognition*, pp. 445–448, 1996.
9. A. W. Senior, Detecção de erros grosseiros em pontos de controle planialtimétricos para aerotriangulação. *Revista Brasileira de Cartografia*, Nb 40, 27–35, 1986.
10. M. Morita. Study to enhance the baseline skew correction of handwritten words. *Master thesis, Cefet-Pr, Brazil*,march 1998.
11. P.Soille. Grey Scale Convex Hulls, Definition, Implementation and Application. *Mathematical Morphology and its Applications to Image and Signal Processing, H.J.A.M.Heijmans, J.B.T.M. Roerdink, Kluwer Academic Publishers*, 83–90, 1998.
12. A. El Yacoubi. Modélisation markovienne de l'écriture manuscrite. Application à la reconnaissance des adresses postales. *PhD Thesis*, Université de Rennes 1, September 1996.

MORPHOLOGICAL SEGMENTATION OF TEXT AND FIGURES IN RENAISSANCE BOOKS (XVI CENTURY)

MICHELE MENGUCCI and ISABEL GRANADO
CVRM / Centro de Geo-Sistemas, Instituto Superior Técnico
Av. Rovisco Pais, 1049-001 Lisboa, PORTUGAL

Abstract. This paper presents a mathematical morphology application for the segmentation of greyscale images of ancient books of the renaissance (16th century), followed by the distinction of its elements in two classes: figures and text characters. The images were obtained by scanning photographs of pages of ancient Portuguese books. Due to the book's age and its preservation state, some irregularities appeared in the resultant scanning process, such as different image's size, distorted pages, noise/spots, among many others. The application attempts to go beyond those issues, in order to obtain the best binary image, reflecting more precisely the original one, and to separate the image's elements into figures and text characters, obtaining finally clean and clear results.

Key words: Document Processing, Geodesic Reconstruction, Segmentation.

1. Introduction

If the information contained in very old books, and therefore delicate to handle, could be translated to a digital format, their very content would be of much wider access. In the image analysis laboratory of the CVRM, it has been developed a methodology that takes the printed part of these books content, cleaning it from all the not useful "signs" appearing on their pages, and afterwards, classifying it in two types: text and figures. An approach, followed by S. Beucher [1] in a similar case, has inspired the way to solve the first part of this problem: the pages' segmentation into binary images. For the second part of our task, the separation between figures and text, we kept using the Mathematical Morphology tools. These are indeed a proper approach [2], while other methodologies [3] could have been applied but would have probably been more time consuming. In fact, the work developed consists in two principal tasks, one being consequent to the other. First, the task we faced is to obtain the correspondent binarized images where the text and the figures are represented (level 1) on a homogeneous background (level 0), and next, to distinguish and separate the text from the figures.

The outcomes presented hereafter pertain to the first phase of an European project still in progress, whose main objective is to build an indexed digital library, containing three sets of books: Portuguese, French and Italian ones, all from the XVI century. This work has, of course, been applied to the Portuguese set. The sequence of mathematical morphology operators involved in the images treatment is applied through a script to be run in the Visilog software, C-interpreter. The scripts developed during the research in the initial part of the project, are two: one for the segmentation and one for the separation between text and figures. They could be easily linked in a single one but, for a better and clearer explanation, they will be illustrated separately.

2. Inputs

The inputs are digital grey level images, produced by scanning photographs of old Portuguese books' pages. Though the efforts of the scanning team have been appreciated, the input images are far from being "standard": this means that the set of images was not homogeneous in their general characteristics. This was due to both the author's imperfections in printing (a few years after Gutenberg's invention), and to the most actual problems in the photo/scanning operation. Because of the imperfection in the "strength" of the printing, some part of figures and letters, are less marked than others: if we consider the topographic analogy of the grey level images, it can be said that the "furrow" of these lines is not as deep as the rest. Due to the non optimal scanning conditions we have had other not ideal situations. For instance the image's size is varying both in the linear dimensions and in the proportion between width and height. Apart from this, which was not a great deal because the software handled it, the variations in the background colour (white) represented the major cause of the few imperfections in the results. The basic input is then a grey level image of about 2600 x 3900 pixels, with dark letters and figures, over a background mainly white with some darker shades in a varying proportion, presenting some dirt spots and occasionally handwriting characters.

3. Application Developed

The two main operations of the application, and the respective sequence of operations, are hereafter discussed. The notations and terminology are the same as in Soille's book [4].

3.1. Segmentation

The concept guiding the methodology to segment is simple, referring to the topographic analogy. The signs (text and figures) to detect are like "valleys" or "furrows" carved on a basically plane background. To be highlighted, they are filled and the original image is subtracted to the one with the filled up furrows, leaving just the filling in the output image. Thus we arranged a mask to cover, in the most accurate way, the text and the figures. To locate the interesting zone of each image, in order to build an appropriate mask, we exploited the morphological gradient information. This would leave out of the process all the objects (dirt spots, etc) with "soft slope" contours. We have also used basic directional morphology operators to suite the text's shape and the figures as well as possible. The mask is therefore obtained through directional dilations of the gradient, in the horizontal and vertical directions. Our intention was to make the minimum number of dilations needed to cover the "furrow" of any sign (text or figure) to be detected, and not more, otherwise we would be in risk of covering the "valleys" of the surrounding noise. Also, it was assumed that the main directions of the text characters were 0° and 90°, which is close enough to the reality to get effective results. This method gives an output image with a dark background and the extracted objects in grey level. It is, basically, a sequence in 3 steps:

1. Dilating the original grey level image.
2. Reconstruction of the original image with the image created in step one, which is the "covering" or "filling" operation.
3. Subtraction of the original image to the reconstructed one.

It was considered convenient to repeat the operation once more, as letters or figures lightly marked were badly segmented with the first gradient mask. The second "covering" is somehow complementary to the first one. The reason is the following. After the dilated gradient is processed, it was created an image which is the multiplication of the binary mask, obtained thresholding the dilated gradient (see step 2 and 3 of the operations sequence), by the maximum value of the gradient (this was called meaningfully "plateau"). Then the gradient was subtracted to this "plateau". The result is an image with minimums located where the dilated gradient presented maximums and vice versa. Next, this "complementary dilated gradient" is raised until its maximum meets the value of the first gradient's maximum. This allows the second reconstruction to cover and fill the "valleys" of light marks, this being a problem especially in areas with a very white, or high in topographic terms, background.

Segmentation operations sequence

The process for the image treatment is explained hereafter, following the sequence of the operations performed on an input image by the algorithm:

1. Previous filtering with a low pass Gaussian kernel (Fig.1A). This was considered the best one as it gives a weight, to the central pixel, which is 4 times the surrounding ones, so it filters the image preserving enough contrasts.
2. Directional dilations, with size 20 in the vertical direction and 10 in the horizontal, on the gradient creating the image "grdD", dilated gradient (Fig.1B).
3. Thresholding of "grdD" with threshold level 2.
4. FIRST RECONSTRUCTION: multiplying the former image described, by the dilated gradient "grdD", obtaining the image used as mask (referring to Soille's terminology in [4]) for the reconstruction of the original filtered image.
5. Subtracting the original filtered image to the reconstructed: the result being an image that presents the printed part of the manuscript in grey level on a black background. Thresholding this with a threshold level 2, producing the first binary image, part of the final output (Fig.1C).
6. SECOND RECONSTRUCTION: multiplying the output of step 3 by the gradient maximum, creating the "plateau" image. Subtracting the "grdD" to it, to attain the complementary gradient image, "grdD_C". Before using

this as a mask for the second reconstruction of the original filtered image (Fig.1D), raising of the image "grdD_C" to the "plateau" level (adding to it the difference between the maximums of the complementary and of the normal gradient).

7. Like in step 5, similar operation between the second reconstruction and the original image generating the second binary image (Fig.1E).
8. Union of the two binary outputs (from step 5 and step 7), producing the final image (Fig.1F), ready to be submitted to the separation algorithm.

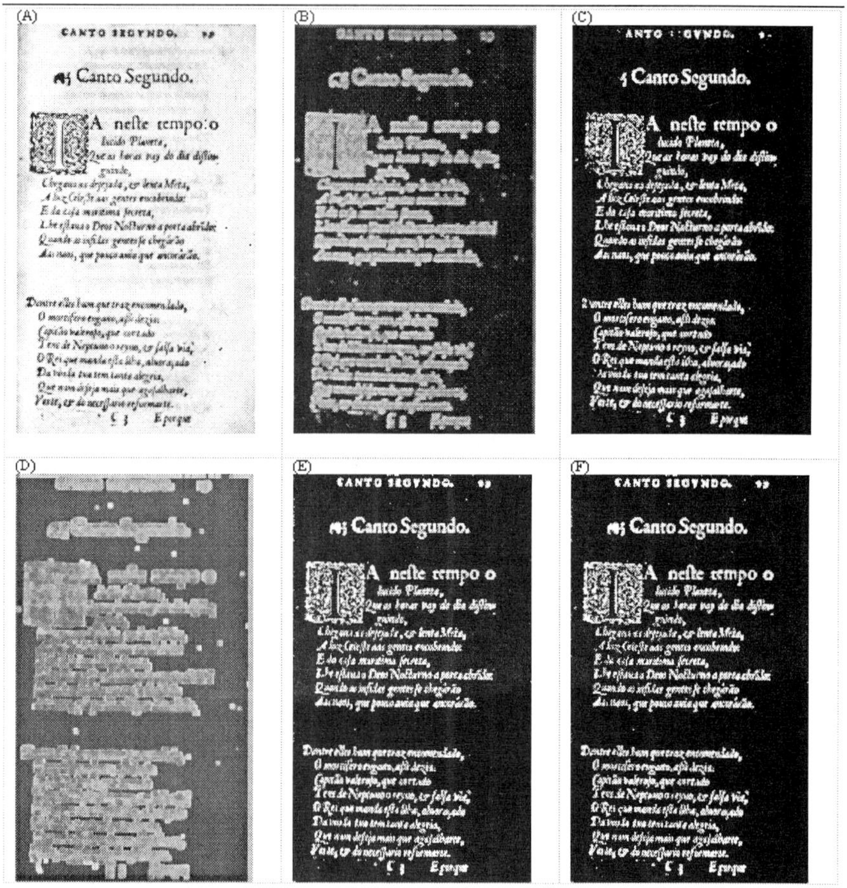

Fig. 1. Images resultant from the algorithm's steps: (A) initial image; (B) image "grdD";(C) first binary output; (D) image "grD_C" after the "raising" operation; (E) second binary output; (F) final image, union of (C) and (E).

3.2. Separation Between Figures and Text

The objective is to separate figures from text characters, using the binary images outputs of the segmentation algorithm. Fundamentally, we used the main directions in which the image's elements evolve [5], i.e., it has been assumed that the majority of lines composing the text characters has an orientation of

0° (length) and 90° (height). A text example is presented in the next figure (Fig.2).

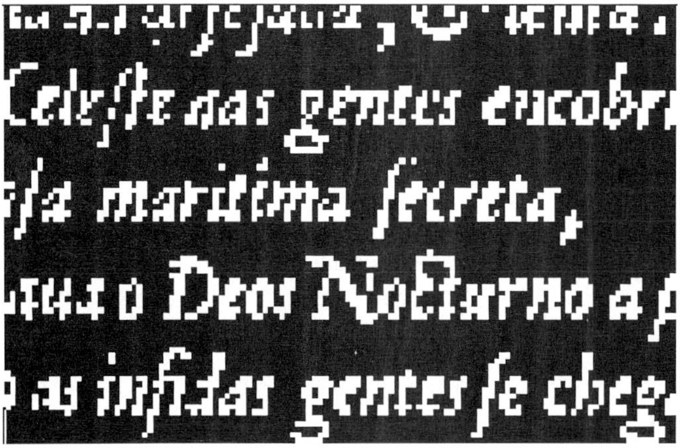

Fig. 2. Example of text orientation

Most of the figures are composed of several small lines, which main directions do not include the ones concerning text characters. So, the figures were obtained using the directions of 30°, 60°, 120° and 150°. All the images were converted to a hexagonal grid, in order to have smoother shapes in the results. The skew of the images was not taken into account because viewing our image set we noticed a regular behaviour, with no perceptible rotation [6]. Concerning the algorithm's implementation, it was used a combination of mathematical morphology operations/primitives to obtain the expected segmentation. It starts with the figure's extraction followed by the separation of the text.

Figures operations sequence

This is the sequence of the algorithm steps:

1. After the conversion to a hexagonal grid, directional closings at 30°, 60°, 120° and 150°, with structuring element size of 5 pixels, in order to connect the figures elements. This value was chosen to avoid the connection between characters, being 7 pixels their average spacing.
2. Directional openings of size 15, at 0° and 90°, eliminating most of the text characters. At this phase, according to the parameter's values all the figures are separated from the rest.
3. Reconstruction of the image resulting from step 1, using as a marker the image resulting from the step 2.
4. Reconstruction of the initial image using as markers the output of step 3.
5. Directional closings applied to the output of step 4 with a structuring element of size 3.
6. Hole fill operation to the output of step 5, closing all the figures elements.

7. Opening of size 22 is applied to the output image of step 6, excluding all the remaining text elements.
8. Reconstruction of the image resultant from step 6, using as marker the output of step 7.
9. Reconstruction of the initial image, using as markers the output of step 8, obtaining the figures.

Text operations sequence

10. Geometrical difference operation, XOR (exclusive or), between the image resultant from step 9 and the initial one. This way it is obtained the respective text, as well as some small particles that should be cleaned.
11. Delete the elements touching the image's border.
12. Directional openings over the output of step 11, at 30°, 60°, 120° and 150° with size 2, and directions 0° and 90° with size 1, eliminating the small particles.
13. Directional closing to the output of step 12, at 0° with size 10, and directions: 30°, 60°, 90°, 120° and 150° with size 7, in order to connect the text characters.
14. Reconstruction of output of step 11 using as markers the output of step 13.
15. Initial image reconstruction using as markers the output of step 14, obtaining the text cleaned from noisy particles.

Basically the algorithm for all the Portuguese books follows the same structure as the one presented above. However, it was necessary to adapt to each type of book, some of the size parameters, due to the mentioned conditions of scanning and preservation state.

4. Results

In this section we show two examples of the output of the methodology, with the respective input image (see figures 3 and 4). In both cases the figures and the text are clearly extracted, even if, in the "Lusiadas" case, a letter "Q" got classified as part of the image. Errors like this one are consequence of a deficient binarization, which is usually [7] a very critical point. In the other example, it can be seen that the evident variations of the background grey level did not prevent from a good binarization.

5. Conclusions

This whole application has been trained mainly on the images taken from the book "Os Lusiadas", by Luis de Camões. Afterwards we tested their application to another Portuguese book: "Obras do Doutor F. Miranda". We found that the changes to be made in the parameters values were not many and, beside of this, they were not hard to come up to. Still, this is a fundamental check to be made before any other upcoming use of the scripts. Considering the project aims the results were in general satisfactory and useful. The mathematical

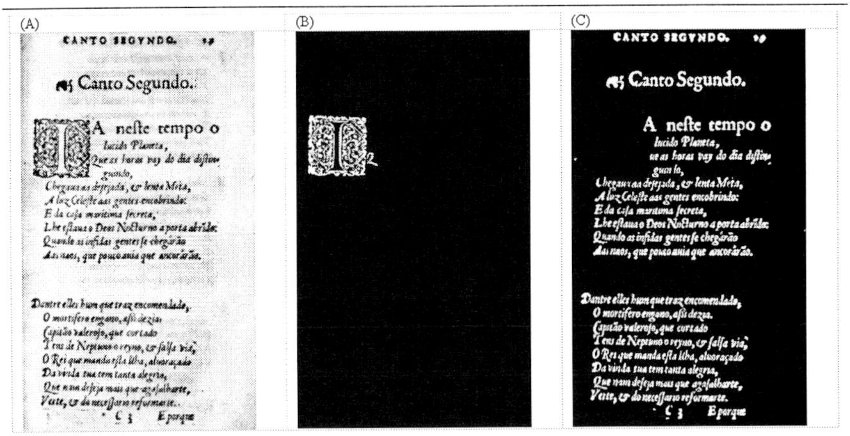

Fig. 3. Algorithm's application to the "Lusiadas" book. (A) initial image; (B) figure segmentation; (C) text segmentation.

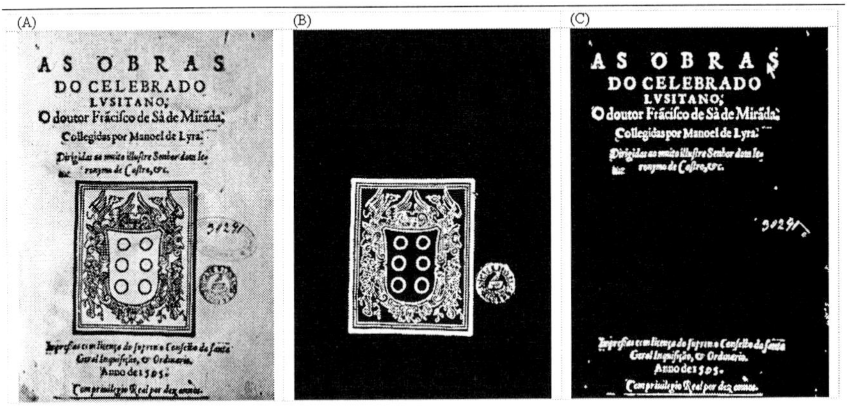

Fig. 4. Algorithm's application to the "Obras do Doutor F.Miranda" book. (A) initial image; (B) figure segmentation; (C) text segmentation.

morphology operators involved in this methodology approach the images in a global way, and not locally. Therefore the major problems derived from the wide shades in the background. Just an example: as the gradient operation is a measure of a slope, the importance of a fixed height surrounding the valleys represented by the letters is evident. In future it will be considered to apply a background normalization, for example with an alternating sequential filter [7]. The same methodology will be applied to other European books from the same age, expecting to find them effective, as long as the kind of characters will not change significantly. In the case of the Coran book, which is also going to be analysed, it will be probably necessary to adapt the algorithms. However, this application will surely contribute to the digitalization of the information contained in the historic books that will be submitted to it. It will also provide an easier way for their consultation and, hence, for their better preservation.

Acknowledgements

The authors are grateful to DGXIII / Telematics Programme/ LB-5608/A for supporting the project's development.

References

1. S. Beucher. *Pré-traitement morphologique d'images de plis postaux* In 4éme Colloque National Sur L'ecrit Et Le Document-Cned'96, Nantes-Juillet, 1996.
2. P. Montolio, T. Gasull, L. Corbera, F. Marqués. *Character Recognition and Document Analysis by Morphological Techniques* In International Workshop on Mathematical Morphology and its Applications to Signal Processing, Barcelona (1993).
3. S. He and N. Abe. *A Clustering-Based Approach to the Separation of Text Strings from Mixed Text/Graphics Documents* Proceedings of ICPR '96, Vienna - Austria.
4. P. Soille. *"Morphological Image Analysis"* pp. 53-89-163, Springer 1999.
5. G. Agam and I. Dinstein. *Adaptive Directional Morphology with Application to Document Analysis* in "Mathematical Morphology and its Applications to Image and Signal Processing", p. 401, edited by Kluwer Academic Publishers, 1996.
6. P. Parodi, G. Piccioli. *An Efficient Pre-Processing of Mixed-Content Document Images for OCR Systems* In Proceedings of ICPR '96, Vienna - Austria.
7. M. Cumplido, P. Montolio and A. Gasull. *Morphological Preprocessing and Binarization for OCR Systems* in "Mathematical Morphology and its Applications to Image and Signal Processing", p. 393, edited by Kluwer Academic Publishers, 1996.

APPLICATION OF MATHEMATICAL MORPHOLOGY AND MARKOV RANDOM FIELD THEORY TO THE AUTOMATIC EXTRACTION OF LINEAR FEATURES IN AIRBORNE IMAGES

A. KATARTZIS, V. PIZURICA and H. SAHLI
ETRO/IRIS
Vrije Universiteit Brussel
Pleinlaan2, 1050 Brussels, Belgium.

Abstract. In this paper we present a model-based approach to the automatic extraction of linear features, like roads and paths, from aerial optical images. The proposed method consists of two steps. The first step utilizes local information related to the geometry and radiometry of the structures to be extracted. It consists of a series of morphological filtering stages. The resulting image (response) serves as input to a line-following algorithm, which produces a set of line segments. In the second step, a segment linking process is carried out incorporating contextual, *a priori* knowledge about the road shape, with the use of Markov random field (MRF) theory. In this approach the extracted line segments, produced by the morphological operators, are organized as a graph. The linking of these segments is then achieved through assigning labels to the nodes of the graph, using domain knowledge, extracted line segments measurements and spatial relationships between the various line segments. The interpretation labels are modeled as a MRF on the corresponding graph and the linear feature identification problem is formulated as a maximum a posteriori (MAP) estimation rule. The proposed approach has been successfully applied to airborne images of different profile.

Key words: Mathematical Morphology, Markov Random Fields, Linear Features, Airborne Images.

1. Introduction

Several approaches for linear feature extraction have been proposed in the literature, most of them dealing with the problem of road identification by using either synthetic aperture radar (SAR) images or optic (visible range) images. The literature contains a variety of schemes, which are mainly based on a *local* criterion, involving the use of local operators, or a *global* criterion, incorporating additional knowledge about the structure of the objects to be detected. The methods based on local criteria evaluate local properties on the image by using either an edge or line detector [4] [6] [12] [13] or morphological operators [5]. The performance of these methods can be greatly increased by using techniques that introduce some global constraints in the image analysis process. These techniques lead to an optimal solution through the minimization of a cost function by using either dynamic programming [3] [6] [10], tracking methods [7] or the Bayesian framework [2] [9] [12].

In this paper, we propose a combined mathematical morphology and MRF technique for road identification. It is a model-based approach that combines both local and global criteria about the geometry and radiometry of the linear structures of interest. During the local analysis step, the detection of elongated structures is performed by applying a series of morphological filters, similar to those proposed by Chanussot et al. [5]. The main axis of the extracted elongated structures is determined by applying the watershed transformation on the response of the morphological filtering. The response values along the watershed lines, together with information about orientation, is then used as an input to a line-following algorithm that produces a set of line segments. The procedure related to the global analysis step is inspired by the earlier work of Tupin et al. [12]. A segment linking process is performed by using contextual *a priori* information with the help of a Markovian model of road like objects.

The use of MRF theory succeeds in extending the results of the morphological filtering towards a better reconstruction of the road network. The proposed Markovian model can be considered as a refined, more robust version of the one of Tupin et al.[12], as it involves a fewer number of parameters, and describes more efficiently the properties of the linear features of interest. Our approach has a high detection performance in heavily textured environments and is able to identify elongated structures of different size.

The paper is organized as follows: In Section 2 the proposed morphological approach for linear feature detection is described. In Section 3 we present the MRF model-based formulation for road identification and show how domain knowledge can be organized into clique functions associated with the MRF model. We illustrate the results of our approach for various airborne images in Section 4. Finally a discussion and directions for future research are given in Section 5.

2. Local Analysis

The features we search for (roads and paths) are characterized by their geometry and image appearance (radiometry). The roads appear on an optical airborne image as thin, elongated structures with a maximum width w_{max}. They are locally rectilinear, with each road pixel belonging to a line segment that is longer than a minimum length ℓ_0 and each road segment is considered as a bright structure with respect to its surrounding. All this information can be integrated and extracted using mathematical morphology. A series of morphological operators, adapted to the geometrical characteristics of the objects we want to identify, are successively applied to the input image. A line-following algorithm is then applied to the resulting image in order to produce a set of line segments. As an example, we will use the airborne image of Fig. 1(a).

2.1. Linear feature detection

1. **Removing non-flat valleys**
   We want to remove dark structures from the image, without influencing the shape of the bright elongated structures of interest. For this reason, we apply a morphological *closing by reconstruction*, using a square flat structuring element (SE) of size equal to $w_{max}/4$.

2. **Removing non-linear bright structures**
Road segments correspond to elongated bright regions with a certain width. In order to eliminate bright structures that do not belong to any line segment with minimum length ℓ_0, we apply on the reconstructed image of step 1 a morphological *opening* by using ℓ_0 pixels long linear structuring elements successively oriented in 32 possible directions. The resulting value at each pixel is the *supremum* of all these directional openings.
3. **Removing bright structures of large width**
In this step, we eliminate very wide linear bright structures that correspond to objects larger than the roads. Initially, we perform a morphological *closing* operation with a flat square SE of size $w_{max}/4$, in order to remove remaining dark spots from the image. After this, we retain only bright structures with widths less than w_{max}, by applying an *opening top-hat* operator with a flat square SE of size w_{max}. The remaining structures correspond to the roads that we want to extract. Finally, we apply once again *closing* with a flat square SE of size $w_{max}/4$, in order to make the regions inside the roads more uniform. We will consider this result as the response (I) of our morphological road detector. The negative image of this response is shown in Fig. 1(b).

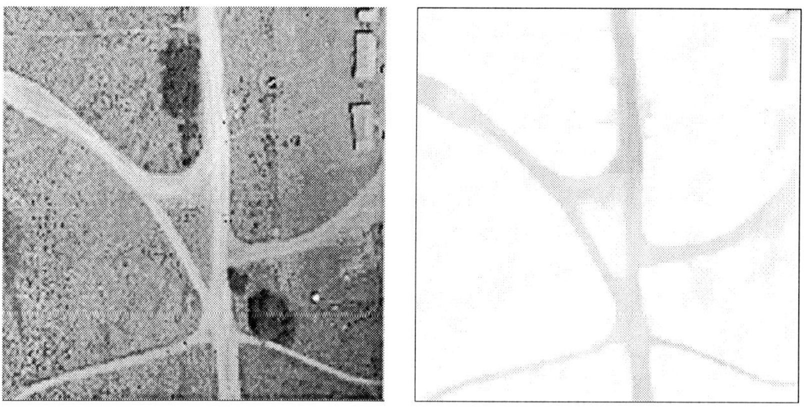

(a) (b)

Fig. 1. (a) Original airborne image. (b) Morphological road detection response (I)

2.2. Line segment extraction

The final result of step 3 (Section 2.1) gives a higher response at the points belonging to the roads compared with the surrounding background. An easy way to extract the road regions would be the application of a threshold to the resulting image as in [5]. Unfortunately, this leads to partial detection of the roads (disconnected segments), together with some spurious results corresponding to false alarms. In order to overcome this problem, and to produce one pixel width line segments, we extract, from the resulting image of step 3, the pseudo-medial axis of the roads, over which we apply a line-following algorithm using the orientation of the SE which produces the supremum of the directional openings (step 2 in Section 2.1). The pseudo-medial axis of the roads is extracted by performing the *watershed* transformation on the response image I. The result of this transformation is shown in Fig. 2(a).

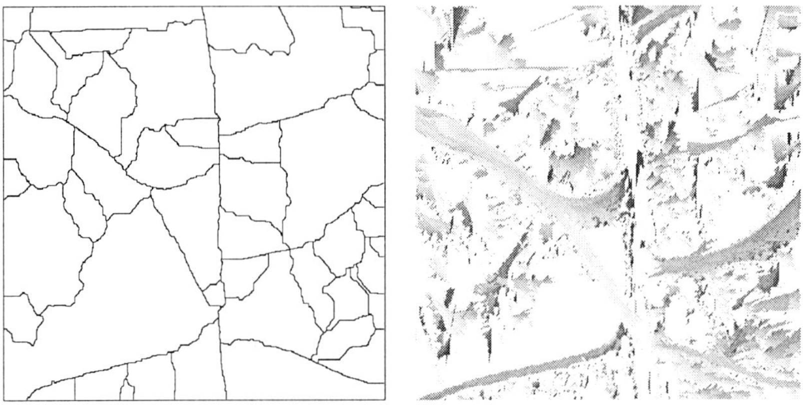

(a) (b)

Fig. 2. (a) Watershed lines. (b) Orientation image.

An orientation image is obtained by assigning to each pixel the direction of the SE which produces the supremum of the directional openings operator (step 2 in section 2.1). An example is given in Fig. 2(b) where a white gray value represents an angle of 0° and black an angle of 180°. We then apply a line-following algorithm along the watershed lines of Fig. 2(a), with each medial-axis pixel considered as a starting point. The tracking is performed in both directions along a line, by taking into account angular information provided by the orientation image of Fig. 2(b). We retain only line segments satisfying an angular deviation of maximum $d\alpha$ degrees and having length greater than a predefined value ℓ_{min}. Fig. 3(a) shows the final result of the line-following algorithm, applied on the image of Fig. 2(a), using the orientation image of Fig. 2(b). The angle offset $d\alpha$ was set equal to 10°. A small value for ℓ_{min}, equal to $\ell_0/5$, is used in order to avoid discontinuities especially in regions close to road bifurcation.

3. Global Analysis

The global analysis step of our approach is based on the earlier work of Tupin et al. [12] and is carried out on the level of the road segments. A graph is built, which contains all possible connected line segments that are created by using some connectivity criteria. The road identification process is then treated as an optimum graph labeling problem. This is carried out by associating an energy function to the line segments, based on a Markovian model of road like objects. Given the observation process, the minimization of this energy function will produce the best configuration of the line segments.

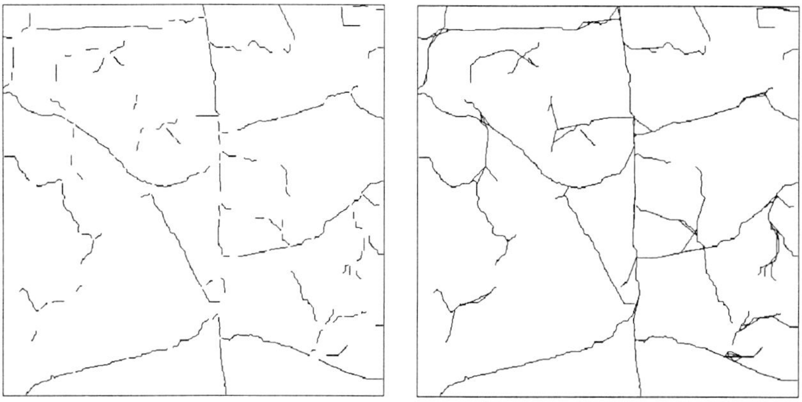

Fig. 3. (a) Detected line segments ($\mathbf{S_{det}}$). (b) Augmented set of line segments ($\mathbf{S}$).

3.1. GRAPH CREATION

We will denote by $\mathbf{S_{det}}$ the set of the detected line segments (Fig. 3(a)). Each segment of $\mathbf{S_{det}}$ is defined by its endpoints. Two line segments of $\mathbf{S_{det}}$ can be connected if their distance is less than a fixed threshold, and if the angle between them is less than a specified value. We create a new set $\mathbf{S_{con}}$, which corresponds to all possible connections between the elements of $\mathbf{S_{det}}$. Let $\mathbf{S} = \{\mathbf{S_{det}} \cup \mathbf{S_{con}}\}$, with its cardinality denoted by N. The elements of the set $\mathbf{S}$ are presented in Fig. 3(b).

For each line segment $i \in \mathbf{S}$ we assign a saliency measure r_i defined as:

$$r_i = \overline{I}/(|\theta - \overline{\alpha}| + 1) \tag{1}$$

where $\overline{I}$ is the mean value of the morphological road detection response, along the line segment (Fig. 1(b)), θ is the line segment orientation, and $\overline{\alpha}$ is the mean value of the morphological road detection orientation response, along the line segment (Fig. 2(b)).

We associate a graph structure ($\mathbf{G}$) to the set $\mathbf{S}$, each segment i (belonging either to $\mathbf{S_{det}}$ or $\mathbf{S_{con}}$) being one of its nodes, and two nodes i and j being linked by an arc if they share a common endpoint. In order to introduce contextual knowledge with the use of a Markovian model, we must define a neighbourhood system. The neighbourhood $\mathcal{N}_i$ of each node i contains all the line segments that have a common endpoint with i. For each segment $i \in \mathbf{S_{det}}$ we define 2 cliques that correspond to both of its end points. Each of these cliques contains all the segments that share the specific extremity. If N_{det} denotes the number of elements of $\mathbf{S}_{\mathbf{det}}$, then the total number of cliques equals $2N_{det}$. An example of such a neighbourhood system is shown in Fig. 4.

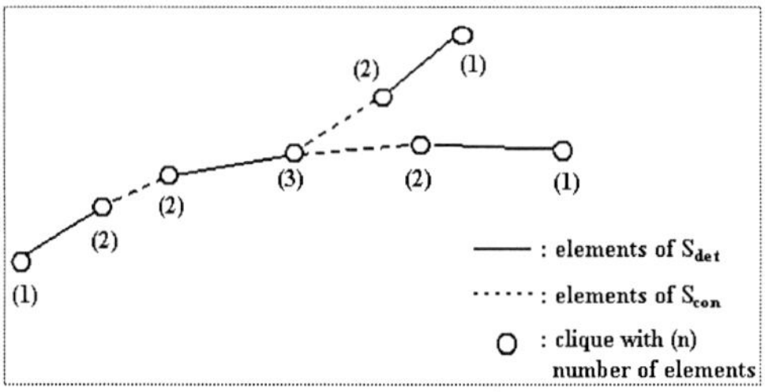

Fig. 4. Neighbourhood system.

After the definition of our neighbourhood system, we attach attributes to the nodes and arcs of $\mathbf{G}$. The arc between nodes i and j is associated with a value θ_{ij} representing the angle between the two segments. For each node $i \in \mathbf{S}$, we associate a normalized length ℓ_i and an observation value d_i, that reflects the probability of this segment belonging to the road. d_i should increase when, adjacent to it, segments belong also to the road, something that rises from the continuity that characterizes our region of interest. For this reason we express d_i as a function of the saliency measures r_k:

$$d_i = \max_{j \in \mathcal{N}_i} \{(r_i + r_j)/2\}$$

The identification of the road will be carried out with an appropriate labeling of the graph. A label l_i is associated to each node i with $l_i = 1$ if i is a part of the road and $l_i = 0$ otherwise. The optimum configuration $\mathbf{L} = (l_1, l_2, ..., l_N)$ of the segments of $\mathbf{S}$, given the observation process $\mathbf{D} = (d_1, d_2, ..., d_N)$, can be estimated with a MAP criterion that maximizes the posterior probability distribution given by:

$$P(l|d) = \frac{p(d|l)P(l)}{p(d)} \tag{2}$$

where $P(l)$ is the prior probability of labelings l, $p(d|l)$ is the conditional probability distribution function (p.d.f.) of the observations d, also called the *likelihood function* of l for d fixed, and $p(d)$ is the density of d which is a constant when d is given.

3.2. Energy definition

3.2.1. *Conditional Probability Distribution*

We consider that the conditional probability distribution $p(d|l)$ corresponds to a Gibbs distribution. By assuming independence between the different observations (d_i) and supposing that the conditional probability distribution of d_i

only depends on the labelings l_i, we can write

$$p(d|l) = \prod_{i=1}^{N} p(d_i|l_i) \propto exp\left(-\sum_{i=1}^{N} V(d_i|l_i)\right) \tag{3}$$

where $V(d_i|l_i)$ denotes the potential of segment i. This type of potential can be deduced from the observation field **D** and reflects the likelihood of every segment as belonging or not to a road.

The conditional probability distributions $p(d_i|l_i)$ are learned from an experiment after a manual segmentation of the roads, performed by a human observer. After this experiment, we notice that road segments may have almost any observation value d, while non-road segments have observations with values greater than a threshold t. Based on this heuristic, the following linear conditional potentials have been chosen:

$$V(d|0) = \begin{cases} \frac{d}{t} & \text{if } d < t \\ 1 & \text{otherwise} \end{cases} \qquad \text{and} \qquad V(d|1) = 0, \forall d$$

In order for the potentials to correspond to a probability distribution, we normalize the values $V(d|l)$ so that: $\int_0^1 p(d = x|l)dx = \int_0^1 exp[-V(d = x|l)]dx = 1$. This condition holds for the potentials that correspond to road segments, as they are equal to zero. For the non-road segments, potentials of the form: $V(d_i|0) + logZ_0$ are used; Z_0 denotes a normalization factor given by: $Z_0 = (1 - t)(1/e) - t(1/e - 1)$, with $e = exp(1)$.

3.2.2. *Prior Probability of Labelings*

Our *prior* road model is based on the assumptions that roads are long structures with low curvature and that intersections between them are rare. By considering the label field **L** as a MRF, we can use once again the MRF-Gibbs field equivalence in order to introduce *a priori* knowledge to the road identification task. The prior probability of labelings $P(l)$ can be expressed in terms of an energy function $U(l)$ as:

$$P(L = l) = \frac{1}{Z_1} exp(-U(l)) \tag{4}$$

where Z_1 is the partition function and $U(l) = \sum_{c \in C} V_c(l)$. The clique potentials $V_c(l)$ carry *a priori* information about the geometrical characteristics of the features to be extracted. Every clique c contains one segment belonging to $\mathbf{S_{det}}$ (with length ℓ^{det}), along with the segments of $\mathbf{S_{con}}$ (with length ℓ^{con}) that share the same extremity. Based on the main assumptions of our road model, we have chosen the following potentials for every clique c:

$$\forall i \in c, l_i = 0 \Rightarrow V_c(l) = 0 \tag{5}$$

$$\exists! i \in c/l_i = 1 \Rightarrow V_c(l) = K_1 + 1 - \ell_i^{det} + logZ_0 \tag{6}$$

$$\exists!(i,j) \in c^2/l_i = l_j = 1 \Rightarrow V_c(l) = sin(\theta_{ij}) + 1 - \ell_i^{det} + \ell_j^{con} + 2logZ_0 \qquad (7)$$

in all other cases,

$$V_c(l) = K_2 \sum_{i/i \in c} l_i \qquad (8)$$

Equation 5 describes a *null* situation, which does not have to be penalized or favored with respect to the *a priori* assumptions about the road structure. In equation 6, by choosing $K_1 > 0$ we penalize short roads: i.e. the clique potential is high for a clique with only one isolated segment, except when this isolated segment has a high normalized length ℓ_i^{det} (close to 1). High values of K_1 favor more connected configurations. Equation 7 imposes the constraint of low curvature and at the same time penalizes configurations with short detected and long connecting segments. Finally, $K_2 > 0$, in equation 8, makes less probable the appearance of crossroads.

The additional factors $logZ_0$ and $2logZ_0$, in equations 6 and 7 respectively, facilitate the comparison between the clique potential values and the conditional potentials of the *null* configurations (where all the segments of the current clique are labeled as 0). In the case of a clique with one segment labeled as 1, the factor $K_1 + 1 - \ell_i^{det}$ in equation 6 is directly compared with the conditional potential component $V(d_i|0)$ of the current segment i. In the case of a clique with two segments i, j labeled as 1, the factor $sin(\theta_{ij}) + 1 - \ell_i^{det} + \ell_j^{con}$ of equation 7 is compared with the sum of the conditional potential components $V(d_i|0)$, $V(d_j|0)$.

3.2.3. *Posterior Probability*

The posterior probability $P(l|d)$ can be also expressed in terms of a global energy function $U(l|d)$, which can be deduced from the potentials described in the previous two sections:

$$P(l|d) = \frac{1}{Z_2} exp(-U(l|d)) \ , \ U(l|d) = \sum_{i=1}^{N} V(d_i|l_i) + \sum_{c \in C} V_c(l) \qquad (9)$$

The MAP configuration of the line segments can be estimated by minimizing the energy function $U(l|d)$.

4. Results

For the minimization of the energy function, we use a simulated annealing scheme with a polynomial-time cooling schedule [1]. By comparing the energy components of different configurations of three adjacent segments, we derived the following accepted range for the parameter K_1: $K_1 < \frac{2+logZ_0}{2}$. Recall that the factor Z_0 is a function of the parameter t. Finally, the parameter K_2 has been empirically set to a value around 0.1. In Fig. 5 we present the results of our method applied on the image (a) of Fig. 1, using a different value for the parameter K_1.

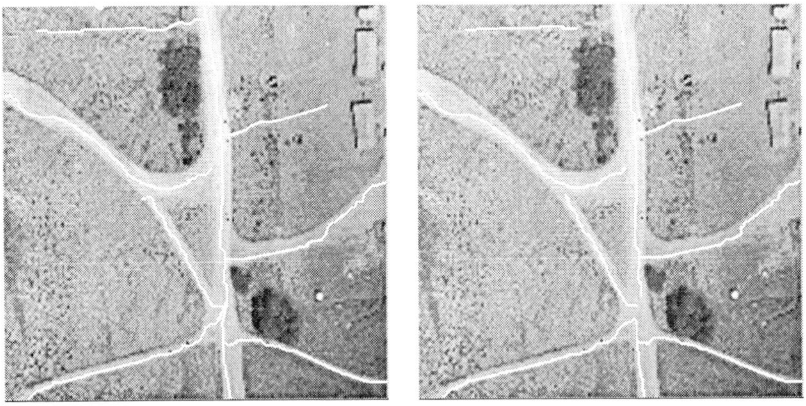

(a) (b)

Fig. 5. Results of the road extraction process for different parameter values. (a) $t = 0.15$, $K_1 = 0.5$, $K_2 = 0.1$ - Optimal result. (b) $t = 0.15$, $K_1 = 0.3$, $K_2 = 0.1$ - Partial reconstruction of the road network.

5. Discussion - Conclusions

We proposed a model-based technique for linear feature extraction, in digitized airborne images, which combines both local and global criteria, and presented its application on the problem of road and path detection. Its main advantage is the high detection performance in heavily textured environments along with its ability of identifying elongated structures independently of their size. Concerning the local analysis step, we utilized the morphological operators proposed by Chanussot et. al. [5] in order to identify road structures with specific geometrical properties. Additionally, we extracted the road main axis and applied a line-following algorithm. This process produced a set of line segments with meaningful orientation properties and eliminated a sufficient number of false alarms. In the next step of our work, we created a Markovian road model, similar to the one proposed by Tupin et. al. [12] in order to introduce contextual knowledge to our analysis. At the same time, we proposed some necessary modifications in order to incorporate additional information about the nature of the line segment candidates. These include a discrimination between the initially detected segments ($\mathbf{S_{det}}$) and the ones corresponding to the possible connections ($\mathbf{S_{con}}$), the introduction of a new observation measure (d_i) that reflects more efficiently the likelihood value of each segment and the use of fewer number of potential parameters (t, K_1, K_2).

One of the most important limitations of our method is that it is not entirely unsupervised, due to the setting of five parameters, two of them concerning the local analysis step (w_{max}, ℓ_0) and three (t, K_1, K_2) influencing the linking process. The parameters w_{max}, ℓ_0 are based on *a priori* knowledge about the size of the roads. On the other hand, the proposed ranges of the parameters t, K_1, K_2, give optimal results for this type of environments, independently of the size of the linear features of interest. Further analysis should be carried out towards the problem of identifying road segments with high curvature, especially when this is higher that the maximum road width found in the image and in the choice of a more efficient skeletonization process for the extraction of the road main axis. Finally, improvements could be obtained during the connection step, by searching for the best path between extremities of the segments we want to connect, instead of assuming that all roads may be found by connecting a set of initially detected segments.

Acknowledgements

This research is part of the European Pilot Project: "Airborne Minefield Detection in Mozambique" (ITC(N), RMA(B), Geograph(P), EOS(UK), Eurosense(B), Aerodata(B), CAE Aviation(L), Recon Optical(UK), Satelitbild(S), Zeiss Eltro Optronic(G), NPA(N), CND(Moz), I.G.I LTD(G), VUB(B)).

References

1. E. H. L. Aarts. *Simulated annealing and Boltzmann machines.* Chapman and Hall, 1993.
2. M. Barzohar and D. B. Cooper. Automatic finding of main roads in aerial images using geometric-stochastic models and estimation. *IEEE Trans. Pattern Anal. Machine Intell.*, 18:707–721, Jul. 1996.
3. M. Buckley and J. Yang. Regularised shortest-path extraction. *Pattern Recognition Letters*, 18:621–629, 1997.
4. J. Canny. A computational approach to edge detection. *IEEE Trans. Pattern Anal. Machine Intell.*, 8:679–698, Nov. 1986.
5. J. Chanussot and P. Lambert. An application of mathematical morphology to road network extraction on SAR images. *Proc. International Symposium on Mathematical Morphology, Amsterdam*, pages 399–406, 1998.
6. M. A. Fischler, J. M. Tenenbaum, and H. C. Wolf. Detection of roads and linear structures in low resolution aerial imagery using a multisource knowledge integration technique. *Comput. Graph. Image Processing*, 15:201–223, no 3 1981.
7. D. Geman and B. Jedynak. An active testing model for tracking roads in satellite images. *IEEE Trans. Pattern Anal. Machine Intell.*, 18:1–14, Jan. 1996.
8. S. Krishnamachari and R. Chellapa. Delineating buildings by grouping lines with MRFs. *IEEE Trans. on Image Processing*, 5:164–168, Jan. 1996.
9. J. L. Marroquin. A Markovian random field of piecewise straight lines. *Biological Cybern.*, 61:457–465, 1989.
10. N. Merlet and J. Zerubia. New prospects in line detection by dynamic programming. *IEEE Trans. Pattern Anal. Machine Intell.*, 18:426–431, Apr. 1996.
11. N. Metropolis, A. Rosenbluth, M. Rosenbluth, A. Teller, and E. Teller. Delineating buildings by grouping lines with MRFs. *J. Chem. Phys.*, 21:1087–1091, 1953.
12. F. Tupin, H. Maitre, J. M. Nicolas, and E. Pechersky. Detection of linear features in SAR images: Application to road network extraction. *IEEE Trans. on Geoscience and Remote Sensing*, 36:434–453, Mar. 1998.
13. Y. T. Zhou, V. Venkateswar, and R. Chellapa. Edge detection and linear feature extraction using a 2-d random field model. *IEEE Trans. Pattern Anal. Machine Intell.*, 11:84–95, Jan. 1989.

TESTING SOME MORPHOLOGICAL APPROACHES TO FACE LOCALIZATION*

B. RADUCANU and M. GRANA
Dept. CCIA, UPV/EHU
e-mail:ccbradub@si.ehu.es, ccpgrrom@si.ehu.es

Abstract. In this paper we present the results of the use of some morphological approaches to feature extraction for face localization in gray level images. Namely we have applied the Morphological Multiscale Fingerprints (MMF), and two grayscale Hit-or-Miss transforms. The morphological feature extraction techniques tested belong to the class of global image feature extraction approaches. They can be combined with others to ensure a more robust face localization. No structural relationships between face elements are taken into account. We compare these results with those obtained using a standard PCA approach.

Key words: Face Localization, Morphological Multiscale Analysis, Generalized Hit-or-Miss Transform.

1. Introduction

The field of face recognition [1] has reached a high degree of maturity, however the problem of face localization remains a research issue. Face localization is the task of determining the position of a face in the image. It is a needed preprocessing step for face recognition. There are several approaches to face localization which have shown some degree of success. The early approaches followed the path of Principal Component Analysis (PCA) which was very successful for face recognition [2],[3]. However, PCA shows a high sensitivity to changes in illumination and in pose and scale. The Local Feature Analysis [4] combines the PCA approach with some structural approaches. The PCA-based localization has been extended to Independent Component Analysis in [5]. Other works based on global picture processing use neural network approaches [6, 7, 8, 9, 10, 11]. These approaches are quite sensitive to the training samples employed, and their tuning is a quite laborious process. On the other hand, approaches based on the color processing [12, 13] are very easy to realize, although again very sensitive to the training data. Structural approaches are based on the detection on face elements and the testing of their relative distances [14, 15]. They can be made robust to occlusions and pose changes, but again their tuning is very tricky. A sensible approach to more robust face localization is the combination of several methods into a multi-cue system.

In this paper we explore the performance of some morphological operators to extract features for the task of face localization in grayscale images. No

* Work supported by project PI-1998-21 from the Gobierno Vasco and project TAP-98-0294-C02-02 from the CICYT. Bogdan Raducanu has a predoctoral grant from the UPV/EHU

structural model is proposed, although the morphological methods could be embedded into structural approaches performing the detection of face elements. The first method is based on the extraction of morphological features from the image. The features are the reduced fingerprint of the image obtained from a multiscale morphological analysis [16]. Two different generalizations of the Hit-or-Miss transform for grayscale images constitute the second and third methods tested. The first generalization was originally proposed as a grayscale operator with the same structure as the binary image Hit-or-Miss transform: a superposition of two erosions [17]. It was called the Graylevel Hit-or-Miss Transform (GHMT). The second generalization is a new proposition based on the decomposition in level sets of the grayscale images. We call it Level Based Hit-or-Miss Transform (LBHMT) to differentiate from GHMT. It consists of the application of the binary Hit-or-Miss transform on each of the level sets of the image, and combining them with the sup operator. The objective of the present work is to compare the performance of the morphological analysis with the PCA approach, taken as a representative of the global feature extraction techniques. We have used a small custom database of images and computed the mean Receiver Operation Curve (ROC) for the techniques.

Section 2 presents the Multiscale Morphological Fingerprint. Section 3 presents the generalizations of the Hit-or-Miss Transform. Section 4 reviews the PCA approach to face localization. Section 5 presents our experimental results. Section 6 presents our conclusions.

2. Multiscale Morphological Fingerprints

Scale-space theory deals with the formal definition of the concept 'scale' in terms of signals/images, i.e how we represent the data at a given scale and how we relate image features from one scale to another. A very important basic requisite for a particular scale-space is the so-called 'causality': Every feature/extremum in coarse scale (large σ) has to have a cause in fine scale (small σ). The linear scale-space results from the convolution with Gaussian kernels of increasing variance.

Another way to generate a scale-space is using mathematical morphology. Dilation and erosion, are the basic operations of morphological scale-space. In the following definitions, we assume that f is the original grayscale image and g is the structuring function, namely $f : D \subset R^n \to R$ and $g : E \subset R^n \to R$. The dilation is defined as: $(f \oplus g)(x) = \sup_{t \in E} \{f(x-t) + g(t)\}$ and the erosion as: $(f \ominus g)(x) = \inf_{t \in E} \{f(x+t) - g(t)\}$. In all the assumptions, in order to dismiss the 'lateral shifting effect', [16] impose for the origin of the structuring function the following conditions: $\sup_{t \in E} \{g(t)\} = 0$ and $g(0) = 0$. A suitable scale-space structuring function is the 'sphere' function defined by the following equation: $g_\sigma(x) = |x| \left(\left(1 - \|x/\sigma\|^2\right)^{1/2} - 1 \right)$, $\|x\| \leq \sigma$. Now, the multiscale dilation-erosion is defined as:

$$(f * g_\sigma)(x) = \begin{cases} (f \oplus g)(x) & \text{if} \sigma > 0 \\ f(x) & \text{if} \sigma = 0 \\ (f \ominus g)(x) & \text{if} \sigma < 0 \end{cases} \tag{1}$$

For positive scales ($\sigma > 0$), the operation corresponds to a dilation, and for negative scales ($\sigma < 0$), the operation corresponds to an erosion. As $|\sigma|$ increases, the image tends to have less 'structure'. When $|\sigma| \to 0$, the image converges to the original one. The features preserved by the multiscale analysis are the local extrema. This is stated by the following theorem, proved in [16]:

Theorem 1 *Considering f the original function and g the structuring function with the property that it has a local maxima at the origin, we define the following sets: $E_{\max} = \{x : f(x)$ is a local maximum$\}$ and $E_{\min} = \{x : f(x)$ is a local minimum$\}$. Then, for any $\sigma_1 < \sigma_2 < 0 < \sigma_3 < \sigma_4$ we have the following relations:$E_{\min}(f*g_{\sigma_1}) \subseteq E_{\min}(f*g_{\sigma_2}) \subseteq E_{\min}(f)$ and $E_{\max}(f*g_{\sigma_4}) \subseteq E_{\max}(f*g_{\sigma_3}) \subseteq E_{\max}(f)$.*

The fingerprints of a scale-space are plots of the point sets of the signal extrema over the scales, i.e. $E^*(\sigma) = E_{\max}(f * g_\sigma) \cup E_{\min}(f * g_\sigma)$. In practice, for computational reasons, is used the so-called 'reduced fingerprint', defined as follows:

$$E_r^*(\sigma) = \begin{cases} E_{\max}(f \oplus g_\sigma) & \text{if} \sigma > 0 \\ E_{\max}(f) \cup E_{\min}(f) & \text{if} \sigma = 0 \\ E_{\min}(f \ominus g) & \text{if} \sigma < 0 \end{cases} \tag{2}$$

The preservation of the local extrema produces a relative insensitivity of the morphological multiscale fingerprint to variations of illumination. Figure 1 shows the reduced fingerprint (second row), the local minima (third row) and maxima (fourth row) of a face image (center) after darkening (left) and brightening (right) with the application of monotonic increasing gamma functions. The localization of the local maxima and minima is not changed by the monotonic transformation of the grayscales. The fingerprint is affected by the brightness transformation in the following way: the darkening of the image produces a reduction and an increase of the scale of the local maxima and the local minima, respectively. The brightening works in the dual direction. However, for moderate brightness variations the MMF is much more insensitive than the PCA described in the next section.

In the experiments reported in section 4, we have computed the distance between two fingerprints as the sum of the difference of the number extreme points at each scale: $d(f_1, f_2) = \sum_\sigma |\mathrm{Card}(\mathrm{E_{f_1}}(\sigma)) - \mathrm{Card}(\mathrm{E_{f_2}}(\sigma))|$, where by Card($A$) we denote the cardinality of the set A. This distance is obviously invariant to affine transformations, and relatively robust against small illumination changes. In the experiments reported below, we have used the sphere structural function defined over a finite set of scales: $\{\sigma = \pm 2^i, i = 1, \ldots, 6\}$. To decide the 'faceness' of an image we compute the nearest neighbor relative to a set of face patterns. The minimum distance is kept as a measure of the similitude of the image to a face. Despite the naivety of the definition of the distance, the results were very good.

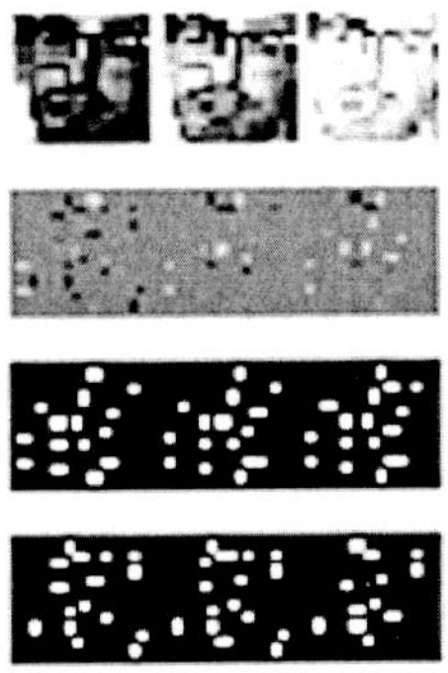

Fig. 1. Fingerprint of a face image after darkening and brightening.

3. Generalizations of Hit-or-Miss Transform to Grayscale Images

In this section we review the definition of the Hit-or-Miss Transform (HMT) morphological operator and one extension to grayscale images. Furthermore, we present our own definition of this operator for grayscale images, based on level sets. The HMT can be seen like a template matching technique, thus it selects pixels that have certain geometrical properties (e.g. isolated points, edge points, corner points or T-junctions, as reported in [18]). This operator is not a morphological filter, because it's not endowed with the two properties: (i) increasingness and (ii) idempotence.

3.1. Binary Hit-or-Miss

Given a window X, consisting of the image J and the background $(X\backslash J)$, and two structuring elements K and L, satisfying the condition $K \cap L = \oslash$, we define the 'Hit-or-Miss' as follows:

$$J \otimes (K, L) = (J \ominus K) \cap ((X\backslash J) \ominus L) \tag{3}$$

Without any loss of the generality of the above definition, we can consider the special case in which K and L form a partition of the template window, Y. In this situation, K represents the actual image, and L, the background, i.e. $L = Y\backslash K$. The required condition, $K \cap L = \oslash$, is thus satisfied. Then, the interpretation of the above formula is: a pattern matches a feature in the original image if the intersection between the erosion of the image with the template and the erosion of the image background with the template background is not empty.

3.2. Grayscale HMT: GHMT

The previous definition of the HMT has been extended [17] to gray-level images. In this case, the sets are substituted by functions, and binary erosions by grayscale erosions. In the following definitions, we assume that f is the original grayscale image and g is the structuring function, namely $f : D \subset R^n \to R$ and $g : E \subset R^n \to R$, as defined in section 2. The 'Hit-or-Miss' transform in this case becomes [17]:

$$(f \otimes g)(x) = [(f \ominus g)(x)] + [((-f) \ominus (-g))(x)] \tag{4}$$

Using the definition of erosion, the above relation can be rewritten into the form:

$$(f \otimes g)(x) = \min_{t \in E} \{f(x+t) - g(t)\} + \min_{t \in E} \{-f(x+t) + g(t)\} \tag{5}$$

The first erosion ensures that the template matches the image from 'above', and the second one, from 'below'. The 'perfect matching' occurs when both matches happen simultaneously. Making use of the dualism between erosion and dilation, the last relation can be written in the following form:

$$(f \otimes g)(x) = \min_{t \in E} \{f(x+t) - g(t)\} - \max_{t \in E} \{f(x+t) - g(t)\} \tag{6}$$

The last relation says that the result of the GHMT is always negative or at most equal to 0. It can be shown that $f \otimes g$ takes value 0, iff $f(x+t) = g(x) + k$, for every $x \in E$, where k is a constant. This property demonstrates the insensitivity of the GHMT to some variations of the image/template. In the experiments reported below, the face patterns become the templates. The GHMT with each template was computed and the maximum result for each pixel gave the measure of faceness of the pixel.

3.3. Grayscale HMT Based on Level Sets: LBHMT

We can interpret the image as a topographical map where local elevation corresponds to the gray value in the image [19]. The image is formed by staking the level sets, with lowest level at the bottom and the highest on the top. Using this analogy, we can analyze the image at level $k, k = 0, \ldots, N_{max} - 1$, independently from the other ones, where by N_{max} we denote the maximum gray-level from the image. Let us consider a gray-level image as a function of its coordinates versus intensity, i.e. $f : D \subset R^n \to \{0, 1, \ldots, N_{max} - 1\}$. The image f can be decomposed into its grayscale sets as given by the formula:

$$S_k(f) = \{x \in D \,|\, f(x) \geq k\} \tag{7}$$

Level sets correspond to binarizations by threshold k. In a similar way, having this level sets, we can reconstruct the original image, by taking the supremum of all of them

$$f(x) = \sup_{k=0,\ldots,N_{\max}-1} \{x \in S_k(f)\} \tag{8}$$

With this interpretation, we can consider both the original image and the structuring function as topographical maps. We can apply the binary HMT locally at each level set. The grayscale Hit-or-Miss Transform can be expressed as the reconstruction of an image whose level sets are given by the binary HMT computed at the corresponding level sets of the original image and pattern. Therefore, the result of this operation is given by the supremum of all local results. In other words,

$$(f \otimes_g g)(x) = \sup_{k=0,\ldots,N_{\max}-1} \{x \in S_k(f) \otimes S_k(g)\} \tag{9}$$

where $S_k(g)$ is the two element partition formed by the foreground (1valued pixels) and the background (0 valued pixels). We call this operator Level Based Hit-or-Miss Transform to differentiate from the GHMT described previously. In the face localization problem the templates correspond to face patterns. The measure of faceness of a pixel corresponds to the distance between the original image and the result of the LBHMT of all the face patterns. The combination of the result of the face patterns is done with the sup operator. Despite of the non-increasing nature of the Hit-or-Miss Transform on binary images, there are no missed threshold sets when the result images are restacked, we are currently working on the formal proof of this assertion. Although it can be proved that the resulting grayscale operator is not invariant to grayscale transformations of the image or the pattern, the experimental results show that it is relatively robust against illumination changes.

4. Review of the Eigenface Approach

For the sake of completeness we review the PCA approach to face localization [3], sometimes referred to as the subspace approach. Given an initial set of M face images, each one of size $N \times N$ pixels. Let them be $\Gamma_1, \Gamma_2, \ldots, \Gamma_n$. The average face of the set is denoted by Ψ. Each face in the set differs from the average by the quantity $\Phi_i = \Gamma_i - \Psi$, with $i = 1, \ldots, M$. From the covariance matrix of the original set: $C = \frac{1}{M}\sum_{n=1}^{M} \Phi_n \Phi_n^T$, we retain the first significant M eigenvectors, denoted by v_l , with $l = 1, \ldots, M$. These eigenvectors are used to create the M eigenfaces in the following way: $u_l = \sum_{k=1}^{M} v_{lk}\Phi_k$, $l = 1, \ldots, M$. To test the faceness of an image Γ, we project it on the subspace defined by the eigenfaces: $\omega_k = u_k^T(\Gamma - \Psi)$, $k = 1, \ldots, M$. The reconstruction from this projection is given by $\Phi_f = \sum_{i=1}^{M} \omega_i u_i$. The reconstruction error $\varepsilon^2 = \|\Phi - \Phi_f\|^2$can be used as a "faceness criterion", in order to decide if at the current location in the image there is a face present or not, where $\Phi = \Gamma - \Psi$. This can be achieved by simply thresholding the parameter ε, i.e. by stating that a region in the image corresponds to a face iff the reconstruction error is below a certain threshold. In the experiments reported in the next section, we have computed this reconstruction error for each pixel neighborhood. This error image is then normalized in the interval [0,1] before thresholding to determine the face localizations.

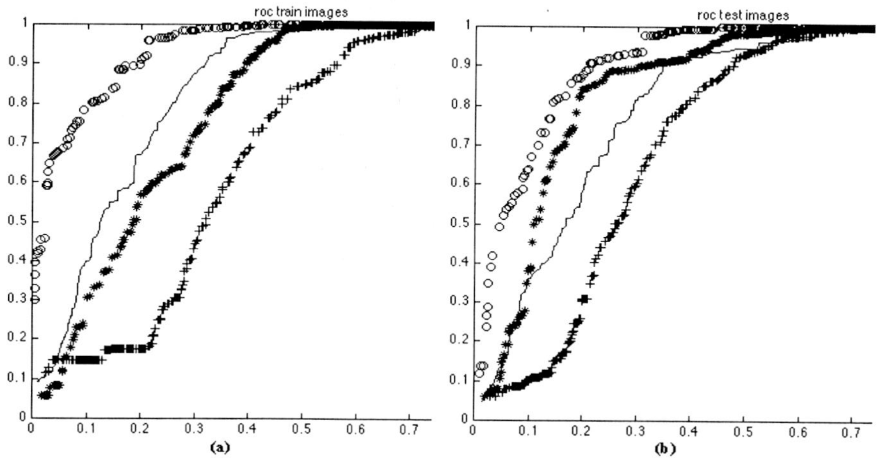

Fig. 2. ROC of the ratio of true positive pixels, PCA ('*'),MMF ('o'), and Generalized Hit-or-Miss: GHMT('+') and LBHMT ('.'). (a) train images (b) test images.

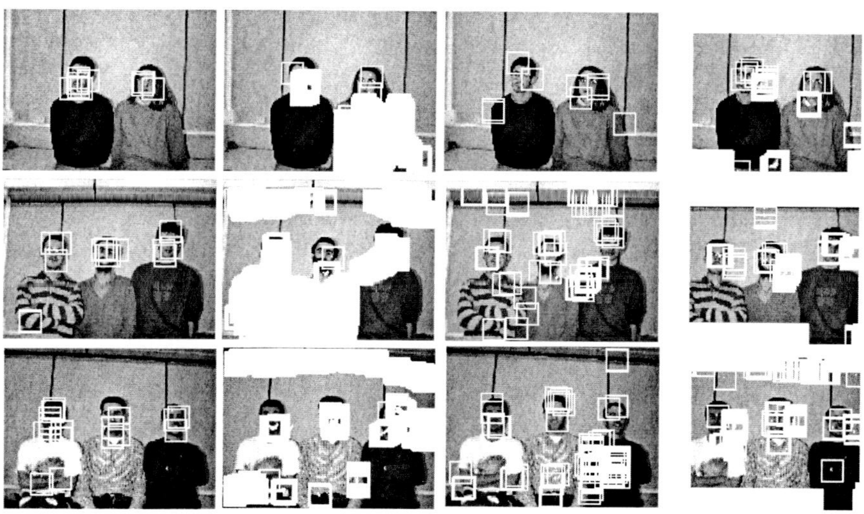

Fig. 3. Face localization on some training images with the constraint of 90% localization of face pixels: MMF, PCA, LBHMT and GHMT.

5. Experimental Results

We have performed the experimental comparison of PCA and MMF, GHMT, LBHMT over a set of 19 images. The faceness measure is normalized in all cases

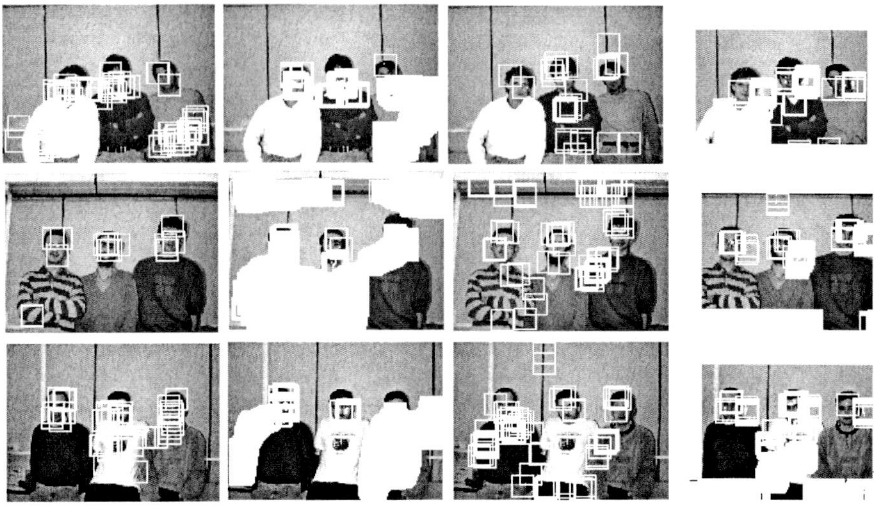

Fig. 4. Face localization on some test images, constrained to 90% localization of face pixels: MMF, PCA, LBHMT, GHMT.

and a threshold is applied to determine the faces. We have extracted 40 sample faces from 10 images, and exactly the same faces have been used to compute both the eigenfaces of the PCA transformation, the MMF prototypes, and as the templates for GHMT and LBHMT. The images used for the face extraction are considered the training set, and the remaining as the test set. The ground truth is given by a set of hand defined rectangles that include most of the faces. Note that these rectangles do not coincide with the face patterns. Under this conditions the training images also require some generalization abilities.

We have computed the Receiving Operator Curve (ROC) (the ratio of the true positive pixels to false positive pixels) by varying the faceness threshold value for all approaches when applied to face localization on the training and test images. In figure 2 we plot the mean of the normalized ratios r1= (true positive / positive ground truth) versus r2 = (false positive / negative ground truth) . The r1 ratio approaches 1 when there is no false positive responses, The r1 ratio approaches 1 when all the faces in the image are detected, the r2 ratio approaches 0 when there are no false positives.

It can be appreciated in figure 2, that the MMF consistently performs better than the other techniques, giving better recognition rates for the same false positive ratio. This is more clear in the training images. All the approaches eventually detect all the labeled faces in the images at the cost of increasing the false positives, but MMF does it faster. The worst results are given by the GHMT. The LBHMT and the PCA alternate their performance. While LBHMT performs better in the training set, the PCA improves relatively in the test set. The exploratory experiments with both GHMT and LBHMT showed that both techniques were able to localize the exact patterns in image. As noted before the ground truth does not match exactly to the set of face patterns, some extrapolation abilities are required of the classification criteria realized by the operators. This produces a great decrease of the accuracy of the response of the Hit-or-Miss techniques.

The results in figure 2 may be misleading because r2 is normalized against the figure background, so that a small variation corresponds to a big increase in false positives. In figure 3 and 4 we show the localizations performed by (ordered by columns) the MMF (left column), PCA, LBHMT, and GHMT (right column) on some images when the process is constrained to localize 90% of the face pixels in the image. This constraint is realized by relaxing the corresponding faceness threshold for each method until the required percentage of true positives is reached. A face localization is given by a white square. Overlapping squares produce white patches in some cases. Namely PCA and GHMT results show this effect.

6. Conclusions and Further Work

In this paper we make an empirical comparison of several techniques for face localization based on morphological operators. We propose the use of Multiscale Morphological Fingerprint (MMF) to obtain a set of global features for face recognition. We also tested the grayscale Hit-or-Miss Transform (GHMT) and a new definition for grayscale HMT based on level sets. The Principal Components Analysis (PCA) has been used for benchmarking. The ROCs computed show that MMF performs consistently better than the other techniques. Further work will be addressed to the search of more elaborate definitions of the similarity between fingerprints that could lead to increased discriminant power. Also we want to explore alternative extensions of the Hit-or-Miss operator with improved classification generalization properties.

References

1. R. Chelappa, C. L. Wilson and S. Sirohey. Human and Machine Recognition of Faces: A Survey. *Proceedings of the IEEE*, 83(5):705–740, May 1995.
2. M. Kirby and L. Sirovich. Application of the Karhunen-Loève Procedure for the Characterization of Human Faces. *IEEE Transactions on Pattern Analysis and Machine Intelligence*, 12(1):103-108, January 1990
3. M. Turk and A. Pentland. Eigenfaces for Recognition. *Journal of Cognitive Neuroscience*, 3(1):71-86, January 1991
4. P. S. Penev and J. J. Atick. Local Feature Analysis: A General Statistical Theory for Object Representation. *Network: Computation in Neural Systems*, Vol. 7:477-500, 1996
5. M. S. Bartlett and T. J. Sejnowski. Viewpoint Invariant Face Recognition Using Independent Component Analysis and Attractor Networks. In M. Mozer, M.Jordan and T. Petsche, editors, *Neural Information Processing Systems-Natural and Synthetic*, volume 9, MIT Press, Cambridge, MA., 1997
6. H. A. Rowley, S. Baluja and T. Kanade. Human Face Detection in Visual Scenes. *Technical Report CMU-CS-95-158R*, Computer Science Dept., Carnegie-Mellon University, November 1995
7. H. A. Rowley, S. Baluja and T. Kanade. Rotation Invariant Neural Network-Based

Face Detection. *Technical Report CMU-CS-97-201*, Computer Science Dept., Carnegie-Mellon University, December 1997
8. H. A. Rowley, S. Baluja and T. Kanade. Neural Network-Based Face Detection. *IEEE Transactions on Pattern Analysis and Machine Intelligence*, 20(1):23-38, January 1998
9. K. K. Sung and T. Poggio. Exemple-Based Learning for View-Based Human Face Detection. *A. I. Memo No. 1521*, MIT, December 1994
10. S. H. Lin, S. Y. Kung and L. J. Lin. Face Recognition/Detection by Probabilistic Decision-Based Neural Networks. *IEEE Transactions on Neural Networks*, 8(1):114-132, January 1997
11. P. Juell and R. Marsh. A Hierarchical Neural Network for Human Face Detection. *Pattern Recognition*, 29(5):781-787, May 1996
12. C. H. Lee, J. S. Kim and K. H. Park. Automatic Human Face Location in a Complex Background Using Motion and Color Information. *Pattern Recognition*, 29(11):1877-1889, November 1996
13. G. Yang and T. S. Huang. Human Face Detection in a Complex Background. *Pattern Recognition*, 27(1):53-63, January 1994
14. T. K. Leung, M. C. Burl and P. Perona. Finding Faces in Cluttered Scenes Using Random Labeled Graph Matching. *Fifth International Conference on Computer Vision*, 1995, http://HTTP.CS.Berkeley.EDU/~leungt/Research/ICCV95_final.ps.gz
15. K. C. Yow and R. Cipolla. Finding Initial Estimates of the Human Face Location. *Technical Report TR-239*, University of Cambridge, 1995
16. P. T. Jackway and M. Deriche. Scale-Space Properties of the Multiscale Morphological Dilation-Erosion. *IEEE Transactions on Pattern Analysis and Machine Intelligence*, 18(1):38-51, January 1996
17. M. Khosravi and R. W. Schafer. Template Matching Based on a grayscale Hit-or-Miss Transform. *IEEE Transactions on Image Processing*, 5(6):1060-1066, June 1996
18. H. J. A. M. Heijmans. Aspects of the Theory of Morphological Operators and Filters. *Workshop on Design Methodologies for Microelectronics and Signal Processing*, pages 377-387, Gliwice-Cracow, Poland, 20-23 October 1993
19. F. Guichard and J.-M. Morel. Image Iterative Smoothing and P.D.E.'s. *Notes de Cours du Centre Émile Borel (Institut Henri Poincaré)*, 14 September 1998-18 December 1998

QUANTITATIVE DESCRIPTION OF TELECOMMUNICATION NETWORKS BY SIMULATION

F. TOURNOIS, C. LANTUÉJOUL and M. SCHMITT
Ecole des Mines de Paris, Centre de Géostatistique
35 rue Saint-Honoré
77305 Fontainebleau
France

Abstract. A stochastic model is used to analyse cellular communication network performance through the average number of users connected to a base station. This average is studied in two different ways : from the servers' point of view and the customers', leading to different results. Analytic limitations appear in the non-stationary case. Simulation seems to be a solution in order to obtain realisations as well as non intuitive results.

Key words: Cellular Networks, Mobile Communications, Macroscopic Modeling, Poisson Process.

1. Introduction and Description of the Network Model

With new telecommunication systems based on cellular communication networks, companies have to make strategic plans to build new networks by setting up servers according to the customers' distribution with appropriate capacity, to modify existing networks by adding new servers or increasing server capacity and to evaluate the performance of these systems (i.e. saturation probability). These problems are very complex and in order to reduce the parameter number, we use a probabilistic model. The main idea consists in considering the spatial configuration of the network objects (customers and servers) as realizations of stochastic point processes [6].

In this paper we consider a simple model for describing cellular communication network [1, 2]. This model is made up of two kinds of objects, the base stations and the users. For each station, we define an area (*cell*). Together these cells form a partition of space. Users in a cell are served by the corresponding base station (see Figure 1). We use two independent non stationary Poisson processes to model the location of servers and users in the system, because of its tractability. Let $\mathcal{P}(\lambda)$ denote a Poisson process with intensity function λ (in other words, λ is not necessarily constant but may vary through space). λ_s stands for the server intensity and λ_c for the user intensity. This model is examined via the number of users connected to each base station in two different ways, first analytically and second using simulations.

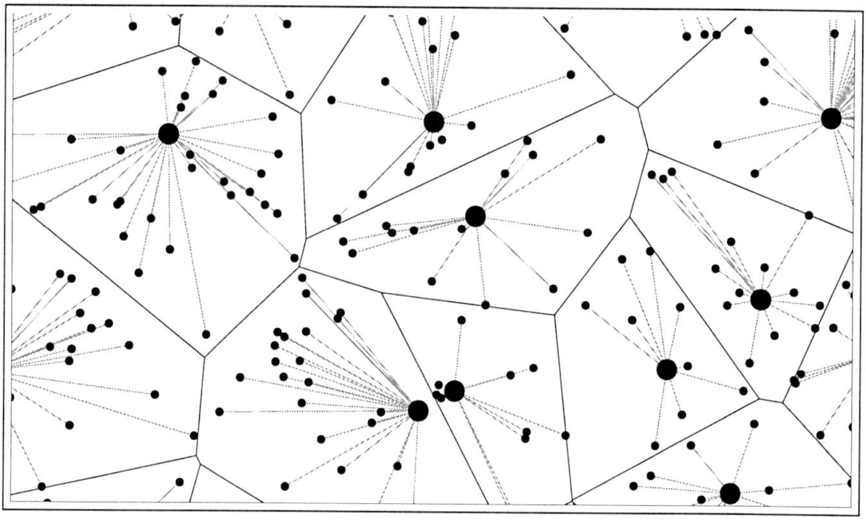

Fig. 1. Basic model with users (small discs), stations (big discs), cells and communication paths.

2. Customers' and Servers' Point of View

Our aim is to study the average number of users connected to one server. There are two ways to answer this question. In the first one, we consider things from the customer's point of view. He wants to know how many users are connected to his server. In the second one, we consider the server's point of view. He wants to know how many customers are connected. Let $N_c(x)$ be the number of users connected to the base station serving a customer located at point x and let $\overline{N_c(x)} = \mathbb{E}(N_c(x))$ be its expectation. We use $N_s(x)$ to denote the number of users connected to the server located at point x and $\overline{N_s(x)} = \mathbb{E}(N_s(x))$ for its expectation. In the following formulae, we can see important differences between these two approaches.

Proposition 1 *If λ_c is the user's process intensity and λ_s the server's process intensity, then for all $x \in \mathbb{R}^n$ the generating functions [8] of $N_c(x)$ and $N_s(x)$ are :*

$$\mathbb{E}\left(z^{N_c(x)}\right) = \int_{R^n} \mathbb{E}\left[e^{-\lambda_c\left[\mathcal{C}(\S,\mathcal{P}(\lambda_\int . \infty_{\overline{\mathcal{B}}(\dagger,|\S-\dagger|)})\cup\{\dagger\})\right](z-1)}\right] \lambda_s(y) e^{-\lambda_s(B(y,|x-y|))} dy$$

$$\mathbb{E}\left(z^{N_s(x)}\right) = \mathbb{E}\left[e^{-\lambda_c(\mathcal{C}(\S,\mathcal{P}(\lambda_\int)\cup\{\S\}))(z-1)}\right]$$

where $\mathcal{C}(\dagger, \{\S_\rangle\}_{\rangle\in\mathcal{I}})$ is the cell of the Voronoi diagram [7, 5] constructed with $\{x_i\}_{i\in I}$ which contained the point y, $B(x,r) = \{\, y \mid |x-y| < r\}$, $\lambda_s(B) = \int_B \lambda_s$, $\lambda_c(B) = \int_B \lambda_c$ and $\overline{B} = \{\, x \mid x \notin B\,\}$.

The second formula is straightforward. Regarding the first one, it can be obtained by randomizing the location of the server of x.

Corollary 1 *If λ_c is the user's process intensity and λ_s the server's process intensity, then for all $x \in \mathbb{R}^n$*

$$\overline{N_c(x)} = \mathbb{E}\left[\int_{R^n} \lambda_c\left(\mathcal{C}(\int, \mathcal{P}(\lambda_\int . \infty_{\overline{\mathcal{B}}(\dagger,|\S-\dagger|)}) \cup \{\dagger\})\right) \lambda_s(y) e^{-\lambda_s(B(y,|x-y|))} dy\right]$$

$$\overline{N_s(x)} = \mathbb{E}\left[\lambda_c\left(\mathcal{C}(\S, \mathcal{P}(\lambda_\int) \cup \{\S\})\right)\right]$$

In general, these formulae are not mathematically tractable. In contrast to this, simulations are possible. Things are particularly simple in the case of a Poisson process. To simulate such a process that contains point x, it is sufficient to add x to any simulation.

$$\{\mathcal{P}(\lambda)(\omega) \,|\, \omega \in \otimes, \, \S \in \mathcal{P}(\lambda)(\omega)\} = \{\mathcal{P}(\lambda)(\omega) \cup \{\S\} \,|\, \omega \in \otimes\}$$

To show that these two formulae lead to different evaluations, let us study the constant case, that is, where for all $x \in \mathbb{R}^n$ $\lambda_c(x) = \Lambda_c$ and $\lambda_s(x) = \Lambda_s$. In that case, in one dimension applying the two formulae gives $\overline{N_s} = \frac{\Lambda_c}{\Lambda_s}$, $\overline{N_c} = \frac{3\Lambda_c}{2\Lambda_s}$ whereas in two dimensions $\overline{N_s} = \frac{\Lambda_c}{\Lambda_s}$, $\overline{N_c} = 1.280\frac{\Lambda_c}{\Lambda_s}$ [3, 1].

3. One-dimensional Study

In this section, we let $n = 1$ and study some specific cases of intensities. We begin with the polynomial case to point out differences between the two points of view. Then, periodic functions are considered.

Proposition 2 *If $\lambda_c \in \mathbb{R}[X]^*$ is the users' process intensity and $\lambda_s \in \mathbb{R}_0[X]^\dagger$ the servers' process intensity, we define two series*

$$\begin{cases} \alpha_{2p} = \frac{3(3^{2p+2}-1)}{2^{4p+4}\lambda_s^{2p+1}} \\ \alpha_{2p+1} = 0 \end{cases} \quad \textit{and} \quad \begin{cases} \beta_{2p} = \frac{1}{2^{2p}\lambda_s^{2p+1}} \\ \beta_{2p+1} = 0 \end{cases}$$

and two linear mappings from $\mathbb{R}[X] \mapsto \mathbb{R}[X]$ by

$$\Phi_c\left(\tfrac{X^n}{n!}\right) = \sum_{j=0}^{n} \alpha_{n-j} \frac{X^j}{j!} \quad \textit{and} \quad \phi_s\left(\tfrac{X^n}{n!}\right) = \sum_{j=0}^{n} \beta_{n-j} \frac{X^j}{j!}$$

then we have :

$$\overline{N_c} = \Phi_c(\lambda_c) \quad \textit{and} \quad \overline{N_s} = \Phi_s(\lambda_s)$$

These formulae are obtained by simple calculus using the first corollary.

Let us choose $\lambda_s(x) = \lambda \in \mathbb{R}$ and $\lambda_c(x) = \lambda(x^2-1)^2$. Figure 2 shows that when λ_s and λ_c increase proportionally, we have the constant case locally $\overline{N_c}(x) = \frac{3\lambda_c(x)}{2\lambda_s(x)}$ and $\overline{N_s}(x) = \frac{\lambda_c(x)}{\lambda_s(x)}$. In fact, the more λ_s increases, the cell size decreases, the calculation becomes more local and then the customers' intensity

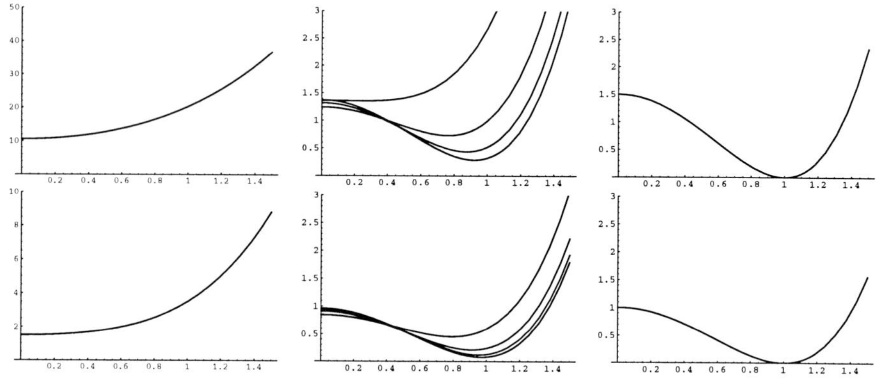

Fig. 2. $\overline{N_c}$ (top) and $\overline{N_s}$ (bottom) with $\lambda_s(x) = \lambda \in R$ and $\lambda_c(x) = \lambda(x^2 - 1)^2$, $\lambda = 1$ (left), $\lambda \in \{2, 3, 4, 5\}$ (middle) and $\lambda = 100$ (right).

variation decreases in the cell. A low intensity variation is near the constant case. For $\lambda = 1$, $\overline{N_c}$ and $\overline{N_s}$ are far from the constant case and when λ increases ($\lambda \in \{2, 3, 4, 5\}$) the constant case appears gradually. Note also that for $\lambda < 10$ the estimated values are far from the constant case. So when optimizing a network, the rule of thumb $\frac{\lambda_c(x)}{\lambda_s(x)}$ is not optimal. We also see a real difference between the customers' and the server' points of view.

Now, we can propose the same formulae for a periodic function.

Proposition 3 *If* $\lambda_c = \sum_{n \in \mathbb{Z}} a_n e^{inx}$ *is the users' intensity process and* $\lambda_s \in \mathbb{R}_0[X]$ *the servers' intensity process then :*

$$\overline{N_c} = \frac{3a_0}{2\lambda_s} + \sum_{n \in \mathbb{Z}^*} \frac{2a_n\lambda_s(12\lambda_s^2 + n^2)}{(4\lambda_s^2 + n^2)^2} e^{inx}$$

and

$$\overline{N_s} = \frac{a_0}{\lambda_s} + \sum_{n \in \mathbb{Z}^*} \frac{4a_n\lambda_s}{4\lambda_s^2 + n^2} e^{inx}$$

For example, if we choose $\lambda_s = \lambda$ and $\lambda_c = 2\lambda + \lambda\cos(x) + \frac{\lambda}{2}\cos(20x)$, then we have a high $(\pi/10)$ and a low (2π) variation frequency. Figure 3 shows that the high frequencies become more and more visible as λ increases.

Now, we choose another simple case in order to show the border effects. We define $\alpha \in \mathbb{R}$, $\lambda_s = \lambda_c = \alpha.1_{]0,1[}$ for this case. Firstly, from Figure 4, we see the convergence to the constant case according to α. Secondly, we notice bumps near the edges corresponding to peaks of the numbers of customers. This result appears all the more interesting as it is unexpected. These peaks correspond in fact to places where base stations cover average-sized areas, located at the end parts of the 1D.

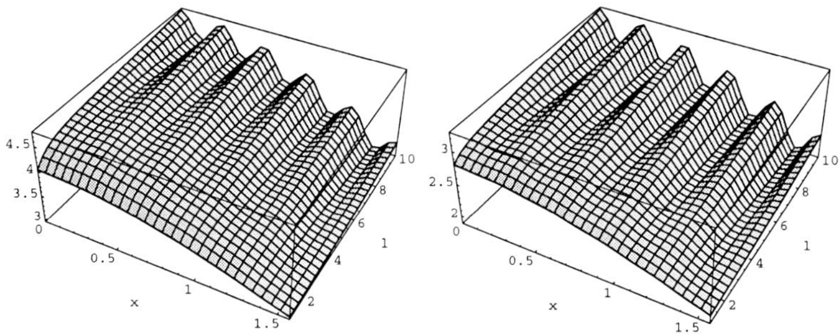

Fig. 3. $\overline{N_c}$ (left) and $\overline{N_s}$ (right) with $\lambda_s = \lambda$, $\lambda_c = 2\lambda + \lambda\cos(x) + \frac{\lambda}{2}\cos(20x)$, $x \in]0, \pi/2[$ and $\lambda \in]1, 10[$.

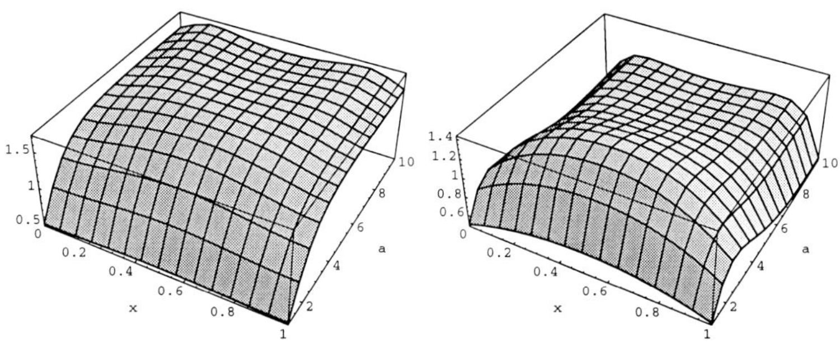

Fig. 4. $\overline{N_c}$ (left) and $\overline{N_s}$ (right) with $\lambda_s = \lambda_c = \alpha.1_{]0,1[}$, $x \in]0, 1[$ and $\alpha \in]1, 10[$

4. Two-dimensional Study

We are now going to study a realistic two-dimensional case. In 2D all the calculations get more complex and require the use of simulations. We only obtain analytical formulae for the average number of users in the constant case (see section 2).

Firstly, similarly to the one-dimensional case, let us choose $\lambda_s = \lambda_c = \alpha.1_{]0,1[^2}$. Figures 5 and 6 show the simulation results for three cases $\alpha = 5$, $\alpha = 20$ and $\alpha = 50$. As in the one-dimensional case, we see that the convergence to the constant case[‡] is observed first in the middle of the square. Peaks appear near the edges and near the corners. In this case, like λ in the one-dimensional study, α defines a scale or precision factor to watch $\alpha.1_{]0,1[^2}$. When α increases, the shape of the square $]0,1[^2$ appears. We can follow the evolution in the three pictures : at first, we cannot see the square ($\alpha = 5$) and finally the square appears ($\alpha = 50$). In the general case λ_s acts as a scale factor to watch or describe λ_c. We can say from these figures that, peaks are better revealed by

‡ $\overline{N_c}(x) = \frac{3\lambda_c(x)}{2\lambda_s(x)}$ and $\overline{N_s}(x) = \frac{\lambda_c(x)}{\lambda_s(x)}$

the customers' point of view than the servers'.

Moreover, in 2D, the number of users is not constant outside the definition set (see the edge of the surface) because a Voronoi diagram defines cells in $\mathbb{R}^2$ outside $]0,1[^2$ and these infinite area cells do not necessarily cover the same area in contrast to the 1D case.

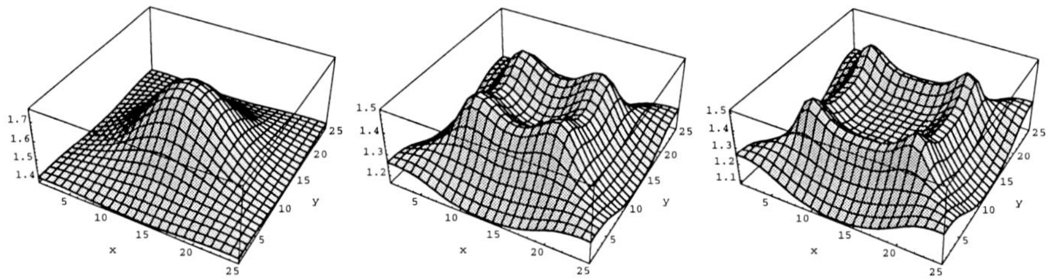

Fig. 5. $\overline{N_c}$ with $\lambda_s = \lambda_c = \alpha.1_{]0,1[^2}$, $x \in]-0.2, 1.2[^2$, $\alpha = 5$ (left), $\alpha = 20$ (middle) and $\alpha = 50$ (right).

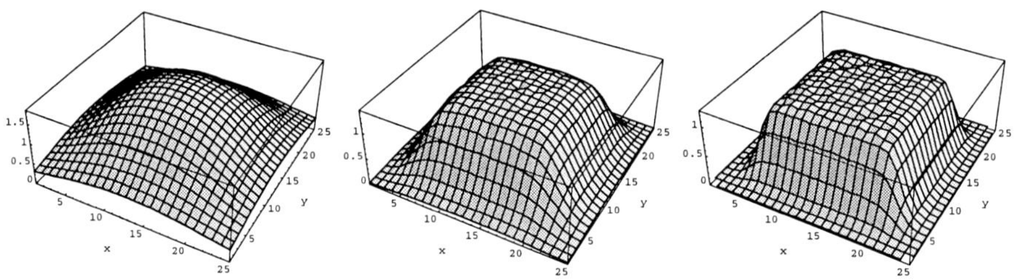

Fig. 6. $\overline{N_s}$ with $\lambda_s = \lambda_c = \alpha.1_{]0,1[^2}$, $x \in]-0.2, 1.2[^2$, $\alpha = 5$ (left), $\alpha = 20$ (middle) and $\alpha = 50$ (right).

5. Two-dimensional non-Euclidian Study

Let us now consider an obstacle deflecting the transmission. The shortest path between two points is not necessarily a line (see Figure 9). The Voronoi diagram becomes a geodesic Voronoi diagram [4, 7] and we see a variation of the number of users, compared to the previous case. We choose $\lambda_s = \lambda_c = \alpha.1_\Delta$ with $\Delta = \{x \in]0,1[^2 \mid x \notin]0.4, 0.6[^2\}$. Figures 7 and 8 show the results of the simulation for three cases $\alpha = 5$, $\alpha = 20$ and $\alpha = 50$. As before, α is a scale factor. As it increases, the square $]0.4, 0.6[^2$ becomes more and more visible.

This experiment makes us realize that it must be very difficult to obtain analytical results. However by using simulations, non intuitive results can be obtained.

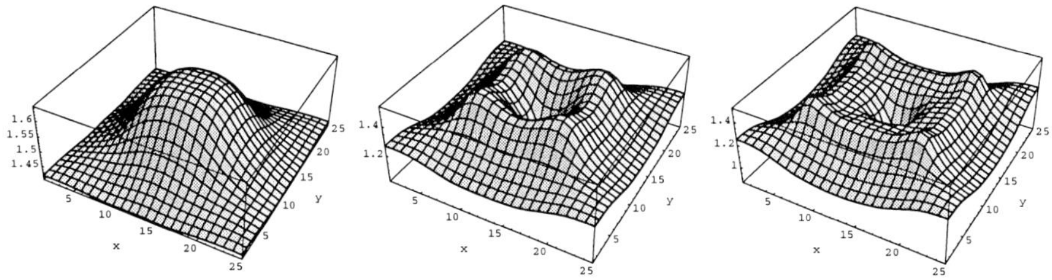

Fig. 7. $\overline{N_c}$ with $\lambda_s = \lambda_c = \alpha.1_\Delta$, $x \in]-0.2, 1.2[^2$, $\alpha = 5$ (left), $\alpha = 20$ (middle) and $\alpha = 50$ (right).

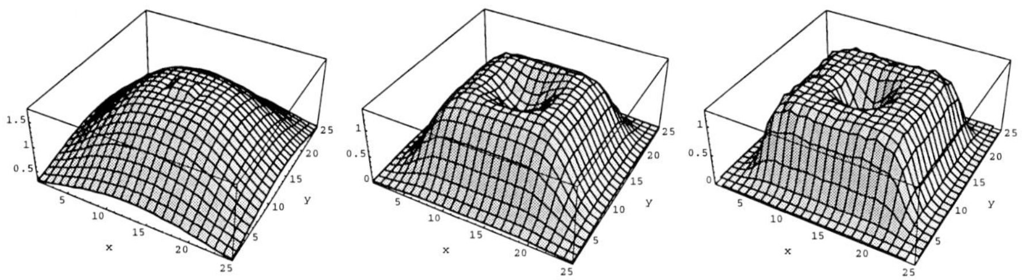

Fig. 8. $\overline{N_s}$ with $\lambda_s = \lambda_c = \alpha.1_\Delta$, $x \in]-0.2, 1.2[^2$, $\alpha = 5$ (left), $\alpha = 20$ (middle) and $\alpha = 50$ (right)

6. Conclusion

The servers' and customers' points of view lead to quite different results. Analytic solutions cannot be obtained even in simple cases. Simulation seems to be a viable alternative to analytical studies for mobile communication network studies. Our preliminary simulations give interesting new results but we could examine more complex cases by using conditional simulations [9]. For example, we could consider the real network state at one point in time and begin the simulation with this data.

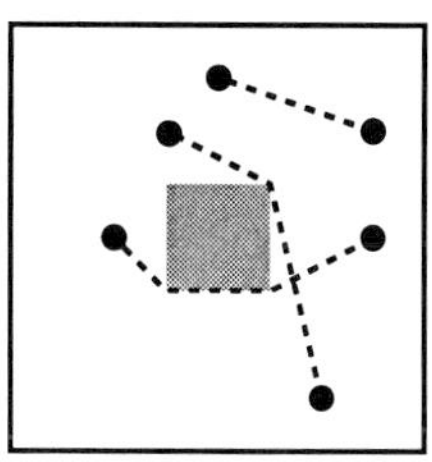

Fig. 9. Some shortest paths in Δ.

References

1. F. Baccelli, M. Klein, M. Lebourges and S.Zuyev. Géométrie aléatoire et architecture de réseaux de communication INRIA, 1995 , Rapport de recherche 2542.
2. F. Baccelli and S. Zuyev. Poisson-Voronoï Spanning trees. INRIA, 1996 , Rapport de recherche 3040.
3. G. Foss and S.Zuyev. On a Voronoï Aggregative Process Related to a Bivariate Poisson Process. *Adv. Appl. Proba.*, 1996
4. C. Lantuéjoul and F. Maisonneuve. Geodesic methods in quantitative image analysis. *Pattern Recognition Journal*, 17-2:177–187, 1984
5. D. Stoyan, W. Kendall and J. Mecke. Stochastic geometry and its applications. *John Wiley & Sons*, 1995
6. G. Matheron. Random sets & integral geometry. *John Wiley & Sons*, 1975
7. J. Serra. Image analysis and mathematical morphology. *Academic press*, 1982
8. W. Feller. An introduction to probability theory and its applications. *John Wiley & Sons*, 1950
9. C. Lantuéjoul. Conditional simulation of random sets. *Proceeding of the Section on statistics and the environment*, Annual Meeting of the American Statistical Association, 37–45, 1998

Author Index

Subject Index